Stars and Stellar Processes

This textbook offers a modern approach to the physics of stars, assuming only undergraduate-level preparation in mathematics and physics, and minimal prior knowledge of astronomy. It starts with a concise review of introductory concepts in astronomy, before covering the nuclear processes and energy transport in stellar interiors, and stellar evolution from star formation to the common stellar endpoints as white dwarfs and neutron stars. In addition to the standard material, the author also discusses more contemporary topics that students will find engaging, such as neutrino oscillations and the MSW resonance, supernovae, gamma-ray bursts, advanced nucleosynthesis, neutron stars, black holes, cosmology, and gravitational waves. With hundreds of worked examples, explanatory boxes, and problems with solved problems, this textbook provides a solid foundation for learning either in a classroom setting or through self-study.

Mike Guidry is Professor of Physics and Astronomy at the University of Tennessee. His current research is focused on the development of new algorithms to solve large sets of differential equations, and applications of Lie algebras to strongly-correlated electronic systems. He has written five textbooks and authored more than 120 journal publications on a broad variety of topics. He previously held the role of Lead Technology Developer for several major college textbooks in introductory physics, astronomy, biology, genetics, and microbiology. He has won multiple teaching awards and is responsible for a variety of important science outreach initiatives.

Stars and Stellar Processes

MIKE GUIDRY
University of Tennessee, Knoxville

Shaftesbury Road, Cambridge CB2 8EA, United Kingdom

One Liberty Plaza, 20th Floor, New York, NY 10006, USA

477 Williamstown Road, Port Melbourne, VIC 3207, Australia

314–321, 3rd Floor, Plot 3, Splendor Forum, Jasola District Centre, New Delhi – 110025, India

103 Penang Road, #05–06/07, Visioncrest Commercial, Singapore 238467

Cambridge University Press is part of Cambridge University Press & Assessment, a department of the University of Cambridge.

We share the University's mission to contribute to society through the pursuit of education, learning and research at the highest international levels of excellence.

www.cambridge.org
Information on this title: www.cambridge.org/9781107197886
DOI: 10.1017/9781108181914

© Mike Guidry 2019

This publication is in copyright. Subject to statutory exception and to the provisions of relevant collective licensing agreements, no reproduction of any part may take place without the written permission of Cambridge University Press & Assessment.

First published 2019

A catalogue record for this publication is available from the British Library

Library of Congress Cataloging-in-Publication data
Names: Guidry, M. W., author.
Title: Stars and stellar processes / Mike Guidry (University of Tennessee, Knoxville).
Description: Cambridge, United Kingdom ; New York, NY :
Cambridge University Press, 2018. | Includes bibliographical references and index.
Identifiers: LCCN 2018034170 | ISBN 9781107197886 (hardback : alk. paper)
Subjects: LCSH: Stars–Structure. | Stars–Evolution.
Classification: LCC QB808 .G85 2018 | DDC 523.8/6–dc23
LC record available at https://lccn.loc.gov/2018034170

ISBN   978-1-107-19788-6   Hardback

Cambridge University Press & Assessment has no responsibility for the persistence or accuracy of URLs for external or third-party internet websites referred to in this publication and does not guarantee that any content on such websites is, or will remain, accurate or appropriate.

For

Delphine, Milo,
Jack, Tethys, and Zelda
The best grandchildren ever!

Brief Contents

1	Some Properties of Stars	*page* 3
2	The Hertzsprung–Russell Diagram	32
3	Stellar Equations of State	53
4	Hydrostatic and Thermal Equilibrium	86
5	Thermonuclear Reactions in Stars	105
6	Stellar Burning Processes	131
7	Energy Transport in Stars	153
8	Summary of Stellar Equations	188
9	The Formation of Stars	201
10	Life and Times on the Main Sequence	228
11	Neutrino Flavor Oscillations	253
12	Solar Neutrinos and the MSW Effect	271
13	Evolution of Lower-Mass Stars	297
14	Evolution of Higher-Mass Stars	324
15	Stellar Pulsations and Variability	337
16	White Dwarfs and Neutron Stars	346
17	Black Holes	378
18	Accreting Binary Systems	401
19	Nova Explosions and X-Ray Bursts	421
20	Supernovae	429
21	Gamma-Ray Bursts	460
22	Gravitational Waves and Stellar Evolution	478

Contents

Preface — *page* xxiii

Part I Stellar Structure — 1

1 Some Properties of Stars — 3
- 1.1 Luminosities and Magnitudes — 3
  - 1.1.1 Stellar Luminosities — 3
  - 1.1.2 Photon Luminosities — 4
  - 1.1.3 Apparent Magnitudes — 5
  - 1.1.4 The Parsec Distance Unit — 6
  - 1.1.5 Absolute Magnitudes — 8
  - 1.1.6 Bolometric Magnitudes — 8
- 1.2 Stars as Blackbody Radiators — 9
  - 1.2.1 Radiation Laws — 9
  - 1.2.2 Effective Temperatures — 10
  - 1.2.3 Stellar Radii from Effective Temperatures — 11
- 1.3 Color Indices — 12
- 1.4 Masses and Physical Radii of Stars — 13
- 1.5 Binary Star Systems — 14
  - 1.5.1 Motion of Binary Systems — 15
  - 1.5.2 Radial Velocities and Masses — 17
  - 1.5.3 True Orbit for Visual Binaries — 18
  - 1.5.4 Eclipsing Binaries — 19
- 1.6 Mass–Luminosity Relationships — 20
- 1.7 Summary of Physical Quantities for Stars — 22
- 1.8 Proper Motion and Space Velocities — 22
- 1.9 Stellar Populations — 23
  - 1.9.1 Population I and Population II — 23
  - 1.9.2 Population III — 24
- 1.10 Variable Stars and Period–Luminosity Relations — 25
  - 1.10.1 Cepheid Variables — 25
  - 1.10.2 RR Lyra Variables — 26
  - 1.10.3 Pulsational Instabilities — 27
  - 1.10.4 Pulsations and Free-Fall Timescales — 28
- Background and Further Reading — 29
- Problems — 29

2 The Hertzsprung–Russell Diagram — 32
- 2.1 Spectral Classes — 32
  - 2.1.1 Excitation and the Boltzmann Formula — 32
  - 2.1.2 Ionization and the Saha Equations — 33
  - 2.1.3 Ionization of Hydrogen and Helium — 35
  - 2.1.4 Optimal Temperatures for Spectral Lines — 36
  - 2.1.5 The Spectral Sequence — 38
- 2.2 HR Diagram for Stars Near the Sun — 41
  - 2.2.1 Solving the Distance Problem — 41
  - 2.2.2 Features of the HR Diagram — 42
- 2.3 HR Diagram for Clusters — 43
- 2.4 Luminosity Classes — 45
  - 2.4.1 Pressure Broadening of Spectral Lines — 46
  - 2.4.2 Inferring Luminosity Class from Surface Density — 47
- 2.5 Spectroscopic Parallax — 48
- 2.6 The HR Diagram and Stellar Evolution — 49
- Background and Further Reading — 49
- Problems — 49

3 Stellar Equations of State — 53
- 3.1 Equations of State — 53
- 3.2 The Pressure Integral — 54
- 3.3 Ideal Gas Equation of State — 54
  - 3.3.1 Internal Energy — 56
  - 3.3.2 The Adiabatic Index — 57
- 3.4 Mean Molecular Weights — 58
  - 3.4.1 Concentration Variables — 59
  - 3.4.2 Partially Ionized Gases — 59
  - 3.4.3 Fully-Ionized Gases — 60
  - 3.4.4 Shorthand Notation and Approximations — 61
- 3.5 Polytropic Equations of State — 63
  - 3.5.1 Polytropic Processes — 63
  - 3.5.2 Properties of Polytropes — 63
- 3.6 Adiabatic Equations of State — 65
- 3.7 Equations of State for Degenerate Gases — 66
  - 3.7.1 Pressure Ionization — 66
  - 3.7.2 Distinguishing Classical and Quantum Gases — 69
  - 3.7.3 Nonrelativistic Classical and Quantum Gases — 70
  - 3.7.4 Ultrarelativistic Classical and Quantum Gases — 72
  - 3.7.5 Transition from a Classical to Quantum Gas — 72
- 3.8 The Degenerate Electron Gas — 74
  - 3.8.1 Fermi Momentum and Fermi Energy — 74
  - 3.8.2 Equation of State for Nonrelativistic Electrons — 75

		3.8.3	Equation of State for Ultrarelativistic Electrons	76
	3.9		High Gas Density and Stellar Structure	77
	3.10		Equation of State for Radiation	78
	3.11		Matter and Radiation Mixtures	79
		3.11.1	Mixtures of Ideal Gases and Radiation	79
		3.11.2	Adiabatic Systems of Gas and Radiation	79
		3.11.3	Radiation and Gravitational Stability	80
	Background and Further Reading			81
	Problems			81
4	**Hydrostatic and Thermal Equilibrium**			**86**
	4.1		Newtonian Gravitation	86
	4.2		Conditions for Hydrostatic Equilibrium	86
	4.3		Lagrangian and Eulerian Descriptions	88
		4.3.1	Lagrangian Formulation of Hydrostatics	88
		4.3.2	Contrasting Lagrangian and Eulerian Descriptions	89
	4.4		Dynamical Timescales	91
	4.5		The Virial Theorem for an Ideal Gas	92
	4.6		Thermal Equilibrium	94
	4.7		Total Energy for a Star	95
	4.8		Stability and Heat Capacity	96
		4.8.1	Temperature Response to Energy Fluctuations	96
		4.8.2	Heating Up while Cooling Down	97
	4.9		The Kelvin–Helmholtz Timescale	97
	Background and Further Reading			101
	Problems			101
5	**Thermonuclear Reactions in Stars**			**105**
	5.1		Nuclear Energy Sources	105
		5.1.1	The Curve of Binding Energy	105
		5.1.2	Masses and Mass Excesses	107
		5.1.3	Q-Values	108
		5.1.4	Efficiency of Hydrogen Burning	109
	5.2		Thermonuclear Hydrogen Burning	110
		5.2.1	The Proton–Proton Chains	110
		5.2.2	The CNO Cycle	111
		5.2.3	Competition of PP Chains and the CNO Cycle	113
	5.3		Cross Sections and Reaction Rates	114
		5.3.1	Reaction Cross Sections	114
		5.3.2	Rates from Cross Sections	115
	5.4		Thermally Averaged Reaction Rates	115
	5.5		Parameterization of Cross Sections	116
	5.6		Nonresonant Cross Sections	117
		5.6.1	Coulomb Barriers	117

		5.6.2	Barrier Penetration Factors	118
		5.6.3	Astrophysical S-Factors	119
		5.6.4	The Gamow Window	120
	5.7	Resonant Cross Sections	121
	5.8	Calculations with Rate Libraries	123
	5.9	Total Rate of Energy Production	123
	5.10	Temperature and Density Exponents	123
	5.11	Neutron Reactions and Weak Interactions	124
	5.12	Reaction Selection Rules	127
		5.12.1	Angular Momentum Conservation	127
		5.12.2	Isotopic Spin Conservation	127
		5.12.3	Parity Conservation	127
	Background and Further Reading	128
	Problems	129

6 Stellar Burning Processes — 131

6.1	Reactions of the Proton–Proton Chains	131
	6.1.1 Reactions of PP-I	131
	6.1.2 Branching for PP-II and PP-III	133
	6.1.3 Effective Q-Values	134
6.2	Reactions of the CNO Cycle	135
	6.2.1 The CNO Cycle in Operation	136
	6.2.2 Rate of CNO Energy Production	137
6.3	The Triple-α Process	138
	6.3.1 Equilibrium Population of ^{8}Be	139
	6.3.2 Formation of the Excited State in ^{12}C	140
	6.3.3 Formation of the Ground State in ^{12}C	141
	6.3.4 Energy Production in the Triple-α Reaction	142
6.4	Helium Burning to C, O, and Ne	143
	6.4.1 Oxygen and Neon Production	143
	6.4.2 The Outcome of Helium Burning	146
6.5	Advanced Burning Stages	147
	6.5.1 Carbon, Oxygen, and Neon Burning	147
	6.5.2 Silicon Burning	148
6.6	Timescales for Advanced Burning	151
Background and Further Reading	151
Problems	151

7 Energy Transport in Stars — 153

7.1	Modes of Energy Transport	153
7.2	Diffusion of Energy	154
7.3	Energy Transport by Conduction	156
7.4	Radiative Energy Transport	157

		7.4.1	Thomson Scattering	157
		7.4.2	Conduction in Degenerate Matter	158
		7.4.3	Absorption of Photons	158
		7.4.4	Stellar Opacities	159
		7.4.5	General Contributions to Stellar Opacity	160
	7.5	Energy Transport by Convection	162
	7.6	Conditions for Convective Instability	163
		7.6.1	The Schwarzschild Instability	164
		7.6.2	The Ledoux Instability	165
		7.6.3	Salt-Finger Instability	166
	7.7	Critical Temperature Gradient for Convection	167
		7.7.1	Convection and the Adiabatic Index	168
		7.7.2	Convection and the Pressure Gradient	169
	7.8	Stellar Temperature Gradients	170
		7.8.1	Choice between Radiative or Convective Transport	170
		7.8.2	Radiative Temperature Gradients	171
	7.9	Mixing-Length Treatment of Convection	171
		7.9.1	Pressure Scale Height	172
		7.9.2	The Mixing-Length Philosophy	173
		7.9.3	Analysis of Solar Convection	174
	7.10	Examples of Stellar Convective Regions	175
		7.10.1	Convection in Stellar Cores	175
		7.10.2	Surface Ionization Zones	177
	7.11	Energy Transport by Neutrino Emission	178
		7.11.1	Neutrino Production Mechanisms	178
		7.11.2	Classification and Rates	182
		7.11.3	Coherent Neutrino Scattering	184
	Background and Further Reading	185
	Problems	185

8 Summary of Stellar Equations — 188

	8.1	The Basic Equations Governing Stars	188
		8.1.1	Hydrostatic Equilibrium	188
		8.1.2	Luminosity	189
		8.1.3	Temperature Gradient	189
		8.1.4	Changes in Isotopic Composition	189
		8.1.5	Equation of State	190
	8.2	Solution of the Stellar Equations	190
	8.3	Important Stellar Timescales	191
	8.4	Hydrostatic Equilibrium for Polytropes	192
		8.4.1	Lane–Emden Equation and Solutions	193
		8.4.2	Computing Physical Quantities	195
		8.4.3	Limitations of the Lane–Emden Approximation	196

8.5	Numerical Solution of the Stellar Equations	196
Background and Further Reading		197
Problems		197

Part II Stellar Evolution — 199

9 The Formation of Stars — 201

9.1	Evidence for Starbirth in Nebulae	201
9.2	Jeans Criterion for Gravitational Collapse	203
9.3	Fragmentation of Collapsing Clouds	204
9.4	Stability in Adiabatic Approximation	206
	9.4.1 Dependence on Adiabatic Exponents	206
	9.4.2 Physical Interpretation	207
9.5	The Collapse of a Protostar	207
	9.5.1 Initial Free-Fall Collapse	208
	9.5.2 A Little More Realism	209
9.6	Onset of Hydrostatic Equilibrium	209
9.7	Termination of Fragmentation	212
9.8	Hayashi Tracks	212
	9.8.1 Fully Convective Stars	212
	9.8.2 Development of a Radiative Core	213
	9.8.3 Dependence on Composition and Mass	214
9.9	Limiting Lower Mass for Stars	214
9.10	Brown Dwarfs	215
	9.10.1 Spectroscopic Signatures	216
	9.10.2 Stars, Brown Dwarfs, and Planets	217
9.11	Limiting Upper Mass for Stars	217
	9.11.1 Eddington Luminosity	218
	9.11.2 Estimate of Upper Limiting Mass	218
9.12	The Initial Mass Function	220
9.13	Protoplanetary Disks	221
9.14	Exoplanets	222
	9.14.1 The Doppler Spectroscopy Method	223
	9.14.2 Transits of Extrasolar Planets	224
Background and Further Reading		224
Problems		224

10 Life and Times on the Main Sequence — 228

10.1	The Standard Solar Model	228
	10.1.1 Composition of the Sun	229
	10.1.2 Energy Generation and Composition Changes	229
	10.1.3 Hydrostatic Equilibrium	229

		10.1.4 Energy Transport	230
		10.1.5 Constraints and Solution	230
	10.2	Helioseismology	233
		10.2.1 Solar p-Modes and g-Modes	233
		10.2.2 Surface Vibrations and the Solar Interior	233
	10.3	Solar Neutrino Production	236
		10.3.1 Sources of Solar Neutrinos	236
		10.3.2 Testing the Standard Solar Model with Neutrinos	237
	10.4	The Solar Electron-Neutrino Deficit	238
		10.4.1 The Davis Chlorine Experiment	238
		10.4.2 The Gallium Experiments	239
		10.4.3 Super Kamiokande	239
		10.4.4 Astrophysics and Particle Physics Explanations	241
	10.5	Evolution of Stars on the Main Sequence	242
	10.6	Timescale for Main Sequence Lifetimes	243
	10.7	Evolutionary Timescales	245
	10.8	Evolution Away from the Main Sequence	246
		10.8.1 Three Categories of Post Main Sequence Evolution	247
		10.8.2 Examples of Post Main Sequence Evolution	247
	Background and Further Reading		250
	Problems		250

11 Neutrino Flavor Oscillations 253

	11.1	Overview of the Solar Neutrino Problem	253
	11.2	Weak Interactions and Neutrino Physics	254
		11.2.1 Matter and Force Fields of the Standard Model	254
		11.2.2 Masses for Particles of the Standard Model	256
		11.2.3 Charged and Neutral Currents	257
	11.3	Flavor Mixing	259
		11.3.1 Flavor Mixing in the Quark Sector	259
		11.3.2 Flavor Mixing in the Leptonic Sector	259
	11.4	Implications of a Finite Neutrino Mass	260
	11.5	Neutrino Vacuum Oscillations	260
		11.5.1 Mixing for Two Neutrino Flavors	261
		11.5.2 The Vacuum Oscillation Length	262
		11.5.3 Time-Averaged or Classical Probabilities	263
	11.6	Neutrino Oscillations with Three Flavors	265
		11.6.1 CP Violation in Neutrino Oscillations	266
		11.6.2 The Neutrino Mass Hierarchy	267
		11.6.3 Recovering 2-Flavor Mixing	267
	11.7	Neutrino Masses and Particle Physics	268
	Background and Further Reading		268
	Problems		268

12 Solar Neutrinos and the MSW Effect — 271
- 12.1 Propagation of Neutrinos in Matter — 271
  - 12.1.1 Matrix Elements for Interaction with Matter — 271
  - 12.1.2 The Effective Neutrino Mass in Medium — 272
- 12.2 The Mass Matrix — 274
  - 12.2.1 Propagation of Left-Handed Neutrinos — 274
  - 12.2.2 Evolution in the Flavor Basis — 275
  - 12.2.3 Propagation in Matter — 276
- 12.3 Solutions in Matter — 276
  - 12.3.1 Mass Eigenvalues for Constant Density — 277
  - 12.3.2 The Matter Mixing Angle θ_m — 277
  - 12.3.3 The Matter Oscillation Length L_m — 278
  - 12.3.4 Flavor Conversion in Constant-Density Matter — 279
- 12.4 The MSW Resonance Condition — 280
- 12.5 Resonant Flavor Conversion — 282
- 12.6 Propagation in Matter of Varying Density — 285
- 12.7 The Adiabatic Criterion — 286
- 12.8 MSW Neutrino Flavor Conversion — 287
  - 12.8.1 Flavor Conversion in Adiabatic Approximation — 287
  - 12.8.2 Adiabatic Conversion and the Mixing Angle — 288
  - 12.8.3 Resonant Conversion for Large or Small θ — 289
  - 12.8.4 Energy Dependence of Flavor Conversion — 290
- 12.9 Resolution of the Solar Neutrino Problem — 290
  - 12.9.1 Super-K Observation of Flavor Oscillation — 291
  - 12.9.2 SNO Observation of Neutral Current Interactions — 291
  - 12.9.3 KamLAND Constraints on Mixing Angles — 292
  - 12.9.4 Large Mixing Angles and the MSW Mechanism — 294
  - 12.9.5 A Tale of Large and Small Mixing Angles — 294
- Background and Further Reading — 295
- Problems — 295

13 Evolution of Lower-Mass Stars — 297
- 13.1 Endpoints of Stellar Evolution — 297
- 13.2 Shell Burning — 298
- 13.3 Stages of Red Giant Evolution — 300
- 13.4 The Red Giant Branch — 302
  - 13.4.1 The Schönberg–Chandrasekhar Limit — 303
  - 13.4.2 Crossing the Hertzsprung Gap — 303
- 13.5 Helium Ignition — 304
  - 13.5.1 Core Equation of State and Helium Ignition — 304
  - 13.5.2 Thermonuclear Runaways in Degenerate Matter — 305
  - 13.5.3 The Helium Flash — 305
- 13.6 Horizontal Branch Evolution — 306

		13.6.1	Life on the Helium Main Sequence	306
		13.6.2	Leaving the Horizontal Branch	306
	13.7	Asymptotic Giant Branch Evolution		307
		13.7.1	Thermal Pulses	308
		13.7.2	Slow Neutron Capture	310
		13.7.3	Development of Deep Convective Envelopes	314
		13.7.4	Mass Loss	314
	13.8	Ejection of the Envelope		315
	13.9	White Dwarfs and Planetary Nebulae		316
	13.10	Stellar Dredging Operations		317
	13.11	The Sun's Red Giant Evolution		319
	13.12	Overview for Low-Mass Stars		321
	Background and Further Reading			321
	Problems			321

14 Evolution of Higher-Mass Stars — 324

	14.1	Unique Features of More Massive Stars		324
	14.2	Advanced Burning Stages in Massive Stars		325
	14.3	Envelope Loss from Massive Stars		326
		14.3.1	Wolf–Rayet Stars	326
		14.3.2	The Strange Case of η Carinae	327
	14.4	Neutrino Cooling of Massive Stars		327
		14.4.1	Local and Nonlocal Cooling	329
		14.4.2	Neutrino Cooling and the Pace of Stellar Evolution	329
	14.5	Massive Population III Stars		330
	14.6	Evolutionary Endpoints for Massive Stars		330
		14.6.1	Observational and Theoretical Characteristics	331
		14.6.2	Black Holes from Failed Supernovae?	331
		14.6.3	Gravitational Waves and Stellar Evolution	333
	14.7	Summary: Evolution after the Main Sequence		333
	14.8	Stellar Lifecycles		333
	Background and Further Reading			335
	Problems			335

15 Stellar Pulsations and Variability — 337

	15.1	The Instability Strip		337
	15.2	Adiabatic Radial Pulsations		337
	15.3	Pulsating Variables as Heat Engines		340
	15.4	Non-adiabatic Radial Pulsations		340
		15.4.1	Thermodynamics of Sustained Pulsation	340
		15.4.2	Opacity and the κ-Mechanism	342
		15.4.3	Partial Ionization Zones and the Instability Strip	342
		15.4.4	The ε-Mechanism and Massive Stars	344
	15.5	Non-radial Pulsation		344

Background and Further Reading 345
Problems 345

16 White Dwarfs and Neutron Stars 346
16.1 Properties of White Dwarfs 346
    16.1.1 Density and Gravity 347
    16.1.2 Equation of State 347
    16.1.3 Ingredients of a White Dwarf Description 348
16.2 Polytropic Models of White Dwarfs 349
    16.2.1 Low-Mass White Dwarfs 349
    16.2.2 High-Mass White Dwarfs 350
    16.2.3 Heuristic Derivation of the Chandrasekhar Limit 352
    16.2.4 Effective Adiabatic Index and Gravitational Stability 354
16.3 Internal Structure of White Dwarfs 355
    16.3.1 Temperature Variation 355
    16.3.2 An Insulating Blanket around a Metal Ball 356
16.4 Cooling of White Dwarfs 356
16.5 Crystallization of White Dwarfs 358
16.6 Beyond White Dwarf Masses 359
16.7 Basic Properties of Neutron Stars 359
    16.7.1 Sizes and Masses 360
    16.7.2 Internal Structure 361
    16.7.3 Cooling of Neutron Stars 362
    16.7.4 Evidence for Superfluidity in Neutron Stars 364
16.8 Hydrostatic Equilibrium in General Relativity 365
    16.8.1 The Oppenheimer–Volkov Equations 366
    16.8.2 Comparison with Newtonian Gravity 366
16.9 Pulsars 367
    16.9.1 The Pulsar Mechanism 367
    16.9.2 Pulsar Magnetic Fields 368
    16.9.3 The Crab Pulsar 369
    16.9.4 Pulsar Spindown and Glitches 369
    16.9.5 Millisecond Pulsars 370
    16.9.6 Binary Pulsars 372
16.10 Magnetars 374
Background and Further Reading 375
Problems 375

17 Black Holes 378
17.1 The Failure of Newtonian Gravity 378
17.2 The General Theory of Relativity 379
    17.2.1 General Covariance 379
    17.2.2 The Principle of Equivalence 379

	17.2.3	Curved Spacetime and Tensors	380
	17.2.4	Curvature and the Strength of Gravity	381
17.3	Some Important General Relativistic Solutions		381
	17.3.1	The Einstein Equation	382
	17.3.2	Line Elements and Metrics	382
	17.3.3	Minkowski Spacetime	383
	17.3.4	Schwarzschild Spacetime	384
	17.3.5	Kerr Spacetime	385
17.4	Evidence for Black Holes		386
	17.4.1	Compact Objects in X-ray Binaries	387
	17.4.2	Causality Constraints	389
	17.4.3	The Black Hole Candidate Cygnus X-1	389
17.5	Black Holes and Gravitational Waves		392
17.6	Supermassive Black Holes		392
17.7	Intermediate-Mass and Mini Black Holes		393
17.8	Proof of the Pudding: Event Horizons		394
17.9	Some Measured Black Hole Masses		396
Background and Further Reading			396
Problems			397

Part III Accretion, Mergers, and Explosions   399

18 Accreting Binary Systems   401

18.1	Classes of Accretion		401
18.2	Roche-lobe Overflow		402
	18.2.1	The Roche Potential	402
	18.2.2	Lagrange Points	403
	18.2.3	Roche Lobes	404
18.3	Classification of Binary Star Systems		405
18.4	Accretion Streams and Accretion Disks		406
	18.4.1	Gas Motion	406
	18.4.2	Initial Accretion Velocity	406
	18.4.3	General Properties of Roche-Overflow Accretion	408
	18.4.4	Disk Dynamics	408
18.5	Wind-Driven Accretion		410
18.6	Classification of X-Ray Binaries		411
	18.6.1	High-Mass X-Ray Binaries	411
	18.6.2	Low-Mass X-Ray Binaries	411
	18.6.3	Suppression of Accretion for Intermediate Masses	412
18.7	Accretion Power		412
	18.7.1	Maximum Energy Release in Accretion	412
	18.7.2	Limits on Accretion Rates	413
	18.7.3	Accretion Temperatures	413

	18.7.4	Maximum Efficiency for Energy Extraction	414
	18.7.5	Storing Energy in Accretion Disks	415
18.8		Some Accretion-Induced Phenomena	415
18.9		Accretion and Stellar Evolution	416
	18.9.1	The Algol Paradox	416
	18.9.2	Blue Stragglers	418

Background and Further Reading — 418
Problems — 418

19 Nova Explosions and X-Ray Bursts — 421

19.1		The Nova Mechanism	421
	19.1.1	The Hot CNO Cycle	423
	19.1.2	Recurrence of Novae	425
	19.1.3	Nucleosynthesis in Novae	425
19.2		The X-Ray Burst Mechanism	425
	19.2.1	Rapid Proton Capture	426
	19.2.2	Nucleosynthesis and the rp-Process	426

Background and Further Reading — 427
Problems — 427

20 Supernovae — 429

20.1		Classification of Supernovae	429
	20.1.1	Type Ia	430
	20.1.2	Type Ib and Type Ic	431
	20.1.3	Type II	433
20.2		Thermonuclear Supernovae	434
	20.2.1	The Single-Degenerate Mechanism	435
	20.2.2	The Double-Degenerate Mechanism	435
	20.2.3	Thermonuclear Burning in Extreme Conditions	437
	20.2.4	Element and Energy Production	438
	20.2.5	Late-Time Observables	439
20.3		Core Collapse Supernovae	440
	20.3.1	The "Supernova Problem"	441
	20.3.2	The Death of Massive Stars	441
	20.3.3	Sequence of Events in Core Collapse	442
	20.3.4	Neutrino Reheating	446
	20.3.5	Convection and Neutrino Reheating	447
	20.3.6	Convectively Unstable Regions in Supernovae	448
	20.3.7	Remnants of Core Collapse	449
20.4		Supernova 1987A	450
	20.4.1	The Neutrino Burst	450
	20.4.2	The Progenitor was Blue!	451
	20.4.3	Radioactive Decay and the Lightcurve	453

		20.4.4	Evolution of the Supernova Remnant	454
		20.4.5	Where is the Neutron Star?	455
	20.5	Heavy Elements and the r-Process		456
	Background and Further Reading			458
	Problems			458

21 Gamma-Ray Bursts — 460

	21.1	The Sky in Gamma-Rays		460
	21.2	Localization of Gamma-Ray Bursts		463
	21.3	Generic Characteristics of Gamma-Ray Burst		464
	21.4	The Importance of Ultrarelativistic Jets		467
		21.4.1	Optical Depth for a Nonrelativistic Burst	467
		21.4.2	Optical Depth for an Ultrarelativistic Burst	467
		21.4.3	Confirmation of Large Lorentz Factors	468
	21.5	Association of GRBs with Galaxies		468
	21.6	Mechanisms for the Central Engine		469
	21.7	Long-Period GRB and Supernovae		470
		21.7.1	Types Ib and Ic Supernovae	470
		21.7.2	Role of Metallicity	470
	21.8	Collapsar Model of Long-Period Bursts		471
	21.9	Neutron Star Mergers and Short-Period Bursts		474
	21.10	Multimessenger Astronomy		476
	Background and Further Reading			476
	Problems			476

22 Gravitational Waves and Stellar Evolution — 478

	22.1	Gravitational Waves		478
	22.2	Sample Gravitational Waveforms		480
	22.3	The Gravitational Wave Event GW150914		482
		22.3.1	Observed Waveforms	483
		22.3.2	The Black Hole Merger	484
	22.4	A New Probe of Massive-Star Evolution		486
		22.4.1	Formation of Massive Black Hole Binaries	486
		22.4.2	Gravitational Waves and Massive Binary Evolution	487
		22.4.3	Formation of Supermassive Black Holes	489
	22.5	Listening to Multiple Messengers		490
	22.6	Gravitational Waves from Neutron Star Mergers		491
		22.6.1	New Insights Associated with GW170817	493
		22.6.2	The Kilonova Associated with GW170817	495
	22.7	Gravitational Wave Sources and Detectors		497
	Background and Further Reading			497
	Problems			497

Appendix A	*Constants*	499
Appendix B	*Natural Units*	502
Appendix C	*Mean Molecular Weights*	505
Appendix D	*Reaction Libraries*	507
Appendix E	*A Mixing-Length Model*	516
Appendix F	*Quantum Mechanics*	519
Appendix G	*Using arXiv and ADS*	522
References		524
Index		534

Preface

This book contains material used in an advanced undergraduate astronomy course on stellar structure and stellar evolution that I teach regularly at the University of Tennessee. The goal of the course and of the book is to provide an introduction that is topically current and accessible to a reader with some physics but minimal astrophysics background.

Specifically, the reader is expected to have physics experience commensurate with that of a third or fourth year US undergraduate physics major, and to be familiar with the material typically covered in an introductory descriptive course in astronomy. The first two chapters are a concise review of introductory concepts in astronomy, so this latter requirement is useful but not essential for the diligent.

I don't assume any special knowledge of nuclear, atomic, or elementary particle physics beyond that usually covered in introductory physics courses. Likewise, I assume that readers are conversant with special relativity, general relativity, and quantum mechanics only at the level typically covered in first or second year university introductions to modern physics. Mathematically I assume the reader to be familiar with basic algebra, geometry, calculus, and differential equations. I strongly encourage the use of programming tools such as MatLab, Mathematica, or Maple, or more formal programming languages like C/C++ or Java where appropriate in solving problems. However, none of these tools is essential for working the problems.

To aid in comprehension, many worked examples and boxes containing supplementary information are scattered throughout each chapter. These show how to solve problems and serve to set the subject matter in context by providing a broader perspective. A total of 240 problems of varying complexity and difficulty may be found at the ends of the chapters, each chosen to illustrate important points, fill in details, or prove assertions made in the text. The solutions for all 240 problems are available from the publisher at www.cambridge.org/GuidryStars as PDF files in typeset book format for instructors, and a subset of 101 problem solutions is available to students from the publisher in the same format. Those problems with solutions available to students are marked by *** at the end of the problem.

Many articles referenced in this book are published in journals with limited free public access. To help ensure broad availability to these references for readers who may not have easy access to these journals, I have included where possible for journal articles information allowing free access through the preprint server *arXiv* or the *ADS Astronomy Abstract Service*. More details may be found at the beginning of the Bibliography in Ref. [1] and instructions for using arXiv and ADS may be found in Appendix G.

Any book dealing with astrophysics at an intermediate level must grapple with the issue of units. One is encouraged to standardize units and in introductory astronomy it makes sense to use the SI (MKS) system of units. However, professionals in the field routinely employ the CGS (centimeter-gram-second) system and more specialized units that are defined such that fundamental constants like the speed of light, gravitational constant, Planck's constant, or Boltzmann constant take the value of one. Since one of the purposes of the present material is to encourage students to use and explore the relevant literature, I have adopted a policy of generally using the CGS system or natural units.

Let me comment on use of this material in teaching courses, based on my experience teaching it for a number of years to senior-level undergraduates and beginning graduate students. As noted above the first two chapters are intended to be a review of introductory astronomy. I generally do not cover this material in class, but assign it as reading and require the students to do about 15 problems from these chapters in the first week or so of class to ensure that everyone is up to speed on introductory astronomy. (I usually have students taking the class who have not had introductory astronomy; this material permits them to catch up on the essentials.)

This leaves 20 chapters. That is too much to cover in depth for a one-semester course, leaving two choices: (1) cover all the material, but assign some as reading and homework only, or (2) cover a select set of topics in more depth. Which topics to emphasize is a clearly a matter of personal choice. If one desires to teach a more traditional course, then topics like neutrino oscillations and the MSW resonance, gamma-ray bursts, black holes, gravitational waves, supernovae, and accretion can be omitted, or treated in much more cursory fashion than in the book chapters. However, I have found from experience that these are precisely the chapters that most excite my students!

Therefore, I submit that the most effective and imaginative use of the material in this book for teaching is to build a course on it that broadens the appeal of stellar structure and evolution to a new set of students beyond the traditional astronomy/astrophysics majors. Most of those condensed matter, materials science, nuclear physics, particle physics, computer science, Earth and planetary science, chemistry, and mathematics majors will not be overly excited about stars as heat engines, but they can be attracted by a meaningful discussion of some of the most exciting topics in all of contemporary science: black holes, gravitational waves, neutrino oscillations, supernovae, and gamma-ray bursts, set within the context of a modern view of stellar evolution.

For those wishing to teach from this book, several additional resources are available from the publisher for instructors and for students:

1. *Instructor Solutions Manual for Stars and Stellar Processes*, which is a PDF file typeset in the format of the book that presents the solutions for all 240 problems at the ends of chapters. This manual is available only to instructors.
2. *Student Solutions Manual for Stars and Stellar Processes*, which is a PDF file typeset in the format of the book that contains the solutions for a subset of 101 of the 240 problems at the ends of chapters. This manual is available to students and instructors. As noted above, the problems contained in this solutions manual for students are marked by *** at the end of the problem in the text.

3. *Stars and Stellar Processes in Lecture Notes*, which is a PDF file, one for each chapter of the book, that presents a synopsis appropriate for projection and presentation of the essential material in each chapter. Individual slides are organized in a presentation format suitable for teaching with text formatted in larger fonts and in color. These are the slides that I use myself when teaching this material.

We conclude this list by noting that the inclusion of DOI or arXiv numbers for all journal references–which allows easy browser access through *arXiv* and *ADS* for most articles (see Appendix G)–may be viewed as an additional resource permitting creative literature-based projects to be assigned with minimal bother, if an instructor is so-inclined.

Finally, I would like to extend my thanks to the many students and colleagues whose questions and comments sharpened this presentation, to Nicholas Gibbons, Ilaria Tassistro, Jon Billam, and Dominic Stock at Cambridge University Press for all their help in shepherding this book to publication, and especially to my wife Jo Ann for her patience and support over many years.

PART I

STELLAR STRUCTURE

1 Some Properties of Stars

The fundamental building blocks of visible matter in the Universe are stars.[1] This chapter will discuss some of the basic properties of stars such as luminosity, radius, mass, color, and temperature as a prelude to a more detailed exploration of their structure and evolution. Let's begin with a discussion of their most obvious characteristic: that they are visible in our sky, and that even casual observation indicates that there is a substantial variation in brightness between different stars.

1.1 Luminosities and Magnitudes

The apparent brightness of a star is a combination of an intrinsic brightness, which is related to the internal structure of the star, and the effect of distance, since the intensity falls off as the square of the distance. To make much headway in understanding stars these two factors must be separated. This requires a direct or indirect measurement of the distance to the star, or comparison of stars that are known to be at equivalent distances (even if the distance itself is not known). Measuring the distance to stars is difficult and can be accomplished directly only for more nearby stars. The effect of the distance scale can be factored out if stars are compared that are members of physical (gravitationally bound) groupings called *clusters*, which come in two types: *open or galactic clusters* containing tens to hundreds of stars that are found preferentially in the plane of the galaxy, and *globular clusters* containing as many as hundreds of thousands of stars that are found preferentially in the galactic halo. Comparison of stars in a cluster makes it certain that they lie at almost the same distance. From the variation in brightness for stars in clusters, it is found that stellar luminosities L vary over some 10 orders of magnitude, $10^{-4} L_\odot < L < 10^6 L_\odot$, where $L_\odot$ represents the luminosity of the Sun.

1.1.1 Stellar Luminosities

A flux is defined to be the amount of energy crossing a unit surface area per unit time. The luminosity L of a star is the power required to sustain the total energy flux across a

---

[1] Chapters 1 and 2 review material normally covered in introductory astronomy courses. For readers without an introductory astronomy background they serve as an overview of concepts that will be important for later discussion. These chapters may be skipped if you are familiar with the basic properties of stars and with the relationship of luminosity to surface temperature for stars captured in the Hertzsprung–Russell diagram.

> **Box 1.1 — Ejection of Mass by Stars**
>
> The presence of the solar wind in our own Solar System suggests that all stars lose at least some mass continuously. However, many stars appear to have periods of very large mass loss early in their lives (*T Tauri winds* from stars just settling to the main sequence and strong mass flows from young, massive main sequence stars) and late in their lives (*red giant winds*, *planetary nebula*, and related phenomena). In addition, explosions such as novae and supernovae associated with dead and dying stars eject mass into interstellar space, sometimes in large amounts.
>
> Systematics of white dwarf populations give generic observational evidence that prior to the white dwarf stage many stars must undergo substantial mass loss [176]. In the solar neighborhood white dwarfs with accurately determined masses $\sim 0.4 M_\odot$ are found. Since the study of stellar evolution in clusters indicates that there has been insufficient time for stars formed with that little mass to have evolved to the white dwarf stage, these white dwarfs must have come from main sequence stars that have shed considerable mass since their formation. In addition, direct observation indicates the presence of white dwarfs in some clusters with masses less than the main sequence stars in the cluster, again indicating that they must have evolved from stars that underwent considerable mass loss in their evolution.

closed surface surrounding the star. It has units of energy per unit time and is a sum of three primary components, $L = L_\gamma + l_\nu + L_{\Delta m}$, which are associated with emission of photons, emission of neutrinos, and surface mass loss, respectively.

Photon emission: The total luminosity associated with the photon flux is L_γ. This flux is emitted primarily from the thin *photosphere* at the surface of the star; it is the principal luminosity source for most young stars, and is most often what is meant when speaking loosely of stellar luminosity.

Neutrino emission: The quantity L_ν is the total luminosity associated with neutrino emission from the star. Cooling by neutrino emission becomes important in massive stars late in their life and the energy of a core collapse supernova, which represents the death of a massive star, is radiated primarily in the form of neutrinos.

Surface mass loss: Most stars have mechanisms by which they lose mass from their surfaces (see Box 1.1). Since ejected matter must be lifted in a gravitational field, mass loss subtracts from the energy budget of the star and is a source of luminosity according to our general definition. The term $L_{\Delta m}$ accounts for this source.

1.1.2 Photon Luminosities

Henceforth, unless otherwise specified, by "luminosity" we will mean the photon luminosity. For a spherical star the luminosity is given by

$$L_\gamma = 4\pi R^2 \int_0^\infty F_\lambda \, d\lambda, \tag{1.1}$$

where F_λ is the net outgoing energy flux at wavelength λ and R is the radius. The corresponding flux f_λ detected at the surface of the Earth is reduced according to the inverse square law,

$$f_\lambda = (R/r)^2 F_\lambda, \qquad (1.2)$$

where r is the distance of the star from Earth. (Generally, the observed flux at the surface of the Earth must be corrected for absorption in the interstellar medium and in the Earth's atmosphere; those corrections will be discussed below.) Therefore, $F_\lambda R^2 = f_\lambda r^2$, and a measurement of $f_\lambda r^2$ at all λ determines $F_\lambda R^2$ at all wavelengths and hence the photon luminosity L_γ through Eq. (1.1). The most difficult part is likely to be the determination of the distance r, unless the star is nearby so that trigonometric methods may be used to measure the distance with confidence.

1.1.3 Apparent Magnitudes

It is useful to express actual and apparent brightness in terms of logarithmic *magnitude scales*. Two general classes of magnitudes may be distinguished: *apparent magnitudes*, which are associated with the apparent brightness of objects in our sky, and *absolute magnitudes*, which define brightness with the dependence on distance scales factored out. Apparent magnitudes will be discussed in this section and absolute magnitudes will be discussed in Section 1.1.5 below.

The apparent magnitude m is defined such that for two stars labeled 1 and 2 with observed fluxes f_1 and f_2, respectively,

$$m_2 - m_1 = 2.5(\log f_1 - \log f_2) = 2.5 \log\left(\frac{f_1}{f_2}\right), \qquad (1.3)$$

where log means the base-10 logarithm[2] and the normalization of the magnitude scale is discussed in Box 1.2. This definition implies that a difference of five orders of magnitude corresponds exactly to a factor of 100 in brightness, and that algebraically smaller magnitudes are associated with brighter objects. It is often useful to define a set of apparent magnitudes that are restricted to a limited range of frequencies (for example, by the use of telescopic filters; see Fig. 1.3 below). Some common ones are

1. The *visual magnitude* m_v, determined from the flux in the range of frequencies to which the human eye is sensitive (peaking in the yellow–green part of the spectrum).
2. The *blue-sensitive magnitude* m_B, which is the magnitude determined if the light is collected using a blue filter.
3. The *photovisual magnitude* m_V, which is the magnitude determined if the light is collected using a yellow filter to make the resulting magnitude correspond more closely to the visual magnitude defined above.
4. The *ultraviolet magnitude* m_U, which is determined using filters to emphasize the UV part of the spectrum.

Various other apparent magnitudes can be introduced by using other filters that emphasize different parts of the spectrum but the ones described above are common and representative. More will be said about such magnitudes when color indices are discussed in Section 1.3.

---

[2] In this book we use $\log \equiv \log_{10}$ to denote the base-10 logarithm and $\ln \equiv \log_e$ to denote the base-e or natural logarithm.

> **Box 1.2** **Normalization of Magnitude Scales**
>
> The definition of individual apparent magnitudes in Eq. (1.3) requires setting the scale by an arbitrary choice. The scales for the magnitudes m_v and m_V are set conventionally so that modern visual magnitude scales coincide well with the subjective scales of ancient astronomy. The convention in use employs the magnitudes of a set of stars to fix a scale but practically it is such that a spectral class A0 star such as Vega has visual magnitude zero (spectral classes are discussed in Chapter 2). In the resulting scale, the brightest stars in the sky have apparent visual or photovisual magnitudes near zero. A representative set of apparent visual magnitudes for common objects is displayed in Table 1.1. The magnitude zeros for other apparent magnitude scales are set by similar conventions. For example, the blue-sensitive magnitude scale is defined so that $m_B = m_V$ for a spectral class A0 star.

Table 1.1 Some apparent visual magnitudes

Object	m_V
Sirius (brightest star)	-1.5
Venus at brightest	-4.4
Full Moon	-12.6
The Sun	-26.8
Faintest naked-eye stars	$+6-7$
Faintest object visible from Earth with largest conventional telescopes	$\sim +25$
Faintest object visible from the Hubble Space Telescope	$\sim +30$

The apparent magnitudes discussed above conflate intrinsic properties (energy output) with geometric effects. It is desirable to factor out the distance dependence to address issues of stellar structure. This is done formally by introducing *absolute magnitude scales*, which will be defined in the next section. Before doing that, it is convenient to introduce a unit called the *parsec* that is a preferred unit of distance for astronomers.

1.1.4 The Parsec Distance Unit

The apparent relative positions of stars on the celestial sphere shift by small amounts over a six-month period because of the parallax effect as the Earth goes around its orbit. The angle p defined in Fig. 1.1, which is equal to half the angular size of the Earth's orbit as viewed from the star, is called the *parallax angle*; it is related trigonometrically to the distance d to the star through

$$\tan p = \frac{1\,\text{AU}}{d}, \qquad (1.4)$$

where the *astronomical unit* AU is the average separation of the Earth and Sun (the length of the Earth's semimajor orbital axis; $1\,\text{AU} \sim 1.5 \times 10^8$ km). A small-angle approximation

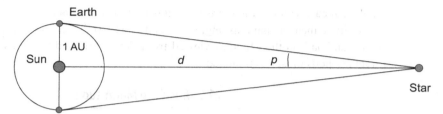

Fig. 1.1 The parallax angle p for a star as observed from Earth.

is justified and p (radians) $= 1\mathrm{AU}/d$. Converting the angular measure to seconds of arc (1 degree = 3600 arcsec $\equiv$ 3600" and 1 radian = 2.06×10^5 arcsec), permits writing

$$\frac{d}{1\,\mathrm{AU}} = \frac{2.06 \times 10^5}{p''}, \qquad (1.5)$$

where the notation indicates that p is to be given in seconds of arc.

The relationship between parallax angle and distance given by Eq. (1.5) suggests defining a natural distance unit equal to the distance at which a star would have a parallax angle of 1". This unit is termed the *parsec* (from concatenating "parallax" and "seconds"), and is abbreviated by the symbol pc. With these units the distance in parsecs is just the inverse of the parallax angle in seconds of arc:

$$d\,(\mathrm{pc}) = \frac{1}{p''}. \qquad (1.6)$$

From this equation the relationship of the parsec to other common distance units is easily found. For example,

$$1\,\mathrm{pc} = 2.06 \times 10^5\,\mathrm{AU} = 3.09 \times 10^{18}\,\mathrm{cm} = 3.26\,\mathrm{ly},$$

where a lightyear (ly) is the distance light travels in a year. Parallax angles for even the nearest stars are tiny, as illustrated in Example 1.1.

Example 1.1 The nearby star α Centauri has a measured parallax of 0.742" [2], corresponding to a distance of $d = 1/0.742 = 1.348\,\mathrm{pc}$, or 4.4 ly. To set this parallax angle in perspective, 1" is the angle subtended by a 2-cm diameter coin at a distance of 4 km.

Parallax angles can be measured reliably down to about 0.01" for ground-based telescopes without adaptive optics, so the traditional parallax method is useful for distance measurements out to about 100 pc (though the uncertainty becomes substantial for larger distances). Observations with the *Hipparcos satellite* could measure a parallax of 0.001" and extended the parallax range to about 1000 pc, allowing determination of high-precision parallaxes for more than 100,000 new stars (and 2.5 million additional stars at low precision). More recently the European Space Agency *Gaia* mission was launched in 2013 with a goal of mapping precisely the position, brightness, and variations in brightness, color, velocities, and evidence for a companion for more than 10^9 stars by 2018, including parallax measurements for more than 200 million new stars. To enable this it has the capability to measure parallax angles as small as 5×10^{-6} arcseconds (the angle subtended by a thumbnail on the Moon as viewed from Earth). Beyond these distances, other less-direct methods must be employed.

1.1.5 Absolute Magnitudes

By convention, the absolute magnitude, denoted by M to distinguish it from the apparent magnitude m, is the apparent magnitude that a star would have if it were placed at a standard distance of 10 pc = 32.6 ly. Using previous expressions for the apparent magnitude, it is then easy to show (Problem 1.3) that the absolute and apparent magnitudes are related by

$$m - M = 5 \log\left(\frac{d}{10\,\text{pc}}\right), \tag{1.7}$$

where the quantity $m - M$ is termed the *distance modulus*. Thus, the absolute magnitude is the apparent magnitude minus the distance modulus, and is easily calculated from (1.7) if the distance d to the star is known.

1.1.6 Bolometric Magnitudes

The *bolometric magnitude* is the magnitude that a star would have if the detector could collect the entire spectrum of emitted radiant energy. Realistic detectors cannot do this because of inherent detector limitations and losses in the atmosphere and interstellar medium, so it is necessary to apply a *bolometric correction* to raw magnitudes; this correction is designed to add back flux that is absorbed in the atmosphere or otherwise not detected. Then the absolute bolometric magnitude is

$$M_{\text{bol}} = M_{\text{v}} + \text{BC}, \tag{1.8}$$

where BC is the *bolometric correction*. (*Note*: Some authors define instead $M_{\text{bol}} = M_{\text{v}} - \text{BC}$, so be mindful of the sign for BC.) The bolometric correction is large for very hot and very cool stars because they output a substantial portion of their radiation at UV and IR wavelengths, respectively, and these wavelengths are absorbed strongly in the atmosphere. Even above the atmosphere there may be significant corrections for absorption in the interstellar medium.

Example 1.2 Because the Sun emits small amounts of UV and IR radiation relative to visible light, its bolometric correction is small. The Sun has an absolute bolometric magnitude of $M_{\text{bol}}^{\odot} = 4.74$, corresponding to a luminosity of $L_{\odot} = 3.828 \times 10^{33}$ erg s^{-1} (this luminosity will be estimated from observations below).

For calculations we often find it convenient to write for an arbitrary star

$$M_{\text{bol}} - M_{\text{bol}}^{\odot} = -2.5 \log \frac{L}{L_{\odot}}, \tag{1.9}$$

which corresponds to expressing the absolute bolometric magnitude M_{bol} and luminosity L for an arbitrary star in units of the corresponding quantities for the Sun.

1.2 Stars as Blackbody Radiators

A temperature can be defined for an object that is in thermodynamical equilibrium. In particular, we may introduce a temperature self-consistently for a star if it is a *blackbody radiator*. Stars are often assumed to be blackbody radiators. They generally are not perfectly so, but this is a sufficiently good approximation to be a very useful starting point.

1.2.1 Radiation Laws

A blackbody radiator has a radiation field that is isotropic, homogeneous, randomly polarized, and independent of the walls of the container. If a body satisfies these conditions, several important *radiation laws* apply.

Planck law: The *Planck radiation law* defines the intensity of emitted radiation for a blackbody. The Planck function $B_\lambda(T)$ giving the power emitted per unit surface area of a blackbody per unit wavelength into unit solid angle is given by

$$B_\lambda(T) = \frac{2hc^2}{\lambda^5} \frac{1}{e^{hc/\lambda kT} - 1}, \qquad (1.10)$$

where λ is the wavelength, T is the temperature, h is Planck's constant, c is the speed of light, and k is Boltzmann's constant. Blackbody spectra for several temperatures are illustrated in Fig. 1.2, where the total area under the curve is seen to grow rapidly with temperature and the distribution exhibits a single peak that shifts to shorter wavelengths as the temperature increases. Two other important laws governing this behavior may be derived from the Planck law, the *Stefan–Boltzmann law* and the *Wien displacement law* (see Example 1.3 and Problem 1.4). The first governs the total energy radiated at all wavelengths and the second governs the wavelength at which the peak intensity is emitted.

Stefan–Boltzmann law: The law of Stefan and Boltzmann says that the total energy E radiated per unit time per unit surface area at all wavelengths varies as the fourth power of the temperature,

$$E = \tfrac{1}{4}acT^4 = \sigma T^4, \qquad (1.11)$$

where a is the *radiation density constant* and σ is the *Stefan–Boltzmann constant*. Multiplication by the surface area then gives the luminosity. Thus for a spherical blackbody $L = 4\pi R^2 \sigma T^4$, where R is the radius.

Wien law: The Wien displacement law states that for a blackbody radiator the maximum in the radiation distribution as a function of wavelength occurs at

$$\lambda_{\max} = 2.90 \times 10^7 \left(\frac{\text{K}}{T}\right) \text{Å}, \qquad (1.12)$$

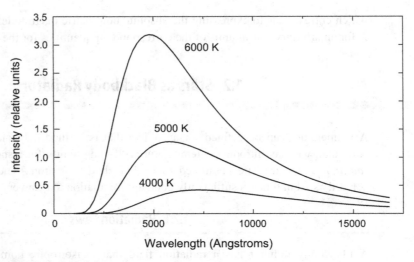

Fig. 1.2 Planck distribution for several temperatures.

where 1 angstrom (Å) = 10^{-8} cm. The Stefan–Boltzmann law explains the increase in total luminosity with temperature seen in Fig. 1.2, while the Wien law accounts for the shift of these distributions to shorter wavelengths as the temperature increases.

Example 1.3 The Stefan–Boltzmann and Wien laws follow from the more general Planck law. We may illustrate by outlining the derivation of the Stefan–Boltzmann law (you are asked to provide the details in Problem 1.4). The total energy flux emitted by a blackbody at temperature T can be expressed as an integral over Eq. (1.10), and using the substitutions

$$u = hc/\lambda kT \qquad \lambda = hc/ukT \qquad d\lambda = \frac{-hc}{u^2 kT}\, du,$$

the integral can be evaluated to give the Stefan–Boltzmann law, $E = \sigma T^4$, where σ is the Stefan–Boltzmann constant.

The Wien law results from differentiating Eq. (1.10) to find the maximum, as you are asked to show in Problem 1.13.

1.2.2 Effective Temperatures

If a star is assumed to be a blackbody radiator, we may use the Stefan–Boltzmann law to define an *effective surface temperature* T_e through the relation

$$L = 4\pi \sigma R^2 T_e^4. \tag{1.13}$$

That is, the effective temperature T_e is the temperature that a perfect blackbody of radius R would need in order to radiate the observed luminosity of the star. This is an integral condition that requires the total luminosity of the star and the fictitious blackbody that

> **Box 1.3** — **Luminosity and Effective Temperature of the Sun**
>
> The preceding discussion may be illustrated by determining the surface flux, total luminosity, and effective surface temperature of the Sun — a star for which the required quantities are known rather well. The total radiant flux of the Sun on our upper atmosphere (the *solar constant*) has a value of 1.36×10^6 erg cm^{-2}s^{-1}, and the Sun subtends an average angle of 32 minutes of arc in our sky at an average distance of 1 AU $\sim 1.496 \times 10^8$ km. Then from geometry the radius of the Sun is given by
>
> $$R_\odot = \tan(16') \, \text{AU} = 0.00465 \, \text{AU} = 6.96 \times 10^{10} \, \text{cm},$$
>
> and if the distance to the Sun is r, the flux at the solar surface is related to the flux on our upper atmosphere by the inverse square intensity law,
>
> $$\text{Flux (solar surface)} = \left(\frac{r}{R_\odot}\right)^2 \times \text{Flux (Earth)} = 6.28 \times 10^{10} \, \text{erg cm}^{-2}\text{s}^{-1}.$$
>
> The total solar luminosity follows from integrating this flux over the surface area of the Sun,
>
> $$L_\odot = 4\pi R_\odot^2 \times \text{Flux (solar surface)} \simeq 3.82 \times 10^{33} \, \text{erg s}^{-1},$$
>
> and this allows an effective surface temperature for the Sun to be calculated as
>
> $$T_e^\odot = \left(\frac{L_\odot}{4\pi \sigma R_\odot^2}\right)^{1/4} = 5770 \, \text{K}.$$
>
> A more careful analysis yields the standard value of $T_e^\odot \sim 5777$ K. Using this temperature in the Wien law (1.12) indicates that the Sun's spectrum peaks at about 5020 Å, which is the yellow–green part of the spectrum. More generally, R is known for only a few nearby bright stars; for others, R must be estimated in some model-dependent way, and an estimate of the distance to the star is required in order to determine the luminosity.

approximates it to be equivalent, but does not constrain whether the detailed wavelength distribution of emitted radiation for the star and the fictitious blackbody are equivalent. Box 1.3 illustrates the use of Eq. (1.13) to determine an effective temperature for the Sun.

1.2.3 Stellar Radii from Effective Temperatures

The (limited) direct ways available to determine the physical sizes of stars will be discussed later in this chapter. However, if a star of known luminosity is assumed to be a spherical blackbody and the effective temperature T_e is estimated from the spectrum (see Chapter 2), Eq. (1.13) can be used to solve for a radius. For calculations, it is often useful to relate the radius R and effective temperature T_e of a star to corresponding quantities for the Sun through

$$\frac{R}{R_\odot} = \left(\frac{T_e^\odot}{T_e}\right)^2 \left(\frac{L}{L_\odot}\right)^{1/2}, \tag{1.14}$$

where $T_e^\odot = 5777\,\text{K}$ and $L_\odot = 3.828 \times 10^{33}\,\text{erg}\,\text{s}^{-1}$ (see Box 1.3). Of course, radii determined in this way are effective radii that are only approximations to physical radii since real stars are not perfect blackbodies. Nevertheless, in many cases this may be expected to yield a rather good estimate of the true radius of the star.

1.3 Color Indices

As noted in Section 1.1.3, magnitudes associated with specific wavelength regions of the electromagnetic spectrum may be defined. This is typically accomplished by using filters that allow light in only a range of wavelengths to pass. Three common filters are illustrated in Fig. 1.3; these filters are termed ultraviolet (U), blue (B), and photovisual (V) because of the wavelength regions that they emphasize. Then quantities called *color indices* (CI) may be defined by taking the difference of the apparent magnitudes in these localized regions of the spectrum. For the filters illustrated in Fig. 1.3 there are three possible independent combinations, conventionally taken to be

$$\text{CI}_{\text{UB}} \equiv U - B \qquad \text{CI}_{\text{BV}} \equiv B - V \qquad \text{CI}_{\text{UV}} \equiv U - V, \qquad (1.15)$$

where we let the symbols (U, V, B) stand for the apparent magnitudes. For example,

$$U - B = -2.5 \log \left(\frac{\int F_\lambda S_\lambda^U\, d\lambda}{\int F_\lambda S_\lambda^B\, d\lambda} \right) + \text{constant}, \qquad (1.16)$$

where F_λ is the monochromatic flux at wavelength λ, the functions S_λ^U and S_λ^B encode the full response of the detectors and filters as a function of wavelength, and the constant is set by convention. These continuous color indices are indicators of the surface temperature for stars, as suggested in Figs. 1.2 and 1.4, because they indicate the slope

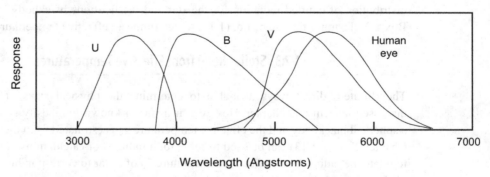

Fig. 1.3 Some common wavelength filters used in astronomical spectroscopy. Each filter has an acceptance width of about 1000 Å. The V filter acceptance is similar to that of the human eye but the U and B filters emphasize much shorter wavelengths.

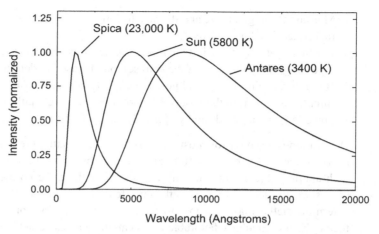

Fig. 1.4 Blackbody intensity distributions for three stars having different surface temperatures. Unlike for Fig. 1.2, each distribution has been normalized to unity at its peak. In reality, the area under the curve for Spica would be 2094 times larger than for Antares, by virtue of Eq. (1.11).

of the blackbody radiation distribution over a range of wavelengths, which is related to the surface temperature.

Example 1.4 Consider the hot star Spica, with a radiation distribution illustrated in Fig. 1.4. For Spica $U = -0.20$, $B = 0.74$, and $V = 0.97$ [2]. The corresponding $B - V$ and $U - B$ color indices are

$$B - V = 0.74 - 0.97 = -0.23 \qquad U - B = -0.20 - 0.74 = -0.94.$$

The negative signs of the color indices in this case are an indication that Spica is a very hot star, as is apparent from Fig. 1.4 and the Wien law. On the other hand, for the star Antares $B = 2.75$ and $V = 0.91$ [2], giving a $B - V$ color index of

$$B - V = 2.75 - 0.91 = 1.84.$$

The positive value of the color index in this case is an indication that Antares is a cool star, as is also apparent from Fig. 1.4 and the Wien law.

1.4 Masses and Physical Radii of Stars

In considering the essential properties of a star, certainly the radius and total mass are quantities of fundamental interest. However, it is very difficult to determine either of these quantities and they are known directly for only a small subset of stars. There are three ways in which the radius of a star can be determined:

1. Measure the angular size and the distance of the star and use trigonometry to deduce the radius, as was done in Box 1.3.
2. For binary star systems in which one star eclipses another, occultation methods (based on timing of the duration of the eclipse) may be used to determine a size.
3. If the distance to the star is known, the blackbody luminosity relations may be used to infer the radius from the observed distribution in wavelength. This is less direct than the preceding two methods and yields only approximate radii.

Determining the mass for most stars is also problematic. Generally we can determine a mass by pushing something (inertial mass) or by watching the gravitational interaction with another object (gravitational mass).[3] Since there are no methods at our disposal to push stars, the second method is the only feasible one. Typically, the only reliable way to determine stellar masses is to observe the interactions in a binary star system, and even then only in particularly favorable cases can the masses be determined reliably.

1.5 Binary Star Systems

From our parochial perch in the Solar System it is easy to conclude that single stars like our Sun are the norm in the Universe. However, most heavier stars, and a significant fraction of all stars, appear to be parts of multiple star systems; most common among those are *binary systems*, in which two stars orbit around the common center of mass for the system. Figure 1.5 shows the binary system Castor, which was deduced by Herschel to be binary in the 1790s. This is an example of a *visual binary*, where both stars can be seen from Earth. If only one of the stars can be seen, it is often possible to infer that a star has an unseen companion through the influence of the orbiting companion on the star's proper motion,

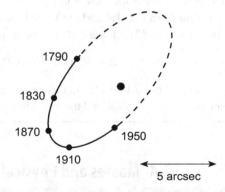

Fig. 1.5 The visual binary Castor, the first binary star system discovered. Numbers give the year of observation. The orbit coordinate system is that illustrated in Fig. 1.9(b).

---

[3] The gravitational and inertial masses of an object are equal, by virtue of the equivalence principle of general relativity.

1.5 Binary Star Systems

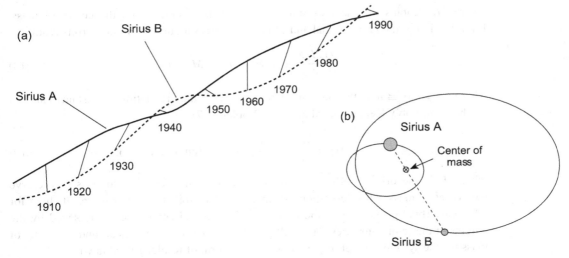

Fig. 1.6 (a) Perturbation of proper motion on the celestial sphere for Sirius A caused by companion star Sirius B from 1910 through 1990. (b) Apparent orbits (projection of true orbit on the celestial sphere) of the Sirius binary system with relative locations of the two stars in 1990. Orbits are to scale but the sizes of the stars are not.

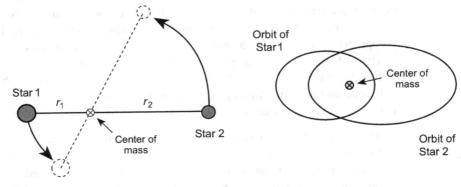

Fig. 1.7 True orbits for binary star systems.

as illustrated in Fig. 1.6(a). If the translational (proper) motion of the binary system is subtracted out, the projection of the orbital motion is recovered, as shown in Fig. 1.6(b).

1.5.1 Motion of Binary Systems

The schematic geometry of a binary system is illustrated in Fig. 1.7. The orbits are ellipses in the most general case and each star revolves around the center of mass for the binary, with the separation between the star and center of mass given by r. From the definition of the center of mass

$$M_1 r_1 = M_2 r_2, \tag{1.17}$$

where M_i denotes the mass of star i and r_i is its distance from the center of mass. Equilibration of the gravitational and centrifugal forces assuming circular orbits requires

$$\frac{GM_1M_2}{(r_1+r_2)^2} = M_1\omega^2 r_1 = M_2\omega^2 r_2, \tag{1.18}$$

where $\omega_1 = \omega_2 \equiv \omega$ is the angular velocity and G is the gravitational constant. This can be shown to imply Kepler's third law (see Problem 1.7):

$$M_1 + M_2 = \frac{4\pi^2}{G}\frac{(r_1+r_2)^3}{P^2} \quad \text{(circular orbits)}, \tag{1.19}$$

where P is the orbital period, which must be the same for the two stars to preserve the center of mass. This derivation assumed circular orbits but the same results hold for elliptical orbits, provided that the orbital radii for the circular case are replaced by the lengths of the semimajor axes for the elliptical motion of the two stars around the center of mass for the system. Therefore, the most general form of Kepler's third law is

$$M_1 + M_2 = \frac{4\pi^2}{G}\frac{(a_1+a_2)^3}{P^2}, \tag{1.20}$$

where a_1 is the length of the semimajor axis (half the long axis of the ellipse) for the orbit of star 1 about the center of mass and a_2 is the corresponding quantity for the second star. As shown in Box 1.4, this equation simplifies for typical applications in the Solar System if appropriate units are chosen and one of the masses is small compared with the other.

Thus, if a_1, a_2, and the period P can be measured, we may use Eq. (1.20) to deduce the total mass $M_1 + M_2$ and then use Eq. (1.17) to determine the masses of the individual components. Although valid in principle, there are some basic difficulties with implementing this in practice:

1. The distance to most binary systems is too large to permit separating the binary pair images observationally.
2. Even if the components of the binary system can be distinguished, what is seen is the *projection* of the elliptical orbits in 3-dimensional space on the 2-dimensional celestial sphere [see Fig. 1.9(a) below]; the 3D orientation of the ellipse is unknown without further information.

Box 1.4	Reduction to Simplified Form of Kepler's Third Law

In the Solar System the mass of a planet is small relative to the Sun and Eq. (1.20) can be simplified. As shown in Problem 1.8, the factor $4\pi^2/G$ is numerically equal to unity if units are chosen such that distance is measured in astronomical units, time in Earth years, and mass in solar masses. Then if the mass of the planet is neglected relative to that of the Sun, the familiar $P^2 = a^3$ proposed by Kepler for planetary motion is obtained. However, in binary star systems the masses of the two stars are often comparable and the effect of the center of mass implies a large modification for this original simple form of the third law proposed by Kepler.

3. If the components can be distinguished and their motion followed in a binary system, the separations observed are in angle; to convert those to a physical length requires knowing the distance to the binary, and this might not be known with precision.

Despite these problems, for a limited number of binary systems it is possible to obtain precise information on the masses and diameters of the components, as will now be described. However, for most binary systems one can obtain (at best) only limits on these quantities.

1.5.2 Radial Velocities and Masses

The preceding difficulties may be circumvented if we can obtain information about the orbits in a binary system by means other than direct measurement of the geometry. One possibility is to use the Doppler effect to determine radial velocities for the two stars and then to use this information to estimate masses. The orbital velocities of binary stars are small enough to justify use of the nonrelativistic Doppler formula

$$\frac{v_\mathrm{r}}{c} = \frac{\Delta \lambda}{\lambda_0} = \frac{\lambda_v - \lambda_0}{\lambda_0}, \tag{1.21}$$

where $\Delta \lambda$ is the shift in wavelength of a spectral line that is normally at a wavelength λ_0 and v_r is the radial component of the velocity. As illustrated schematically in Fig. 1.8 for a simple binary system having stars with circular orbits and the same mass and spectra, and an observer assumed to be in the plane of the binary orbits, the Doppler effect leads to periodic doubling of the spectral lines if the light from both stars is collected simultaneously. This periodic doubling gives a way to determine that a system is binary, even if the two components cannot be resolved observationally. Binaries that are inferred from details of their spectra are called *spectroscopic binaries,* as discussed further in Box 1.5.

If for simplicity of illustration we assume circular orbits with their plane along the line of sight, because the period is the orbital circumference divided by the orbital velocity (which is constant for circular orbits),

$$v_1 = \frac{2\pi r_1}{P} \qquad v_2 = \frac{2\pi r_2}{P}. \tag{1.22}$$

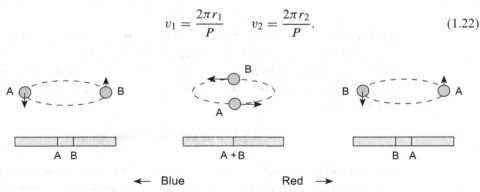

Fig. 1.8 A double-line spectroscopic binary. Periodic doubling of spectral lines because of Doppler shifts indicates that the system is binary, even if the two stars can't be resolved.

> **Box 1.5**   **Spectroscopic Binaries**
>
> A binary system that is not resolved telescopically but is inferred to be binary from Doppler shifts of spectral lines is termed a *spectroscopic binary*. In fact, most known binary systems are spectroscopic binaries because only when binaries are relatively nearby and/or have large orbital separations can the components be resolved telescopically. Spectroscopic binaries may be further subdivided into *double-line* spectroscopic binaries, where lines from both stars are seen in the spectrum and the schematic picture is as in Fig. 1.8, and *single-line* spectroscopic binaries, where the lines from one star are too faint to see but it is still possible to infer the binary nature of the system from the periodic shift of the lines observed for the other star.

Therefore, if the radial velocities for *both Stars* are measured, the mass ratio may be obtained as

$$\frac{v_1}{v_2} = \frac{r_1}{r_2} = \frac{M_2}{M_1}, \qquad (1.23)$$

where (1.17) has been used. In addition, since Eq. (1.22) implies that $r_i = Pv_i/2\pi$,

$$r_1 + r_2 = \frac{P}{2\pi}(v_1 + v_2).$$

Inserting this result into Kepler's law (1.20) with $r_i = a_i$ for circular orbits yields

$$M_1 + M_2 = \frac{P(v_1 + v_2)^3}{2\pi G}, \qquad (1.24)$$

and Eqs. (1.23) and (1.24) can be solved simultaneously for the individual masses. Unfortunately, in the realistic case the preceding analysis is complicated by two issues:

1. The orbits are ellipses, not circles.
2. The line of sight for the observer is usually not the orbital plane of the binary system.

Since the orbital plane will in the general case be tilted by some (often unknown) angle i with respect to the observer, the realistic situation is as illustrated in Fig. 1.9(a) and the observer measures only a component

$$v_r = v \sin i \qquad (1.25)$$

of the orbital velocity v. Therefore, masses cannot be determined unless the angle i can be ascertained. However, even if i is unknown Eq. (1.25) may allow a limit to be set on masses; this can still be very useful information. Finally, spectral lines often are visible from only one component of the binary. This more complex case will be discussed in Section 17.4.1.

1.5.3 True Orbit for Visual Binaries

Conventionally, for visual binaries the orbit of the less bright component is plotted relative to the position of the brighter component using the coordinate system that is illustrated in Fig. 1.9(b). This represents the *apparent relative orbit*. The *true relative orbit* is usually tilted by an angle i relative to the line of sight, as illustrated in Fig. 1.9(a). The true orbit

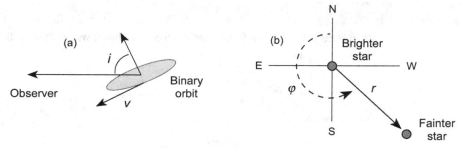

Fig. 1.9 (a) Tilt angle *i* of a binary orbit. (b) Coordinate system for visual binaries. Compass directions on the celestial sphere are indicated on the axes.

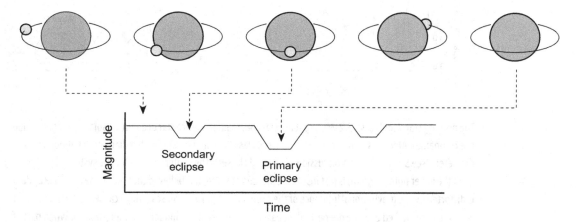

Fig. 1.10 Schematic representation of an eclipsing binary system.

results from "untilting" the apparent orbit. Detailed observation of a visual binary often can tell us something about the tilt angle. For example, if the true orbit is elliptical the projected orbit will also be elliptical but the primary star will not be at the focus of the ellipse if the tilt angle is non-zero.

In favorable cases the distance to a visual binary may be found by noting from Eq. (1.22) that $r_1 + r_2$ can be inferred if v_1 and v_2 can be measured by Doppler methods. Then observation of the orbit allows determination of the angular semimajor axis α and the distance can be computed then from

$$d = \frac{r_1 + r_2}{\alpha}. \tag{1.26}$$

Thus, precise distances to visual binaries can be determined without using parallax, but the number of binary systems for which the required observations are possible is limited.

1.5.4 Eclipsing Binaries

For a small subset of binary systems the stars periodically totally or partially eclipse each other, as illustrated in Fig. 1.10. These *eclipsing* binaries have some important features.

Box 1.6 — Winking Demons

Algol (β Persei or the "Winking Demon Star") is an eclipsing binary system. Its properties are illustrated in the following diagram.

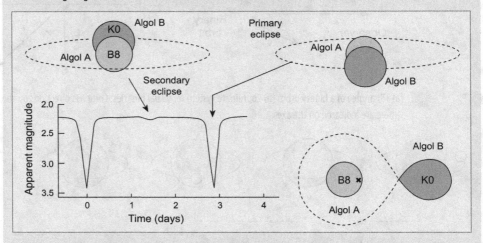

The primary star, Algol A, is a 3.2 $M_\odot$, spectral class B8, main sequence star of about 2.7 solar diameters, while the companion Algol B is a cooler 0.7 $M_\odot$, spectral class K0, subgiant star with a diameter of about 3.5 solar diameters (see Section 2.1 for a discussion of spectral classes). The center of mass for the system (denoted by × in the lower right diagram) is just inside the primary star, and it is believed that the shape of the secondary is distorted by the gravitational influence of the primary. The partial eclipses (which cause light variation that is visible to the naked eye) are deepest when part of the hotter star is hidden and more shallow when part of the cooler star is hidden in eclipse; these eclipses occur with a period of 68.8 hours.

1. For eclipsing binaries the angle i defined in Fig. 1.9(a) is approximately $\frac{\pi}{2}$; otherwise the eclipses would not be visible. Furthermore, from the details of the lightcurves (for example, the depth of the eclipses), the value of i can often be fixed even more precisely.
2. The period can be determined directly from the lightcurve, whether the individual components can be resolved or not.
3. If the eclipse is approximately total, precise timing of the lightcurve (duration of eclipses) can give the radii of the stars relative to the size of the orbit.

Notice that because the tilt angle is known rather well the true orbital velocity can be determined; then the time for an eclipse gives the true radius without needing to know the distance to the binary. A well-known eclipsing binary is described in Box 1.6.

1.6 Mass–Luminosity Relationships

Masses and radii for some visual binary systems are summarized in Table 1.2 (naming systems for the stars in this table are discussed in Box 1.7), and the corresponding quantities

1.6 Mass–Luminosity Relationships

Table 1.2 Masses and radii for some visual binaries [52, 173]

Star	Spectral/luminosity class	$R/R_\odot$	$M/M_\odot$	Parallax (arcsec)
α CMa A	A1 (V)	1.68	2.20	0.377
α CMi	F5 (IV–V)	2.06	1.77	0.287
ξ Her A	G0 (IV)	2.24	1.25	0.104
ξ Her B	K0 (V)	0.79	0.70	0.104
α Cen	G2 (V)	1.27	1.14	0.743
γ Vir	F0 (V)	1.35	1.08	0.094
η Cas A	G0 (V)	0.98	0.91	0.172
η Cas B	M0 (V)	0.59	0.56	0.172
ξ Boo	G8 (V)	0.77	0.90	0.148

Table 1.3 Main sequence masses and radii [52, 173, 196]

Spectral class	CI (B–V)	$R/R_\odot$	$M/M_\odot$
O8	−0.32	10	23
B0	−0.30	7.5	16
A0	−0.01	2.4	2.8
F0	+0.30	1.4	1.6
G0	+0.58	1.1	1.1
K0	+0.81	0.85	0.82
M0	+1.40	0.60	0.54
M5	+1.62	0.27	0.20

Box 1.7 — Naming the Stars

In Table 1.2 stars are named using the *Bayer system*, in which brighter stars are designated by an abbreviation for the Latin possessive of the constellation and a Greek letter giving the (approximate) order of brightness within the constellation. Stars may be named also using the *Flamsteed system*, which employs the Latin possessive of the constellation and an Arabic numeral indicating the order of the star's location with respect to the western edge of the constellation. Thus α Car is α Carinae, the brightest star in the constellation Carina, and 31 Leo is 31 Leonis, the 31st star from the western edge of the constellation Leo. The brighter stars also have common names; for example, α Ori is Betelgeuse and α Car is Canopus. Letter suffixes indicate a component of a multiple star system. Thus, α CMa A is the brightest component of the Sirius binary star system.

---

for various main sequence spectral types are summarized in Table 1.3. Systematics indicate that there is a strong correlation between the mass of a star and its luminosity. Generally, it is found that a *mass–luminosity relationship* expressed in the form of a power law,

$$L \simeq M^\alpha, \qquad (1.27)$$

with $\alpha \sim 3.5$ is valid for many main sequence stars (not white dwarfs and red giants). This mass–luminosity relationship is illustrated in Fig. 1.11.

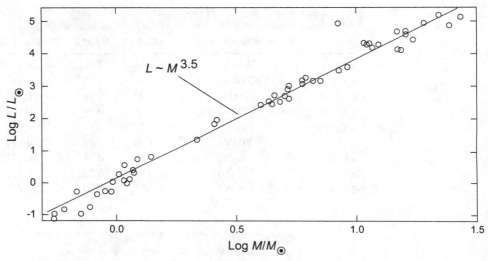

Fig. 1.11 Mass–luminosity relationship. Main sequence stars with $M \geq 1 M_\odot$ obey $L \simeq M^{3.5}$.

1.7 Summary of Physical Quantities for Stars

Let us summarize our findings to this point for the basic physical properties of stars. In units of the solar quantities $M_\odot = 1.99 \times 10^{33}$ g, $L_\odot = 3.83 \times 10^{33}$ erg s^{-1}, and $R_\odot = 6.96 \times 10^{10}$ cm, it is found that for most stars the masses M, luminosities L, radii R, and effective surface temperatures T_e lie in the ranges

$$10^{-1} M_\odot < M < 100 M_\odot \qquad 10^{-4} L_\odot < L < 10^6 L_\odot$$
$$10^{-2} R_\odot < R < 10^3 R_\odot \qquad 2 \times 10^3 \text{ K} < T_e < 10^5 \text{ K}.$$

Later we will address *why* stars should have physical properties respecting these limits.

1.8 Proper Motion and Space Velocities

Although their apparent motion is slow because of their great distance, the stars change their relative position on the celestial sphere over time. The rate of change in angular position is called *proper motion*, which is usually expressed in seconds of arc per year. The star with the largest known proper motion, Barnard's Star, changes its relative position on the celestial sphere by about 10.3 seconds of arc per year (which extrapolates to the angular diameter of the Moon over 175 years). Hundreds of stars are known with proper motions of more than one second of arc per year.

The velocity of a star with respect to the Sun is termed the *space velocity* v_s. This velocity may be resolved into a component perpendicular to the line of sight termed the *tangential*

Fig. 1.12 The space velocity v_s and its components for a star.

velocity v_t (which is responsible for the proper motion), and a component along the line of sight termed the *radial velocity* v_r, as illustrated in Fig. 1.12. The radial velocity can be measured from shifts in spectral lines using the Doppler formula (1.21),

$$\frac{v_r}{c} = \frac{\Delta \lambda}{\lambda_0} = \frac{\lambda_v - \lambda_0}{\lambda_0}.$$

The tangential component of velocity requires knowledge of the distance of the star to convert the proper motion (in angle) into a tangential velocity. The tangential velocity is

$$v_t = 4.74\, \mu d \text{ km s}^{-1}, \tag{1.28}$$

where μ is the rate of proper motion in arcsec per year and d is the distance in parsecs. The space velocity magnitude is then given by

$$v_s^2 = v_r^2 + v_t^2, \tag{1.29}$$

and the direction is given by trigonometry from Fig. 1.12. Typical space velocity magnitudes for stars are 20–100 km s^{-1}.

1.9 Stellar Populations

Much of the initial discussion in this chapter has focused on the properties of individual stars but properties shared by large groups of stars also are significant in understanding stellar structure and stellar evolution. A set of individual stars sharing a similar set of group characteristics is termed a *stellar population*.

1.9.1 Population I and Population II

Consider the following history of our galaxy. The original galaxy formed in a Universe dominated by hydrogen with some helium, and only traces of any other elements, because that is what was produced in the big bang. The original galaxy is thought to have been more spherical, with the subsequent flattening into a rotating disk resulting from conservation of angular momentum as the galaxy contracted. Thus, the first generations of stars were poor in elements heavier than hydrogen and helium ("metal-poor stars"), and these stars were formed with a more spherical distribution than the present disk. They constitute the globular clusters presently found in the halo surrounding the galaxy and in the central

Table 1.4 Some features of stellar populations I and II [55]		
Characteristic	Population I	Population II
Heavy element concentration	2–3%	< 1%
Dominant spectral classes	O, B, A	K, M
Most luminous stars	Blue supergiants	Red giants
Location	Galactic disk	Nucleus and halo
Age	< 1.5×10^9 years	> 1.5×10^9 years

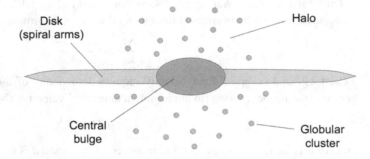

Fig. 1.13 A schematic picture of our galaxy.

bulge of the galaxy (Fig. 1.13). This set of old, metal-poor stars concentrated in the globular clusters and galactic core is called *Population II* (or just "Pop II" for short).

In contrast, younger stars must have formed in the galactic disk because that is where the star-forming material (gas and dust) presently is found. Over its history, the galaxy has been enriched in heavier elements by stellar processing and distribution; thus, this younger population of stars is richer in metals. These stars typically form in open clusters in the spiral arms, and constitute what is called *Population I* (or "Pop I" for short).

To summarize, Pop I is a mix of stars like that found in open clusters or in the vicinity of the Sun, while Pop II is that mixture of stars typical for globular clusters. These populations have different characteristics, as will be explored more extensively in conjunction with the Hertzsprung–Russell diagram (Chapter 2). For example, the most luminous stars in Pop I are blue supergiants, but the most luminous stars in Pop II are much fainter red giants. Some principal characteristics of these two populations are summarized in Table 1.4.

1.9.2 Population III

Theory and simulations provide strong evidence for an extremely metal-poor population of stars that is believed to represent the very first generation of stars formed in the Universe. This is termed *Population III*. Since the elements beyond helium are not produced in significant amounts in the big bang and must be synthesized in stars, this first generation of stars contained almost no metals. It is believed that Pop III stars grew to hundreds of solar masses or more because of a complex set of processes associated with their low metal

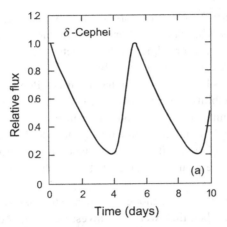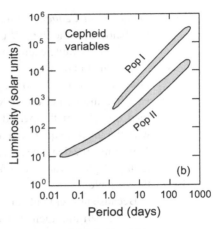

Fig. 1.14 (a) Lightcurve for δ-Cephei. (b) Period–luminosity relations for the two classes of Cepheid variables.

content and that these stars quickly exploded as *supernovae* (often of the *pair instability* type; see Box 20.2), seeding the Universe with heavier elements up to iron. It is highly unlikely that Pop III stars survive in the nearby Universe. In principle Pop III stars might be detected in high redshift galaxies dating to the early Universe; there are hints of Pop III stars in galaxies at very large redshift, but thus far no such interpretation has been confirmed. The study of Pop III is important because of the light it sheds on initial structure formation in the Universe.

1.10 Variable Stars and Period–Luminosity Relations

As discussed in Section 1.5.4, eclipsing binary stars vary in brightness because of the eclipses but many other stars vary their light output because of changes in their intrinsic properties. Variable stars may be classified broadly into three categories: (1) eclipsing variables, (2) pulsating variables, and (3) eruptive variables. Examples of pulsating variables include Cepheid variables, RR Lyra variables, and long-period red variables. Examples of eruptive variables are novae, supernovae, and X-ray bursters. As will be illustrated later in Fig. 15.1, pulsating variable stars tend to occur in localized regions of temperature–luminosity space (the *Hertzsprung–Russell diagram* to be discussed in Chapter 2).

1.10.1 Cepheid Variables

Cepheid variables are named for the prototype, δ-Cephei, whose lightcurve is shown in Fig. 1.14(a). They are yellow supergiant stars that vary their light output with a well-defined period typically lying in the range 1–100 days. Because they are luminous they can be seen in other nearby galaxies as well as our own. There are two classes of Cepheid variables, the *classical Cepheids*, which are Population I stars (the prototype δ-Cephei is a classical Cepheid), and the *Type II Cepheids*, which are Population II stars. These two classes of Cepheids resemble each other except for subtle spectral effects.

As illustrated in Fig. 1.14(b), Cepheid variables obey striking period–luminosity relations. This means that once it has been established that a star is a Pop I or Pop II Cepheid variable and its period for variability has been measured, its absolute magnitude can be read directly off the period–luminosity graph, and this can then be used in conjunction with the inverse square intensity law and the apparent magnitude to determine its distance. Thus, once the distance scale is calibrated, the period can be used to determine the distance to any Cepheid variable and hence to any grouping of stars (such as a galaxy) containing the variable.

Because no Cepheid variables were close enough to Earth to use standard parallax methods to determine their distance when they were first investigated, the period–luminosity relations for Cepheid variables were calibrated originally using indirect methods such as *statistical parallax* (a method that determines the mean parallax for a population of stars by analyzing their proper motion with the effect of the Sun's motion approximately removed) or spectroscopic parallax, which is described in Section 2.5).[4] The Cepheid variable period–luminosity relation was used by Henrietta Swan Leavitt (1868–1921) in 1917 to demonstrate conclusively that the Magellanic Clouds were too far away to be part of our own Milky Way Galaxy and thus constituted separate galaxies in their own right. Later, in 1925 Edwin Hubble (1889–1953) used Cepheids to show that the "spiral nebulae" such as Andromeda were too distant to be part of our own galaxy and thus were also external galaxies.

1.10.2 RR Lyra Variables

Another class of pulsating variable star that may be used to determine distances is that of the RR Lyra variables, with the class named after the prototype, RR Lyra.[5] The RR Lyra variables are of much shorter period than the Cepheid variables (typically 0.5–1 day); they are Pop II blue giants (spectral class $\sim$ F0), and all are of approximately the same luminosity. Thus, they are confined to a small region of temperature–luminosity space, as will be illustrated in Fig. 15.1, and the observed brightness of an RR Lyra variable indicates rather directly its distance. Because the RR Lyra stars are giants, they are much less luminous than the supergiant Cepheid variables and cannot be used to measure distances reliably beyond our galaxy. All known RR Lyra variables are telescopic stars since the brightest, RR Lyra, is at the unaided visual limit (apparent magnitude 7).

---

[4] More recently, space-based observations have extended the range of the parallax method considerably and have permitted the distances to larger numbers of Cepheids to be measured directly by parallax. For example, the parallax distances to several hundred Cepheid variables were determined using the Hipparcos data set and further improvements are expected with the newer Gaia data.

[5] This is an example of a common nomenclature in naming variable stars where a one- or two-letter sequence of capital letters (or the letter V with a number appended) is followed by a Latin possessive for the constellation. The sequence of letters gives the temporal order of discovery for variable stars in the constellation (with the convention being rather esoteric). Two well-known examples of this naming system are RR Lyra and T Tauri. Brighter variable stars (for example, Betelgeuse) usually go by their common or Bayer name.

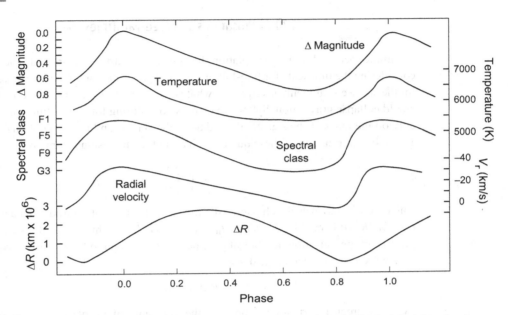

Fig. 1.15 Variation in brightness, temperature, spectral class, surface radial velocity V_r, and radius for δ-Cephei as a function of the phase of the pulsation period.

1.10.3 Pulsational Instabilities

Cepheid and RR Lyra variability is associated with pulsational instability, as illustrated in Fig. 1.15, which indicates that the pulsation in brightness of a Cepheid variable is correlated with a variation in the spectral class, surface temperature, and surface radial velocity of the star (the first two determined from the pattern of spectral lines and the last from Doppler shift of those spectral lines). The periodic sign reversal of the radial velocity (referenced to a constant space velocity) is a direct indicator that the radius is pulsating for such stars.

The change in radius ΔR may be measured by using the Doppler shift to determine the surface velocity corrected for motion of the center of mass relative to the Earth; then from $dR = v dt$ and Eq. (1.21) [52],

$$\Delta R(t) = R(t) - R(t_0) = \int_{t_0}^{t} v\, dt = c \int_{t_0}^{t} \frac{\lambda(t) - \lambda_0}{\lambda_0}\, dt. \qquad (1.30)$$

We may derive an expression for the total radius of the star by comparing magnitudes at two points in the oscillation corresponding to equivalent surface temperatures. Since the surface temperatures are the same, it may be assumed that the variation in brightness is caused entirely by the change in surface area and therefore

$$m_V(t_1) - m_V(t_2) = -2.5 \log \frac{R^2(t_1)}{R^2(t_2)}. \qquad (1.31)$$

Equations (1.30) and (1.31) give two equations to solve for R and ΔR. From these relations it is found that for typical Cepheids the pulsations lead to a variation in radius $\Delta R/R \simeq 0.20$.

1.10.4 Pulsations and Free-Fall Timescales

A simple period–luminosity relation for pulsating variable stars may be derived by considering the timescale for free-fall in the star's gravitational field: if the pressure support for the star were suddenly taken away, what is the characteristic timescale on which the star would collapse gravitationally? Since there is no restoring force, this timescale can depend only on the mass of the star M, the radius of the star R, and the gravitational constant G. The only combination of these quantities having the dimension of time is

$$t_{\text{ff}} = \left(\frac{R^3}{GM}\right)^{1/2}, \tag{1.32}$$

which may be taken as the characteristic timescale for free-fall. Introducing a mean density $\bar{\rho} \sim M/R^3$, this may be written as $t_{\text{ff}}(\bar{\rho})^{1/2} \simeq G^{-1/2}$. If this timescale is equated with a period for one pulsation, the pulsational period P is found to be inversely proportional to the square root of the average density:

$$P \simeq (G\bar{\rho})^{-1/2}. \tag{1.33}$$

(A more precise derivation supplying the constants of proportionality is requested in Problem 4.1.) Now if the surface temperature is fixed, the larger the star the more luminous it is. But $\bar{\rho}$ is much smaller for larger stars because they are so diffuse. For example, the average density of the Sun is 1.4 g cm^{-3} but that of a typical supergiant may be only 10^{-7} g cm^{-3}, and that of a white dwarf is approximately 10^6 g cm^{-3}. Thus, Eq. (1.33) predicts a period–luminosity relation in which the period goes up as the luminosity goes up – *in reality the period–luminosity relation is a period–density relation*, and there is generally an inverse correlation between average stellar density and the luminosity.

This result provides a qualitative explanation for why the period of Cepheid variables is much longer than that of RR Lyra variables. Cepheids are supergiants, with much larger radii and much smaller average densities than the RR Lyra stars, which are giants. Thus, the periods for Cepheids should be considerably longer than for RR Lyra variables, as observed. Another important class of pulsating variables is the long-period red variables (see the long-period variable region in Fig. 15.1). They are red supergiants, which are the largest stars, and their periods can be hundreds of days – even longer than Cepheids.

Example 1.5 If the luminosity and average density are assumed to be related by a power law $L \simeq k(\bar{\rho})^{-\alpha}$ [55], then (1.33) may be invoked to yield a period–luminosity relation

$$P \simeq L^{1/2\alpha}, \tag{1.34}$$

which can be expressed in terms of absolute magnitude M as

$$\log P = -\frac{1}{5\alpha} M + \text{constant}. \tag{1.35}$$

Many variable stars (for example, Cepheids) are observed to obey a period–luminosity relation of this form.

Background and Further Reading

Good introductions to the material of this chapter may be found in Böhm-Vitense [52]; Clayton [71]; Hansen, Kawaler, and Trimble [107]; Carrol and Ostlie [68]; and Tayler [211], or any good introductory astronomy text. The book by Hansen, Kawaler, and Trimble [107] contains an annotated guide to selected literature at the end of each chapter that will prove useful for those wishing to pursue many of the topics discussed here in more depth.

Problems

1.1 The star Sirius is 8.6 lightyears away. What is its parallax angle?***

1.2 The star Gliese 710 is presently 62 lightyears from Earth, but Hipparcos data suggest that in about a million years it will pass within one lightyear of Earth. Its apparent visual magnitude is +9.7. What is its absolute magnitude? What will its apparent visual magnitude be in a million years as viewed from Earth (assume that the intrinsic brightness remains the same as it is today)? What is the present radial velocity of Gliese 710?

1.3 Show that the absolute magnitude M and apparent magnitude m are related by

$$M = m - 5 \log \frac{d}{10},$$

where d is the distance to the star in parsecs.

1.4 Fill in the steps of deriving the Stefan–Boltzmann law for blackbody radiation from the Planck Law (1.10) that are outlined in Example 1.3.***

1.5 The average density of stars in the vicinity of the Sun is about 0.08 stars per cubic parsec. If a ground-based telescope can detect reliably a parallax shift as small as 0.02 arcsec, estimate the upper limit on the number of stars for which the distance could be determined by parallax. If space-based observations such as those of the Hipparcos satellite permit parallax shifts as small as 0.001 arcsec to be measured, how many stars could in principle have their distances determined by parallax in this case? Repeat this analysis for Gaia, assuming that it can measure a parallax of 5×10^{-6} arcsec.

1.6 The first star for which a parallax was measured was 61 Cygni, reported by Bessel in 1838. The currently accepted parallax shift for 61 Cygni is 0.286 arcsec (which is relatively close to Bessel's original measurement of 0.316 arcsec). What is the distance to 61 Cygni in parsecs and lightyears?

1.7 Starting from Eq. (1.18), show that for the special case of circular orbits the motion is governed by

$$M_1 + M_2 = \frac{4\pi^2}{G} \frac{(r_1 + r_2)^3}{P^2},$$

where r denotes orbital radii. As noted in the text, this equation holds also for elliptical orbits provided the orbital radii for circular orbits are replaced by the semimajor axes of the corresponding ellipses.***

1.8 Show that Eq. (1.20) reduces to the usual simple form of Kepler's third law for planetary motion ($P^2 = a^3$) if time is measured in years, masses in units of $M_\odot$, and distances in AU, and the mass of a planet is neglected relative to that of the Sun.***

1.9 Ellipses are of fundamental importance in astronomy because keplerian orbits are elliptical. The properties of an ellipse are determined by two quantities, the length of the semimajor axis a and the eccentricity ε, which is defined by requiring that the distance from the center to either of the two foci of the ellipse be $a\varepsilon$. In terms of the distances r_1 and r_2 from the respective foci to a point on the ellipse, the equation of the ellipse is $r_1 + r_2 = 2a$. Use this to show that

(a) The length of the semiminor axis b is related to a and ε through

$$b^2 = a^2(1 - \varepsilon^2).$$

(b) In terms of a polar coordinate system centered on one of the foci, the distance from that focus to any point on the ellipse is given by

$$r(\theta) = \frac{a(1 - \varepsilon^2)}{1 + \varepsilon \cos\theta},$$

where θ is the polar angle between the semimajor axis and the line joining the focus to the point on the ellipse, and where $0 \leq \varepsilon < 1$.

(c) The area of an ellipse is $A = \pi ab$. Use the preceding results and Kepler's second law for planetary motion to show that in a time interval dt the planet travels an angular distance measured from the focus

$$d\theta = \frac{2\pi}{P} \frac{a^2}{r(\theta)^2} \sqrt{1 - \varepsilon^2}\, dt,$$

where P is the period for the orbit and $r(\theta)$ is given in part (b).

1.10 The Balmer H_α absorption line in the spectrum for the star Vega is observed to be shifted to 6562.5 Å, the star is observed to change its angular position on the celestial sphere by 0.35" per year, and the parallax shift of Vega is found to be 0.130". What are the radial, tangential, and space velocities for Vega?

1.11 The position of the Moon with respect to the background stars is observed at moonrise. When the Moon crosses the celestial meridian $6^h 12^m$ later, it is observed to have shifted 148' to the east relative to the stars. How far is the Moon from the Earth? *Hint:* The shift in apparent angle is caused both by parallax and the motion of the Moon on its orbit between observations.

1.12 (a) Use the orbital properties of the Moon to estimate the mass of the Earth.
(b) Use the orbital properties of the Earth to determine the mass of the Sun (neglect the mass of the Earth relative to that of the Sun).

(c) Look up the orbital properties of the four Galilean moons of Jupiter (Io, Europa, Ganymede, and Callisto). Use this to determine the mass of Jupiter, assuming that it is much larger than the mass of a Galilean moon. *Hint*: Plot log P versus $\frac{3}{2}\log a$, where P is the period and a the semimajor axis, and apply Kepler's third law.

1.13 Use the Planck Law (1.10) to derive the Wien law for blackbody radiation. *Hint*: You will obtain an equation that must be solved numerically.***

1.14 Halley's Comet has a very elliptical orbit with a period of 75.32 years and a closest approach to the Sun of 0.586 AU. What is the greatest distance of the comet from the Sun?

2 The Hertzsprung–Russell Diagram

Perhaps the most important observation concerning stellar structure is that there are strong correlations between the luminosities of stars and their surface temperatures.[1] A few stars exhibit emission lines but the dominant spectral feature for most normal stars is a set of absorption lines associated with various atoms, molecules, and ions superposed on a continuum. It was realized by the late nineteenth century that stars exhibited regular patterns in these absorption spectra. This led to a classification with a letter sequence A, B, C, ... used to denote the relative strength of hydrogen absorption lines. It was believed at first that this reflected differing elemental compositions so that, for example, A stars had the strongest hydrogen lines because they contained more hydrogen than other stars. We understand now that *all stars have similar composition* and that the spectral sequence is not primarily an indicator of composition but one of conditions in the surface region of the star where the absorption lines are produced (the *chromosphere*). In particular, the spectral sequence results from differing levels of excitation and ionization that depend strongly on the surface temperature and more weakly on the surface density. These ideas are codified in what is perhaps the most important theoretical construct in the stellar astronomer's toolkit, the *spectrum–luminosity* or *Hertzsprung–Russell (HR) diagram*.

2.1 Spectral Classes

The absorption spectrum for a given element depends crucially on the atomic and chemical properties: whether the atoms are parts of molecules, whether they are excited, and whether they are ionized. To illustrate these issues the hydrogen atom often will be used, with atomic level structure and transition properties summarized in Fig. 2.1. This is appropriate both because hydrogen is the simplest atom and because hydrogen is by far the most abundant element in the Universe, so it occupies a central place in astrophysics.

2.1.1 Excitation and the Boltzmann Formula

If we assume approximate thermal equilibrium, the degree of electronic excitation for a species at temperature T is given by the *Boltzmann formula*,

---

[1] Chapters 1 and 2 review material normally covered in introductory astronomy courses. These chapters may be skipped if you are familiar already with basic stellar properties and with the relationship of luminosity to surface temperature that is captured in the Hertzsprung–Russell diagram.

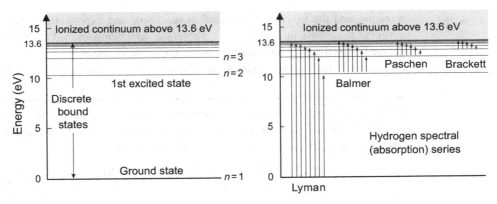

Fig. 2.1 Hydrogen energy levels and some absorption spectral transition series. The corresponding emission series result from reversing the direction of the arrows. The ground state has been placed at zero energy. It is also common to place the zero of the energy scale at the ionization threshold, so that all bound states have negative energy and all continuum states have positive energy. Then the ground state is at -13.6 eV and the first excited state is at -3.4 eV. Energy differences are the relevant quantities, so choice of the energy zero has no physical consequences.

$$\frac{n_2}{n_1} = \frac{g_2}{g_1} \exp[-(E_2 - E_1)/kT], \qquad (2.1)$$

where i labels a state at energy E_i having statistical weight (degeneracy factor) g_i and number density n_i, and k is the Boltzmann constant. The exponential dependence on energies measured in units of kT in Eq. (2.1) strongly favors population of low-lying states at temperatures of interest in most astrophysical applications.

2.1.2 Ionization and the Saha Equations

Equations governing the degree of ionization may be derived in a manner similar to that leading to the Boltzmann formula (2.1) by extending the available states to include, with appropriate weights, the continuum states that are populated by ionization (the gray region above 13.6 eV in Fig. 2.1). The resulting equations are known in astrophysics as the *Saha equations*. For single ionization, the ratio of singly ionized to neutral atom populations is

$$\frac{n_+}{n_0} = 2\left(\frac{u_+}{u_0}\right) \frac{(2\pi m_e kT)^{3/2}}{h^3 n_e} \exp(-E_i/kT), \qquad (2.2)$$

where n_+ is the number density of $+1$ ions, n_0 is the number density of neutral atoms, n_e is the number density of free electrons, m_e is the electron mass, E_i is the ionization energy, and the partition functions u_+ for ions and u_0 for atoms are given by sums of the form

$$u = \sum_n g_n e^{-E_n/kT}, \qquad (2.3)$$

where the sum is over ground and excited states available to the species. The degree of ionization thus depends linearly on statistical (state-degeneracy) factors that are often of order one, inversely on the electron density (because n_e influences the competition of ionization and the inverse recombination reaction), and exponentially on the ionization energy as measured in units of kT. Therefore, ionization caused by thermal effects is expected to be determined primarily by the temperature and more weakly by pressure and atomic energy-level effects (but see the discussion of pressure ionization in Section 3.7.1).

Example 2.1 By taking the (base-10) logarithm of both sides, exchanging the electron pressure P_e for the electron number density n_e using $P_e = n_e kT$, and evaluating the constants, Eq. (2.2) may be expressed in the convenient form [52]

$$\log \frac{n_+}{n_0} = \log \frac{u_+}{u_0} + \frac{5}{2} \log T - E_i \,(\mathrm{eV}) \frac{5040}{T} - \log P_e - 0.179, \qquad (2.4)$$

where energy is given in eV, temperature in K, and pressure in dyn cm^{-2}.

For double ionization the Saha equation takes the form

$$\frac{n_{++}}{n_+} = 2 \left(\frac{u_{++}}{u_+} \right) \frac{(2\pi m_e kT)^{3/2}}{h^3 n_e} \exp(-E_i^+/kT), \qquad (2.5)$$

where E_i^+ is the energy for the second ionization $A^+ \rightarrow A^{++}$ of the species A, the partition function associated with the $+2$ ion is u_{++}, and n_+ is the number density of $+1$ ions calculated from Eq. (2.2) for the first ionization step. Equations (2.2) and (2.5) are generalized in a straightforward way to third and higher degrees of ionization.

Assuming ideal gas behavior (see Section 3.3), the electron pressure P_e appearing in Eq. (2.4) and the gas pressure P_g are related to the corresponding number densities through

$$P_e = n_e kT \qquad P_g = nkT, \qquad (2.6)$$

where the electron number density n_e is related to the ion number densities through

$$n_e = n_+ + 2n_{++} + 3n_{+++} + \dots,$$

and n is the number density for all particles in the gas. Typical free-electron pressures in stellar surfaces are of order 1 dyn cm^{-2} for cooler stars and 1000 dyn cm^{-2} for hotter stars. For the Sun a free-electron pressure of about 30 dyn cm^{-2} is commonly used in stellar atmosphere calculations.

Example 2.2 If a gas of hydrogen is completely ionized,

$$n_e = n_+ = \tfrac{1}{2} n \qquad P_e = (n_e/n) P_g = \tfrac{1}{2} P_g,$$

and for a pure helium atmosphere with all helium twice ionized, $P_e/P_g = \tfrac{2}{3}$.

From simple algebra (see Problem 2.15), for two stages of ionization we may express the fractional abundances of neutral, $+1$, and $+2$ ions, respectively, as

$$\frac{n_0}{n_T} = \frac{1}{1 + \frac{n_+}{n_0}\left(1 + \frac{n_{++}}{n_+}\right)} \qquad \frac{n_+}{n_T} = \frac{1}{1 + \frac{n_0}{n_+} + \frac{n_{++}}{n_+}}$$

$$\frac{n_{++}}{n_T} = \frac{1}{1 + \frac{n_+}{n_{++}}\left(1 + \frac{n_0}{n_+}\right)}, \qquad (2.7)$$

where the total number densities of atoms plus ions, $n_T = n_0 + n_+ + n_{++}$, is assumed constant, and where the ratios n_+/n_0 and n_{++}/n_+ are given by the Saha equations as a function of temperature and electron density.

2.1.3 Ionization of Hydrogen and Helium

The ionization of hydrogen and helium calculated using Eq. (2.7) and the Saha formulas is illustrated in Figs. 2.2 and 2.3 as a function of temperature for three different pressures. In these plots we have employed the common astrophysics convention of using roman numerals to indicate the degree of ionization: I denotes no ionization (neutral atoms), II denotes the first state of ionization, and so on; thus He^{++} is He III in this notation. Notice that the dependence on temperature is much stronger than the dependence on pressure, and that the ionization transitions as a function of T are rather sharp. For example, the partial ionization of hydrogen is confined to a region less than about 3000 K in extent.[2] Notice also that helium is appreciably harder to ionize than hydrogen: its first excited state is at

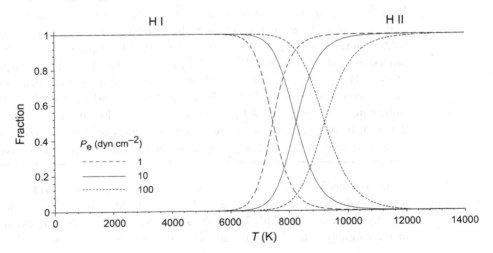

Fig. 2.2 Fractional abundance for ionic forms of hydrogen as a function of temperature for three different electron pressures P_e.

---

[2] Regions of a star where atoms are partially ionized are called *partial ionization zones*. Such zones are important in stellar structure and dynamics (for example they are central to understanding many pulsational instabilities) because ionization is an energy sink that changes the number of particles in the gas and therefore alters the effective equation of state. Figure 2.2 suggests that partial hydrogen ionization may be expected in stellar regions having temperatures in the vicinity of 7,000–10,000 K.

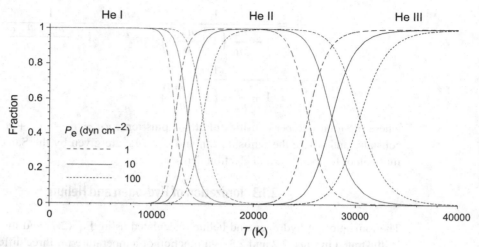

Fig. 2.3 Fractional abundance for ionic forms of helium as a function of temperature for three different electron pressures P_e.

about 20 eV, it first ionizes at 24.6 eV, and an additional 54.4 eV is required to remove the second electron; conversely, hydrogen's first excited state is at 10.2 eV above the ground state and it requires only 13.6 eV to liberate the single hydrogen electron.

Example 2.3 As another example of applying the Saha equations, consider the degree of ionization for hydrogen in the surface of the Sun, taking for this estimate a temperature of 6000 K and an electron pressure of 30 dyn cm^{-2} [52]. For neutral hydrogen the ground state has a statistical factor $g_0 = 2J + 1 = 2(\frac{1}{2}) + 1 = 2$ because of the $2J + 1$ spin degeneracy of the spin $J = \frac{1}{2}$ ground state, while excited states make no significant contribution to the partition function because they lie at higher energy and are strongly suppressed by the Boltzmann factor $\exp(-E_n/kT)$. Therefore, for neutral hydrogen, $u_0 = \sum_n g_n e^{-E_n/kT} \simeq 2$. For the hydrogen ion things are even simpler: there is no electron so there are no excited electronic states and $u_+ = 1$. Inserting these results in Eq. (2.4) gives $n_+/n_0 = 10^{-4}$

It may be concluded from the preceding example that the solar surface is dominated by neutral hydrogen, with only about 1 of every 10,000 atoms of hydrogen ionized. This changes quickly even a little below the solar surface where the temperature and density increase rapidly to levels causing almost total ionization of hydrogen.

2.1.4 Optimal Temperatures for Spectral Lines

To understand spectral properties it is necessary to use in concert both the Boltzmann formula, which determines the states that are likely to be populated in a given species (neutral atoms or ions), and the Saha equations, which determine the abundances of different ionic species.

The Balmer series: As an example, let's estimate the optimal surface temperature to produce strong Balmer series absorption lines in a star. The Balmer absorption series for hydrogen illustrated in Fig. 2.1 occurs in the visible part of the spectrum and is generated by transitions where an electron already in the first excited ($n = 2$) state of hydrogen is promoted to a higher level when a photon is absorbed. For the Balmer absorption series to be produced *there must be a population of hydrogen atoms already in the first excited state.* Therefore, the relevant quantity is the ratio n_2/n_T, where n_2 is the number density of neutral hydrogen atoms in the first excited electronic state and n_T is the total number density of hydrogen atoms plus ions.

Denoting the number density of neutral hydrogen atoms by n_I and the number density of hydrogen ions by n_{II}, we have $n_I \sim n_1 + n_2$ and $n_T = n_I + n_{II}$, where it is assumed in the first equation that only the ground and first excited state are likely to be populated in the temperature range of interest because of the Boltzmann factor. Therefore, from the identity $n_2/n_T = (n_2/n_I)(n_I/n_T)$ and the above expression for n_I,

$$\frac{n_2}{n_T} = \left(\frac{n_2}{n_1 + n_2}\right)\left(\frac{n_I}{n_I + n_{II}}\right) = \left(\frac{n_2/n_1}{1 + n_2/n_1}\right)\left(\frac{1}{1 + n_{II}/n_I}\right), \tag{2.8}$$

and it is only necessary to evaluate the ratios n_2/n_1 and n_{II}/n_I.

1. The ratio n_2/n_1 may be calculated from the Boltzmann formula (2.1) once the degeneracy factors are evaluated. For a hydrogen atom $g_n = 2n^2$, where n is the principal quantum number ($n = 1$ for the ground state and $n = 2$ for the first excited state). Therefore, the ground state of the hydrogen atom at $-13.6\,\text{eV}$ relative to the ionization threshold has degeneracy factor $g_1 = 2$ and the first excited state at $-3.4\,\text{eV}$ has $g_2 = 8$.
2. The ratio $n_{II}/n_I = n_+/n_0$ can be calculated from the Saha equation (2.2) or (2.4) using an ionization energy of 13.6 eV.

Using these results in Eq. (2.8) predicts a Balmer absorption spectrum strength peaking near 10,000 K (depending weakly on the surface electron pressure), falling off rapidly at

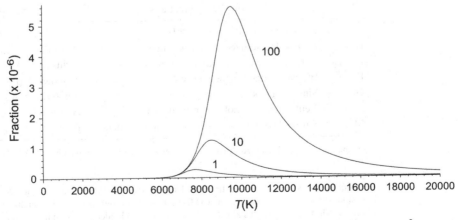

Fig. 2.4 Fraction of Balmer series absorption versus temperature at electron pressures of 1, 10, and 100 dyn cm^{-2}.

lower or higher temperatures, as shown in Fig. 2.4 for three different electron pressures. Observations indicate that indeed the strongest hydrogen Balmer absorption lines occur for stars with surface temperatures around 10,000 K (see Fig. 2.6).

Calcium H and K lines: Similar considerations govern the temperatures at which the absorption lines associated with other species have their greatest intensity. For example, you might be surprised to learn that the strongest lines in the solar spectrum are absorption lines associated with states in first-ionized calcium (called the Ca H and K lines), which are hundreds of times stronger than the solar Balmer lines. As you are asked to show in Problem 2.9, this is not because the Sun is particularly calcium-rich (hydrogen is about a million times more abundant than calcium at the Sun's surface). Rather, it is because an analysis similar to that for the Balmer series above indicates that almost all the small amount of calcium present in the solar surface is in the ground state of first-ionized calcium, which is optimal for absorption to produce the calcium H and K lines. In contrast, only about one in a billion hydrogen atoms in the solar surface is in the first excited state that can lead to Balmer series absorption. Therefore, although hydrogen is 10^6 times more abundant than calcium on the Sun, each calcium is almost 10^9 times more likely to absorb a visible photon than each hydrogen; as a consequence, the calcium H and K lines are much stronger than the hydrogen lines in the solar spectrum.

2.1.5 The Spectral Sequence

The preceding examples suggest that the spectral classification scheme is not a sequence in abundance, as originally thought, but rather is one in *surface temperature*. The modern spectral classification utilizes (largely for rather illogical historical reasons) the sequence O B A F G K M, with the associated characteristics given in Table 2.1.[3] A qualitative illustration of dominant excitations and ionizations as a function of temperature for some species relevant for the spectral classification is shown in Fig. 2.5, and representative

Table 2.1 Spectral classes and their characteristics

Class*	Distinguishing features	Examples
O	Ionized He and metals; weak H	θ^1 Orionis C (O6)
B	Neutral He, ionized metals, stronger H	Rigel (B8), υ Spica (B1)
A	Balmer H dominant, singly ionized metals	Sirius (A1), Vega (A0)
F	H weaker, neutral and singly ionized metals	τ Boötes A (F6)
G	Singly ionized Ca, H weaker, neutral metals	Sun (G2), β Aquilae (G8)
K	Neutral metals, molecular bands appear	Arcturus (K1.5), Pollux (K0)
M	Strong Ti oxide molecular lines, neutral metals	Betelgeuse (M2)

*Standard spectral classes. Additional specialized classes are discussed in Box 2.1.

---

[3] The sequence is associated with a traditional mnemonic: Oh Be A Fine Girl/Guy Kiss Me. Most of our discussion will assume the basic sequence O B A F G K M but some additional specialized classes are discussed in Box 2.1. These classifications also may be decimally subdivided by appending a number 0–9 to give even finer classification of details. Thus a G8 star is much closer to K0 than to G0 in its spectral characteristics.

> **Box 2.1** **Special Spectral Classes**
>
> It often proves useful to add to the traditional spectral classes in Table 2.1 new classes and subclasses specialized for particular types of stars. These may reflect discovery of newer categories of stars, or deeper understanding of stars previously classified under standard spectral classes. Here are a few examples.
>
> **Hot Blue Stars with Strong Emission Lines**
> Hot stars exhibiting strong emission lines are of large current interest. An example is Wolf–Rayet stars (Section 14.3.1): hot, massive, stars exhibiting large luminosity, rapid mass loss, almost no hydrogen lines, and broad emission lines from helium, carbon, nitrogen, and oxygen. Wolf–Rayet stars are given a special spectral class designation W, with subclasses reflecting the details of their spectra (particularly the relative strength of carbon and nitrogen emission lines). For example, WN is a Wolf–Rayet star exhibiting strong emission lines from helium in ionization states I–II and nitrogen in ionization states III–V, and WC is a Wolf–Rayet star with strong carbon (II–IV) emission lines. Another similar classification is that of "Slash" stars (so-named because of a slash in their designation), which are O stars by their absorption spectrum but with emission lines similar to WN Wolf–Rayet stars.
>
> **White Dwarfs**
> White dwarfs are designated by D, with appended letters indicating spectral features. For example DA indicates strong Balmer lines, implying hydrogen-rich outer layers, while DQ white dwarfs display evidence of a carbon-rich atmosphere through atomic or molecular carbon lines. An appended V indicates pulsation: DAV is a pulsating spectral-class DA white dwarf (also known as a ZZ Ceti variable).
>
> **Infrared Objects**
> Red dwarfs (low-mass main sequence stars) and brown dwarfs (Section 9.10) are faint in the visible and often studied in the infrared. New spectral classes have been created for these IR objects. Class L objects are cooler than class M main sequence stars. Some are powered by hydrogen burning but most are brown dwarfs. Class T are cooler brown dwarfs with strong methane spectral features. Class Y are even cooler brown dwarfs that may contain some objects on the boundary between planets and brown dwarfs.
>
> **Carbon Stars**
> Traditionally, R, N, and S were optional sub-branches for M. The R and N classes are now subsumed under *carbon stars* (designated by C), which have high surface carbon abundance. Carbon stars are further subdivided according to specific properties. For example, C-R and C-N have the characteristics of the former R and N classes, respectively. The older class S is now viewed as interpolating between the properties of M and C stars.

absorption spectra for different spectral classes are shown in Fig. 2.6. Once it is established that the spectral sequence is a temperature sequence, we may conclude that it is also a sequence in color index, with more negative color indices lying toward the O end and more positive color indices lying toward the M end of the sequence. This also establishes that the spectral sequence is a color sequence, with red stars lying near the cool M end of

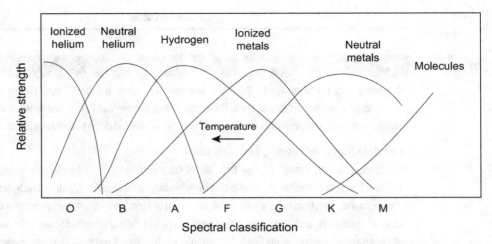

Fig. 2.5 Dominant spectral line strength as a function of spectral class (qualitative). These can be understood primarily in terms of the temperature dependence for electronic excitation and ionization of various species. From the description of the spectral classes in Table 2.1 it is clear that the spectral sequence is a temperature sequence, with the hottest stars lying near O and the coolest stars lying near M. For example, only in O stars are ionized helium lines strong because cooler stars cannot ionize helium. On the other hand, only in the K and M stars are strong molecular lines observed because only these stars are cool enough for molecules to exist in their atmospheres.

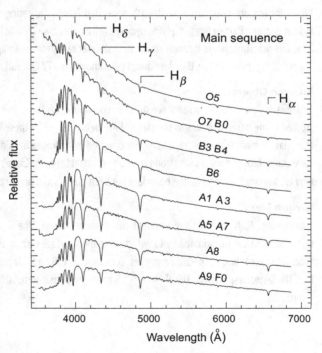

Fig. 2.6 Absorption spectra at optical wavelengths for some main sequence spectral classes (see Table 2.1) [3, 202]. They are characterized by a smoothly varying continuum punctuated by sharp absorption dips (*line absorption*). The positions of hydrogen Balmer series lines are marked. Note the maximal strength of the Balmer series near spectral class A, corresponding to a surface temperature of $\sim$ 10,000 K, and its decreasing strength in spectral classes corresponding to higher and lower temperatures. Constructed from spectra at http://zebu.uoregon.edu/spectra.html; from Silva and Cornell, *Astrophysical Journal Supplement Series* (ISSN 0067-0049), **81**(2), 1992, 865–881. Research supported by University of Michigan.

the sequence, yellow stars in the middle, and blue and blue-white stars lying near the hotter end of the spectral sequence. For (now discredited) historical reasons stars near the O end of the spectral sequence are sometimes referred to as "early" and those near the M end as "late." Although arcane, it is still common terminology in astronomy.

2.2 HR Diagram for Stars Near the Sun

The stars maintain their luminosities by virtue of their enormous energy production and the surface temperatures of stars also are ultimately linked to their internal energy production and energy transport. Hence, we might expect that there is a relationship between luminosity (or equivalently absolute magnitude) and spectral class (or equivalently surface temperature or color index). This could be checked by plotting these quantities against each other for a set of stars and determining whether any correlations were evident. However, determining the luminosity or absolute magnitude for a star requires knowing its distance, and this is problematic for all but the nearest stars.

2.2.1 Solving the Distance Problem

There are two ways in which we might circumvent this distance problem without appeal to more uncertain methods:

1. Use only nearby stars for the sample, for which the distance is known by parallax.
2. Use a population of stars in an open or globular cluster. Then even if the distance to the cluster is unknown, it is certain that it is essentially the *same* unknown distance for each member of the cluster. In this case the apparent magnitude may be used instead of the (unknown) absolute magnitude. This corresponds to an arbitrary global shift of the luminosity axis and preserves relationships in a plot of luminosity versus temperature that would indicate correlations.[4]

The first choice yields a diagram of the form shown in Fig. 2.7, where absolute magnitude is plotted versus the $B - V$ color index. Such a diagram is termed a Hertzsprung–Russell (HR) or spectrum–luminosity diagram (though some purists reserve the term Hertzsprung–Russell for a spectral class–luminosity diagram, as originally investigated by Hertzsprung and by Russell). As suggested above, the horizontal axis of such a diagram can be a color index, effective surface temperature, or spectral class, while the vertical axis is typically either luminosity or absolute magnitude, or in the case of clusters at unknown distances, the apparent magnitude.

---

[4] In practice there is a correction for interstellar reddening: because of the characteristic size of the dust particles that light encounters as it passes through the interstellar medium, shorter wavelengths are preferentially scattered relative to longer wavelengths and light becomes more dim and more red as it passes through the interstellar medium. This correction is of large practical importance but it will be ignored for much of the present introductory discussion.

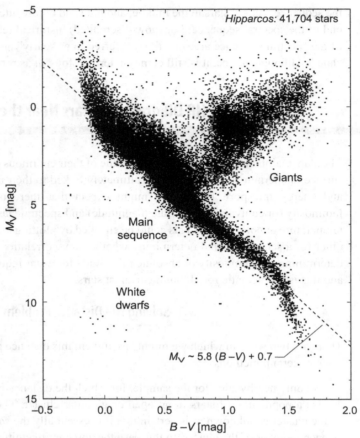

Fig. 2.7 HR diagram for stars near the Sun from selected Hipparcos data [4]. There are almost no supergiants (which would be above the giants around absolute magnitude −5) in this sample. The main sequence for stars near the Sun is seen to be reasonably well described by the linear relationship $M_V \simeq 5.8(B - V) + 0.7$, where M_V is the absolute visual magnitude and $B - V$ is a color index. This diagram does not represent the entire Hipparcos data set but rather is based on 41,704 stars for which measurements were judged to be more precise. © ESA.

2.2.2 Features of the HR Diagram

The dominant feature of the HR diagram for the stars near the Sun is the strong clustering into particular regions of temperature–luminosity space:

1. Approximately 90% of the stars are concentrated in the narrow band cutting diagonally across the center of the diagram called the *main sequence*. The Sun itself is a main sequence star of absolute magnitude +4.8 and spectral class G2. The concentration of many stars in this band implies that their properties may be parameterized in terms of a single quantity. Later it will be shown that this quantity is the mass of the star.

2. A much smaller group of stars is concentrated at higher temperatures but at luminosities far below the corresponding main sequence values. These are the *white dwarfs*.
3. Another smaller group consists of stars having considerably higher luminosities than main sequence stars of the same spectral class; these are the *giants* and the *supergiants*.

The preceding size-specific terminology arises because a vertical cut in the HR diagram selects stars having the same spectral class and thus (approximately) the same surface temperatures. Hence, substantial variation in the luminosity along a vertical line can come only from different surface areas for the stars in question, with stars below the main sequence having smaller surface areas and stars above the main sequence having larger surface areas than a corresponding main sequence star.

2.3 HR Diagram for Clusters

Now let us inquire whether the HR diagram for stars in clusters exhibit the same correlations seen for the group of stars in the immediate vicinity of the Sun. There are two types of clusters to consider: *open clusters* and *globular clusters*, with examples displayed in Fig. 2.8. The HR diagrams for open clusters resemble those for the stars near the Sun but the HR diagrams for globular clusters are substantially different: there are no luminous main sequence stars and no supergiants. Instead, the most luminous stars in globular clusters are giants. We may understand these observations as resulting from an age

Fig. 2.8 (a) The open cluster Trumpler 14, which contains some of the most luminous stars in our galaxy. It is located in the Carina Nebula, about 8000 ly distant. © NASA and ESA, Jesús Maíz Apellániz (Instituto de Astrofísica de Andalucia). (b) The globular cluster M3, which is about 30,000 ly away in the constellation Canes Venatici. It contains about 500,000 stars and is some 200 ly in diameter. NASA © Karel Teuweu. (c) Two merging open clusters in the 30 Doradus Nebula (Tarantula Nebula), which is a massive star-forming region some 170,000 ly away in the Large Magellanic Cloud. The two merging clusters appear to differ in age by about a million years. NASA, ESA, and E. Sabbi (ESA/STScI).

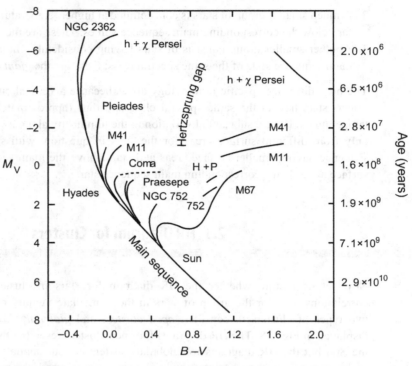

Fig. 2.9 HR diagram for some open clusters with the color index $B - V$ a surrogate for temperature or spectral class (relationships among the three are given in Tables 2.2–2.4). The right axis relates the age of the cluster to the turnoff point for that cluster. Clusters such as M11 or M41 exhibit two branches, with no stars in the region between the branches. As will be discussed in Section 13.4, this region corresponds to the *Hertzsprung gap*, through which stars evolve rapidly and thus are unlikely to be found at any one time. Adapted from figure in Ref. [55].

difference between open and globular clusters, which represents our first clear indication that the HR diagram is a snapshot in time of an evolving population of stars. The open clusters are much younger than the globular clusters. Thus, the original more-luminous main sequence and supergiant stars in the globular clusters have long since evolved beyond those stages, while the much younger open clusters still contain luminous main sequence and supergiant stars that have not yet had time to evolve to later stages in their lives. The point in the HR diagram for a cluster that marks the most luminous main sequence stars is called the *turnoff point* for the cluster. The location of the turnoff point is a direct indicator of the age of the cluster; this is illustrated in Fig. 2.9 for a set of open clusters.

Example 2.4 From Fig. 2.9 the turnoff point for the Pleiades Cluster is at $B - V \sim 0$ and $M_V \sim 0$. From the right axis, this indicates an age of about 1.6×10^8 yr for the Pleiades. On the other hand, M67 has $B - V \sim 0.5$ and $M_V \sim +4$ at its turnoff point, indicating that M67 is a much older cluster with an age of about 7×10^9 y.

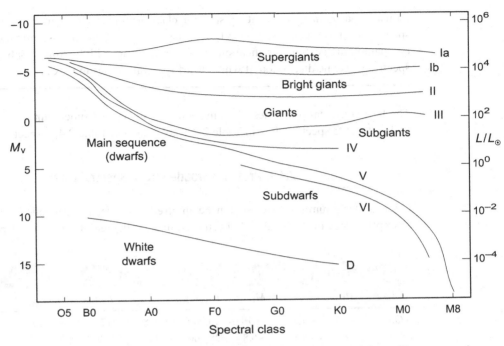

Fig. 2.10 Primary luminosity classes for stars. The modern classification using letters for spectral class and Roman numerals for luminosity class is called the Morgan–Keenan (MK) system.

2.4 Luminosity Classes

The qualitative groupings according to luminosity discussed above may be made more quantitative by introducing the *luminosity classes* illustrated in Fig. 2.10. In this classification, main sequence stars (sometimes called "dwarfs" – not to be confused with white dwarfs or brown dwarfs) are in luminosity class V, there is a class VI of stars somewhat less luminous than the main sequence called subdwarfs, white dwarfs have their own luminosity class D, giants are distributed in three classes (II, III, IV) according to their relative luminosities, and supergiants are divided into two categories (Ia and Ib) according to their relative luminosities. Classification of a star according to both its spectral class and luminosity class then confines it to a rather localized region of the HR diagram.

Example 2.5 The bright blue star Rigel (β-Orionis) may be classified as B8Ia, which means that it is a luminous blue supergiant of spectral class B8 and luminosity class Ia (its absolute magnitude is an impressive -7.8). Properties of many astronomical objects, such as spectrum and luminosity classes for individual stars, may be found in the online SIMBAD database [2]. For example, entering "Sirius" as an object identifier in SIMBAD returns (among other things) that it is a multiple star system with spectral class A1V+DA, indicating a main component of spectral classification A1V (a main sequence star of spectral sequence classification A1, so it has a strong Balmer absorption series), and a companion of spectral classification DA (a white dwarf with hydrogen Balmer lines in its spectrum; see the discussion of white dwarf spectral classes in Box 2.1).

The basic physical properties of main sequence, giant, and supergiant stars are tabulated with respect to spectral class in Table 2.2, Table 2.3, and Table 2.4, respectively.

2.4.1 Pressure Broadening of Spectral Lines

Often a star's luminosity class can be inferred from density-dependent features of its absorption spectra. It has already been seen that the degree of ionization depends weakly

Table 2.2 Main sequence parameters according to spectral type [55, 75]

Class	M_V	B–V	M_{bol}	T_{eff} (K)	$M/M_\odot$	$R/R_\odot$	$L/L_\odot$	$\langle\rho\rangle$ (g cm^{-3})
O5	−5.8	−0.35	−10	40,000	39.8	17.8	5×10^5	0.01
B0	−4.1	−0.31	−6.8	28,000	17.8	7.41	2×10^4	0.06
B5	−1.1	−0.16	−2.6	15,500	6.45	3.80	794	0.16
A0	0.7	0.0	0.1	9,900	3.20	2.51	79.4	0.28
A5	2.0	0.13	1.7	8,500	2.09	1.74	20	0.55
F0	2.6	0.27	2.6	7,400	1.70	1.35	6.3	0.98
F5	3.4	0.42	3.4	6,580	1.29	1.20	2.5	1.07
G0	4.4	0.58	4.3	6,030	1.10	1.05	1.25	1.35
G5	5.1	0.70	5.0	5,520	0.93	0.93	0.8	1.58
K0	5.9	0.89	5.8	4,900	0.78	0.85	0.4	1.78
K5	7.3	1.18	6.7	4,130	0.69	0.74	0.16	2.40
M0	9.0	1.45	7.8	3,480	0.47	0.63	0.06	2.51
M5	11.8	1.63	9.6	2,800	0.21	0.32	0.0079	10
M8	16	1.8		2,400	0.10	0.13	0.00079	63

Table 2.3 Parameters according to spectral type for giant stars [55, 75]

Class	M_V	B–V	M_{bol}	T_{eff} (K)	$M/M_\odot$	$R/R_\odot$	$L/L_\odot$	$\langle\rho\rangle$ (g cm^{-3})
G0	1.1	0.65	1.1	5,600	2.5	6.3	31.6	1.6×10^{-2}
G5	0.7	0.85	0.5	5,000	3.2	10	50	4.0×10^{-3}
K0	0.5	1.07	0.2	4,500	4.0	15.8	79.4	1.3×10^{-3}
K5	−0.2	1.41	−1.0	3,800	5.0	25	200	4.0×10^{-4}
M0	−0.4	1.60	−1.8	3,200	6.3		398	1.0×10^{-4}
M5	−0.8	1.85	−3				1000	

Table 2.4 Parameters according to spectral type for supergiant stars [55, 75]

Class	M_V	$B-V$	M_{bol}	T_{eff} (K)	$M/M_\odot$	$R/R_\odot$	$L/L_\odot$	$\langle\rho\rangle$ (g cm^{-3})
B0	−6.4	−0.25	−9	30,000	50.1	20.0	2.5×10^5	7.9×10^{-3}
A0	−6.2	0.0	−7	12,000	15.8	39.8	2.0×10^4	3.2×10^{-4}
F0	−6	0.25	−6	7,000	12.6	63.0	7.9×10^3	6.3×10^{-5}
G0	−6	0.70	−5.2	5,700	10.0	100	6.3×10^3	1.3×10^{-5}
G5	−6	1.06	−5.2	4,850	12.6	126	6.3×10^3	6.3×10^{-6}
K0	−5	1.39	−5.4	4,100	12.6	200	7.9×10^3	2.0×10^{-6}
K5	−5	1.70	−6	3,500	15.8	398	1.6×10^4	4.0×10^{-7}
M0	−5	1.94	−7		15.8	501	3.2×10^4	2.0×10^{-7}
M5		2.14						

on the electron density in the stellar atmosphere. Further information concerning surface density may be obtained from spectral linewidths, which are finite for three basic reasons:

1. There is a *natural linewidth* ΔE associated with the uncertainty principle, $\Delta E \cdot \Delta t \sim \hbar$, that is inversely related to the lifetime Δt of the state producing the line.
2. Otherwise sharp spectral lines are *Doppler broadened* by turbulent motion of the gas.
3. Collisions of atoms and ions lead to broadening of spectral lines. This *pressure broadening* is larger for higher densities because collision frequency increases with density.

These sources of line broadening have different characteristic line profiles so it is often possible to disentangle their effects by a careful study of spectral line shapes. The source of immediate interest here is the pressure broadening, which is expected to have the form

$$\Delta\lambda \simeq \text{constants} \times n\lambda^2\sigma\sqrt{kT}, \tag{2.9}$$

where λ is the wavelength, σ is the collisional cross section, and n is the number density. Therefore, the amount of pressure broadening is approximately linear in the density and the width of key spectral lines may be used to estimate the surface density of a star.

2.4.2 Inferring Luminosity Class from Surface Density

Once the surface density is estimated, it is then possible to make a luminosity classification because of the inverse correlation noted earlier between average density and absolute luminosity. For example, pressure broadening of spectral lines is expected to be larger for a main sequence star of a given spectral class than for a supergiant star of the same spectral class because the surface density of the main sequence star is orders of magnitude higher than for the supergiant star. By similar reasoning, the linewidths associated with white dwarfs are more broadened still because they have much higher densities than main sequence stars.

2.5 Spectroscopic Parallax

The considerations of the preceding paragraph suggest an indirect method for determining the distance to a star. If the spectral class can be determined from general features of the absorption spectrum and the luminosity class can be determined from subtle density-dependent effects in the absorption spectrum, then the HR diagram may be used to infer the approximate absolute magnitude of the star. Once the absolute magnitude is known, a comparison with the apparent magnitude and invocation of the $1/r^2$ intensity law yields the distance. This method of determining distances is termed *spectroscopic parallax*; it has nothing to do with trigonometric parallax but astronomers are fond of using the term "parallax" for any measurement of distance.

Example 2.6 Let's estimate the distance to α Centauri using spectroscopic parallax and data in the SIMBAD database [2]. From its spectrum, α Centauri is spectral class G2 and luminosity class V; that is, G2V. Interpolating by eye from Fig. 2.10, for spectral class G2 on the main sequence the absolute magnitude should be approximately 4, while the apparent visual magnitude for α Centauri is observed to be $m = -0.1$. From Eq. (1.7),

$$d = 10^{(m-M+5)/5}. \qquad (2.10)$$

Inserting the numbers gives approximately 1.5 pc for the distance to α Centauri. Actual parallax measurements [2], which are quite reliable for a star this close since the parallax angle of $0.742''$ is relatively easy to measure, indicate that it is 1.35 pc away. In this case, spectroscopic and trigonometric parallax agree to within about 11%. If a similar analysis is applied to Arcturus, the apparent magnitude of -0.05 and spectral–luminosity classification $\sim$K2III [2] imply a distance by spectroscopic parallax of 9.8 pc. This is about 13% less than the distance of 11.25 pc inferred from its trigonometric parallax of $0.0888''$ [2].

Spectroscopic parallax yields only approximate distances because of uncertainties in determining spectral and luminosity classes. These uncertainties typically are 15–20%, largely independent of distance if the absorption spectrum can be measured accurately and corrections for reddening by interstellar dust can be performed reliably. In contrast, uncertainties for trigonometric parallax grow rapidly with distance because of the difficulty in measuring very small angles. Trigonometric parallax with traditional Earth-based telescopes can be used out to distances of 50–100 parsecs but uncertainties associated with spectroscopic parallax often are smaller than those for trigonometric parallax once distances exceed $\sim$ 20–30 pc. Newer space-based observations are greatly extending the range of direct parallax: Hipparcos pushed the direct parallax scale out to greater than 1000 pc and the Gaia spacecraft has a goal of measuring distances to 20 million stars with 1% uncertainty and distances to stars as far away as the galactic center with 20% uncertainty. This will permit parallax to compete with spectroscopic parallax at much larger distances and allow a direct check of many star distances that formerly were known only through spectroscopic parallax.

2.6 The HR Diagram and Stellar Evolution

As mentioned above in the discussion of clusters, the HR diagram is a snapshot in time of the relationship between surface temperature and luminosity for stars in a group of stars. Since it has been suggested that HR diagrams change in time for cluster populations, this indicates that individual stars change their position on the HR diagram with time and it is meaningful to speak of an *evolutionary track* on the HR diagram for a star (see Fig. 10.8). Thus eventually the Sun will evolve from the main sequence into the red giant region, will emit a planetary nebula as it sheds mass late in its life, and will finally become a white dwarf. The entire timescale for this evolution is of order 10 billion years. In contrast, the fate of a 20 solar mass star is more spectacular, ending in a supernova explosion on an evolutionary timescale orders of magnitude smaller than that of the Sun. The single most important factor governing the evolution of a star is its mass. The more massive the star, the more rapidly it evolves through all phases of its life. Furthermore, the reason that most stars are found on the main sequence in Fig. 2.7 is that for typical stars the major portion of their lives (specifically, the hydrogen core-burning period) is spent on the main sequence.

Background and Further Reading

The material in this chapter is discussed well in Böhm-Vitense [52], Bowers and Deeming [55], Carrol and Ostlie [68], and Tayler [211].

Problems

2.1 Estimate the color index $B - V$ for a star assumed to be a blackbody with a surface temperature of 18,000 K. In making this estimate, assume the appropriate color filters to have a δ-function (that is, very sharply peaked) response of the actual response for the filter.

2.2 In the Sirius binary system, which has a parallax of 0.38″ as observed from Earth, the companion Sirius B has apparent visual magnitude 8.44 and its spectrum suggests an effective temperature of $\sim 30,000$ K. In addition, the semimajor axis for the primary orbit and companion orbit are observed to be 6.54 and 13.26 AU, respectively, and the period for the binary is found to be 50.13 years.

(a) What are the absolute visual magnitude and absolute bolometric magnitude if the bolometric correction is $BC = -3.3$?

(b) What is the luminosity of Sirius B based on this bolometric magnitude?

(c) What is the radius of Sirius B, assuming it to be a spherical blackbody?
(d) What is the total mass of the binary and the individual mass of the companion Sirius B?
(e) Assuming Sirius B to be spherical, what is its average density?
(f) What region of an HR diagram does Sirius B occupy? What kind of star is it, based on the information deduced in this exercise?

2.3 A certain star has an absolute bolometric magnitude of -5 but its spectrum suggests that the surface temperature is only about 3000 K. Assuming the star to radiate as a spherical blackbody, estimate the luminosity and the radius of this star in comparison with the Sun. How much of the inner Solar System would this star encompass if placed at the position of the Sun? Assuming the mass of this star to be no more than 10 solar masses, estimate its maximum average density. What region of the HR diagram does this star occupy, based on the information just deduced?***

2.4 Deneb (the tail of Cygnus the Swan) has apparent visual magnitude 1.25 and is of spectral–luminosity class A2Ia. Use spectroscopic parallax to estimate its distance from Earth.

2.5 Assume that apparent visual magnitude ~ 28 is the limit at which such systems can reliably determine light variability and spectrum in order to use the Cepheid variable method (the spectrum is necessary to ensure that the star has been identified correctly as a Cepheid variable). Estimate the maximum distance that space-based or ground-based adaptive effects systems can determine by using Cepheids. Assume the limit for standard ground-based telescopes to be apparent visual magnitude ~ 23. What is the maximum distance in that case?

2.6 Suppose that the main sequence of a cluster of stars at unknown distance is observed and is found to be approximately described by a linear relation $m \simeq 5.8 \cdot CI + 15$, where m is the apparent magnitude and CI is the $B - V$ color index. Assuming the main sequence of stars in the cluster to be similar to the main sequence observed for stars near the Sun, estimate the distance to the cluster.

2.7 The Hubble Space Telescope has been used to study Cepheid variables in the galaxy M100. Assume the period–luminosity relation to be parameterized by [see Eq. (1.35)]

$$M_V = -2.76 \log P - 1.4,$$

with the period P expressed in days. What is the distance to M100 indicated by an observed Cepheid having a period of 51 days and average apparent visual magnitude of 24.9? In estimating the distance, assume that interstellar absorption between us and M100 has dimmed the Cepheid light by 0.15 magnitudes.

2.8 A cluster of stars contains a type-II Cepheid variable with apparent magnitude 20 and a period of 10 days. How far away is the cluster?

2.9 In the solar spectrum there are strong calcium absorption lines (the Ca II H and K lines), corresponding to excitations from the ground state of singly ionized calcium. These lines are hundreds of times stronger than the hydrogen Balmer series absorption lines, even though calcium is about a million times less abundant than hydrogen on the Sun. Estimate the expected strength of the calcium K line and Balmer series lines at the solar surface temperature assuming that the partition functions for Ca I and Ca II are 1.32 and 2.30, respectively, the electron pressure is $\log P_e = 1.5 \, \text{dyn cm}^{-2}$, the ionization energy of calcium is 6.11 eV, the K line corresponds to a transition from the Ca II ground state to an excited state at 3.12 eV with statistical factors $g_1 = 2$ for the Ca II ground state and $g_2 = 4$ for the excited state, and that the ratio of calcium abundance relative to hydrogen abundance in the solar surface is 2.2×10^{-6}. *Hint:* You can solve this problem using the Saha equations to determine the population of ionic species and the Boltzmann formula to determine the relative population of ground and excited states within each ionic species.***

2.10 Absorption in the surface of a star involves both discrete lines, such as the Balmer absorption series for hydrogen, and continuum absorption. An important source of continuum absorption is bound–free transitions where an electron is excited from a bound state to an unbound state (ionization). Since the final state contains an essentially free electron, bound–free transitions produce continuous absorption for wavelengths shorter than the threshold wavelength corresponding to the energy difference between the bound state and the ionization threshold. Show that Balmer bound–free continuum absorption (corresponding to ionization from the $n = 2$ hydrogen level at 10.2 eV) and Lyman bound–free continuum absorption (corresponding to ionization from the $n = 1$ ground state of hydrogen) both occur at UV wavelengths where the Sun outputs little light. Show that Paschen bound–free absorption (corresponding to ionization from the $n = 3$ level of hydrogen) occurs primarily in the visible spectrum where the solar light output peaks and therefore is expected to dominate the contribution of neutral hydrogen to continuum absorption at visible wavelengths. (However, note from Problem 2.11 that this is not the dominant overall contribution to continuum absorption for the Sun.)

2.11 Negative ions are generally difficult to produce and are of less consequence in astrophysics than are positive ions. One negative ion that is important in astrophysics is H^-, which is formed by addition of an electron to neutral hydrogen and has a binding energy of only 0.7 eV. Negative hydrogen ions have very low abundance in the solar surface but they are thought to be the dominant source of continuum absorption at visible wavelengths. Work through the following considerations to see why this is so.

(a) Show that under conditions expected in the solar surface the abundance of H^- ions relative to neutral hydrogen atoms is only about 3×10^{-8}.

(b) Show that bound–free absorption involving ionization of H^- occurs at wavelengths shorter than about 17,000 nm and therefore overlaps completely the visible spectrum of the Sun. Thus, essentially every H^- ion in the solar surface is in a state that can contribute to visible continuum absorption.

(c) Neutral hydrogen is by far the most abundant species in the solar surface. In Problem 2.10 it was demonstrated that the dominant neutral hydrogen contribution to continuum absorption is expected to originate in ionization of electrons from the $n = 3$ atomic hydrogen level (Paschen continuum bound–free transitions). Compare the relative abundance of H^- ions in the solar surface with the relative abundance of neutral hydrogen atoms in the $n = 3$ level (the only ones that can contribute significantly to continuum absorption), and show that — even though neutral hydrogen atoms are of order 10 million times more abundant than negative hydrogen ions — negative hydrogen ions are of order 100 times more abundant than neutral hydrogen in the $n = 3$ level and therefore negative hydrogen ions are expected to dominate neutral hydrogen in producing continuum solar absorption. Since separate problems in this chapter demonstrate that other sources of visible continuum absorption (for example, from helium and metals) are less important, it is expected that ionization of H^- ions is the dominant source of visible continuum absorption for the Sun.

2.12 Neutral helium has an ionization energy of 24 eV. Show that, although helium is the second most abundant element in the Sun, ionization from either the ground or first excited state at 20 eV is not expected to be a significant source of continuum absorption for the Sun. Order of magnitude estimates are sufficient for this problem and the effect of statistical factors may be ignored by setting them all to one.***

2.13 Estimate the possible contribution of metals (elements with atomic number greater than that of helium, which have a total abundance of several percent in the Sun) to solar continuum absorption. Take as a representative example iron, which has an ionization energy of 7.9 eV and a sufficiently high density of bound states that one may assume as a rough approximation that for any energy below the ionization threshold there is a nearby bound state. The relative abundance of iron in the solar surface is about 10^{-4} that of hydrogen and, for order of magnitude estimation, you may assume that the statistical factor ratios are of order 1.

(a) Consider continuum absorption at a wavelength of 4000 Å. Show, by considering the expected abundance of iron in a level appropriate to give absorption at that wavelength in the solar surface, that neutral iron is expected to be much less important than H^- ions in contributing to continuum absorption at 4000 Å (see the results of Problem 2.11 and don't forget the effect of ionization on neutral iron abundance).

(b) Repeat the considerations of part (a) but for a wavelength of 2000 Å. Show that at these short wavelengths iron is expected to dominate both H^- and neutral hydrogen in its contribution to continuum absorption through bound–free transitions.

2.14 For stellar atmospheres the free-electron pressures P_e typically range from of order 1 dyn cm^{-2} for the coolest stars to of order 1000 dyn cm^{-2} for the hottest stars. Estimate the range of free-electron number densities for stellar atmospheres from the hottest to coolest main sequence stars.

2.15 Use the condition $n_T = n_0 + n_+ + n_{++}$ to derive Eqs. (2.7) for the fractions of ionized species.***

2.16 Suppose a pure atomic hydrogen gas is found to be 70% ionized at a temperature of 14,000 K. What is the corresponding electron number density, assuming the ionization to be caused entirely by thermal effects?

3 Stellar Equations of State

Our fundamental initial task in astrophysics is to understand the structure of stars. In this chapter and the next four the basic equations that govern stellar structure and stellar evolution will be developed. At a minimum, an understanding of stars will require

1. A set of equations describing the behavior of stellar matter in gravitational fields. These often are well approximated by hydrodynamics in the general case, reducing to hydrostatics in the simplest cases.
2. A set of equations governing energy production and associated composition changes driven by thermonuclear reactions.
3. A set of equations describing how energy is transported from the energy-producing regions deep in the star to the surface.
4. Equations of state that carry information about the microscopic physics of the star and that relate macroscopic thermodynamic variables to each other.

These sets of equations are coupled to each other in highly non-trivial ways. For example, the hydrodynamics is influenced by the energy production in the thermonuclear processes and the thermonuclear processes are in turn strongly dependent on variables such as temperature and density that are controlled by the hydrodynamical evolution and equation of state. The full problem will correspond to a set of coupled, nonlinear, partial differential equations that can be solved only through large-scale numerical computation, but in many cases assumptions can be made that allow simpler solutions illustrating many basic stellar features. We begin the discussion by considering typical equations of state in this chapter, with the other topics enumerated above to be addressed in subsequent chapters.

3.1 Equations of State

An equation of state is a relationship among thermodynamic variables for a system carrying information about the basic physical properties of the system that goes beyond what is known on purely thermodynamic grounds. Schematically, an equation of state is of the form

$$P = P(T, \rho, X_i, \ldots), \qquad (3.1)$$

where P is the pressure, T is the temperature, ρ is the density, the X_i are concentrations variables, and so on. Equation (3.1) is intended to be highly schematic at this point, since an equation of state can take many forms. It need not even be specified analytically.

For example, equations of state employed in numerical astrophysics simulations often use direct interpolation in multidimensional tables that have been constructed numerically. It was fortunate for the development of quantitative stellar models that in many physically relevant instances the minimal equation of state is relatively simple and can be approximated by an analytical function of the thermodynamical variables.

We now consider some specific equations of state that may be important for the physics of stars. In addressing this issue it is important to ask whether the equation of state can be described in terms of classical thermodynamics, or whether quantum physics is required. We will find that for many applications a classical treatment is adequate but for some – in particular those involving high densities – a quantum prescription will become necessary. Let us first address cases where a classical treatment is adequate before tackling the more interesting issue of equations of state for quantum gases. The starting point for a classical description will be to write down a general expression for the pressure of a gas in terms of an integral over the momentum distribution of its particles.

3.2 The Pressure Integral

Except at extremely high densities where liquid or even crystalline phases may be found, our primary concern in astrophysics is with equations of state for gases. If quantum-mechanical effects can be neglected, the pressure in a gas may be expressed in terms of the *pressure integral* derived in Problem 3.6,

$$P = \tfrac{1}{3} \int_0^\infty vpn(p)dp, \tag{3.2}$$

where P is the pressure, v is the velocity, p is the momentum, and $n(p)$ is the number density of particles with momentum in the interval p to $p + dp$. This formula represents a very general result that can be shown to be valid for gas particles with any velocity, up to and including $v = c$ (see Problem 3.18), as long as quantum effects can be neglected.

3.3 Ideal Gas Equation of State

If a gas consists of point particles in random motion with weak interactions, the gas obeys the *ideal gas* equation of state, which may be expressed in a variety of equivalent forms:

$$P = nkT = \frac{N}{V}kT = \frac{NM_u}{V}RT = \rho\frac{kT}{\mu M_u}, \tag{3.3}$$

where P is the pressure, n is the number density of gas particles, V is the volume, $N = nV$ is the number of particles contained in a volume V, the Boltzmann constant is k, the temperature is T, the number of moles in V is NM_u, the universal gas constant is $R = kN_A = k/M_u$ (where the Avogadro constant is $N_A = M_u^{-1}$, with M_u the atomic mass unit), $\mu = \rho/nM_u$ is the mean molecular weight for the gas particles in atomic mass units,

> **Box 3.1**   **Masses, Moles, and Mean Molecular Weights**
>
> The terminology *mean molecular weight* is a misnomer, since molecules are unlikely to be found in stars except for the surface layers of the coolest ones. The mean molecular weight μ is actually the average mass of a particle in the gas (which consists mostly of atomic ions and electrons), as will be discussed further in Section 3.4. Sometimes μ will be expressed in units of amu (see below) and sometimes in physical mass units like grams; it should be clear from dimensional analysis which units are being used.
>
> **Masses and the Atomic Mass Unit**
>
> Masses are commonly given in terms of the atomic mass unit (amu), which will be denoted by M_u. The amu is defined to be $\frac{1}{12}$ of the mass of ^{12}C (in its neutral ground state), so that
>
> $$M_u = 1.66054 \times 10^{-24} \text{ g}.$$
>
> For many (not all) applications in astrophysics it is sufficiently accurate to take the mass of a hydrogen atom or proton to be 1 amu; thus, $m_H \simeq m_p \simeq 1$ amu. Such approximations introduce errors that are less than 1%: in reality the atomic mass of hydrogen is 1.007825 amu and the mass of a proton is 1.007277 amu. For our purposes it will usually be sufficient to ignore the distinction between nuclear mass and atomic mass (which includes electron masses and binding energies in addition to the nuclear mass), as will be discussed further in Chapter 5.
>
> **Moles and the Avogadro Constant**
>
> A *mole* (abbreviation mol) is defined to be a quantity of matter that contains as many constituent objects (atoms, ions, ...) as the number of atoms in 12 grams of ^{12}C. A mole of atoms is equal to a number given by the *Avogadro constant*, which takes the value $N_A = 6.022 \times 10^{23}$. The terminology 'Avogadro's constant is often employed rather than the also-common 'Avogadro's number' to distinguish current usage from historical usage with a different definition that gave essentially the same numerical value of N_A. The distinction will not be important for us except to note that Avogadro's number is just a number but Avogadro's constant carries dimension of mol^{-1}, since the mole is now officially recognized as a unit. Formally,
>
> $$N_A = 6.022 \times 10^{23} \text{ mol}^{-1}.$$
>
> Then, since N_A and $1/M_u$ are equivalent numerically, in equations N_A in units of mol^{-1} can be interchanged with $1/M_u$ in units of g^{-1}. Finally, if so inclined, one may pay homage to the exponent in Avogadro's constant by celebrating October 23 (10/23) as *Mole Day*.

and ρ is the mass density (see Box 3.1 for a discussion of these quantities). The ideal gas equation (3.3) follows from the more general Eq. (3.2) evaluated for a Maxwellian velocity distribution,

$$n(v)dv = 4\pi n \left(\frac{m}{2\pi kT}\right)^{3/2} \exp(-mv^2/2kT)v^2 dv, \quad (3.4)$$

where the total particle number density is $n = \int_0^\infty n(v)dv$ (see Problem 3.7). This distribution is illustrated in Fig. 3.1 for hydrogen gas at several temperatures. Some authors make

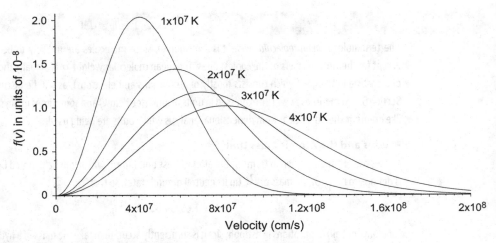

Fig. 3.1 Maxwell velocity distribution $f(v) = n(v)/n$ from (3.4) for hydrogen gas at various temperatures.

subtle distinctions between ideal gases and perfect gases but that will not be important here and any gas described at a reasonable level of approximation by the equation of state (3.3) will qualify as an ideal gas for our purposes.

3.3.1 Internal Energy

For a gas described by the equation of state (3.3) the *internal energy* U is given by

$$U = \int_0^T C_V(T)\, dT, \tag{3.5}$$

with the *heat capacity at constant volume* $C_V(T)$ defined by[1]

$$C_V(T) = \left(\frac{\partial U}{\partial T}\right)_V = \left(\frac{\partial Q}{\partial T}\right)_V = T\left(\frac{\partial S}{\partial T}\right)_V, \tag{3.6}$$

where Q is heat, S is the entropy, and the first law of thermodynamics,

$$dU = \delta Q - P\, dV = T\, dS - P\, dV, \tag{3.7}$$

has been used (see Box 3.2 for the distinction between δQ and dQ). The corresponding internal *energy density* u is given by

$$u = \frac{U}{V} = \frac{1}{V}\int_0^T C_V\, dT, \tag{3.8}$$

where V is the volume and Eq. (3.5) was used.

---

[1] The heat capacity has units in the CGS system of erg K^{-1}. A related quantity is the specific heat, which is the heat capacity per unit mass and has CGS units erg K^{-1} g^{-1}. In this book we will use an upper-case C to denote heat capacities and a lower-case c to denote specific heats.

Box 3.2   Exact and Inexact Differentials

In Eq. (3.7) we wrote the change in heat as δQ instead of dQ to make explicit that the change in heat is *not an exact differential* (or *not a perfect differential*). The change in internal energy dU is an exact differential because there is a function U such that $U = \int dU$. Thus the change in U evaluated over a path depends only on the endpoints of the path and U is a *state function*: it characterizes the state of the system at a given time and a change in U depends only on initial and final states. Conversely, change in heat δQ and change in work δW are not state functions and cannot be expressed as exact differentials. They are *inexact differentials* and cannot be integrated without full knowledge of the integration path.

State Functions and Non-state Functions

We may think of state functions as the integrals of perfect differentials that characterize an equilibrium state of the system, independent of how the system got there. On the other hand, non-state functions may be thought of as being associated with imperfect differentials that characterize a system in transition between equilibrium states. For example, the internal energy U characterizes an equilibrium state and is a state function but the work ΔW done in moving the system from a previous state to the present one depends on the path followed. Thus it characterizes a path-dependent change in energy that is not a state function because its value is not determined by the present state.

A Simple Analogy

Consider a banking account where either checks or cash may be deposited [50]. The total money in the account is the analog of U, check deposits are the analog of δQ, and cash deposits are the analog of δW. At any time the state of the account can be specified by the total amount of money, but that information is insufficient to tell us how much of the total corresponds to check deposits and how much to cash deposits. Only a specification of the detailed history of deposits (the "integration path") can tell us that, since many possible "paths" lead to the same total amount in the account. Likewise, if an equilibrium state is changed into another in a thermodynamic system the final state is uniquely characterized by its internal energy, irrespective of the path followed to get there, but there are many paths leading from the initial to final state corresponding to different relative contributions of δQ and δW to the change in total energy, and state information alone cannot distinguish them.

3.3.2 The Adiabatic Index

For the special case of a monatomic, nonrelativistic ideal gas,

$$C_V = \tfrac{3}{2}Nk \qquad U = C_V T = \tfrac{3}{2}NkT. \tag{3.9}$$

Expressing the internal energy relation (3.5) in differential form, $dU = C_V(T)dT$, introducing the heat capacity at constant pressure,

$$C_P = \left(\frac{\partial U}{\partial T}\right)_P = T\left(\frac{\partial S}{\partial T}\right)_P, \tag{3.10}$$

and using the first law (3.7) gives for an ideal gas (see Problem 3.1)

$$C_P = C_V + Nk. \tag{3.11}$$

The *adiabatic index* γ is defined by

$$\gamma \equiv \frac{C_P}{C_V}. \tag{3.12}$$

For an ideal gas the heat capacities are independent of temperature and if the gas is monatomic,

$$\gamma = \frac{C_P}{C_V} = \frac{C_V + Nk}{C_V} = \frac{\frac{3}{2}Nk + Nk}{\frac{3}{2}Nk} = \frac{5}{3}. \tag{3.13}$$

As we shall see in Eq. (7.37), the adiabatic index of an ideal gas is directly related to the number of degrees of freedom for each particle in the gas. From Problem 3.21, the relationship between the pressure P and energy density u for an ideal gas may be expressed in terms of the adiabatic index:

$$P = (\gamma - 1)u. \tag{3.14}$$

This equation may be used to *define* an effective adiabatic index γ for the general case, but only in the ideal gas limit is γ given by Eq. (3.13). The adiabatic speed of sound v_s in an ideal gas depends on γ through

$$v_s = \sqrt{\gamma P/\rho}, \tag{3.15}$$

where ρ is the density and P is the pressure, as shown in Problem 3.10.

3.4 Mean Molecular Weights

Realistic gases in astrophysics usually consist of more than one atomic or molecular species, and each may be partially or totally ionized. For example, the gas in a star may contain hydrogen atoms and ions, helium atoms and ions, various heavier elements in atomic, molecular, or ionic form, and the electrons produced by ionization. In many situations it may be possible to treat a mixture of different gases as if it were a single gas with an effective molecular weight. For example, as long as the density is not too large a mixture of hydrogen ions, fully ionized helium ions, and electrons will behave as a mixture of three ideal gases, each contributing a partial pressure to the total pressure of the system (Dalton's law of partial pressures). Under these conditions the system can be treated formally as a single gas with an effective molecular weight representing the relative contributions of each individual gas to the system properties. The following sections introduce some formalism and terminology associated with concentrations and densities of various species in a gas that allow such a treatment to be implemented.

3.4 Mean Molecular Weights

3.4.1 Concentration Variables

The mass density ρ_i of an ionic species i is given by

$$\rho_i = n_i A_i M_u = n_i \frac{A_i}{N_A}, \qquad (3.16)$$

where A_i is the atomic mass number, M_u is the atomic mass unit, n_i is the number density, and $N_A = 1/M_u$ is the Avogadro constant (recall the discussion in Box 3.1). The *mass fraction* X_i for the species i may be introduced by the definition

$$X_i \equiv \frac{\rho_i}{\rho} = \frac{n_i A_i M_u}{\rho} = \frac{n_i A_i}{\rho N_A}, \qquad (3.17)$$

where ρ is the total mass density and the label i may refer to ions, atoms, or molecules; by definition the mass fractions sum to unity: $\sum_i X_i = 1$. It also will be useful to introduce the *abundance* Y_i,

$$Y_i \equiv \frac{X_i}{A_i} = \frac{n_i}{\rho N_A}. \qquad (3.18)$$

Generally, the sum of the Y_i will not be unity.

3.4.2 Partially Ionized Gases

Let's address first the most general case of a gas with multiple species, each with arbitrary degrees of ionization. In Appendix C the definitions of the previous section are used to demonstrate that if the energy density and the pressure of radiation can be neglected the average mass of a particle (atoms, ions, or electrons) in the gas is given in atomic mass units by the expression

$$\mu = \left(\sum_i (1 + y_i Z_i) Y_i \right)^{-1}, \qquad (3.19)$$

where the sum is over isotopic species, y_i is the fractional ionization of the species i ($y_i = 0$ for no ionization and $y_i = 1$ if the species i is completely ionized), and Z_i is the atomic number for isotopic species i.

Therefore, with this formalism the actual gas, which is a mixture of electrons and different atomic, possibly molecular, and ionic species, has been replaced with a gas containing a single kind of fictitious particle having an effective mass μ (often termed the *mean molecular weight*) that is given by Eq. (3.19). In very hot stars the momentum and energy density carried by photons are non-trivial and this will modify further the effective mean molecular weight of the gas. This will be discussed in Section 3.11.1.

Example 3.1 For completely ionized hydrogen gas there is a single ionic species and $y_i = Z_i = Y_i = 1$. Thus from Eq. (3.19)

$$\mu = \frac{1}{(1+1) \times 1} = \frac{1}{2}$$

(in atomic mass units). This is just the average mass of a particle in a gas having equal numbers of protons and electrons, if the mass of the electrons is neglected relative to that of the protons. As a second example, the composition of many white dwarfs may be approximated by a completely ionized gas consisting of equal parts ^{12}C and ^{16}O by mass. The mass fractions are $X_{12C} = X_{16O} = 0.5$, so the abundances are

$$Y_{12C} = \frac{X_{12C}}{12} = 0.04167 \qquad Y_{16O} = \frac{X_{16O}}{16} = 0.03125.$$

From Eq. (3.19), this gives $\mu = 1.745$ amu if complete ionization ($y_i = 1$) is assumed. From Eq. (C.4) in Appendix C the average number of electrons produced per ion is $\bar{z} = 6.86$ (the concentration-weighted average of six electrons from each completely ionized carbon and eight electrons from each completely ionized oxygen), so in a typical white dwarf the gas contains nearly seven electrons for every ion.

As will be seen later, the preceding example implies that the mass of the white dwarf is carried almost entirely by the ions but the pressure is dominated by the electrons.

3.4.3 Fully-Ionized Gases

We will show later in this chapter that stars are generally completely ionized except relatively near the surface. Hence complete ionization will often be assumed in this discussion, which simplifies the considerations of Section 3.4.2. Assume a completely ionized gas for which both the ions and electrons obey the ideal gas law. From Eq. (C.2) of Appendix C the mean weight of the ions μ_I is given by

$$\frac{1}{\mu_I} = \sum_i Y_i = \sum_i \frac{X_i}{A_i}, \qquad (3.20)$$

and from the ideal gas law the pressure contributions from the ions is

$$P_I = \frac{\rho}{\mu_I M_u} kT. \qquad (3.21)$$

Now consider the electrons. For the fully ionized gas the electron number density is

$$n_e = \frac{\rho}{M_u} \sum_i X_i \frac{Z_i}{A_i} = \frac{\rho}{\mu_e M_u}, \qquad (3.22)$$

where [see Eq. (C.3)] μ_e is defined by

$$\frac{1}{\mu_e} = \sum_i X_i \frac{Z_i}{A_i}. \qquad (3.23)$$

The electronic contribution to the pressure is then

$$P_e = n_e kT = \frac{\rho}{\mu_e M_u} kT. \qquad (3.24)$$

The total pressure of the gas is the sum of contributions from the ions and the electrons,

$$P_{\text{gas}} = P_I + P_e = \left(\frac{1}{\mu_I} + \frac{1}{\mu_e}\right) \frac{\rho kT}{M_u} = \frac{\rho kT}{\mu M_u}, \qquad (3.25)$$

where the mean molecular weight of the gas μ (averaged over all ionic species and the electrons) is defined by

$$\frac{1}{\mu} = \frac{1}{\mu_I} + \frac{1}{\mu_e}. \qquad (3.26)$$

Thus, the completely ionized gas has the same equation of state as an ideal gas of particles having an effective mass μ given by Eq. (3.26).

3.4.4 Shorthand Notation and Approximations

It is common to define a shorthand notation

$$X \equiv X_{\text{hydrogen}} \qquad Y \equiv X_{\text{helium}} \qquad Z \equiv X_{\text{metals}} \qquad (3.27)$$

where "metals" refers to the aggregate of all elements other than hydrogen and helium[2] and $X + Y + Z = 1$. For the Sun the composition suggested by a study of the solar atmosphere is [28]

$$X = 0.738 \qquad Y = 0.249 \qquad Z = 0.013, \qquad (3.28)$$

which is similar to that expected for a typical Pop I star just entering the main sequence (commonly termed a *Zero Age Main Sequence* or *ZAMS* star). The metal fraction Z will be less than this in Pop II stars. As an alternative to the mass fraction Z, astronomers often specify the metal content of a star in terms of *metallicity*, which is defined in Box 3.3.

Simplified versions of the preceding equations and those of Appendix C that are adequate for many applications may be obtained by approximations assuming the contributions of metals to be small and to come from symmetric isotopes (equal numbers of protons and neutrons). As an example, for the mean ionic mass of a completely ionized gas

$$\frac{1}{\mu_I} \simeq X + \tfrac{1}{4}Y + \frac{1 - X - Y}{\langle A \rangle}, \qquad (3.29)$$

where $\langle A \rangle$ is the average atomic mass number for the metals in the star, and for the average electronic mass per nucleon of that gas

$$\frac{1}{\mu_e} \simeq X + \tfrac{1}{2}Y + (1 - X - Y)\left\langle \frac{Z}{A} \right\rangle \simeq \frac{1 + X}{2}, \qquad (3.30)$$

where angle brackets indicate an average over metals (see Problem 3.20) and the last form results from approximating $\langle Z/A \rangle \sim \tfrac{1}{2}$, since the most abundant metals in stars correspond to isotopes with approximately equal numbers of protons and neutrons (symmetric matter).

---

[2] Hence the joke that it is much easier for an astronomer to memorize the periodic table than for a chemist, since the astronomer's version contains only three elements. However, the astronomer's definition is not so strange for the astrophysical context in which differing chemical properties of elements (which are determined by the electronic structure of atoms) play no role because most of a star is completely ionized. Then it makes sense to single out hydrogen and helium, since they are the lightest and by far most abundant elements, and to view all other elements to first approximation as donors of (often many) electrons to the plasma that can be lumped into one category. Under normal Earth conditions nitrogen is a gas and iron is a (true) metal, implying quite different physical characteristics, but with complete ionization these chemical distinctions are no longer operative, which greatly simplifies the periodic table of elements!

> **Box 3.3 Metallicity**
>
> The metal content of stars is a few percent or less by mass, but even in small doses metals are important because they produce copious electrons when ionized. For example, ionization of a hydrogen atom produces one electron but complete ionization of silicon and iron produces 14 and 26 electrons, respectively. Since increasing the number of free electrons increases the photon opacity and this in turn strongly influences energy transport (see Chapter 7), even a small enhancement of metal concentration can have significant influence on stellar structure and stellar evolution.
>
> **Definition of Metallicity**
>
> The metal mass fraction Z can be used to specify metal concentration, as in Eq. (3.28). However, the metal content of a star also may be specified in terms of a related logarithmically defined quantity called the *metallicity*. Typically the metallicity is designated by the symbol [Fe/H] and is defined as
>
> $$[\text{Fe/H}] = \log\left(\frac{n_{\text{Fe}}}{n_{\text{H}}}\right)_{\text{star}} - \log\left(\frac{n_{\text{Fe}}}{n_{\text{H}}}\right)_{\text{Sun}},$$
>
> where log is the base-10 logarithm and n_{Fe} and n_{H} are the iron and hydrogen number densities, respectively.[a] Thus stars with solar metal content have [Fe/H] $= 0$, those with more metals than the Sun have a positive [Fe/H], and those with fewer metals than the Sun have a negative [Fe/H]. For example, a star with 1% of the iron found in the Sun has [Fe/H] $= -2$. Typical observed metallicities are as large as $+1$ for metal-rich stars and as small as -6 for metal-poor stars. This orders-of-magnitude difference in metal content for stars makes the logarithmic definition of the metallicity particularly convenient.
>
> **Stellar Age and Metallicity**
>
> The early Universe had no metals to speak of. They were manufactured later by stars and distributed in the interstellar medium by winds and explosions. Hence newer generations of stars incorporate on average more metals than older generations when they form, and metallicity is (approximately) correlated with age: other things being equal, high metallicity tends to indicate a young star and low metallicity tends to indicate an old star.
>
> [a] The metallicity 'unit' is termed the *dex* ("decimal exponent") in older literature. Fe is the usual metallic reference because its spectral lines are easily identified but other metals may be used. For example, an oxygen-based metallicity [O/H] may be defined in complete analogy with the Fe-based one.

Example 3.2 For the Sun, assuming complete ionization and the composition (3.28), $\mu_e \simeq 2/(1+X) = 1.15$, and neglecting the small contribution of metals

$$\mu_I \simeq \frac{1}{X + \frac{1}{4}Y} = 1.25.$$

Then the total mean molecular weight characterizing the gas in the Sun is

$$\mu = \left(\frac{1}{\mu_I} + \frac{1}{\mu_e}\right)^{-1} = 0.60,$$

with the assumption of complete ionization.

An approximate formula for the mean molecular weight of a completely ionized gas is

$$\mu \simeq \frac{4}{5X+3}, \tag{3.31}$$

which follows from Eqs. (3.29), (3.30), and (3.26) if metals are neglected so that $Z \sim 0$ and $Y \sim 1 - X$. This often is an acceptable approximation since most stars have a metal content of several percent or less. Applying (3.31) using the solar composition (3.28) gives $\mu \sim 0.60$, as expected from Example 3.2.

3.5 Polytropic Equations of State

An ideal gas equation of state with an effective mean molecular weight μ is a realistic approximation for many astrophysical processes, but there are other equations of state that are important in particular contexts. Generally, systems in thermodynamic equilibrium may be described in terms of three state variables P, T, and V or ρ. The most general equation of state provides a constraint that reduces the number of independent variables to two. If an additional constraint is placed on the system, a specialized equation of state is obtained that allows only *one* state variable to be varied independently.

3.5.1 Polytropic Processes

An example of imposing such an additional constraint is a *polytropic equation of state*. Generally, we may define a polytropic process by the requirement

$$\frac{\delta Q}{\delta T} = c, \tag{3.32}$$

where Q is the heat, T is the temperature, and c is a constant that may be interpreted as the heat capacity. A number of specific processes may be defined by Eq. (3.32), depending on the choice of c [224]. (1) If $c = 0$ the process is *adiabatic*, as discussed in Section 3.6. (2) If $\delta T = 0$ the process is *isothermal* and $c \to \infty$. (3) If $c = C_V$ the process is *isometric* (occurs at constant volume V). (4) If $c = C_P$ the process is *isobaric* (occurs at constant pressure P).

3.5.2 Properties of Polytropes

Important properties of polytropes are explored in Problem 3.11. For example, the first law (3.7) and Eq. (3.32) may be used to prove that for polytropic processes in ideal gases

$$\frac{dT}{T} = (1-\gamma)\frac{dV}{V}, \tag{3.33}$$

where the *polytropic* γ is defined by

$$\gamma \equiv \frac{C_P - c}{C_V - c}, \tag{3.34}$$

which reduces to the ideal gas law adiabatic parameter γ only if the constant $c = 0$. By inspection, three classes of *polytropic equations of state* satisfy Eq. (3.33):

$$PV^\gamma = \text{constant} \qquad P^{1-\gamma}T^\gamma = \text{constant} \qquad TV^{\gamma-1} = \text{constant}. \qquad (3.35)$$

The most common form of a polytropic equation of state employed in astrophysics is

$$P(r) = K\rho^\gamma(r) = K\rho^{1+1/n}(r) \qquad (3.36)$$

(this is of the form $PV^\gamma = \text{constant}$, since $\rho \propto 1/V$), where the *polytropic index n* is related to γ by

$$n = \frac{1}{\gamma - 1}. \qquad (3.37)$$

Equation (3.36) implies physically that the pressure is independent of the temperature, depending only on density. A polytropic approximation for the equation of state often makes solution of the equations for stellar structure easier because it decouples the differential equations describing hydrostatic equilibrium from those governing energy transfer and the temperature gradients (see Section 8.4). Box 3.4 discusses some cases where polytropic equations of state are relevant in astrophysics applications.

Box 3.4   **Examples of Polytropic Equations of State**

For at least three situations of interest in stellar structure the appropriate equation of state may be approximated by a polytrope:

1. For a completely ionized star that is fully mixed by convective motion of the gas and has negligible radiation pressure, the equation of state is

$$P = K_1 \rho^{5/3},$$

which corresponds to a polytrope with $\gamma = \frac{5}{3}$ and $n = \frac{3}{2}$. The parameter K_1 is constant for a given star but is phenomenological and differs from star to star.

2. For a completely degenerate gas of nonrelativistic ($v \ll c$) fermions (degeneracy is defined in Box 3.6 below) the appropriate equation of state is (see Section 3.8)

$$P = K_2 \rho^{5/3},$$

again corresponding to a polytrope with $\gamma = \frac{5}{3}$ and $n = \frac{3}{2}$. However, in this case the parameter K_2 is determined by fundamental constants and is not adjustable.

3. For a completely degenerate gas of ultrarelativistic fermions ($v \sim c$), the equation of state is (see Section 3.8)

$$P = K_3 \rho^{4/3},$$

corresponding to a polytrope with $\gamma = \frac{4}{3}$ and $n = 3$. As in the nonrelativistic degenerate example, the parameter K_3 is fixed by fundamental constants.

These three polytropic equations of state will be relevant for homogeneous stars that are completely mixed by convection, low-mass white dwarfs, and neutron stars and high-mass white dwarfs, respectively.

3.6 Adiabatic Equations of State

In terms of the heat Q and the entropy S, *adiabatic processes* are a specific type of polytropic process defined by the requirement that

$$\delta Q = T dS = 0. \tag{3.38}$$

From the first law (3.7), in adiabatic processes the change in internal energy comes only from PdV work. Adiabatic processes do not exchange heat with their environment, which makes them fully reversible ($dS = 0$). Realistic phenomena in astrophysics are not adiabatic but many are at least approximately so; therefore the adiabatic approximation is often a useful one. Some properties of adiabatic processes for ideal gases are explored in Problem 3.10 and considerably more will be said about them in subsequent material.

It is standard practice to introduce three *adiabatic exponents* Γ_1, Γ_2, and Γ_3 through the definitions

$$\Gamma_1 \equiv \left(\frac{\partial \ln P}{\partial \ln \rho}\right)_S \qquad \frac{\Gamma_2}{\Gamma_2 - 1} \equiv \left(\frac{\partial \ln P}{\partial \ln T}\right)_S \qquad \Gamma_3 - 1 \equiv \left(\frac{\partial \ln T}{\partial \ln \rho}\right)_S, \tag{3.39}$$

where the subscripts S are a reminder that adiabatic processes occur at constant entropy, and where the logarithmic derivatives are equivalent to $\partial \ln A = \partial A/A$. As may be verified directly by substitution, these definitions imply equations of state that take one of the three forms

$$PV^{\Gamma_1} = c_1 \qquad P^{1-\Gamma_2}T^{\Gamma_2} = c_2 \qquad TV^{\Gamma_3-1} = c_3, \tag{3.40}$$

Box 3.5   **A Plethora of Gammas**

For ideal gases the three adiabatic exponents defined through Eq. (3.39) are all equal and are also equivalent to the ideal gas adiabatic coefficient γ,

$$\Gamma_1 = \Gamma_2 = \Gamma_3 = \gamma \qquad \text{(ideal gas)}.$$

But for more general cases Γ_1, Γ_2, and Γ_3 are distinct and carry information emphasizing different aspects of the gas thermodynamics:

1. Because Γ_1 specifies how changes in pressure are related to changes in density, it enters into dynamical properties of the gas such as the speed of sound (see Problem 3.17).
2. Γ_2 relates changes in pressure to changes in temperature; this, for example, is important in understanding convective gas motion (see Chapter 7).
3. Γ_3 depends on the derivative of temperature with respect to density, which influences the response of the gas to compression.

An example of employing these adiabatic exponents may be found in Section 3.11, where a mixture of ideal gas and photons will be examined in the adiabatic limit.

where the c_n are constants and $\rho \propto V^{-1}$. Equationa (3.39) imply that $(\Gamma_2 - 1)/\Gamma_2 = (\Gamma_3 - 1)/\Gamma_1$, so the three adiabatic exponents are not independent. The relationship among the different adiabatic Γ_i and the ideal gas parameter γ is discussed in Box 3.5.

3.7 Equations of State for Degenerate Gases

Degenerate matter, which is defined in Box 3.6, is important in a variety of astrophysical applications. For example, in white dwarf stars the electrons are thought to be degenerate except very near the surface, while in neutron stars the neutrons are expected to be degenerate. Let us look at this in a little more detail for electrons, beginning with a demonstration that – as a consequence of quantum mechanics – most stars are expected to be completely ionized over much of their volume because ionization can be induced by sufficiently high pressure, even for low temperatures. Ionization by pressure rather than temperature implies the possibility of producing a (relatively) cold gas of electrons, which is the necessary condition for a degenerate electron equation of state.

3.7.1 Pressure Ionization

Consider the schematic diagram shown in Fig. 3.2, where atoms are represented by the darker spheres of radius r and the average spacing between atoms is represented in terms of the lighter spheres with radius d. Although the argument can be generalized easily, it is assumed for simplicity of discussion that the stellar material consists only of ions of a single species and electrons produced by ionizing that species. Electrons confined in the atoms obey Heisenberg uncertainty relations of the form $p \cdot \Delta x \geq \hbar$ (Box 3.6). Taking an

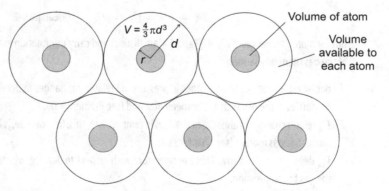

Fig. 3.2 Snapshot of idealized atomic spacing in dense matter. These are slices of 3-dimensional spherical volumes. Darker central spheres represent atomic volumes. The larger unshaded spheres represent the average volume available to each atom at a given density.

Box 3.6 — Quantum Mechanics and Equations of State

Stellar equations of state reflect microscopic properties of the gas. At low densities this gas behaves classically but at higher densities a quantum-mechanical description becomes essential. A brief but more mathematical overview of quantum mechanics is given in Appendix F but the requisite physics can be understood conceptually in terms of four basic ideas.

de Broglie Wavelength

The foundation of a quantum description of matter is particle–wave duality: a microscopic particle takes on wave properties characterized by a *de Broglie wavelength* $\lambda = h/p$, where p is the momentum and h is Planck's constant. Thus in quantum mechanics the location of a particle becomes fuzzy, spread out over an interval comparable to the de Broglie wavelength.

Heisenberg Uncertainty Principle

The uncertainty principle quantifies the fuzziness of particle–wave duality, requiring that $\Delta p \cdot \Delta x \geq \hbar$, where Δp is the uncertainty in momentum, Δx is the uncertainty in position, and $\hbar \equiv h/2\pi$, as well as $\Delta E \cdot \Delta t \geq \hbar$, where ΔE is the uncertainty in energy and Δt is the uncertainty in the time over which the energy is measured.

Quantum Statistics

Elementary particles may be classified as either *fermions* or *bosons*, which characterizes how aggregates of the same type of particle behave. Fermions (such as electrons, or neutrons and protons if their internal quark and gluon structure is neglected) obey *Fermi–Dirac statistics*, which implies the *Pauli exclusion principle*: no two fermions can occupy the same quantum state. All elementary particles of half-integer spin are fermions. Bosons (photons are the most important example for us) obey *Bose–Einstein statistics*, which places no restriction on how many identical particles can occupy the same state. All elementary particles of integer spin are bosons. Matter is made from fermions but forces are mediated by exchange of bosons. For example, electromagnetic forces result from virtual photon exchange between charged particles.

Degeneracy

The exclusion principle requires each fermion to be in a different quantum state, so the lowest-energy state results from filling energy levels from the bottom up. *Degenerate matter* is a many-fermion state with all the lowest energy levels filled and all higher ones unoccupied. It is common at high densities and has an unusual equation of state with various implications for astrophysics. Degenerate matter has many similarities with metals in the solid state.

---

average volume per electron of $V_0 \simeq (\Delta x)^3$, the uncertainty principle relation becomes

$$p \geq \hbar/V_0^{1/3}. \qquad (3.41)$$

The uncertainty principle leads to ionization when the effective volume of the atoms becomes too small to confine the electrons because of Eq. (3.41). The volume per electron V_0 and the volume per ion V_i are related by $ZV_0 = V_i$, since there are Z electrons per ion, allowing (3.41) to be expressed as the inequality $p \geq \hbar Z^{1/3}/V_i^{1/3}$ [55, 134].

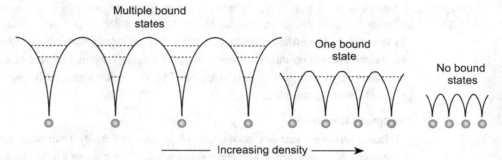

Fig. 3.3 Pressure ionization. At lower density the electrons occupy quantum-mechanical bound states (denoted by dashed lines) in potential wells centered on the ions. With increasing density, fewer locally bound states are possible until at some critical density none remain and the electrons are all ionized. Thus, sufficiently high density can cause complete ionization, even at zero temperature.

From basic atomic physics the atomic radius may be approximated by $r \simeq a_0 Z^{-1/3}$, where $a_0 = 5.3 \times 10^{-9}$ cm is the Bohr radius. From Fig. 3.2, if the star is composed entirely of a single element with atomic number Z and mass number A, there are on average Z electrons in each sphere of radius d and the average number density for the electrons is $n_e = 3Z/4\pi d^3$, which may be solved for the separation d,

$$d \simeq (3Z/4\pi n_e)^{1/3}. \tag{3.42}$$

As illustrated in Fig. 3.3, ionization is likely to result at densities where $d < r$ because no bound electronic states remain in the potential wells centered on the ions at those densities. Since there are A nucleons in each volume of radius d in Fig. 3.2, the mass density is $\rho = 3AM_u/4\pi d^3$ and requiring that $d \simeq r \simeq a_0 Z^{-1/3}$ defines a critical density

$$\rho_{\text{crit}} \simeq 3ZAM_u/4\pi a_0^3. \tag{3.43}$$

We may expect that for densities greater than this there will be almost complete pressure ionization, irrespective of the temperature.

Example 3.3 The condition (3.43) is satisfied rather easily. Consider pure hydrogen. Inserting $Z = A = 1$ gives a critical density of 3.2 g cm^{-3}, only a little larger than that of water.

The critical pressure ionization densities for gases composed of isotopes for a few representative elements are summarized in Table 3.1. These critical densities may be compared with typical actual densities of order 10^2 g cm^{-3} for the center of the Sun, 10^6 g cm^{-3} for a carbon–oxygen white dwarf, and 10^9 g cm^{-3} for the iron core of a massive pre-supernova star. These considerations imply that Saha ionization equations such as Eq. (2.2), which are derived assuming ionization to be caused by thermal effects, are no longer reliable in the deep interior of stars.

Example 3.4 The Saha equations predict approximately 24% of the hydrogen in the core of the Sun to be neutral. However, comparison of Table 3.1 with Table 10.1 for properties of

Table 3.1 Critical pressure-ionization densities

Element	(Z, A)	ρ_{crit} (g cm^{-3})
Hydrogen	(1, 1)	3.2
Helium	(2, 4)	26
Carbon	(6, 12)	230
Oxygen	(8, 16)	410
Silicon	(14, 28)	1254
Iron	(26, 56)	4660

the solar interior indicates that the density is sufficiently high to pressure-ionize hydrogen over the inner 40% of the Sun. Between thermal and pressure effects, the solar interior is almost entirely ionized, in contrast to what would be expected from the Saha equations alone.

3.7.2 Distinguishing Classical and Quantum Gases

We now examine in more depth the distinction between a classical gas and a quantum gas, and the corresponding implications for stellar structure. The difference between classical and quantum gases can be understood in terms of possible statistical distributions for particles in the gas.

The Fermi–Dirac distribution: Identical fermions are described statistically in quantum mechanics by the *Fermi–Dirac distribution*

$$f(\varepsilon_p) = \frac{1}{e^{(\varepsilon_p - \mu)/kT} + 1} \quad \text{(Fermi–Dirac)}, \tag{3.44}$$

where the energy ε_p for nonrelativistic and relativistic regimes is given by

$$\varepsilon_p = mc^2 + \frac{p^2}{2m} \quad \text{(nonrelativistic)}, \tag{3.45}$$

$$\varepsilon_p = \left(p^2 c^2 + m^2 c^4\right)^{1/2} \quad \text{(relativistic)}, \tag{3.46}$$

and the *chemical potential* μ is the energy associated with changing the number of particles in the system. It may be introduced formally by including in Eq. (3.7) a term μdN accounting for a possible change in the particle number N so that the first law generalizes to

$$dU = TdS - PdV + \mu dN, \tag{3.47}$$

which expresses how the internal energy changes with transfer of entropy or heat, with compression or expansion, and through transfer of particles.

The Bose–Einstein distribution: Identical bosons are described statistically in quantum mechanics by the *Bose–Einstein distribution*

$$f(\varepsilon_p) = \frac{1}{e^{(\varepsilon_p - \mu)/kT} - 1} \quad \text{(Bose–Einstein)}, \tag{3.48}$$

which seems similar in form to the Fermi–Dirac distribution but is associated with statistical behavior for bosons that is quite different from that of fermions.

The Maxwell–Boltzmann distribution: Now it is possible to make a concise formal distinction between classical and quantum gases. A gas is a quantum gas if it is described by one of the distributions (3.44) or (3.48), and it is a classical gas if the condition

$$\exp[(mc^2 - \mu)/kT] \gg 1 \tag{3.49}$$

is fulfilled. If the classical condition (3.49) is satisfied, for either fermions or bosons the distribution function becomes well-approximated by *Maxwell–Boltzmann statistics*,

$$f(\varepsilon_p) = e^{-(\varepsilon_p - \mu)/kT} \quad \text{(Maxwell–Boltzmann)}, \tag{3.50}$$

where generally $f(\varepsilon_p) \ll 1$. Thus, in a classical gas the individual energy states are scarcely occupied, quantum effects are minimized, and the gas obeys Maxwell–Boltzmann statistics.

3.7.3 Nonrelativistic Classical and Quantum Gases

It is useful to introduce a *critical (number) density* variable n_c through the definition

$$n_c \equiv \left(\frac{2\pi mkT}{h^2}\right)^{3/2} = \frac{(2\pi)^{3/2}}{\lambda^3}, \tag{3.51}$$

where the deBroglie wavelength λ for nonrelativistic particles is given by

$$\lambda = \frac{h}{p} \simeq \left(\frac{h^2}{mkT}\right)^{1/2}. \tag{3.52}$$

As shown in Box 3.7, the number of particles in the gas is

$$N = \int_0^\infty f(\varepsilon_p) g(p) \, dp, \tag{3.53}$$

where the integration measure is

$$g(p)dp = g_s \frac{V}{h^3} 4\pi p^2 dp, \tag{3.54}$$

with p the momentum, V the volume, and $g_s = 2s + 1 = 2$ the spin-degeneracy factor for electrons. To obtain a rough estimate we may substitute the Maxwell–Boltzmann distribution (3.50) with a nonrelativistic energy (3.45) into (3.53), approximate $g_s \sim 1$, integrate, and rearrange the results to give

$$\frac{n_c}{n} \simeq e^{(mc^2 - \mu)/kT}, \tag{3.55}$$

where the number density is $n = N/V$. Therefore, the condition (3.49) has a simple physical interpretation: satisfying it at a given temperature is equivalent to requiring that $n \ll n_c$, implying that a classical gas is characterized by an actual number density n that is small on a scale set by the critical quantum density n_c.

Box 3.7 — Density of Quantum States

The quantum states of a gas may be enumerated by confining the particles to a cubic box of volume $V = L^3$. The wavefunctions must obey the Schrödinger equation

$$-\frac{\hbar}{2m}\left(\frac{\partial^2}{\partial x^2} + \frac{\partial^2}{\partial y^2} + \frac{\partial^2}{\partial z^2}\right)\psi_k(r) = \varepsilon_k \psi_k(r),$$

where the wavevector $\boldsymbol{k} = \boldsymbol{p}/\hbar$ has been introduced. The boundary conditions (periodic in L_i) imply solutions labeled by the conserved momentum $\boldsymbol{k} = \boldsymbol{p}/\hbar$

$$\psi_k(r) = \exp(i\boldsymbol{k}\cdot\boldsymbol{r}) \qquad \boldsymbol{k} = (k_x, k_y, k_z) = (n_x, n_y, n_z)\frac{\pi}{L},$$

where the n_i ($i = x, y, z$) are positive integers, and the corresponding energy is

$$\varepsilon_k = \frac{\hbar^2}{2m}(k_x^2 + k_y^2 + k_z^2) = \frac{\hbar^2}{2m}k^2.$$

Allowed states may be thought of as points in the k-space labeled by quantum numbers (n_x, n_y, n_z), each with an additional spin degeneracy $g_s = 2s + 1$.

Physical quantities involve integrals over contributions from the particles, so an appropriate integration measure is required. In the interval k_i to $k_i + dk_i$ there are $(L/\pi)dk_i$ distinct values of n_i, so upon defining a 3-dimensional k-space with coordinates $k_x, k_y,$ and k_z a unit volume in this space will contain $(L/\pi)^3$ quantum states. Now consider the quantum states in the k-space lying between the concentric spheres of radius k and $k + dk$, restricting to the octant where all the k_i are positive. The total volume of this region is

$$\tfrac{1}{8}\left(\tfrac{4}{3}\pi(k+dk)^3 - \tfrac{4}{3}\pi k^3\right) = \tfrac{1}{2}\pi k^2 dk.$$

There are $(L/\pi)^3$ discrete states per unit volume, each with additional spin degeneracy g_s, so the density of states is $g(k)dk = \tfrac{1}{2}g_s(L/\pi)^3(\pi k^2 dk)$. Reverting to the momentum variable $\boldsymbol{p} = \hbar\boldsymbol{k}$ and using $V = L^3$ yields the integration measure

$$g(p)\,dp = g_s \frac{4\pi V}{h^3} p^2 dp.$$

For example, if the distribution in ε_p is $f(\varepsilon_p)$, the energy E and particle number N are given by

$$E = \int_0^\infty \varepsilon_p f(\varepsilon_p) g(p)\,dp \qquad N = \int_0^\infty f(\varepsilon_p) g(p)\,dp.$$

For the specific case of fermions, $f(\varepsilon_p)$ is given by Eq. (3.44).

Example 3.5 The preceding result has an instructive alternative interpretation. The average separation between particles is $d \sim n^{-1/3}$, the condition $n \ll n_c$ defining a classical gas implies that $1/n \gg 1/n_c$, and from Eq. (3.51), $n_c \sim \lambda^{-3}$. Thus, the condition $n \ll n_c$ is equivalent to the requirement that $d \gg \lambda$, meaning that for a classical gas the average separation between particles is much larger than the average de Broglie wavelength λ for particles. This makes sense conceptually. The "quantum fuzziness" of a particle extends

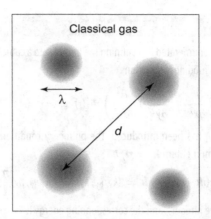 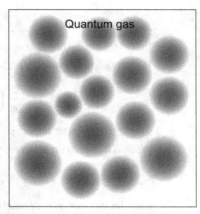

Fig. 3.4 Schematic illustration of classical and quantum gases. The width of each fuzzy ball represents the quantum uncertainty in position (not the physical size) of the gas particle. In the classical gas (left) the average spacing d between gas particles is much larger than their de Broglie wavelengths λ. In the quantum gas (right) d is comparable to λ. The gas particles have a range of λ values because they have a velocity distribution.

over a distance $\sim \lambda$, so if particles are separated on average by distances considerably larger than λ quantum effects are minimized and a classical description becomes valid. This difference between a classical and quantum gas is illustrated in Fig. 3.4.

3.7.4 Ultrarelativistic Classical and Quantum Gases

Proceeding in a manner similar to that for the nonrelativistic case, for ultrarelativistic particles ($v \sim c$) the rest mass of the particle may be neglected and $\varepsilon_p \simeq kT = (m^2c^4 + p^2c^2)^{1/2} \simeq pc$ and Eq. (3.50) may be employed to obtain

$$\frac{n'_c}{n} \sim e^{-\mu/kT}, \tag{3.56}$$

where the relativistic critical quantum density n'_c is defined by

$$n'_c = 8\pi \left(\frac{kT}{hc}\right)^3. \tag{3.57}$$

Comparing Eqs. (3.56) and (3.49) with the mass m neglected suggests that in the ultrarelativistic limit the condition that the gas behave classically is equivalent to a requirement that $n \ll n'_c$. As for the nonrelativistic limit, this implies that the de Broglie wavelength $\lambda = h/p \simeq hc/kT$ is small compared with the average separation of particles in a classical gas.

3.7.5 Transition from a Classical to Quantum Gas

Equations (3.51) and (3.57) imply that we also may view the quantum gas condition as a temperature constraint. For example, from Eq. (3.51) the condition $n \gg n_c$ implies that

3.7 Equations of State for Degenerate Gases

$$kT \ll \frac{h^2 n^{2/3}}{2\pi m}. \qquad (3.58)$$

A quantum gas is a *cold gas*, but cold on a temperature scale set by the right side of Eq. (3.58) – if the density is high enough the gas could be "cold" and still have a temperature of billions of degrees! The precise meaning of a cold fermionic gas is that the fermions are concentrated in the lowest available quantum states, which is the definition of a *degenerate gas* (see Box 3.6).

We may conclude from the preceding results that at high density the classical approximation fails and the gas behaves as a quantum system subject to the quantum statistics (Fermi–Dirac or Bose–Einstein) appropriate for the gas. Notice from (3.51) that with increasing gas density the least massive particles in the gas will be more likely to deviate from classical behavior because the scale set by the critical density is proportional to $m^{3/2}$. Thus photons, neutrinos, and electrons are most susceptible to such effects. The massless photons never behave as a classical gas and the nearly massless neutrinos interact so weakly with matter that they leave the star unimpeded when they are produced.[3] It follows that in normal stellar environments electrons are most susceptible to a transition from classical to quantum gas behavior. On the other hand, the ions are sufficiently massive that they can often be treated as a classical ideal gas, even if the electrons behave as a quantum gas.

Example 3.6 In the center of the Sun the number density for electrons is about 6×10^{25} cm^{-3} and for a temperature of 15×10^6 K the nonrelativistic critical quantum density from Eq. (3.51) is $n_c \sim 1.4 \times 10^{26}$ cm^{-3}. Thus at the center the actual electron density is about half the critical density. A similar analysis at 30% of the solar radius gives that the electron density is about 12% of critical (see Problem 10.14). Thus, the electrons in the Sun are reasonably well approximated by a dilute classical gas and quantum corrections are small. However, the core of the Sun, as for all stars, will contract late in its life as its nuclear fuel is exhausted. The approximate relationship between a star's temperature T and radius R is

$$kT \simeq \frac{GM\bar{\mu}}{3R} \simeq \frac{1}{R},$$

where M is the star's mass, $\bar{\mu}$ is the average mass of a gas particle, and G is the gravitational constant [169]. Combining this equation with Eq. (3.51) yields $n_c \simeq R^{-3/2}$. Since the actual number density as $n \sim R^{-3}$, as the core of the Sun contracts eventually n will exceed n_c in the central regions and the electrons there will begin to behave as a quantum rather than classical gas. Because the present electron density in the core is not very far below the critical density, quantum effects in the gas of the solar core will presumably become important relatively early in this contraction.

---

[3] An exception occurs for a core collapse supernova (Chapter 20), where densities and temperatures become high enough to trap neutrinos for a time that is long compared with the dynamical timescale. Note also that in the very dense environment of a neutron star the neutrons and protons may become degenerate, but they are never degenerate in normal stars.

3.8 The Degenerate Electron Gas

As discussed above, the most important impact of quantum gas behavior in normal stars is for the electron gas. Accordingly, let us apply the preceding rather general discussion specifically to both nonrelativistic and ultrarelativistic degenerate electrons.

3.8.1 Fermi Momentum and Fermi Energy

As illustrated in Fig. 3.5, evaluation of Eq. (3.44) at finite temperature gives an occupation function that drops from one to zero over a region of finite width, with this width decreasing as the temperature is lowered. In the limit that the temperature may be neglected, the Fermi–Dirac distribution (3.44) becomes a step function in energy space,

$$f_{\rm f}(\varepsilon_p) = \frac{1}{e^{(\varepsilon_p - \mu)/kT} + 1} \xrightarrow[T \to 0]{} \begin{cases} f(\varepsilon_p) = 1 & \varepsilon_p \leq \varepsilon_{\rm f} \\ f(\varepsilon_p) = 0 & \varepsilon_p > \varepsilon_{\rm f} \end{cases} \qquad (3.59)$$

where the value of the chemical potential μ at zero temperature is denoted by $\varepsilon_{\rm f}$ and is termed the *Fermi energy*. The corresponding value of the momentum is denoted by $p_{\rm f}$ and is termed the *Fermi momentum*. Thus, $\varepsilon_{\rm f}$ gives the energy of the highest occupied state in the degenerate Fermi gas and $p_{\rm f}$ specifies its momentum. We say that $\varepsilon_{\rm f}$ or $p_{\rm f}$ define a *Fermi surface* that contains all occupied levels within it in either energy or momentum space.[4] Figure 3.6 illustrates the energy and momentum space for a degenerate Fermi gas.

The number of electrons in the degenerate gas at zero temperature is just the number of states with momentum less than the Fermi momentum $p_{\rm f}$,

$$N = \int_0^{p_{\rm f}} g(p)\,dp = 4\pi V \frac{g_{\rm s}}{h^3} \int_0^{p_{\rm f}} p^2\,dp = \frac{8\pi V}{3h^3} p_{\rm f}^3, \qquad (3.60)$$

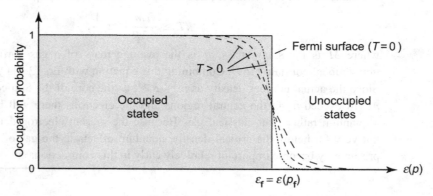

Fig. 3.5 The Fermi–Dirac distribution (3.44) as a function of temperature. Curves with successively shorter dashes represent successively lower temperatures. The solid line defines a step function corresponding to the limit $T \to 0$ in Eq. (3.44). This degenerate-gas limit is illustrated further in Fig. 3.6.

---

[4] It is termed a surface because in the general case the momentum is a 3-component vector so $\varepsilon_{\rm f}$ becomes a surface in a multidimensional space defined by the components of momentum; see Fig. 3.6.

3.8 The Degenerate Electron Gas

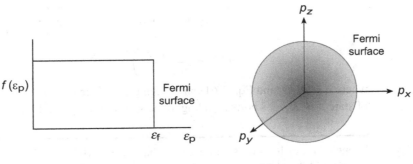

Fig. 3.6 The degenerate Fermi gas with its sharp Fermi surface in energy and momentum. In condensed matter physics the Fermi surface may have a more complex shape but it is assumed to be isotropic in momentum for our discussion of degenerate gases in stars.

where $g_s = 2$ for electrons has been used. Solving for p_f and introducing the number density $n = N/V$, the Fermi momentum and Fermi energy $\varepsilon_f = p_f^2/2m$ are found to be determined completely by the electron number density,

$$p_f = \left(\frac{3h^3}{8\pi} \cdot \frac{N}{V}\right)^{1/3} = h\left(\frac{3n}{8\pi}\right)^{1/3} \qquad \varepsilon_f = \frac{p_f^2}{2m}. \tag{3.61}$$

The interparticle spacing is of order $n^{-1/3}$, so the de Broglie wavelength for an electron at the Fermi surface, $\lambda = h/p_f \sim n^{-1/3}$, is comparable to the average separation of electrons.

3.8.2 Equation of State for Nonrelativistic Electrons

We may construct the equation of state by evaluating the internal energy of the gas. Let's do this first in the nonrelativistic limit and then in the ultrarelativistic limit for degenerate electrons. In the nonrelativistic limit $p_f \ll mc$, which implies that

$$n \ll \left(\frac{1}{\lambda_c}\right)^3 = \left(\frac{mc}{h}\right)^3, \tag{3.62}$$

where $\lambda_c \equiv h/mc$ is the *Compton wavelength* for an electron. In this limit the internal energy density u for the degenerate electron gas is

$$u = \frac{U}{V} = \frac{1}{V}\int_0^\infty \varepsilon_p f(\varepsilon_p) g(p)\, dp \simeq \frac{Nmc^2}{V} + \frac{3N}{10mV} p_f^2, \tag{3.63}$$

where V is the volume and Eqs. (3.45), (3.54), (3.59), (3.60), and $g_s = 2$ have been used. For a nonrelativistic gas the pressure is given by $\frac{2}{3}$ of the kinetic energy density (see Problem 4.10); identifying the second term of Eq. (3.63) as the kinetic energy density yields the equation of state [169]

$$P = \frac{2}{3} \times \text{(kinetic energy density)} = \frac{2}{3}\left(\frac{N}{V}\frac{3p_f^2}{10m}\right)$$

$$= n\frac{p_f^2}{5m} = \frac{h^2}{5m}\left(\frac{3}{8\pi}\right)^{2/3} n^{5/3}, \qquad (3.64)$$

where $n = N/V$ and Eq. (3.61) have been used. Since $n \propto \rho$, this is a polytropic equation of state $\rho \sim K_2 \rho^{5/3}$ corresponding to the second example in Box 3.4.

Example 3.7 For low-mass white dwarfs having $\rho \lesssim 10^6$ g cm^{-3}, the electrons are nonrelativistic and the electron pressure is approximated well by the $\gamma = \frac{5}{3}$ polytrope implied by Eq. (3.64). Utilizing Eq. (3.22) for the electron number density n_e,

$$P_e = \frac{h^2}{5m_e}\left(\frac{3}{8\pi}\right)^{2/3}\left(\frac{\rho}{M_u \mu_e}\right)^{5/3} = \frac{1.0 \times 10^{13}}{\mu_e^{5/3}}\left(\frac{\rho}{\text{g cm}^{-3}}\right)^{5/3} \text{dyne cm}^{-2}. \qquad (3.65)$$

As was already noted in Box 3.4, the constant factor K_2 in this case is fixed by fundamental constants.

3.8.3 Equation of State for Ultrarelativistic Electrons

For ultrarelativistic electrons, $n \gg n'_c$ implies that $n \gg (mc/h)^3$. Utilizing Eq. (3.54) and the ultrarelativistic limit $\varepsilon_p = pc$ of Eq. (3.46), the internal energy density is given by

$$u = \frac{U}{V} = \frac{1}{V}\int_0^\infty \varepsilon_p f(\varepsilon_p) g(p) \, dp$$

$$\simeq \frac{8\pi c}{h^3}\int_0^{p_f} p^3 \, dp = \frac{3}{4}hc\left(\frac{3}{8\pi}\right)^{1/3} n^{4/3}, \qquad (3.66)$$

where (3.59) and (3.61) were used. For an ultrarelativistic gas the pressure is $\frac{1}{3}$ of the kinetic energy density (see Problem 4.10), so from Eq. (3.66)

$$P = \frac{1}{3} \times \text{(kinetic energy density)} = \frac{hc}{4}\left(\frac{3}{8\pi}\right)^{1/3} n^{4/3} \qquad (3.67)$$

for ultrarelativistic electrons.

Example 3.8 For higher-mass white dwarfs having $\rho \gtrsim 10^6$ g cm^{-3}, the electrons are very relativistic and the corresponding degenerate equation of state is well approximated by Eq. (3.67). Utilizing Eq. (3.22) for the electron number density n_e,

$$P_e = \frac{hc}{4}\left(\frac{3}{8\pi}\right)^{1/3}\left(\frac{\rho}{M_u \mu_e}\right)^{4/3} = \frac{1.24 \times 10^{15}}{\mu_e^{4/3}}\left(\frac{\rho}{\text{g cm}^{-3}}\right)^{4/3} \text{dyne cm}^{-2}, \qquad (3.68)$$

which is a polytrope $P = K_3 \rho^{4/3}$ corresponding to the third example in Box 3.4. As for K_2 in Example 3.7, the constant factor K_3 is not adjustable but is fixed by fundamental considerations.

3.9 High Gas Density and Stellar Structure

The preceding discussion implies that increasing the density can have a large impact on the structure of stars by magnifying quantum effects in the gas. In fact, there are a number of important implications for high densities in stellar environments that it is useful to summarize:

1. An increase in the gas density above a critical amount exemplified in Table 3.1 enhances the probability for pressure ionization, thereby creating a gas of electrons and ions irrespective of possible thermal ionization.

Box 3.8 — **Thermal Pressure and Quantum Pressure**

As suggested by preceding discussion, a fermionic gas has an effective pressure of purely quantum-mechanical origin, independent of its temperature [193]. To keep the following discussion simple and qualitative, the gas pressure is assumed to be dominated by nonrelativistic electrons and often factors of order one will be dropped. For an ideal gas of temperature T the average energy of an electron is $E \sim kT = \frac{1}{2}mv^2$, so that the electron may be assumed to have a velocity

$$v_{\text{thermal}} \simeq \left(\frac{kT}{m}\right)^{1/2}$$

of thermal origin. But even at $T = 0$ electrons have a velocity v_{QM} implied by the uncertainty principle. If n is electron number density, $p \sim \Delta p \sim \hbar/\Delta x \sim \hbar n^{1/3}$, and

$$v_{\text{QM}} \simeq \frac{p}{m} \simeq \frac{\hbar n^{1/3}}{m}.$$

Thus, the velocity of particles in the gas may be viewed as having two contributions, one from the temperature and one from quantum fluctuations, with the two logically distinct because the thermal contribution vanishes identically at zero temperature.[a] The pressure contributed by the thermal motion is

$$P_{\text{thermal}} = nkT = nmv_{\text{thermal}}^2$$

and, dropping some constant factors in Eq. (3.64), the quantum pressure is given by

$$P_{\text{QM}} \simeq \frac{\hbar^2}{m} n^{5/3} = nm\left(\frac{\hbar n^{1/3}}{m}\right)^2 = nmv_{\text{QM}}^2.$$

Then a degenerate gas is one for which $P_{\text{QM}} \gg P_{\text{thermal}}$. Thermal pressure is proportional to T and density, but quantum pressure is independent of T and proportional to a power of the density. Hence, degeneracy is favored in low-temperature, dense gases, and a gas can have a high pressure of purely quantum origin, even at $T = 0$. Furthermore, changing T in a degenerate gas will have little initial effect on pressure, which is dominated by a term independent of T (as long as the gas remains degenerate). These properties will have profound consequences for stellar structure and stellar evolution when high densities are encountered.

[a] This is related to the distinction between a *thermal phase transition* (driven by thermal fluctuations that vanish as $T \to 0$), and a *quantum phase transition* (driven by quantum fluctuations that remain in the zero-temperature limit). Such concepts are important in fields like condensed matter physics.

2. By uncertainty principle arguments an increase in the gas density raises the average momentum of gas particles, making them more relativistic.
3. As indicated by Eq. (3.61), an increased density raises the Fermi momentum. This, for example, influences the weak-interaction processes like β-decay that can take place in the star.
4. An increase in the gas density decreases the interparticle spacing relative to the average de Broglie wavelength, making it more likely that the least massive fermions transition from classical to degenerate quantum gas behavior.
5. If high density drives the electron gas into a degenerate state, this will modify the normal relationship between temperature and pressure, with many implications for stellar structure and stellar evolution (see Box 3.8).
6. Increased density enhances the strength of the gravitational field and makes it more difficult to maintain stability of the star against gravitational collapse. Higher density also makes it more likely that general relativistic corrections to Newtonian gravitation become important.
7. Higher density (often implying a higher temperature) changes the rates of thermonuclear reactions and alters the opacity of the stellar material to radiation. The former changes the rate at which energy is produced; the latter changes the efficiency of transporting that energy. Both can have substantial influence on stellar structure and evolution.

These consequences of increased density are highly relevant for understanding stellar structure and stellar evolution because all stars are expected to go through late evolutionary stages that may dramatically increase their central densities.

3.10 Equation of State for Radiation

Unlike the matter, which is fermionic with finite mass and can be relativistic or non-relativistic in different contexts, electromagnetic radiation in stars may be viewed as an ultrarelativistic gas of massless bosons. The photon gas is unusual in that the particles in the gas are massless and they all move at the same speed $v = c$. It is also unusual in that the number of photons is generally not a conserved quantity, since the massless photons can be freely created and destroyed by interactions. Thus the photon number adjusts itself to minimize the free energy of the system. The equation of state associated with radiation follows from using the Planck frequency distribution

$$n(v)\,dv = \frac{8\pi}{c^3} \frac{v^2}{(e^{hv/kT} - 1)}\,dv, \tag{3.69}$$

to evaluate the pressure and energy density (see Problem 3.8). This yields for the radiation pressure of the photon gas,

$$P_{\text{rad}} = \tfrac{1}{3} a T^4, \tag{3.70}$$

where a is the radiation density constant, and for the corresponding energy density

$$u_{\text{rad}} = a T^4 = 3 P_{\text{rad}}, \tag{3.71}$$

which implies an equation of state $P_{\text{rad}} = \tfrac{1}{3} u_{\text{rad}}$ for the photon gas.

3.11 Matter and Radiation Mixtures

For a simple model of a star containing gas and radiation, it is often a good starting point to assume an ideal gas equation of state for the matter (provided that the density is not too high) and a blackbody equation of state for the radiation. In that case, for the pressure P and internal energy U,

$$P = \frac{N}{V}kT + \frac{aT^4}{3}, \tag{3.72}$$

$$U = uV = C_V T + aT^4 V, \tag{3.73}$$

where the first term in each equation is the contribution of the ideal gas and the second term in each equation is the contribution of the radiation.

3.11.1 Mixtures of Ideal Gases and Radiation

In high-temperature stellar environments we will often encounter mixtures of gas and radiation. In that case it is convenient to define a parameter β that measures the relative contributions of gas pressure P_g and radiation pressure P_{rad} to the total pressure P:

$$\beta \equiv \frac{P_g}{P} \quad 1 - \beta = \frac{P_{rad}}{P} \quad P = P_g + P_{rad}. \tag{3.74}$$

Thus $\beta = 1$ implies that all pressure is generated by the gas, $\beta = 0$ implies that all pressure is generated by radiation, and all values in between correspond to situations where pressure receives contributions from both gas and radiation.

Example 3.9 Assuming an ideal gas equation of state, the pressure generated by the gas alone in a mixture of ideal gas and radiation is

$$P_g = nkT = \beta P.$$

Solving this equation for the total pressure,

$$P = \frac{nkT}{\beta} = \frac{\rho kT}{\beta \mu}, \tag{3.75}$$

which is of ideal gas form. Thus, the formal effect of mixing the radiation with the gas is to produce an ideal gas equation of state but with an effective mean molecular weight $\beta \mu$, where μ is the mean molecular weight for the gas alone.

The result of Example 3.9 implies that we may view a mixture of ideal gas and radiation as a modified ideal gas. However, the relative contribution of radiation and gas to the pressure varies through the volume of a star, so the parameter β is a local function of position in the star.

3.11.2 Adiabatic Systems of Gas and Radiation

The preceding discussion of gas and radiation mixtures depends only on the ideal gas assumption. Suppose that discussion is restricted further to adiabatic processes. From the

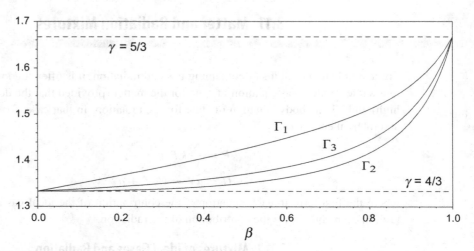

Fig. 3.7 Adiabatic exponents in a mixture of ideal gas and radiation.

adiabatic condition $\delta Q = 0$, the first law of thermodynamics, and the definition of β it may be shown that at constant entropy (see Ref. [55] and Problem 3.16)

$$\frac{d \ln T}{d \ln V} = \frac{-(\gamma - 1)(4 - 3\beta)}{\beta + 12(\gamma - 1)(1 - \beta)}. \tag{3.76}$$

These logarithmic derivatives may then be used to evaluate the adiabatic exponents with the results

$$\Gamma_1 = \frac{d \ln P}{d \ln \rho} = \beta + \frac{(4 - 3\beta)^2(\gamma - 1)}{\beta + 12(1 - \beta)(\gamma - 1)}, \tag{3.77}$$

$$\Gamma_2 = \left(1 - \frac{d \ln T}{d \ln P}\right)^{-1} = 1 + \frac{(4 - 3\beta)(\gamma - 1)}{\beta^2 + 3(\gamma - 1)(1 - \beta)(4 + \beta)}, \tag{3.78}$$

$$\Gamma_3 = 1 + \frac{d \ln T}{d \ln \rho} = 1 + \frac{(4 - 3\beta)(\gamma - 1)}{\beta + 12(1 - \beta)(\gamma - 1)}. \tag{3.79}$$

The adiabatic exponents Γ_1, Γ_2, and Γ_3 are plotted in Fig. 3.7 as a function of the parameter β governing the relative contribution of gas and radiation to the pressure. They have the expected limiting behavior: assuming $\gamma = \frac{5}{3}$ for a monatomic ideal gas and $\beta = 1$ (no radiation contribution to the pressure) gives $\Gamma_1 = \Gamma_2 = \Gamma_3 = \frac{5}{3}$, while for $\beta = 0$ (all pressure generated by radiation), $\Gamma_1 = \Gamma_2 = \Gamma_3 = \frac{4}{3}$. For other values of β the adiabatic exponents are generally not equal to each other and lie between $\frac{4}{3}$ and $\frac{5}{3}$.

3.11.3 Radiation and Gravitational Stability

Problem 3.14 and Fig. 3.7 indicate that the adiabatic exponents (3.39) for a pure radiation field are all equal to $\frac{4}{3}$. As shall be explored in greater detail later (see Sections 9.4 and 16.2.4, for example), an adiabatic exponent with a value less than $\frac{4}{3}$ generally implies an instability against gravitational collapse. Therefore, admixtures of radiation (more generally, of any ultrarelativistic component) often signal decreased gravitational stability for a gas.

Background and Further Reading

Our treatment of equations of state follows the discussions in Bowers and Deeming [55]; Hansen, Kawaler, and Trimble [107]; and the especially clear presentation in Phillips [169]. Equations of state for condensed objects such as white dwarfs and neutron stars are discussed in Shapiro and Teukolsky [200]. Leff [142] gives a pedagogical discussion of the photon gas equation of state. A clearly written general introduction to the thermodynamics and statistical mechanics underlying this chapter may be found in Blundell and Blundell [50].

Problems

3.1 Show that for an ideal gas the heat capacity at constant pressure C_P is related to the heat capacity at constant volume C_V by $C_P = C_V + Nk$, where N is the number of particles and k is the Boltzmann constant.***

3.2 Using the considerations of Section 3.7.1 and the Bohr model of the hydrogen atom, derive an expression for the principal quantum number n of the highest bound state as a function of density. Assume the Sun to be pure hydrogen. Show that even the ground state of the hydrogen atom is expected to be unbound at densities comparable to those at the center of the Sun.

3.3 Define a parameter $\beta = P_g/P$ to be the ratio of the pressure contributed by the gas to the total pressure $P = P_g + P_{\rm rad}$ (gas plus radiation) in a star. Assuming ideal behavior for the gas and a blackbody spectrum for the photons, show that the temperature is given by

$$T = \left(\frac{3k}{\mu a M_u}\frac{(1-\beta)}{\beta}\right)^{1/3}\rho^{1/3},$$

where ρ is the density, M_u is the atomic unit mass, μ is the mean molecular weight, k is the Boltzmann constant, and a is the radiation constant. Show that if β is assumed constant, the system behaves as an ideal gas with an effective molecular weight $\beta\mu$ (where μ is the mean molecular weight of the gas alone), and that the pressure is given by the polytropic equation of state $P = K\rho^\gamma$ with $\gamma = \frac{4}{3}$ and the constant K defined by

$$K = \left[\frac{3}{a}\left(\frac{k}{\mu M_u}\right)^4 \frac{(1-\beta)}{\beta^4}\right]^{1/3}.$$

(This is called the *Eddington model*. The parameter β is not constant in a normal star so these results can be applied only approximately to realistic stars.)

3.4 Hydrogen inside the Sun is almost completely ionized from not too far below the surface to the center. From a typical temperature and density profile of the Sun (for example, see Table 10.1), calculate the percentage hydrogen ionization as a function of solar radius, assuming that the ionization is caused solely by the temperature. Explain any discrepancies between your estimate and the actual degree of ionization in the solar interior.

3.5 Assume the Sun to be completely ionized and assume that the most important interaction between the gas particles is the Coulomb interaction. Derive approximate expressions for the average separation between gas particles in terms of the density, the average Coulomb energy between these (charged) particles as a function of the average separation, and the average kinetic energy of the particles in terms of the temperature. The density and temperature at the solar center and at 50% and 94% of the solar radius are given in the following table (see Table 10.1 for more detail).

$R/R_\odot$	Density (g cm^{-3})	Temperature (K)
0	1.5×10^2	1.6×10^7
0.50	1.35	4×10^6
0.94	1.1×10^{-2}	3.3×10^5

Use the formulas just derived and the data in this table to evaluate the validity of the ideal gas equation of state for describing the solar interior.

3.6 Use Newtonian mechanics to analyze the momentum transfer to an arbitrary flat surface in a gas by a beam of particles incident from one side and demonstrate that the pressure P is given by the *pressure integral* of Eq. (3.2),

$$P = \tfrac{1}{3} \int_0^\infty vpn(p)dp,$$

where v is the velocity, p is the momentum, and $n(p)$ is the number density of particles with momentum in the interval p and $p + dp$. Although derived in this example using Newtonian mechanics, this formula can be shown to be valid for any velocity up to and including $v = c$ (see Problem 3.18).***

3.7 Show that for the Maxwellian velocity distribution,

$$n(v)dv = 4\pi n \left(\frac{m}{2\pi kT}\right)^{3/2} \exp(-mv^2/2kT)v^2 dv,$$

the pressure integral (3.2) leads to the ideal gas equation of state (3.3).***

3.8 (a) Show that for a photon gas described by the Planck distribution (3.69),

$$n(v)dv = \frac{8\pi}{c^3} \frac{v^2}{(e^{hv/kT} - 1)} dv,$$

the energy density of the gas is given by

$$u = \int_0^\infty n(v)\varepsilon_v \, dv = \int_0^\infty n(v) hv \, dv = aT^4,$$

where T is the temperature and

$$a \equiv \frac{8\pi^5 k^4}{15 h^3 c^3} = 7.565 \times 10^{-15} \, \text{erg cm}^{-3} \text{K}^{-4}$$

is the radiation density constant (related to the Stefan–Boltzmann constant σ by $a = 4\sigma/c$). *Hint:* You will find

$$\int_0^\infty \frac{x^3}{e^x - 1} dx = \frac{\pi^4}{15}$$

to be useful for this problem.

(b) Show that the Planck distribution in frequency (3.69) is equivalent to the momentum distribution

$$n(p)dp = \frac{8\pi}{h^3} \frac{p^2}{e^{pc/kT} - 1} dp.$$

(c) Use the result of part (b) and the pressure integral (3.2) to show that the pressure of the photon gas is $P = \frac{1}{3}aT^4$. Hence, the equation of state for the photon gas is $P = \frac{1}{3}u$, as expected generally for ultrarelativistic particles.***

3.9 Assuming the internal energy $U(T, V)$ to be a function of T and V, and using the definition (3.6), show that $dU = C_V dT$ is not valid in the general case but it is true for ideal gases.

3.10 Prove that for adiabatic processes ($\delta Q = 0$, where Q is the heat) in ideal gases

$$\gamma \frac{dV}{V} = -\frac{dP}{P}$$

and that this has solutions

$$PV^\gamma = \text{constant} \qquad P^{1-\gamma} T^\gamma = \text{constant} \qquad TV^{\gamma-1} = \text{constant},$$

where P is the pressure, V is the volume, T is the temperature, and $\gamma = C_P/C_V$ is the ideal gas adiabatic index. Show that the adiabatic sound speed v_s in an ideal gas is given by

$$v_s \equiv \sqrt{B/\rho} = \sqrt{\gamma P/\rho},$$

where ρ is the density and B is the bulk modulus.***

3.11 A polytropic process is specified by the requirement $\delta Q/\delta T = c$, where Q is the heat, T is the temperature, and c is a constant. Prove that for polytropic processes in an ideal gas the temperature T and volume V are related by

$$\frac{dT}{T} = (1 - \gamma) \frac{dV}{V},$$

where the *polytropic gamma* is defined by

$$\gamma = \frac{C_P - c}{C_V - c},$$

with C_P and C_V the heat capacities at constant pressure and volume, respectively. Demonstrate explicitly that

$$P^{1-\gamma} T^\gamma = \text{constant} \qquad PV^\gamma = \text{constant} \qquad TV^{\gamma-1} = \text{constant}$$

are each solutions of this equation and therefore define polytropic equations of state for an ideal gas. Demonstrate that if a process takes place at constant entropy in an ideal gas the polytropic γ reduces to the adiabatic γ, and that for isothermal (constant-temperature) processes in an ideal gas the polytropic γ becomes unity.***

3.12 Show that for the Maxwell distribution (3.4) the most probable speed is $(2kT/m)^{1/2}$. Thus justify the common assumption that the average energy of a particle in a classical gas is $E \simeq kT$.

3.13 Prove that the three adiabatic exponents defined in Eq. (3.39) all become equal to $\frac{5}{3}$ for an ideal gas in adiabatic approximation.

3.14 Show that for pure radiation in adiabatic approximation the three adiabatic exponents defined in Eq. (3.39) all become equal to $\frac{4}{3}$.***

3.15 Demonstrate that the adiabatic equations of state (3.40) imply the relations (3.39).

3.16 Use the first law of thermodynamics, the adiabatic assumption, and the definition of β given in Eqs. (3.74)–(3.75) to derive Eq. (3.76) for a mixture of ideal gas and radiation at constant entropy. Then use Eq. (3.76) to derive the values of the adiabatic exponents given in Eqs. (3.77)–(3.79). ***

3.17 It was asserted in the text that the adiabatic exponent Γ_1 is associated with dynamical responses of the gas. In support of this contention, prove that for adiabatic gases the sound speed is proportional to the square root of Γ_1.***

3.18 In Problem 3.7 the ideal gas equation of state was derived assuming a classical Maxwell–Boltzmann velocity distribution. In this problem the ideal gas equation of state is derived on more general grounds, showing that it is valid even for relativistic velocities as long as the Maxwell–Boltzmann energy distribution function (3.50) is applicable.

(a) Use that the internal energy U of a gas may be written as [see Eqs. (3.63) and (3.66)]

$$U = \int_0^\infty \varepsilon_p f(\varepsilon_p) g(p) dp,$$

where $f(\varepsilon_p)$ is the number of particles with energy ε_p given by Eq. (3.46), to show that the pressure integral (3.2) derived using classical considerations in Problem 3.7 may be expressed in the more general form

$$P = -\frac{\partial U}{\partial V} = -\int_0^\infty \frac{d\varepsilon_p}{dV} f(\varepsilon_p) g(p) dp$$

$$= \frac{1}{3V} \int_0^\infty v_p p f(\varepsilon_p) g(p) dp.$$

Hint: Prove as intermediate steps that

$$\frac{dp}{dV} =\sim \frac{p}{3V} \text{ (uncertainty principle)} \qquad \frac{d\varepsilon_p}{dp} = \frac{pc^2}{\varepsilon_p} = v_p,$$

where v_p is the velocity of the particle with momentum p.

(b) Use the pressure integral derived in the preceding step, the Maxwell–Boltzmann distribution (3.50), the energy expression (3.46), the integration measure (3.54), an integration by parts, and a comparison with the expression (3.53) for the number of particles in the gas to show that the corresponding equation of state has the ideal gas form.

Since this derivation for the ideal gas equation of state utilizes the fully relativistic form of the energy relation (3.46), it is valid for arbitrary gas velocities as long as Eq. (3.50) correctly describes the momentum distribution in the gas.***

3.19 Letting $\delta w \equiv P\,dV$, show that the integral

$$\Delta w = \int_{V_1,P_1}^{V_2,P_2} P\,dV$$

depends on the integration path by showing explicitly that its value for the straight-line integration path from $(V, P) = (0, 0)$ to $(1, 1)$ differs from that on the integration path $(0, 0)$ to $(1, 0)$ to $(1, 1)$. Thus, show that δw is not a perfect differential and therefore that Δw is not a state function. Conversely, prove that $f(V, P) = PV$ is a state function.

3.20 The mean molecular weight per free electron μ_e may be defined by requiring that the total electron number density n_e satisfy

$$n_e = \sum_i n_i^e = \frac{\rho N_A}{\mu_e},$$

where n_i^e is defined in Eq. (C.3). Show that in the general case of an arbitrary ionization fraction y_i for each species

$$\mu_e = \left(\sum_i y_i Z_i Y_i\right)^{-1} \simeq \frac{2}{2X + Y} \simeq \frac{2}{X + 1},$$

where in the last two steps the approximations of complete ionization and negligible metal concentration have been made.***

3.21 Prove that for an ideal gas the pressure P and energy density u are related by $P = (\gamma - 1)u$, where $\gamma = C_P/C_V$ is the adiabatic index.***

3.22 Show that $dU = d(PV)/(\gamma - 1)$ for an ideal gas.

3.23 The ideal gas law assumes that the particles in a gas do not interact with each other. At least three effects known to occur in stars can violate this assumption.

1. In an ionized plasma the electrons and ions will exert attractive Coulomb forces on each other, which tends to reduce the pressure.
2. Photons may contribute to the pressure, enhancing its value over that for the gas alone.
3. The action of the Pauli exclusion principle will increase the pressure if the density is high enough to cause the gas to become degenerate.

The contribution of photons to the pressure was addressed in Section 3.11 and the pressure in a degenerate gas was addressed in Section 3.7. This problem considers the influence of electrostatic interaction between particles in the gas.

(a) Derive an approximate formula for the ratio of electrostatic interaction energy to the thermal energy for an ion–electron pair in a gas assumed to contain only completely ionized hydrogen.
(b) Estimate the magnitude of this effect by applying this formula at the center of the Sun using data from the Standard Solar Model in Table 10.1.
(c) Make the same estimate for a low-mass main sequence star for which the central density is $\sim 400\,\text{g}\,\text{cm}^{-3}$ and the central temperature is $\sim 5 \times 10^6$ K.

4 Hydrostatic and Thermal Equilibrium

A fundamental property of main sequence stars like the Sun is their stability over long periods of time. For example, the geological record indicates that the Sun has been emitting energy at its present rate for several billion years, with relatively small variation. The key to this stability is that for large portions of their lives stars are able to maintain a state of nearly perfect hydrostatic equilibrium, with the pressure gradients produced by thermonuclear fusion and internal heat almost exactly balanced by the gravitational forces. In addition, the long-term stability of stars implies that they must be in approximate thermal equilibrium during the stable stages of their lives. Thus the starting point for an understanding of stellar structure is an understanding of hydrostatic and thermal equilibrium, and departures from that equilibrium.

4.1 Newtonian Gravitation

The Newtonian gravitational field is derived from a gravitational potential Φ that obeys the Poisson equation, which for spherical symmetry takes the form

$$\frac{1}{r^2}\frac{\partial}{\partial r}\left(r^2\frac{\partial \Phi}{\partial r}\right) = 4\pi G\rho. \tag{4.1}$$

The corresponding gravitational acceleration is given by

$$g = \frac{\partial \Phi}{\partial r} = \frac{Gm}{r^2}, \tag{4.2}$$

where $m = m(r)$ is the mass contained within the radius r. Hence, for spherical geometry

$$\Phi(r) = \int_0^r \frac{Gm}{r^2}\,dr + \text{constant}. \tag{4.3}$$

The constant is usually fixed by requiring that $\Phi \to 0$ as $r \to \infty$.

4.2 Conditions for Hydrostatic Equilibrium

The magnitude of the local gravitational acceleration at a radius r is given by Eq. (4.2), where $m(r)$ is the mass contained within a radius r. From Fig. 4.1(a) the mass contained in a thin spherical shell is

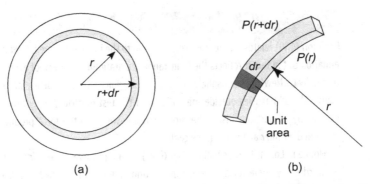

Fig. 4.1 (a) Spherical mass shell. (b) The small shaded volume has height dr and unit area on its inner surface. Therefore its volume is $1 \times dr = dr$ and its mass is $\Delta m = \rho \times 1 \times dr = \rho dr$.

$$dm = m(r + dr) - m(r) = 4\pi r^2 \rho(r) dr. \tag{4.4}$$

Integrating this from the origin to a radius r yields the *mass function* $m(r)$,

$$m(r) = \int_0^r 4\pi r^2 \rho \, dr. \tag{4.5}$$

Now consider the total gravitational force acting on a small volume of unit surface area in the concentric shell of radius r and depth dr illustrated in Fig. 4.1(b). The magnitude of this force (per unit area) will be

$$F_g = -\rho g(r) dr = -\rho \frac{Gm(r)}{r^2} dr, \tag{4.6}$$

where the negative sign indicates that the gravitational force is directed toward the center of the sphere. From Fig. 4.1, the force per unit area arising from the pressure difference between r and $r + dr$ is

$$F_p = P(r) - P(r + dr) = -\frac{\partial P}{\partial r} dr, \tag{4.7}$$

where the negative sign indicates that the pressure-gradient force is directed outward. Thus the inwardly directed gravitational force is opposed by a net outward force arising from the pressure gradient of the gas and radiation, and the total force acting on this volume of unit surface area is

$$F = F_g + F_p = -\frac{\partial P}{\partial r} dr - \frac{Gm(r)}{r^2} \rho \, dr. \tag{4.8}$$

From Newton's second law the equation of motion for the mass element is $F = \Delta ma = \rho dr \partial^2 r / \partial t^2$, which leads to

$$\rho \frac{\partial^2 r}{\partial t^2} = -\frac{\partial P}{\partial r} - \frac{Gm(r)}{r^2} \rho. \tag{4.9}$$

In hydrostatic equilibrium the left side vanishes because the acceleration is zero, giving

$$\frac{dP}{dr} = -\frac{Gm(r)}{r^2} \rho = -g \rho, \tag{4.10}$$

> **Box 4.1** **Hydrostatic Equilibrium and Stellar Interiors**
>
> Equation (4.10) relates momentum conservation to the inverse-square law of gravity. It appears to be a local equation but in fact it reflects the long-range gravitational influence of every particle in the star. As shown originally by Newton, for a spherical mass distribution and a test particle at radius r the gravitational forces exerted by all particles outside the radius r on the test particle exactly cancel, and the gravitational force exerted by all particles inside the radius r is equivalent to the force produced by concentrating all mass contained within r at a point in the center.
>
> Notice in Eq. (4.10) that both ρ and $Gm(r)/r^2$ are positive, so $dP/dr \leq 0$ and pressure must decrease outward everywhere for a spherical gravitating system to be in hydrostatic equilibrium. This will in turn imply that density and temperature must increase toward the center of a star. It follows that the condition of hydrostatic equilibrium alone is sufficient to ensure that stars must be much more dense and hot near their centers than near their surfaces.

where partial derivatives have been replaced with derivatives because by our assumption there is no longer any time dependence. As discussed in Box 4.1, dP/dr is always negative under conditions of hydrostatic equilibrium. Equations (4.4) and (4.10) represent our first two equations of stellar structure. They constitute two equations in three unknowns (P, m, and ρ as functions of r). This system of equations may be closed by specifying an equation of state relating these quantities (see Chapter 3). Before considering that, let us explore some consequences that follow from these equations alone.

4.3 Lagrangian and Eulerian Descriptions

In the study of fluid motion two basic computational points of view may be adopted. We can take a fixed grid and describe the fluid flow through the grid; this is called *Eulerian hydrodynamics*. Alternatively, we can describe the fluid motion in terms of coordinates that are attached to the mass elements and that move with them; this is called *Lagrangian hydrodynamics*.[1] In the limit that accelerations of the fluid can be neglected the Lagrangian and Eulerian descriptions of hydrodynamics reduce to Lagrangian and Eulerian descriptions of hydrostatics, respectively.

4.3.1 Lagrangian Formulation of Hydrostatics

The equations in Section 4.2 represent an Eulerian description of hydrostatics. The Lagrangian approach to hydrostatics may be illustrated by reformulating the preceding

---

[1] To appreciate the difference, consider determining the temperature of the atmosphere over time either by using weather balloons drifting with the wind, or by observing from fixed points on the ground. The first is a Lagrangian point of view, if the balloon is imagined to be tied approximately to the motion of a packet of air. The second is Eulerian, since the air is observed from fixed points as it flows by. Leonhard Euler (1707–1783) is usually credited with the development of both the Eulerian and Lagrangian approaches.

Table 4.1 The equations of hydrostatics

Eulerian coordinates (r,t)	Lagrangian coordinates (m,t)
$\dfrac{dm}{dr} = 4\pi r^2 \rho$	$\dfrac{dr}{dm} = \dfrac{1}{4\pi r^2 \rho}$
$\dfrac{dP}{dr} = -\dfrac{Gm\rho}{r^2}$	$\dfrac{dP}{dm} = -\dfrac{Gm}{4\pi r^4}$

equations with $m(r)$ rather than r as the independent variable. The general result for a change of variables between Eulerian and Lagrangian representations, $(r,t) \to (m,t)$, is specified by [134],

$$\frac{\partial}{\partial m} = \frac{\partial}{\partial r} \cdot \frac{\partial r}{\partial m} \qquad \left(\frac{\partial}{\partial t}\right)_m = \left(\frac{\partial}{\partial t}\right)_r + \frac{\partial}{\partial r} \cdot \left(\frac{\partial r}{\partial t}\right)_m, \qquad (4.11)$$

where the subscripts denote variables that are held constant. Clearly the Lagrangian version of Eq. (4.4) is

$$\frac{dr}{dm} = \frac{1}{4\pi r^2 \rho}. \qquad (4.12)$$

This implies that the first transformation in Eq. (4.11) between the two representations is explicitly

$$\frac{\partial}{\partial m} = \frac{1}{4\pi r^2 \rho} \frac{\partial}{\partial r}, \qquad (4.13)$$

which may be used to convert Eq. (4.9) to

$$\frac{1}{4\pi r^2} \frac{\partial^2 r}{\partial t^2} = -\frac{\partial P}{\partial m} - \frac{Gm(r)}{4\pi r^4}. \qquad (4.14)$$

For hydrostatic equilibrium the acceleration on the left side may be neglected, giving the Lagrangian version of Eq. (4.10),

$$\frac{dP}{dm} = -\frac{Gm}{4\pi r^4}. \qquad (4.15)$$

Table 4.1 summarizes the equations of spherical hydrostatics in Eulerian and Lagrangian form. This will be adequate for our consideration of hydrostatics but a more extensive discussion of the relationship between Eulerian and Lagrangian descriptions of hydrodynamics may be found in Box 4.2.

4.3.2 Contrasting Lagrangian and Eulerian Descriptions

Eulerian and Lagrangian representations are each a valid description of hydrodynamics, with advantages and disadvantages in a particular context. Our observational mindset is often Eulerian: we tend to think of monitoring a river for say water temperature by placing measuring devices at fixed points on the river rather than imagining measuring devices floating down the river with moving packets of water. On the other hand, the microscopic laws of physics are often formulated in Lagrangian form: in describing the collision of

Box 4.2 Lagrangian and Eulerian Derivatives

As shown in Problem 4.9, the Lagrangian time derivative D/Dt and Eulerian time derivative $\partial/\partial t$ are related by

$$\frac{D}{Dt} \equiv \frac{d}{dt} = \frac{\partial}{\partial t} + \boldsymbol{v} \cdot \nabla,$$

where $\boldsymbol{v}$ is velocity. The Lagrangian derivative has two contributions: the local (Eulerian) time derivative $\partial/\partial t$ giving the intrinsic change within the fluid element and $\boldsymbol{v} \cdot \nabla$ giving the contribution from displacement (*advection*) of the fluid element by the flow. The distinction between the Eulerian derivative and the Lagrangian derivative corresponds to the distinction between a partial derivative and a total derivative.

A partial derivative is a derivative with respect to one variable with all other variables held constant. Illustrating for a function $f(x, y)$ of two variables, $\partial f/\partial x$ is the slope in the x direction and $\partial f/\partial y$ is the slope in the y direction. The total derivative is relevant when the independent variables can be related to a single independent variable. A standard example is a function of two variables relevant only along some 1-dimensional path specified by a constraint. Think of a winding road through a range of mountains, with the altitude of the road at any point (x, y) specified by a function $f(x, y)$. What is likely to be relevant is not the (partial) derivatives in the x and y directions at arbitrary points (x, y), but instead the slope of the road at various points (which generally has contributions from slopes in both x and y directions). These points can be parameterized by a single variable s that is the distance along the road measured from some reference point, with $x = x(s)$ and $y = y(s)$.

A *total derivative* $Df/Ds \equiv df/ds$ may be defined that is the slope *in the direction of the road* at the point parameterized by the single independent variable s. By the chain rule, the total derivative and partial derivatives in this example are related by

$$\frac{df}{ds} = \frac{\partial f}{\partial x}\frac{dx}{ds} + \frac{\partial f}{\partial y}\frac{dy}{ds}.$$

The time derivative d/dt is the only total derivative that normally appears in the description of fluids. It is often termed the *material* or *substantial* or *Lagrangian derivative*, to distinguish it from the Eulerian or local derivative $\partial/\partial t$, which is just the partial derivative evaluated at a fixed spatial point. The total time derivative is analogous to the total derivative in the preceding mountain-road example, because it represents the full derivative along a path followed by a Lagrangian particle.

billiard balls it is normal to imagine following each ball and asking what forces are applied to it. It is less common to imagine staking out points on the table and asking how balls move past those fixed points, which would be the Eulerian point of view.

Because the Lagrangian approach is often more simply tied to the underlying physical laws, the advantages of the Lagrangian formulation are most apparent when symmetries and conservation laws are important in the system. For example, imagine a spherical star that is neither gaining nor losing mass but is pulsating in size. The radial distance to the surface (an Eulerian coordinate) is changing with time but the mass contained within the radius (a Lagrangian coordinate) is constant in time. On the other hand, if spherical

4.4 Dynamical Timescales

Dynamical timescales are of particular importance in astrophysics because they set the approximate time required to respond to perturbations of hydrostatic equilibrium. From Problem 4.1, the free-fall timescale t_ff introduced in Section 1.10.4 (now with constants properly evaluated) is

$$t_\mathrm{ff} = \sqrt{\frac{3\pi}{32 G \bar{\rho}}} \simeq \sqrt{\frac{1}{G\bar{\rho}}} \simeq \sqrt{\frac{R}{g}}, \tag{4.16}$$

where $\bar{\rho} = M/(\frac{4}{3}\pi R^3)$ is the average density, G is the gravitational constant, and $g = GM/R^2$ is the gravitational acceleration. This defines a timescale for collapse of a gravitating sphere if it suddenly lost all pressure support. A second dynamical timescale may be introduced by considering the opposite extreme: if gravity were taken away, how fast would the star expand by virtue of its pressure gradients? This timescale can depend only on R, $\bar{\rho}$, and $\bar{P}$, and the only combination of these quantities having time units is

$$t_\mathrm{exp} \simeq R\sqrt{\frac{\bar{\rho}}{\bar{P}}} \simeq \frac{R}{\bar{v}_s}, \tag{4.17}$$

where $\bar{v}_s \sim (\bar{P}/\bar{\rho})^{1/2}$ is the mean sound speed. This result makes physical sense because changes in pressure are mediated by waves propagating at the speed of sound. Hydrostatic equilibrium will be precarious unless the two timescales (4.16) and (4.17) are comparable ($\tau_\mathrm{ff} \sim \tau_\mathrm{exp}$), which suggests defining a *dynamical timescale*

$$\tau_\mathrm{dyn} = \tau_\mathrm{hydro} \simeq \sqrt{\frac{1}{G\bar{\rho}}}, \tag{4.18}$$

which may be used to characterize the timescale for the response of hydrostatic equilibrium to perturbations.

Example 4.1 For the Sun $\bar{\rho} = 1.4\,\mathrm{g\,cm^{-3}}$ and Eq. (4.18) gives $t_\mathrm{hydro}^\odot \simeq 55$ minutes. If hydrostatic equilibrium were not satisfied observable changes should appear in a matter of hours, but the fossil record indicates that the Sun has been extremely stable for billions of years. Hence it must maintain very good hydrostatic equilibrium over such extended timescales.

Dynamical timescales for some other astronomical objects calculated using Eq. (4.18) are displayed in Table 4.2. The dynamical timescales vary widely for these objects because they represent a huge range of densities.

Table 4.2 Characteristic hydrodynamical timescales

Object	$\sim M/M_\odot$	$\sim R/R_\odot$	$\bar{\rho}/\rho_\odot$	τ_{hydro}
Red giant	1	100	10^{-6}	36 days
Sun	1	1	1	55 minutes
White dwarf	1	1/50	10^5	9 seconds

4.5 The Virial Theorem for an Ideal Gas

Many concepts from statistical mechanics are modified in the astrophysical context because of the long range of the gravitational force. In non-astrophysical systems of physical interest the effective interactions often are short-ranged and the energy of the system may be approximated as an extensive variable (roughly, the total is the sum of the parts).[2] For example, if a dilute gas of atoms or molecules is divided spatially into N parts, the total energy is approximately the sum of the energy for the N separate parts because the interactions are short-ranged and thus small between the parts. But for a group of objects interacting gravitationally the total energy of the system is more than the sum of the parts because considerable energy resides in the interactions between the parts, by virtue of the gravitational interaction being always unscreened and of long range. Thus results that depend on assuming the energy to be an extensive variable are not generally valid in gravitating systems. However, there is one powerful and rather general result that holds in gravitating many-body systems and thus is of considerable utility in astrophysics that will now be discussed in the context of stellar structure.

Stars have at their disposal two large energy reservoirs: gravitational energy, which can be released by contraction (and absorbed by expansion), and internal energy, which can be produced by nuclear reactions and conversion of gravitational energy. An important relationship between these energy resources for objects in approximate hydrostatic equilibrium may be derived if we multiply both sides of the Lagrangian hydrostatic equation (4.15) by $4\pi r^3$ and then integrate over dm from 0 to $M \equiv m(R)$ to give

$$\begin{aligned}
\int_0^M \frac{Gm}{r} dm &= -4\pi \int_0^M r^3 \frac{\partial P}{\partial m} dm \\
&= -4\pi r^3 P \Big|_{m=0}^{m=M} + 12\pi \int_0^M r^2 P \frac{\partial r}{\partial m} dm \\
&= 12\pi \int_0^M r^2 P \frac{1}{4\pi r^2 \rho} dm \\
&= \int_0^M \frac{3P}{\rho} dm,
\end{aligned} \qquad (4.19)$$

---

[2] Recall that an *extensive property* is a sum of the properties of separate noninteracting subsystems that compose the entire system; thus, an extensive quantity depends on the amount of material in a system. Mass is an example. An *intensive property* is independent of the amount of material; density of a homogeneous system is an example.

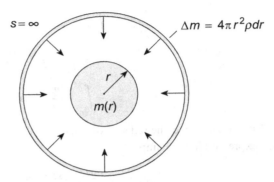

Fig. 4.2 Gravitational assembly of a star by the accretion of concentric shells, each of mass $\Delta m = 4\pi r^2 \rho dr$.

where ρ, r, and P are functions of the independent Lagrangian variable m, an integration by parts has been performed in line two, Eq. (4.12) has been used in going from line two to line three, and the first term in line two is identically zero because r vanishes when $m = 0$ (center of star) and P vanishes when $m = M$ (surface of star).

What is the physical interpretation of the result that we have obtained in Eq. (4.19)? By dimensional analysis the left and right sides define energies of some kind. Assuming an ideal gas, the equation of state is $P/\rho = kT/\mu$ and the factor of kT indicates that the right side of (4.19) is related to the internal energy of the gas, while the appearance of G on the left side of Eq. (4.19) implies that it is a gravitational energy. Indeed, as shown in Problem 4.5, for an ideal monatomic gas the right side of Eq. (4.19) is equal to twice the internal energy U of the gas,

$$\int_0^M \frac{3P}{\rho} dm = 2U.$$

The left side of Eq. (4.19) may be interpreted by calculating the gravitational energy that is released in the formation of a star by accretion of mass. Consider Fig. 4.2, where a shell of mass $\Delta m = 4\pi r^2 \rho dr$ is allowed to fall from infinity onto the surface of a spherical mass of radius r and enclosed mass $m(r)$. The gravitational energy released in adding this shell to the star is then given by,

$$d\Omega = \int_\infty^r F_g \, ds = \int_\infty^r g(s) \Delta m \, ds$$

$$= \int_\infty^r \frac{Gm(r)}{s^2} 4\pi r^2 \rho dr \, ds$$

$$= -\frac{Gm(r)}{s}\bigg|_\infty^r \times 4\pi r^2 \rho dr$$

$$= -4\pi r^2 \rho dr \frac{Gm(r)}{r},$$

and the total gravitational energy Ω released in assembling a star of radius R and mass M from such mass shells is

$$\Omega = \int d\Omega$$
$$= -4\pi \int_0^R r^2 \rho \frac{Gm(r)}{r} dr$$
$$= -\int_0^M \frac{Gm(r)}{r} dm, \qquad (4.20)$$

with $M \equiv m(R)$. Hence, the left side of Eq. (4.19) is just $-\Omega$. Collecting results, Eq. (4.19) may be expressed in the compact form

$$2U + \Omega = 0, \qquad (4.21)$$

where U is the internal energy of the star and Ω is its gravitational energy. Equation (4.21) is termed the *virial theorem* (for an ideal, monatomic gas). It will prove to be one of our most important tools for understanding stellar structure and stellar evolution because it establishes a general relationship between the internal energy and the gravitational energy of a star that is in hydrostatic equilibrium. The virial theorem is of broad applicability because of the non-restrictive conditions under which it was derived, and because it relates the two most important energy reserves for a star.

4.6 Thermal Equilibrium

In addition to being in hydrostatic equilibrium, stars are in approximate thermal equilibrium. Let us investigate this using as a guide the discussion in Prialnik [176]. By energy conservation (first law of thermodynamics) the internal energy of a star can be changed by adding or removing heat, or by PdV work. Assume hydrostatic equilibrium and consider a spherical mass shell as in Fig. 4.3. If the concentric shell is at radius r and of width dr, its volume is $dV = 4\pi r^2 dr$. It will prove convenient to work in Lagrangian coordinates to examine the heat flow, with $dm = \rho dV = 4\pi r^2 \rho dr$, where ρ is the mass density at the radius of the concentric shell.

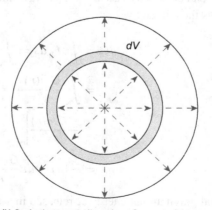

Fig. 4.3 A spherical mass shell of volume dV. Dashed arrows indicate heat flow out of the star.

Let u be the internal energy per unit mass and let δf denote the change of some quantity f within the mass shell over a time t. The change in heat over a time δt is then denoted δQ and the work done in a time δt is denoted by δW. Then the total change in internal energy over a time δt (recall that u is the internal energy per unit mass) is given by

$$\delta(udm) = (\delta u)dm = \delta Q + \delta W, \qquad (4.22)$$

where the first step follows because dm is constant by mass conservation. As you are asked to show in Problem 4.20, the change in heat over a time δt is given by

$$\delta Q = q \, dm \, \delta t - \frac{\partial L}{\partial m} dm \, \delta t \qquad (4.23)$$

and the work done in a time δt is

$$\delta W = -P \delta \left(\frac{1}{\rho}\right) dm, \qquad (4.24)$$

where $L(m)$ is the luminosity associated with the heat flow across the shell, P is the pressure and q the rate of energy release per unit mass in the mass shell, and where $dV/dm = \rho^{-1}$ has been used. Substituting Eqs. (4.23) and (4.24) into Eq. (4.22) and taking the limit $\delta t \to 0$ gives a differential equation specifying the energy balance in a mass shell,

$$\frac{du}{dt} + P \frac{d}{dt}\left(\frac{1}{\rho}\right) = q - \frac{\partial L}{\partial m}, \qquad (4.25)$$

as shown in Problem 4.21. In thermal equilibrium the temporal derivatives on the left side of (4.25) vanish, implying that $q = dL/dm$. Integrating both sides of this equation over the mass coordinate and introducing

$$L_0 \equiv \int_0^M q \, dm \qquad L \equiv \int_0^M \frac{dL}{dm} dm, \qquad (4.26)$$

where L is the total luminosity and L_0 is the luminosity produced by nuclear reactions in the star, leads to $L_0 = L$. For a star in thermal and hydrostatic equilibrium energy is radiated away at the same rate that it is produced by nuclear reactions, which you should hardly find surprising!

4.7 Total Energy for a Star

Integrating Eq. (4.25) over the entire star yields

$$\int_0^M \frac{du}{dt} dm + \int_0^M P \frac{d}{dt}\left(\frac{1}{\rho}\right) dm = \int_0^M q \, dm - \int_0^M \frac{\partial L}{\partial m} dm, \qquad (4.27)$$

while a corresponding integration of Eq. (4.14) multiplied by $\dot{r}$ over the entire star leads to

$$\int_0^M \ddot{r} \dot{r} \, dm = -4\pi \int_0^M r^2 \dot{r} \frac{\partial P}{\partial m} dm - \int_0^M \frac{Gm\dot{r}}{r^2} dm. \qquad (4.28)$$

As you are invited to show in Problem 4.22, Eqs. (4.27)–(4.28) imply an energy-conservation equation

$$\dot{E} = \dot{U} + \dot{\Omega} + \dot{K} = L_0 - L, \tag{4.29}$$

where dots indicate time derivatives and the total energy E is

$$E = U + K + \Omega, \tag{4.30}$$

where U is the total internal energy, Ω is the total gravitational energy, K is the total kinetic energy,

$$K = \tfrac{1}{2} \int_0^M \dot{r}^2 \, dm, \tag{4.31}$$

L is the total luminosity, and L_0 is the luminosity deriving from nuclear reactions. If the star is in thermal equilibrium $\dot{E} = 0$ and if it is in hydrostatic equilibrium $K = 0$. In that limit the properties of the star are governed by the virial theorem relating U to Ω.

4.8 Stability and Heat Capacity

We have argued above that stars are in a hydrostatic equilibrium that balances gravitational forces against pressure-differential forces, and a thermal equilibrium that balances energy production against energy emission, but how *stable* is that equilibrium? A ball sitting in the bottom of a deep valley and a ball balanced on a knife edge are both in equilibrium, but they exhibit very different stability to perturbations. Are stars in a deep valley, or are they balanced on a knife edge? As we shall see, the answer can depend very much on the appropriate equation of state, and is the basis of both the remarkable stability of main sequence stars and of some of the most violent explosions observed in our Universe. We will address a number of instabilities in later chapters; here we illustrate for thermal instability.

4.8.1 Temperature Response to Energy Fluctuations

Consider a star with an ideal gas plus radiation equation of state given by Eq. (3.72), so that

$$P = P_g + P_r = nkT + \tfrac{1}{3}aT^4 = \tfrac{2}{3}u_g + \tfrac{1}{3}u_r,$$

where for the ideal gas an internal energy density $u_g = \tfrac{3}{2}nkT = \tfrac{3}{2}P_g$ is assumed and for the radiation $u_r = aT^4 = 3P_r$ is assumed. From Eqs. (4.20) and (4.19) the gravitational energy is

$$\Omega = -\int_0^M \frac{3P}{\rho} \, dm = -2 \int_0^M \frac{u_g}{\rho} \, dm - \int_0^M \frac{u_r}{\rho} \, dm = -2U_g - U_r,$$

since the total internal energies are given by

$$U_g = 4\pi \int_0^R u_g r^2 \, dr = \int_0^M \frac{u_g}{\rho} \, dm \qquad U_r = 4\pi \int_0^R u_r r^2 \, dr = \int_0^M \frac{u_r}{\rho} \, dm,$$

by virtue of Eq. (4.4). Thus the total energy is

$$E = \Omega + U_r + U_g = -U_g = -\tfrac{3}{2}NkT$$

for an average temperature T. Letting L denote the luminosity of the star and L_0 the energy generation rate, their difference may be written as

$$L_0 - L = \frac{dE}{dt} = -\frac{3}{2}Nk\frac{dT}{dt},$$

where at thermal equilibrium $L_0 - L = 0$. Now suppose a small fluctuation away from equilibrium occurs such that $L_0 - L = \delta L$. Solving the preceding equation for dT/dt gives

$$\frac{dT}{dt} = -\frac{2}{3}\frac{\delta L}{Nk},$$

which governs how the temperature will respond to small fluctuations in energy average. Now consider two situations:

1. If $\delta L > 0$, the nuclear energy generation rate exceeds the luminosity, $L_0 > L$, and $dT/dt < 0$. Thus the response to an increase in energy generation rate is a decrease in temperature, which tends to lower the energy generation rate.
2. If $\delta L < 0$, the nuclear energy generation rate is less than the luminosity and $dT/dt > 0$, so the temperature increases, which increases the rate of energy generation.

This behavior represents the essence of a stable system: an imbalance causes an automatic restorative action that re-establishes the balance. Yet this essential feature of normal stars is quite counterintuitive; we could not have predicted it based on normal experience!

4.8.2 Heating Up while Cooling Down

When $\delta L > 0$ the star is producing more energy than it is radiating, yet it *cools*. Unlike every system dealt with in everyday life, a star has a *negative heat capacity,* as discussed more extensively in Box 4.3. This is a generic feature of a system bound by the long-range, attractive gravitational interaction: adding energy to a bound self-gravitating system does not make it hotter, it makes it colder! The objects around us locally are not bound by long-range interactions (the force of gravity is negligible on an atomic or nuclear scale) and they – except for a few exceptions known for nanoscale systems – exhibit positive heat capacities. This is why our normal intuition can be very misleading when it comes to the behavior of stars, or any bound gravitating system such as a clusters of stars or galaxies.

4.9 The Kelvin–Helmholtz Timescale

Stars go through various phases in which they contract gravitationally, particularly during formation and during the transition between different nuclear energy sources in late stellar evolution. Except in the stellar explosions to be described in Chapters 19–21,

> **Box 4.3** **Stars, Black Holes, and Negative Heat Capacities**
>
> In interpreting U in Eq. (4.21) an ideal gas was assumed, so the internal energy U can be identified with the kinetic energy of the gas particles, $E_{kin} = U$, and the gravitational energy Ω with the potential energy, $E_{pot} = \Omega$. Thus an alternative statement of the virial theorem (4.21) is that for a self-gravitating, spherical distribution of ideal-gas particles in hydrostatic equilibrium,
>
> $$E_{kin} = -\tfrac{1}{2} E_{pot}.$$
>
> Expressing the virial theorem in this form lays bare a property that is quite unusual.
>
> **Heat Capacities of Stars and Clusters**
> For almost all physical systems the heat capacity is positive but if $E = E_{kin} + E_{pot}$ is the total energy then $E_{kin} = -E$, so that decreasing the total energy of the system *increases* the kinetic energy. If the temperature is assumed to be a measure of average particle kinetic energy, this implies that for an object governed by the virial theorem temperature increases as energy is lost: it "heats up as it cools down." Since heat capacity is the ratio of change in heat to change in temperature, a star governed by the virial theorem has a *negative heat capacity*.
>
> This counterintuitive behavior is associated with the role of the gravitational field in the equilibrated system. That it makes physical sense is suggested by considering a gravitationally bound cluster of stars in which the kinetic energy of the stars is balanced by the potential energy of the gravitational attraction of each star in the cluster for all other stars. For a large cluster, the motion of the stars may be viewed as defining a temperature for a "gas of stars." Now imagine removing slowly a small amount of energy. The cluster will re-equilibrate, but with the average velocity of the gas particles (stars) increased because the cluster is now more tightly bound. Thus the temperature characterizing the gas of stars increases as energy is extracted and the gravitating cluster exhibits a negative heat capacity.
>
> **Heat Capacities of Black Holes**
> Using advanced quantum field theory the distribution of energy emitted by a *black hole* as *Hawking radiation* is found to be equivalent to that of a blackbody with temperature given by
>
> $$T = \frac{\hbar c^3}{8\pi kGM}$$
>
> (see Chapter 12 of Ref. [100]), where M is the mass, k is Boltzmann's constant, $h = 2\pi\hbar$ is Planck's constant, and G is the gravitational constant. As Hawking radiation is emitted the black hole loses mass (energy) and the temperature rises: the black hole becomes hotter as it loses energy, so it exhibits a negative heat capacity.

these gravitational contractions are typically under conditions of approximate hydrostatic equilibrium. Gravitational contraction for a short period releases an amount of energy $\Delta\Omega$ and, since the virial theorem must be satisfied for hydrostatic equilibrium to hold, Eq. (4.21) implies that as a star contracts the thermal energy must change by[3]

---

[3] Of course a contracting star cannot be in hydrostatic equilibrium. However, if the collapse is slow the star is at each instant only slightly out of hydrostatic equilibrium and the virial theorem will be satisfied approximately.

4.9 The Kelvin–Helmholtz Timescale

$$\Delta U = -\tfrac{1}{2}\Delta\Omega, \qquad (4.32)$$

and the excess energy must be transported away. Thus, gravitational contraction has three consequences for a star: (1) the star heats up, (2) some energy is radiated into space, and (3) the total energy of the star decreases and it becomes more bound gravitationally. As was discussed in Box 4.3, these steps are mutually consistent only because stars have *negative heat capacities*, with the released gravitational potential energy supplying both the radiated energy and the internal heating [in equal amounts if Eq. (4.32) holds].

If approximate hydrostatic equilibrium is to be maintained, at each infinitesimal step of the contraction the star must wait until half of the released gravitational energy is radiated away before it can continue to contract. This implies that there is a *timescale for contraction* in near hydrostatic equilibrium that is set by the time required to radiate the excess energy. This contraction timescale is called the *Kelvin–Helmholtz timescale* or the *thermal adjustment timescale*. We may estimate it by assuming uniform density ρ and a corresponding mass $m(r) = \tfrac{4}{3}\pi r^3 \rho$ during the gravitational contraction.[4] Then, from Eq. (4.20) the gravitational energy released in collapsing the initial cloud of gas and dust to a radius R is

$$\begin{aligned}
\Omega &= -\int_0^R 4\pi r^2 \rho \frac{Gm(r)}{r}\, dr \\
&= -\frac{16}{3}\pi^2 \rho^2 G \int_0^R r^4 \, dr \\
&= -\frac{16}{15}\pi^2 \rho^2 G R^5 \\
&= -\frac{3}{5}\frac{GM^2}{R},
\end{aligned} \qquad (4.33)$$

where $M = \tfrac{4}{3}\pi R^3 \rho$. Taking $M = M_\odot$ and $R = R_\odot$ gives that $\Omega_\odot = 2.3 \times 10^{48}$ erg of gravitational energy was released in forming the Sun. By the virial theorem, half of this must have been radiated while the protosun contracted:

$$E^\odot_\text{rad} = \tfrac{1}{2}\Omega_\odot \simeq 10^{48} \text{ erg}. \qquad (4.34)$$

The *Kelvin–Helmholtz timescale* t_KH sets the characteristic time required to radiate this energy.

Example 4.2 A rough estimate of the Kelvin–Helmholtz timescale for contraction of the protosun to the main sequence follows from assuming that its present luminosity of $L_\odot \sim 3.8 \times 10^{33}$ erg s^{-1} characterized its luminosity for the longer part of its collapse to the main sequence (see Problem 9.14 for a justification). Then $t_\text{KH} \simeq E^\odot_\text{rad}/L_\odot \simeq 10^7$ years, implying that the Sun contracted to the main sequence on a Kelvin–Helmholtz timescale of about 10 million years. A more sophisticated treatment suggests $t^\odot_\text{KH} = 3 \times 10^7$ yr.

---

[4] The assumption of uniform density is an oversimplification but any more realistic density profile consistent with hydrostatic equilibrium will give $\Omega \propto GM^2/R$ as in Eq. (4.33), with the constant of proportionality differing from $\tfrac{3}{5}$ but still of order one; see Section 9.2 and Problem 9.6.

Box 4.4 — Timescale Set by Random Walk of Photons

At a more microscopic level the contraction timescale may be viewed as being set by the time for photons produced in the core of the star to make their way by a random walk to the surface of the star. For a random walk, the distance traveled after Z scatterings is (see Problem 4.3)

$$\Delta x \simeq \lambda \sqrt{Z},$$

where the mean free path λ is the average distance the photon travels before being absorbed (λ is defined more precisely in Box 7.1). To escape, a photon must undergo approximately

$$Z = \left(\frac{\Delta x}{\lambda}\right)^2 = \left(\frac{R}{\lambda}\right)^2$$

absorptions and random re-emissions. A timescale may be associated with this random walk by estimating the average time to be emitted again once absorbed (a typical estimate is 10^{-8} seconds [50]). This approach is explored further in Problem 10.8 and Problem 10.9.

From the virial theorem and Eq. (9.1), if the luminosity is assumed to derive entirely from gravitational contraction the total radiated energy is

$$E = \frac{\Delta \Omega}{2} = -\frac{fGM^2}{2R},$$

where f is of order one. Then the solution of Problem 9.14 indicates that

$$\frac{dR}{dt} = -\left(\frac{2LR}{fGM^2}\right) R = -\frac{1}{t_{\text{KH}}} R, \qquad (4.35)$$

where a Kelvin–Helmholtz timescale for the star is defined by

$$t_{\text{KH}} \equiv \frac{fGM^2}{2LR} \simeq \frac{GM^2}{LR}, \qquad (4.36)$$

where R is the radius, M the mass, and L the luminosity, and the last step approximates $\tfrac{1}{2} f \sim 1$. Assuming t_{KH} to be constant, Eq. (4.35) has the solution $R = R_0 e^{-t/t_{\text{KH}}}$, so the Kelvin–Helmholtz timescale is (approximately, since t_{KH} is not constant) the time for the radius to decrease by a factor of e^{-1} in the contraction. From Eq. (4.36), the ratio of the Kelvin–Helmholtz timescale for some star relative to that of the Sun is given by

$$\frac{t_{\text{KH}}}{t_{\text{KH}}^{\odot}} = \left(\frac{R_\odot}{R}\right)\left(\frac{L_\odot}{L}\right)\left(\frac{M}{M_\odot}\right)^2, \qquad (4.37)$$

where $t_{\text{KH}}^{\odot} \sim 3 \times 10^7$ yr. The contraction timescale also may be viewed as being set by the timescale for photons to diffuse out of the stellar interior, as discussed in Box 4.4.

Example 4.3 From Table 2.2 an A0 main sequence star has $R = 2.5\, R_\odot$, $M = 3.2\, M_\odot$, and $L = 79.4\, L_\odot$. From Eq. (4.37), the Kelvin–Helmholtz timescale is $t_{\text{KH}} \sim 0.052\, t_{\text{KH}}^{\odot} \sim 1.55 \times 10^6$ yr. This is one of many examples that we shall encounter illustrating that more massive stars evolve more rapidly through all phases of their lives, including periods of gravitational contraction.

The Kelvin–Helmholtz timescale is distinct from the free-fall or dynamical timescale (Section 4.4). The dynamical timescale is characteristic of processes where gravity is not strongly opposed by other forces; the Kelvin–Helmholtz timescale governs the rate at which liberated gravitational energy can be radiated from a system in which gravitational forces are almost exactly balanced by pressure-differential forces. The dynamical timescale is generally much shorter than the Kelvin–Helmholtz timescale. For example, the dynamical timescale for the Sun is of order 1 hour but its Kelvin–Helmholtz timescale is of order 10 million years.

Background and Further Reading

For clear introductions to hydrostatic equilibrium in gravitational fields, see Hansen, Kawaler, and Trimble [107]; Kippenhahn, Weigert, and Weiss [134]; Prialnik [176]; and Böhm-Vitense (Vol. 3) [52]. Our discussion of thermal equilibrium and the total energy of the star follows that of Prialnik [176], as does our general discussion of stability.

Problems

4.1 Use the gravitational equations to show that a free-fall timescale to collapse a gravitating sphere to its center with no pressure to oppose the collapse is given by

$$t_{\text{ff}} = \sqrt{\frac{3\pi}{32G\bar{\rho}}},$$

where G is the gravitational constant and $\bar{\rho}$ is the average density contained in the sphere.***

4.2 Estimate the hydrodynamical timescale for the Sun, a red giant, and a white dwarf star.

4.3 Derive an expression for the time required for a photon to random walk from the center to the surface of a star in terms of an average mean free path λ and stellar radius R. Estimate the random walk time for the Sun, assuming an average mean free path of 0.5 centimeters. *Hint*: Treat each step of the random walk as a vector and consider the vector sum after N steps.***

4.4 Use the equations of hydrostatic equilibrium to place a lower bound on the central pressure of a star in equilibrium and evaluate this quantity for the Sun. *Hint*: Integrate the pressure from the center to the surface and assume that the radius of the surface in the resulting expression is always larger than the radius of any interior point.

4.5 Demonstrate that the right side of Eq. (4.19) is equal to twice the internal energy for an ideal gas.***

4.6 Use the virial theorem to place a lower bound on the mean temperature of a star and estimate this bound for the Sun. Use this result to estimate the ratio of radiation to gas pressure for the Sun. *Hint*: In the first step, if R is the stellar radius, obviously $1/r > 1/R$ for any interior point.

4.7 Can a star simultaneously expand and cool through its entire volume if total energy is conserved and it remains in hydrostatic and thermal equilibrium?

4.8 Assume a star of uniform density and uniform temperature, composed of a monatomic ideal gas.

(a) Show that the internal energy is $U = \frac{3}{2} M N_A k T / \mu$, where M is the mass, N_A is Avogadro's number, T is the temperature, and μ the mean molecular weight of the gas.

(b) Use the result of part (a), the virial theorem, and the gravitational energy of a constant-density sphere to show that

$$T = 4.09 \times 10^6 \mu \left(\frac{M}{M_\odot}\right)^{2/3} \left(\frac{\rho}{\text{g cm}^{-3}}\right)^{1/3},$$

in units of K.

(c) Use the formula derived above in part (b) to plot $\log T$ versus $\log \rho$ for $M/M_\odot = 0.2, 1, 10, 35, 100$.

(d) The preceding derivation assumed an ideal gas. On the plot constructed in part (c) above, place

   (i) A curve above which radiation pressure would dominate the gas pressure.
   (ii) A curve below which electron degeneracy would be important. *Hint*: Set the Fermi energy equal to kT to estimate where degeneracy becomes important.
   (iii) A temperature above which relativistic effects would be important. *Hint*: When does kT become comparable to the rest mass of the lightest particles in the gas?
   (iv) A density above which relativistic effects would be important. *Hint*: Relativity will become important when the momentum divided by Planck's constant becomes comparable to the inverse Compton wavelength of the lightest particles.

(e) Use the preceding results to derive a formula estimating the stellar mass $M/M_\odot$ above which the radiation pressure would be expected to dominate the gas pressure.***

4.9 From the definition of a total derivative in terms of a limit, prove the relationship

$$\frac{D}{Dt} = \frac{\partial}{\partial t} + \boldsymbol{v} \cdot \nabla$$

given in the first equation of Box 4.2 between the total (Lagrangian) derivative and the local (Eulerian) partial derivative with respect to time.***

4.10 The pressure of an ideal gas may be expressed as [see Eq. (3.2) and Problem 3.6]

$$P = \frac{N}{3V} \langle pv \rangle = \frac{1}{3} n \langle pv \rangle$$

where N is the number of particles, V the volume, n the number density, p the momentum, v the velocity, and $\langle \rangle$ signifies an average over all particles in the gas. Show that for a nonrelativistic gas $P = \frac{2}{3}\varepsilon_{\text{kin}}$ and that for an ultrarelativistic gas $P = \frac{1}{3}\varepsilon_{\text{kin}}$, where ε_{kin} is the corresponding kinetic energy density.***

4.11 Show that for a nonrelativistic ideal gas the virial theorem (4.21) may also be expressed in the form $P = -\Omega/3V$, where Ω is the gravitational energy and V is the volume.

4.12 In deriving the virial theorem for a spherical star of radius R the pressure was assumed to vanish at the surface of the star: $P(R) = 0$. However, stars don't have a sharply defined surface but rather a region (atmosphere) where the pressure falls off rapidly. This is often modeled by assuming that the star in hydrostatic equilibrium is immersed in a thin medium (the atmosphere) of finite pressure. How is the virial theorem (4.21) modified if it is assumed that the star has $P(R) = P_0 \neq 0$?

4.13 Prove that for a sphere of radius R and constant density ρ in hydrostatic equilibrium, the pressure at radius $r \leq R$ is given by

$$P(r) = \tfrac{2}{3}\pi G \rho^2 (R^2 - r^2),$$

where G is the gravitational constant.

4.14 In the text the virial theorem was derived starting from the Lagrangian description of hydrostatics. Derive the virial theorem starting from the Eulerian form (4.10) of the hydrostatic equation

$$\frac{dP}{dr} = -\frac{Gm(r)\rho(r)}{r^2}.$$

Hint: Multiply by $4\pi r^3$ and integrate both sides; then use an integration by parts and boundary conditions to simplify the left side.

4.15 Find the functional form of the Lagrangian mass coordinate $m(r)$ and the average value of the density for a star of mass M with density parameterized by

$$\rho(r) = \rho_0 \left(1 - \left(\frac{r}{R}\right)^2\right),$$

where ρ_0 is the central density and R is the radius.

4.16 Three general equations of state $P = P(\rho, T)$ that are relevant for stellar structure have been considered: Eq. (3.3) for an ideal gas, Eq. (3.36) for a polytrope, and Eq. (3.70) for radiation. Show that all can be expressed in the form

$$\frac{dP}{P} = \alpha \frac{d\rho}{\rho} + \beta \frac{dT}{T},$$

where the coefficients α and β are greater than or equal to zero.***

4.17 A sphere of ideal gas has a total mass M. Assuming hydrostatic equilibrium, at a given radius $r(m)$ the gas pressure P_{gas} and the hydrostatic pressure P_{hydro} will be equivalent and by virtue of Eq. (4.15) will be given by

$$P_{\text{gas}} = P_{\text{hydro}} = \int_m^M \frac{Gm}{4\pi r^4} \, dm.$$

Now perturb the gas slightly with a compression so that everywhere $r \to (1 - \delta)r$, where δ is a small constant. Show that in adiabatic approximation the condition for gravitational stability (that the gas is capable of reacting to this perturbation and restoring hydrostatic equilibrium) can be met only if the adiabatic index of the gas satisfies $\gamma > \frac{4}{3}$. *Hint*: The perturbation was a compression, so the gas must expand to restore equilibrium.

4.18 Consider the atmosphere of a spherical planet. Assume that the atmosphere is an ideal gas in hydrostatic equilibrium at a uniform constant temperature containing molecules with mean molecular weight μ, and that the gravitational acceleration may be assumed constant over the extent of the atmosphere. Derive equations describing the variation of atmospheric pressure and the variation of atmospheric mass density with height above the planetary surface.

4.19 Find the functional form of the Lagrangian mass coordinate $m(r)$ for a star of mass M with density parameterized (crudely) by

$$\rho(r) = \rho_0 \left(1 - \frac{r}{R}\right),$$

where ρ_0 is the central density and R is the radius. Use this result to show that the corresponding pressure gradient is

$$\frac{dP}{dr} = -\frac{4\pi G}{3} \rho_0^2 r \left(1 - \frac{3r}{4R}\right) \left(1 - \frac{r}{R}\right).$$

Integrate this to obtain a formula for the pressure as a function of r. This formula will involve the central density ρ_0 and central pressure P_0. Use physical boundary conditions on the star to evaluate ρ_0 and P_0 in terms of the stellar radius R and the stellar mass M, thus finally giving an expression for $P(r)$ depending only on M, R, and r.

4.20 Prove Eq. (4.23) for the change in heat and Eq. (4.24) for the change in work in a time δt for a concentric mass shell.***

4.21 Show that the differential equation (4.25) follows from Eqs. (4.22)–(4.24).***

4.22 Prove that the energy conservation equation (4.29) follows from Eqs. (4.27) and (4.28). *Hint*: Show that $d(\rho^{-1})/dt = \partial \dot{V}/\partial m$ and then evaluate the second integral on the left side of Eq. (4.27) using an integration by parts.***

5 Thermonuclear Reactions in Stars

Stars have three primary sources of energy: (1) heat left over from earlier processes, (2) gravitational energy, and (3) thermonuclear energy. Gravitational energy is important for various stages of star birth and star death, and white dwarfs shine because of heat left over from earlier thermonuclear and gravitational energy generation. However, since the virial theorem indicates that gravity can supply the energy of the Sun only on a 10^7 year timescale, thermonuclear reactions are the only viable long-term source for observed stellar luminosities. In this chapter and the next we address in some depth how thermonuclear reactions influence both the structure and the evolution of stars. To do so we require the use of knowledge from nuclear physics about the reactions that ions in stars can undergo, but with the added twist that these reactions occur in a hot, dense gas having a statistical distribution of velocities for the reacting particles. This leads to the key idea of thermally averaged rates for nuclear reactions, which will provide the basis of a formalism for describing quantitatively the thermonuclear reactions that power stars.

5.1 Nuclear Energy Sources

The luminosity of the Sun is $L_\odot \simeq 3.8 \times 10^{33}$ erg s^{-1} and that of the most luminous stars is about $10^6 L_\odot$. From the Einstein relation $m = E/c^2$, the rate of mass conversion to energy that is required to sustain the Sun's luminosity is

$$\frac{\Delta m}{\Delta t} = \frac{1}{c^2}\frac{\Delta E}{\Delta t} = \frac{L_\odot}{c^2} = 4.2 \times 10^{12} \text{ g s}^{-1}, \tag{5.1}$$

and the most luminous stars require conversion rates a million times larger. Let us now discuss how nuclear reactions in stars can account for mass-to-energy conversion on this scale.

5.1.1 The Curve of Binding Energy

The binding energy for a nucleus of atomic number Z and neutron number N is

$$B(Z, N) \equiv [Zm_\text{p} + Nm_\text{n} - m(Z, N)]c^2, \tag{5.2}$$

where $m(Z, N)$ is the mass of the nucleus, m_p is the mass of a proton, and m_n is the mass of a neutron. The binding energy may be interpreted either as the energy released in assembling a nucleus from its constituent nucleons, or as the energy required to break

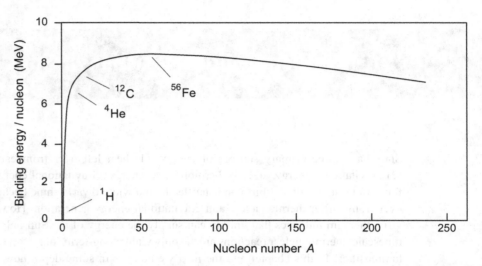

Fig. 5.1 The smoothed curve of nuclear binding energy. Only the average behavior is shown; local fluctuations have been suppressed, as has the isotopic dependence on (Z, N) for a given A. Those details are important in nuclear physics and in practical astrophysics calculations, but only the average behavior is important for the present discussion.

a nucleus apart into its free constituents.[1] The more relevant quantity is often the *binding energy per nucleon*, $B(Z, N)/A$, where $A = Z+N$ is the atomic mass number. The average behavior of binding energy per nucleon as a function of the atomic mass number A is shown in Fig. 5.1. The qualitative behavior of the binding energy curve may be understood from the nuclear physics considerations listed below.

Saturation of nuclear forces: The almost constant binding energy per nucleon over most of the range of A at about 8 MeV per nucleon is a consequence of the saturation of nuclear forces. Because the nuclear forces are short-ranged, vanishing quickly outside a distance of order 10^{-13} cm, nucleons can interact strongly only with their nearest neighbors in a nucleus. Thus, the binding energy per nucleon is approximately constant (except at low A) with added nucleon number.

Surface and volume effects: The rapid increase of the binding energy per nucleon with A for very light nuclei is a surface versus volume effect. The nucleons in the interior of a nucleus can interact more strongly with their neighbors than those at the surface, which don't have other nucleons surrounding them on all sides. But the relative importance of surface to volume decreases with increasing nucleon number. This causes the binding

---

[1] Most mass tables give the total *atomic mass* rather than the *nuclear mass* for $m(Z, N)$. Nuclear masses can be obtained by subtracting from the tabulated atomic masses the rest mass of the Z electrons of the atom and the binding energy of these electrons to the nucleus. Since the latter is typically of order 10 eV per electron, the rest mass of an electron is 511 keV, and one atomic mass unit is $\sim$ 931.5 MeV, the differences in electron binding are small on the scale set by the rest mass of an atom. In most applications the interchange of nuclear and atomic masses in formulas leads to errors of order one keV or less and these may be ignored in all but the most precise considerations.

energy to increase with increasing nucleon number in the light nuclei because it minimizes the importance of the more loosely-bound nucleons at the surface.

Coulomb repulsion: The slow decrease of B/A with nucleon number above $A \sim 60$ is a Coulomb effect. The increased number of protons in the heavier nuclei causes the long-range, repulsive Coulomb force to destabilize the nucleus and lower the binding energy. The Coulomb effect on binding increases with A because the repulsive Coulomb energy is proportional to Z^2, which increases with A.

Maximal stability of iron group nuclei: The competition of the surface versus volume and Coulomb effects establishes a maximum of the binding energy per nucleon in the vicinity of $A = 60$. The isotopes in this region are called the *iron group nuclei*, and they are thermodynamically the most stable isotopes in the Universe. A large amount of stellar physics rests directly or indirectly on this fact.

Symmetry energy: Other things being equal, nuclei with approximately equal numbers of neutrons and protons are more stable than those with an excess of one or the other. This contribution to the binding energy is called the *symmetry energy*. In light nuclei, isotopes with $N \sim Z$ are generally more stable than those with asymmetry between N and Z. In heavy nuclei, the competition between the symmetry energy and Coulomb energy implies that the most stable nuclei have more neutrons than protons (60%–70% more in the heaviest isotopes known). This effect does not show up directly in Fig. 5.1 because only the most stable isotopes for each A are shown in this plot. It would show up if the binding energy of isotopic chains (same Z but different N) were plotted.

Shell effects: The realistic B/A curve exhibits fluctuations around the smoothed average of Fig. 5.1 caused by shell-closure effects at the microscopic nuclear structure level. These shell effects can be significant for the details of many processes important for stellar structure but are not important in our present general discussion.

Correlation energy: Nuclei are stabilized by 2-body interactions between nucleons that make a contribution to nuclear binding called *correlation energy*. The most important correlations are short-range *pairing interactions* and longer-range *quadrupole interactions*. The pairing correlation is less effective if either the number of neutrons or number of protons is odd. Therefore *odd-mass nuclei*, which have an odd number of either protons or neutrons, are generally less stable than *even-mass nuclei*, while *odd–odd nuclei* (odd numbers of both protons and neutrons) are generally even less stable. These pairing effects are not visible in Fig. 5.1 because this plot is a smoothed representation that averages over such fluctuations.

5.1.2 Masses and Mass Excesses

It is conventional to define the *mass excess*, $\Delta(A, Z)$, through

$$\Delta(A, Z) \equiv (m(A, Z) - A)\, M_u c^2, \tag{5.3}$$

where $m(A, Z)$ is measured in atomic mass units (amu), $A = Z + N$ is the atomic mass number, and the atomic mass unit M_u (see Box 3.1) is given by

$$M_u = \frac{1}{N_A} = 1.660539 \times 10^{-24}\,\text{g} = 931.494\,\text{MeV}/c^2, \tag{5.4}$$

where Avogadro's constant is $N_A = 6.022 \times 10^{23}\,\text{mol}^{-1}$. The total number of nucleons is constant in low-energy nuclear reactions, so the atomic mass numbers cancel on the two sides of any equation and sums and differences of masses (large numbers in standard units) may be replaced by the corresponding sums and differences of mass excesses (small numbers in standard units). For example, we may rewrite Eq. (5.2) as

$$\begin{aligned} B(Z, N) &= [Z m_p + N m_n - m(Z, N)]c^2 \\ &= [Z m_p + (A - Z) m_n - m(Z, N)]c^2 \\ &= [Z \Delta_p + (A - Z) \Delta_n - \Delta(A, Z)]c^2, \end{aligned} \tag{5.5}$$

where the mass excesses of the neutron and proton have been abbreviated to $\Delta(1, 0) \equiv \Delta_n$ and $\Delta(1, 1) \equiv \Delta_p$, respectively. Atomic masses and mass excesses are tabulated in Ref. [29].

Example 5.1 To illustrate the use of these formulas, let's calculate the binding energy of ^{4}He using Eq. (5.5). The relevant mass excesses are [29]

$$\Delta_p c^2 = 7.289\,\text{MeV} \qquad \Delta_n c^2 = 8.071\,\text{MeV} \qquad \Delta(4, 2)c^2 = 2.425\,\text{MeV}$$

and the binding energy of ^{4}He is then

$$B(2, 2) = 2 \times 7.289 + 2 \times 8.071 - 2.425 = 28.3\,\text{MeV}.$$

This implies that 28.3 MeV of energy is required to separate ^{4}He into free neutrons and protons, or that 28.3 MeV of energy is released by assembling two free protons and two free neutrons into a ^{4}He nucleus.

5.1.3 Q-Values

The Q-value for a reaction is the total mass of the reactants minus the total mass of the products, which is equivalent to the corresponding difference in mass excesses,

$$\begin{aligned} Q &= \text{Mass of reactants} - \text{Mass of products} \\ &= \text{Mass excess of reactants} - \text{Mass excess of products}. \end{aligned} \tag{5.6}$$

It is common to specify the Q-value in energy units rather than mass units.

Example 5.2 For the nuclear reaction $^2\text{H} + ^{12}\text{C} \rightarrow ^1\text{H} + ^{13}\text{C}$ the mass excesses are

$$\Delta(^2\text{H})c^2 = 13.136\,\text{MeV} \qquad \Delta(^{12}\text{C})c^2 = 0\,\text{MeV}$$

$$\Delta(^1\text{H})c^2 = 7.289\,\text{MeV} \qquad \Delta(^{13}\text{C})c^2 = 3.1246\,\text{MeV}.$$

The Q-value for this reaction is then

$$Q = \Delta(^2\text{H})c^2 + \Delta(^{12}\text{C})c^2 - \Delta(^1\text{H})c^2 - \Delta(^{13}\text{C})c^2 = +2.72\,\text{MeV}.$$

The positive Q indicates that this is an *exothermic reaction:* 2.72 MeV is liberated from binding energy in the reaction, appearing as kinetic energy or internal excitation of the products. Conversely, a negative value of Q would indicate an *endothermic reaction:* additional energy must be supplied to make the reaction viable.

5.1.4 Efficiency of Hydrogen Burning

Examination of the curve of binding energy in Fig. 5.1 suggests two potential nuclear sources of energy: fission of heavier elements into lighter elements or fusion of lighter elements into heavier ones. In either case the reaction products are more bound than the reactants, implying that energy has been released. Since stars are composed mostly of hydrogen and helium, their primary nuclear energy source must be the fusion of lighter elements to heavier ones. Coulomb repulsion between charged nuclei will inhibit fusion, so hydrogen ($Z = 1$) will be easier to fuse than helium ($Z = 2$), and is the primary candidate for a thermonuclear fuel accounting for stellar energy production. In particular, it will be shown below that main sequence stars are powered by thermonuclear processes that convert four ^{1}H into ^{4}He.[2] Before considering the detailed mechanisms by which this conversion takes place in stars, let us first estimate how much energy can be derived from the fusion of light elements into heavier ones, and an efficiency associated with this process.

Example 5.3 The total rest mass energy in one gram of material is

$$E = mc^2 = 9 \times 10^{20}\,\text{erg}, \tag{5.7}$$

and in Problem 5.1 you are asked to show that the energy released in the fusion of one gram of hydrogen into ^{4}He is

$$\Delta E(\text{fusion H} \rightarrow {}^4\text{He}) = 6.3 \times 10^{18}\,\text{erg g}^{-1}. \tag{5.8}$$

Therefore, the ratio is

$$\frac{\Delta E(\text{fusion H} \rightarrow {}^4\text{He})}{\text{total rest-mass energy}} = \frac{6.3 \times 10^{18}\,\text{erg g}^{-1}}{9 \times 10^{20}\,\text{erg g}^{-1}} \simeq 0.007, \tag{5.9}$$

and less than 1% of the initial rest mass is converted into energy in the stellar burning of hydrogen into helium.

These considerations show that thermonuclear burning of hydrogen is a rather *inefficient source of energy.* Furthermore, reaction rates in the cores of lower-mass main sequence

---

[2] As is standard we use the shorthand "hydrogen fusion" for this conversion, but it actually involves multiple reaction steps, not all of which are fusion reactions.

stars like the Sun are actually quite small.[3] As demonstrated in Problem 5.12, the luminosity of the Sun derives from a central source of such low power density that it is equivalent to distributing several 100-watt light bulbs per cubic meter of the solar core. The reason that thermonuclear fusion is able to power stars is not because it is intrinsically efficient at converting mass to energy, or that it has a high reaction rate under conditions found in main sequence stars. Rather, it is because of the enormous mass of stars, which ensures that they have large reservoirs of hydrogen available as thermonuclear fuel in their cores.

5.2 Thermonuclear Hydrogen Burning

The generic energy source for stars on the main sequence is the thermonuclear burning of hydrogen into helium. There are two separate sets of nuclear reactions that can convert hydrogen to helium with the release of energy under conditions found in stellar interiors:

- The *proton–proton chains* (*PP chains*).
- The *CNO (carbon–nitrogen–oxygen) cycle*.

The proton–proton chains are responsible for most of the present energy of the Sun and generally are dominant in main sequence stars of a solar mass or less. The energy production in the CNO cycle quickly surpasses that of the proton–proton chains as soon as the mass of the star exceeds about a solar mass. As will be seen below, the primary reason for this rapid switch is that the CNO cycle has an extremely strong dependence on temperature. This will favor the CNO cycle in more massive main sequence stars because they have higher core temperatures.

5.2.1 The Proton–Proton Chains

The most important reactions of the PP chains are summarized in Fig. 5.2. Calculations within the Standard Solar Model to be described in Chapter 10 indicate that the Sun is producing 98.4% of its energy from the PP chains and only 1.6% from the CNO cycle, and that within the fraction coming from PP chains PP-I produces about 85% of the energy, PP-II produces about 15%, and PP-III produces only about 0.02%. The basic reason for this branching lies in the rates for key nuclear reactions under present conditions in the solar core, and will be described below.

---

[3] The most efficient conversion of mass to energy is matter–antimatter annihilation, which can convert 100% of the mass to energy. However, little antimatter exists naturally in the Universe, so that is not a viable large-scale astrophysical source of energy. The next best conversion efficiencies come from processes involving black holes, particularly rotating black holes, which can in principle give as much as $\sim 30\%$ conversion of mass to energy (see Chapters 13–15 of Ref. [100]). This is why it is generally thought that the gigantic but extremely compact central engines powering quasars must be rotating supermassive ($\sim 10^9 M_\odot$) black holes.

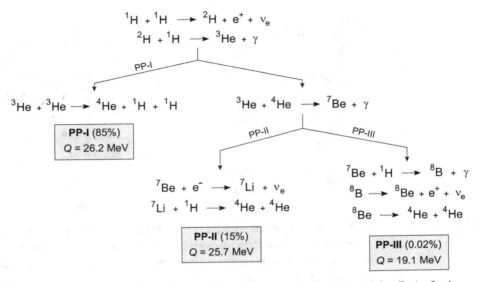

Fig. 5.2 Main branches of the PP chains. The percentage contribution to solar energy production and the effective Q-value are shown for each branch of the chains.

5.2.2 The CNO Cycle

The name of the carbon–nitrogen–oxygen or CNO cycle derives from the role of carbon, nitrogen, and oxygen in the corresponding sequence of reactions. The primary cycle is summarized in Fig. 5.3(a), which corresponds to the following set of nuclear reactions.[4]

$$\begin{array}{lll} {}^{12}C(p,\gamma){}^{13}N & {}^{13}N \to e^+ + \nu_e + {}^{13}C & {}^{13}C(p,\gamma){}^{14}N \\ {}^{14}N(p,\gamma){}^{15}O & {}^{15}O \to e^+ + \nu_e + {}^{15}N & {}^{15}N(p,\alpha){}^{12}C. \end{array} \qquad (5.10)$$

The reactions (5.10) of the CNO cycle are sometimes called the CN cycle. A second set of reactions can occur as a branch from the basic cycle [see Fig. 5.3(b)]:

$$\begin{array}{ll} {}^{15}N(p,\gamma){}^{16}O & {}^{16}O(p,\gamma){}^{17}F \\ {}^{17}F \to e^+ + \nu_e + {}^{17}O & {}^{17}O(p,\alpha){}^{14}N. \end{array} \qquad (5.11)$$

This set of reactions branches from the first set once ^{15}N has been produced, and it feeds back into the first set as ^{14}N is produced in the last step. The rate of branching into the second set of reactions depends on stellar conditions but for temperatures below 10^8 K the branching is typically less than one part in 10,000. The two sets of reactions (5.10)–(5.11), together with some minor side branches that are not shown, constitute the full CNO cycle.

---

[4] A compact nuclear physics notation for reactions will sometimes be used. For example, the 2-body reaction $a + b \to c + d$ is written concisely as b(a, d)c. Thus, $^{12}C(p,\gamma)^{13}N$ denotes a reaction in which a proton is captured on ^{12}C to produce ^{13}N and a γ-ray is emitted, while $^7Be(e^-, \nu_e)^7Li$ denotes an electron capture reaction in which a proton in the ^7Be nucleus absorbs an extranuclear electron and is converted to a neutron, and a neutrino is emitted. A schematic illustration of the types of nuclear reactions important for astrophysics may be found in Fig. D.1 of Appendix D.

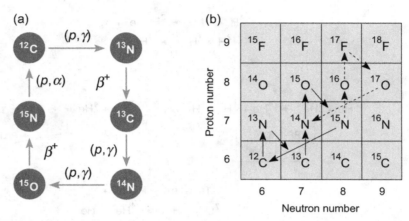

Fig. 5.3 The CNO cycle. (a) The main part of the cycle [Eq. (5.10)] is illustrated schematically. (b) The main part of the cycle is indicated with solid lines and the branch shown in Eq. (5.11) is indicated with dashed lines. The notation (p, i) means a proton capture followed by emission of i; for example $^{12}C(p, \gamma)^{13}N$. The notation β^+ indicates β-decay by positron emission; for example, $^{13}N \rightarrow {}^{13}C + e^+ + \nu_e$.

Box 5.1   **CNO Catalysis**

The reactants and products for each reaction around the CN cycle (5.10) may be summed to obtain

$$^{12}C + 4p \rightarrow {}^{12}C + {}^4He + 2\beta^+ + 2\nu.$$

(The sum neglects γ-rays because there is no conservation law for the number of photons.) Therefore, the CNO cycle converts protons into ^{4}He with a corresponding release of energy and neutrinos, just as for the PP chains, but in a very different way: ^{12}C serves as a *catalyst* for the conversion of four protons to ^{4}He. Its presence is required but it is not consumed because a ^{12}C is returned in the last step of Eq. (5.10). The sequence (5.10)–(5.11) has been written as if the (p, γ) reaction on ^{12}C were the first step but it is a closed cycle and any step may be considered to be the initial one. As illustrated in Problem 5.9 and Box 6.1, this means that any of the carbon, nitrogen, or oxygen isotopes appearing in the cycle (5.10)–(5.11) may be viewed as catalysts that serve to convert protons into helium. The closed nature of the cycle implies also that

1. Any mixture of these CNO isotopes will play the same catalytic role.
2. If any one of the CNO isotopes is present initially a mixture of the others will inevitably be produced by the cycle of reactions in Eqs. (5.10)–(5.11).
3. The sum of abundances for the CNO isotopes is conserved by the cycle.

These properties of the CNO cycle are elaborated further in the results of the numerical calculation displayed in Box 6.1.

(For this reason, it is sometimes called a bi-cycle; see the dashed-line and solid-line paths in Fig. 5.3.) As we discuss in Box 5.1 and illustrate further in Box 6.1, the isotopes of carbon, nitrogen, and oxygen appearing in Fig. 5.3 *catalyze* the CNO cycle: they must be present for it to occur, but they are not consumed by it.

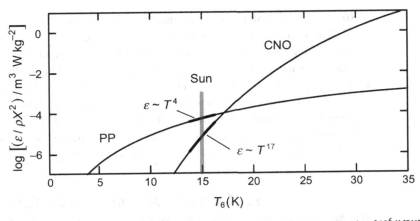

Fig. 5.4 Rate of energy release in the PP chains and CNO cycle as a function of T_6, the temperature in units of 10^6 K [211]. (See Section 5.9 for calculation of the energy production rate.) The present temperature of the Sun is indicated. Near that temperature PP chain energy production varies as $\sim T^4$ and CNO cycle energy production varies as $\sim T^{17}$ (see Section 5.10).

5.2.3 Competition of PP Chains and the CNO Cycle

The rates of energy release from hydrogen burning for the PP chains and CNO cycle are illustrated in Fig. 5.4, where it is clear that the CNO cycle depends more strongly on temperature than do the PP chains. This temperature dependence implies that the star's mass on the main sequence is a crucial factor governing the competition of PP and CNO energy production because it has a strong influence on the central temperature. The PP chains can occur in any star but the CNO cycle requires the presence of carbon, nitrogen, or oxygen isotopes. The CNO cycle might be expected to be relatively more important in Pop I stars because of their higher concentration of heavier elements, but CNO abundances are typically of secondary importance to the temperature for the competition between the PP chains and CNO cycle (except for extreme Pop II stars with essentially no CNO isotopes, where the CNO cycle cannot operate at all).

As an important aside, rates for PP chains and the CNO cycle are of importance in understanding the very first generation of stars that formed in the Universe (*Pop III*; see Section 1.9.2). The first such stars were probably more massive than current stars, perhaps by factors of 5–10. This was largely because of the absence of metals, which inhibited cooling (metals in a gas aid cooling because of the forest of emission lines that they produce), and the resulting higher temperatures favored the collapse of more massive protostars (see the Jeans mass in Section 9.2). The (photon) opacity (Section 7.4.4) for stars in the early Universe was much less than current stellar opacities because there were few metals and metals increase opacity by virtue of the large number of electrons that they release when ionized (which then interact strongly with photons). Thus, as will be discussed in Section 9.11, the low opacity stabilized these more massive nascent Stars against pulsational instabilities and ejection of their envelopes, allowing them to grow even more massive by accretion.

Since no CNO isotopes were produced in the big bang (see Chapter 20 of Ref. [100] for a description of big bang nucleosynthesis), the first stars must have operated by the proton–proton chains until some of them could produce carbon by the triple-α process (Section 6.3) and seed formation of later stars that could operate by the CNO cycle. The CNO cycle produces energy more efficiently in massive stars than the PP chains by virtue of its stronger temperature dependence. Therefore, the pace of early structure evolution in the Universe presumably depended on when the earliest stars produced and distributed (by winds and supernova explosions) sufficient CNO isotopes to allow a succeeding generation of stars to switch to the more efficient CNO cycle for energy production.

5.3 Cross Sections and Reaction Rates

A quantitative analysis of energy production in stars requires the basics of nuclear reaction theory as applied in stellar environments. Let us examine this, using for guidance the discussions in Refs. [71, 107, 128, 188]. Consider a representative nuclear reaction

$$\alpha + X \rightarrow (Z^*) \rightarrow \beta + Y, \tag{5.12}$$

where Z^* denotes an excited compound nucleus as a possible intermediate state [Eq. (5.12) also may be expressed as $X(\alpha, \beta)Y$ in nuclear reaction notation]. A compound nucleus is an excited composite formed in the initial collision that quickly decays into the final products of the reaction. In Eq. (5.12) the left side ($\alpha + X$) is called the *entrance channel* of the reaction and the right side ($\beta + Y$) is called the *exit channel* of the reaction. It is standard to classify nuclear reactions according to the number of (nuclear) species in the entrance channel; thus (5.12) is a *2-body reaction*, while the photodisintegration reaction $\gamma + A \rightarrow B + C$ is a *1-body reaction* (since the photon γ on the left side is not a nuclear species), and $A + B + C \rightarrow D$ is a *3-body reaction*. Ideas will be illustrated primarily with 2-body reactions in the following discussion but 1-body reactions and 3-body reactions also are important in stellar energy production.

5.3.1 Reaction Cross Sections

Imagine first the typical laboratory setting where the reaction (5.12) is initiated by a beam of projectiles α directed onto a target containing nuclei X. The *cross section* $\sigma_{\alpha\beta}(v)$, which generally is a function of the velocity v, is defined as[5]

$$\sigma_{\alpha\beta}(v) \equiv \frac{\rho_{\alpha\beta}}{F(v)} = \left(\frac{\text{reactions per unit time per target nucleus}}{\text{incident flux of projectiles}} \right), \tag{5.13}$$

and has units of area [a commonly-used unit of cross section is the *barn (b)*, which is defined to be a cross section of 10^{-24} cm^2]. The incident particle flux $F(v)$ is given by

---

[5] A shorthand label $\alpha\beta$ will be used to denote the reaction being considered. This should be thought of as a possibly-composite label carrying sufficient information to identify the reaction uniquely.

$$F(v) = n_\alpha v, \qquad (5.14)$$

where n_α is the number density of projectiles α in the beam and v is their relative velocity.

5.3.2 Rates from Cross Sections

The number of reactions per unit time (reaction rate) per target nucleus $\rho_{\alpha\beta}$ is

$$\rho_{\alpha\beta} = n_\alpha v \sigma_{\alpha\beta}. \qquad (5.15)$$

The total reaction rate per unit volume $r_{\alpha\beta}(v)$ then is obtained from multiplying $\rho_{\alpha\beta}$ by the number density n_X of target nuclei X:

$$r_{\alpha\beta}(v) = \rho_{\alpha\beta} n_X = \eta_{\alpha X}\, n_\alpha n_X v\, \sigma_{\alpha\beta}(v) \qquad \eta_{\alpha X} \equiv \frac{1}{1+\delta_{\alpha X}}, \qquad (5.16)$$

and has units of cm^{-3} s^{-1} in the CGS system. The factor $\eta_{\alpha X}$ involving the Kronecker δ_{ab} (which is one if $a=b$ and zero if $a \neq b$) is introduced to prevent overcounting if the colliding particles are identical; this is explained further in Section D.2.3. Normally it is most convenient to work in the center of mass coordinate system, so velocities, energies, momenta, and cross sections will be center of mass quantities,

$$E = E_{\rm CM} = \left(\frac{m_X}{m_\alpha + m_X}\right) E_{\rm Lab} \qquad v = \sqrt{2E/\mu} \qquad \mu \equiv \frac{m_\alpha m_X}{m_\alpha + m_X},$$

unless otherwise noted.

5.4 Thermally Averaged Reaction Rates

The preceding equations assume a beam of monoenergetic particles. In a stellar environment the gas is in approximate thermal and hydrostatic equilibrium and so has a distribution of velocities instead. Assuming that the gas can be described classically (see Chapter 3), at equilibrium it has a Maxwell–Boltzmann distribution $\Phi(E)$ of energies[6]

$$\Phi(E) = \frac{2}{\pi^{1/2}} \frac{E^{1/2}}{(kT)^{3/2}} \exp(-E/kT). \qquad (5.17)$$

We may define a thermally-averaged cross section $\langle \sigma v \rangle_{\alpha\beta}$ for a 2-body reaction by averaging the reaction cross section over the velocity distribution in the gas,

$$\langle \sigma v \rangle_{\alpha\beta} \equiv \int_0^\infty \Phi(E) \sigma_{\alpha\beta}(E) v\, dE,$$

$$= \sqrt{\frac{8}{\pi \mu}} (kT)^{-3/2} \int_0^\infty \sigma_{\alpha\beta}(E) e^{-E/kT} E\, dE, \qquad (5.18)$$

where $v = (2E/\mu)^{1/2}$ was used and the units of $\langle \sigma v \rangle_{\alpha\beta}$ are cm^3 s^{-1} (cross section times velocity). The corresponding thermal average of the reaction rate (5.16) is then given by

---

[6] If the two colliding particles belong to two separate Maxwell–Boltzmann velocity distributions, their relative velocity will also belong to a Maxwell–Boltzmann distribution. A proof may be found in Ref. [71].

$$r_{\alpha\beta} = \eta_{\alpha X} n_\alpha n_X \int_0^\infty \Phi(E)\sigma_{\alpha\beta}(E) v\, dE$$
$$= \eta_{\alpha X} n_\alpha n_X \langle \sigma v \rangle_{\alpha\beta}$$
$$= \eta_{\alpha X} \rho^2 N_A^2 \frac{X_\alpha X_X}{A_\alpha A_X} \langle \sigma v \rangle_{\alpha\beta}$$
$$= \eta_{\alpha X} \rho^2 N_A^2 Y_\alpha Y_X \langle \sigma v \rangle_{\alpha\beta}, \tag{5.19}$$

where the last two lines introduce the mass fractions X_i and the abundances Y_i defined in Eqs. (3.17)–(3.18). The units of $r_{\alpha\beta}$ are cm^{-3} s^{-1} (rate per unit volume), and the intuitively reasonable interpretation of Eq. (5.19) is that the total rate per unit volume for the 2-body reaction (5.12) is the (thermally averaged) rate for a single α to react with a single X to produce $Y + \beta$, multiplied by the number of αs per unit volume and by the number of Xs per unit volume.

5.5 Parameterization of Cross Sections

To proceed further we require cross sections to enable calculation of the thermally averaged rates. These cross sections may be parameterized in the general form[7]

$$\sigma_{\alpha\beta}(E) = \pi g \lambda^2 \frac{\Gamma_\alpha \Gamma_\beta}{\Gamma^2} f(E), \tag{5.20}$$

where the *energy widths* $\Gamma_i \equiv \hbar/\tau_i$ are expressed in terms of the corresponding mean life τ_i for decay of the compound system through channel i, the entrance channel is denoted by α, the exit channel by β, the total width is $\Gamma = \sum_i \Gamma_i$, where the sum is over all open channels i, the probability to decay to channel i is $P_i = \Gamma_i/\Gamma$, the reduced de Broglie wavelength is defined through $\lambda^2 = \hbar^2/2\mu E$, the statistical factor g contains information on the spins of projectile, target, and compound nucleus (and is typically of order unity), and the detailed reaction information resides in the factor $f(E)$. This factor can be complicated to evaluate in general but in two instructive limiting cases it simplifies.

1. The reaction may be *resonant*, in which case the rate is strongly peaked at some energy because of a narrow (quasibound) state in the compound nucleus. If this resonance is well-separated in energy from other resonances (an *isolated resonance*), the cross section can be approximated as described below.
2. The reaction may be *nonresonant* because there are no resonances in the channel of interest, or because the reaction energy lies far from any resonance. The approximate treatment of non-resonant cross sections also will be described below.

---

[7] The parameters Γ appearing in (5.20) have units of $\hbar$ divided by time, which is energy. They are called *energy widths* because states with short lifetimes for decay (that is, with large decay rates) correspond to spectral peaks (resonances) broad in energy, by a $\Delta E \cdot \Delta t \simeq \hbar$ uncertainty principle argument. Conversely, states with long decay lifetimes (small decay rates) correspond to narrow resonances. The limiting case is a state that is completely stable, which then corresponds to a vanishing decay rate and a sharply-defined energy.

In this idealized situation the total rates will be a sum of resonant and nonresonant pieces that can be handled separately. The realistic situation is often more complex than the idealized one imagined above. For example, there may be multiple resonances overlapping each other, or there may be a *sub-threshold resonance,* which peaks below the minimum energy for the reaction to occur but has a tail extending into the region of allowed energies because of its finite width. We shall ignore such complications here (see Ref. [128] for discussion of more general cases), and proceed to describe first the non-resonant contribution, and then the contribution of a single isolated resonance to the cross section.

5.6 Nonresonant Cross Sections

You will not be surprised to learn that most (though not all) reactions important for stellar energy production are exothermic ($Q > 0$). Typically for the reactions of interest $Q \sim 1$ MeV, with this energy going into kinetic energy of particles in the exit channel and any internal excitation of the products. This additional energy leads to a marked asymmetry in the entrance and exit channels for charged particle reactions of interest in stellar energy production. In the entrance channel the thermal energies available are set by the temperatures through $kT = 8.6174 \times 10^{-8} T$ keV, with the temperature expressed in kelvin. Hydrogen, helium, and carbon burning occur in a temperature range 10^7 K $\lesssim T \lesssim 10^9$ K, implying kinetic energies in the plasma of 1 keV $\lesssim kT \lesssim$ 100 keV. In light of the average Q-values noted above, generally $E(\text{entrance}) \ll E(\text{exit})$ for the charged-particle reactions of interest.

5.6.1 Coulomb Barriers

Because of the low energies in the entrance channel, charged-particle reactions are influenced strongly by the Coulomb barrier between colliding ions that is illustrated in Fig. 5.5. The potential energy associated with the Coulomb barrier may be approximated as

$$E_{\text{CB}} = 1.44 \frac{Z_\alpha Z_X}{R(\text{fm})} \text{ MeV}, \tag{5.21}$$

where Z_i is the atomic number of particle i, the characteristic distance R is

$$R \simeq 1.3(A_\alpha^{1/3} + A_X^{1/3}) \text{ fm}, \tag{5.22}$$

A_i is the atomic mass number (in atomic mass units) of particle i, and 1 fm = 10^{-13} cm.

Example 5.4 Consider a proton scattering from ^{28}Si. The Coulomb barrier is

$$E_{\text{CB}} = 1.44 \frac{(1)(14)}{1.3(1^{1/3} + 28^{1/3})} \text{ MeV} = 3.8 \text{ MeV},$$

where Eqs. (5.21) and (5.22) were used.

Table 5.1 Coulomb barriers for p + X

X	$Z_\alpha Z_\beta$	R (fm)	E_{CB} (MeV)
$^1_1$H	1	2.6	0.55
$^{12}_6$C	6	4.3	2.0
$^{28}_{14}$Si	14	5.2	3.8
$^{56}_{26}$Fe	26	6.3	6.0

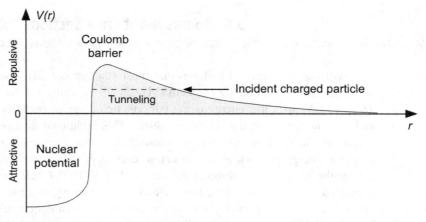

Fig. 5.5 The Coulomb barrier for charged-particle reactions. Average energies in stellar plasmas are far below the top of the barrier, so charged-particle reactions occur by tunneling.

Some typical Coulomb barriers for proton reactions p + X are shown in Table 5.1, where it may be noted that entrance channel energies for hydrogen fusion in stars (which lie in the range 10^{-3} to 10^{-1} MeV) are typically orders of magnitude lower than the Coulomb barrier. This implies a dramatic temperature dependence for hydrogen fusion reactions. On the other hand, exit channel energies (approximately 1 MeV in typical cases) are not so different from the barrier energies for fusion of protons with lighter ions.

5.6.2 Barrier Penetration Factors

Classically, the characteristic particle energies in a stellar plasma are far too small to surmount the Coulomb barrier for charged-particle reactions. However, quantum mechanically it is possible for tunneling to take place at energies that lie below the height of the barrier, albeit with exponentially suppressed probability. Assuming $E_{CB} \gg E$, the quantum-mechanical barrier penetration probability for a collision having zero relative orbital angular momentum (these are termed s-waves in scattering theory) may be expressed as

$$P(E) \propto e^{-2\pi\eta} \qquad \eta \equiv \frac{Z_\alpha Z_X e^2}{\hbar v}, \tag{5.23}$$

where the *Sommerfeld parameter* η is dimensionless. For representative values of the parameters a typical result is that for a barrier penetration in a star $P(E) \sim \exp(-12)$. Thus, the charged particle reactions crucial to the power generation in stars (and to your existence!) turn out to be highly-improbable events. Since the reaction rate will be dominated by the likelihood to penetrate the barrier, it is reasonable to take as an entrance channel width for nonresonant reactions $\Gamma_\alpha \simeq e^{-2\pi\eta}$, which has a very strong energy dependence.

5.6.3 Astrophysical S-factors

In the exit channels for charged-particle reactions the energies are roughly comparable to the barrier energies; thus we may assume that Γ_β is a weakly varying function of E and the nonresonant cross section then may be parameterized as

$$\sigma_{\alpha\beta}(E) = \pi g \lambdabar^2 \frac{\Gamma_\alpha \Gamma_\beta}{\Gamma^2} f(E) \equiv \frac{S(E)}{E} e^{-2\pi\eta}, \qquad (5.24)$$

where $\lambdabar^2 \propto 1/E$ has been used. The factor $S(E)$ is termed (rather prosaically) the *astrophysical S-factor*. It is assumed to vary slowly with E and contains all energy dependence not introduced explicitly through the factors $\lambdabar^2 \sim 1/E$ and $\exp(-2\pi\eta)$. The S-factor may be determined experimentally from Eq. (5.24) by measuring the cross section at a given energy and computing the $\exp(-2\pi\eta)/E$ factor. The measured S-factor for a typical (p, γ) reaction is illustrated in Fig. 5.6.

Because of the low values of kT for typical stellar plasmas reaction cross sections are often needed at energies lower than can be measured in a laboratory. Then, experimental measurements at higher energy must be extrapolated to the lower energy of interest. Usually this is done by assuming no resonances at the lower energy and plotting

$$S(E) \equiv \sigma_{\alpha\beta}(E) E e^{2\pi\eta}, \qquad (5.25)$$

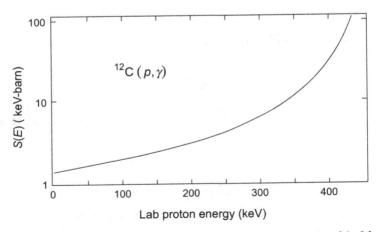

Fig. 5.6 S-factor versus laboratory proton energy for a typical capture reaction [88]. This is the portion of the S-factor for the region below the resonance in Fig. 5.8.

which has smoother behavior than the full cross section. Then, from Eqs. (5.18) and (5.24) the thermally-averaged nonresonant cross section is given in terms of the S-factor as

$$\langle \sigma v \rangle_{\alpha\beta} = \sqrt{\frac{8}{\pi\mu}} (kT)^{-3/2} \int_0^\infty S(E) e^{-E/kT} e^{-bE^{-1/2}} \, dE, \quad (5.26)$$

where $b \equiv (2\mu)^{1/2} \pi Z_\alpha Z_X e^2 / \hbar$.

5.6.4 The Gamow Window

The energy dependence of Eq. (5.26) resides primarily in the quantity

$$F_G \equiv e^{-E/kT} e^{-bE^{-1/2}}. \quad (5.27)$$

Since the first factor in this expression (arising from the Maxwell–Boltzmann velocity distribution) decreases rapidly with energy while the second (arising from the barrier penetration factor) increases rapidly with energy, the product (5.27) defines a strong localization in energy called the *Gamow window*, which is illustrated schematically in Fig. 5.7. Only if the energies fall within the Gamow window will charged particle reactions occur with significant probability. Thus, the Gamow window is of extreme significance in the physics of stars. The maximum of the Gamow peak is found to lie at

$$E_0 = \left(\frac{bkT}{2} \right)^{2/3} \quad (5.28)$$

(see Problem 5.4). Useful approximate expressions for the width of the Gamow peak and for the nonresonant cross section can be obtained by assuming the Gamow peak to be a gaussian having the same peak position and curvature at the peak as the realistic Gamow peak (see Problem 5.5). In this approximation

$$\Delta = \frac{4}{3^{1/2}} (E_0 kT)^{1/2} \quad (5.29)$$

for the $1/e$ width of the Gamow peak.

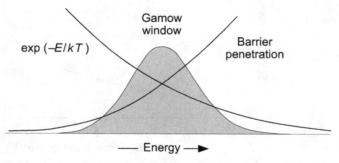

Fig. 5.7 Schematic representation of the Gamow window. In this figure, each curve and the Gamow peak are plotted on different scales to better visualize the location of the peak.

Example 5.5 Equations (5.28) and (5.29) can be evaluated numerically to give

$$E_0 = 1.22(Z_\alpha^2 Z_X^2 \mu T_6^2)^{1/3} \text{ keV} \qquad \Delta = 0.75(Z_\alpha^2 Z_X^2 \mu T_6^5)^{1/6} \text{ keV}. \qquad (5.30)$$

Then for the interaction of two protons at a temperature of $T_6 = 20$ (that is, at a temperature of $T = 20 \times 10^6$ K),

$$kT = 1.7 \text{ keV} \qquad E_0 = 7.1 \text{ keV} \qquad \Delta = 8.1 \text{ keV},$$

and from Table 5.1 the corresponding Coulomb barrier is about 550 keV. Thus, as a consequence of the Coulomb barrier this charged-particle reaction is most likely to occur in an energy window that is centered at about 7 keV with a width of approximately 8 keV.

In gaussian approximation the integral in Eq. (5.26) can be evaluated analytically and the cross section is found to be

$$\langle \sigma v \rangle_{\alpha \beta} \simeq \frac{7.2 \times 10^{-19} S(E_0) a^2}{\mu Z_\alpha Z_X T_6^{2/3}} \exp(-a T_6^{-1/3}) \qquad (5.31)$$

in units of cm^3 s^{-1}, where $a = 42.49(Z_\alpha^2 Z_X^2 \mu)^{1/3}$ and $S(E_0)$ is the S-factor in units of keV barns, evaluated at the energy corresponding to the maximum of the Gamow peak.

5.7 Resonant Cross Sections

A resonant cross section is dominated by strong enhancement within a narrow range of energies. In the simplest case of an isolated resonance, $f(E)$ can be expressed in the *Breit–Wigner* form

$$f(E)_{\text{res}} = \frac{\Gamma^2}{(E - E_{\text{r}})^2 + (\Gamma/2)^2}, \qquad (5.32)$$

where the resonance energy E_{r} is related to a corresponding excitation energy E^* for a quasibound state in the compound nucleus through (see Fig. 6.5)

$$E_{\text{r}} = E^* - Q, \qquad (5.33)$$

where Q is the energy released by the reaction because of differing binding energies in the entrance and exit channels. The corresponding Breit–Wigner cross section for a reaction with orbital angular momentum ℓ is

$$\sigma_{\alpha \beta} = \pi g \lambdabar^2 (2\ell + 1) \frac{\Gamma_\alpha \Gamma_\beta}{(E - E_{\text{r}})^2 + (\Gamma/2)^2}, \qquad (5.34)$$

which will exhibit a strong peak for $E \sim E_{\text{r}}$.

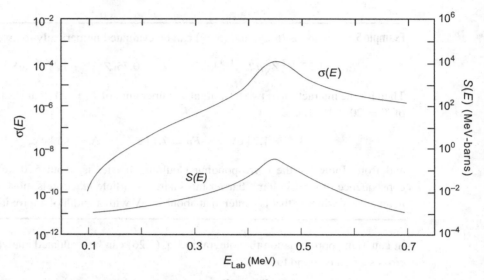

Fig. 5.8 Measured cross section $\sigma(E)$ in barns for the reaction ^{12}C(p, γ)^{13}N [88]. The S-factor $S(E)$ is defined through Eq. (5.24).

Example 5.6 Consider the reaction ^{12}C(p, γ)^{13}N illustrated in Fig. 5.8. It has a resonance corresponding to a state in ^{13}N at an excitation energy of 2.37 MeV that is strongly excited at a laboratory proton energy of 0.46 MeV (see Problem 5.13).

If the widths Γ_i vary slowly over a resonance, it is valid to assume that the dominant contribution to the integral will come for $E \sim E_r$ and approximate by

$$\Phi(E) \rightarrow \Phi(E_r) \qquad \Gamma_\alpha \rightarrow \Gamma_\alpha(E_r) \qquad \Gamma_\beta \rightarrow \Gamma_\beta(E_r),$$

in which case we may write for the resonant velocity-averaged cross section [107]

$$\langle \sigma v \rangle_{\alpha\beta} = \frac{\pi \hbar^2 g (2\ell + 1)}{2\mu} \sqrt{\frac{8}{\pi \mu}} (kT)^{-3/2} e^{-E_r/kT}$$
$$\times \Gamma_\alpha(E_r) \Gamma_\beta(E_r) \int_0^\infty \frac{dE}{(E - E_r)^2 + (\Gamma/2)^2}. \qquad (5.35)$$

The integrand peaks near E_r; extending the lower limit to negative infinity and assuming the widths to be constant allows the integral to be evaluated, giving [107]

$$\langle \sigma v \rangle_{\alpha\beta} = 2.56 \times 10^{-13} \frac{(\omega\gamma)_r}{(\mu T_9)^{3/2}} \exp(-11.605 E_r/T_9) \, \text{cm}^3 \, \text{s}^{-1}, \qquad (5.36)$$

where E_r is in MeV, T_9 indicates the temperature in units of 10^9 K, and the quantity

$$(\omega\gamma)_r \equiv (2\ell + 1) g \frac{\Gamma_\alpha \Gamma_\beta}{\Gamma} \qquad (5.37)$$

is an intrinsic measure of reaction strength in units of MeV that is tabulated in Ref. [88].

5.8 Calculations with Rate Libraries

Rates entering into the reaction formalism of this chapter have strong temperature dependence and a weaker density dependence. Such rates, which are determined by some combination of experiments, theory, and extrapolation, are required in a broad range of phenomena exhibiting temperatures and densities differing by orders of magnitude. A common approach to making these rates available for practical calculations is to parameterize the density and temperature dependence guided by theory and physical intuition, and then to tabulate the corresponding parameters in *rate libraries*. Appendix D describes how to use two such rate libraries that have been employed at various places in this book. These libraries allow rates for a reaction to be computed quickly for any temperature and density.

5.9 Total Rate of Energy Production

The total reaction rate per unit volume $r_{\alpha\beta}$ is given by Eq. (5.19). The corresponding total rate of energy production per unit mass $\varepsilon_{\alpha\beta}$ is then given by the product of the rate and the Q-value, divided by the density:

$$\varepsilon_{\alpha\beta} = \frac{r_{\alpha\beta} Q}{\rho}, \tag{5.38}$$

which has CGS units of $\text{erg}\,\text{g}^{-1}\text{s}^{-1}$. The Q-value entering this expression is defined in Eq. (5.6), but with the proviso that if a reaction produces a neutrino that removes energy from the star without appreciable interaction in the core, this neutrino energy should be subtracted from the total Q-value before it is inserted into Eq. (5.38).

5.10 Temperature and Density Exponents

It can be useful to parameterize the energy production rate in the form of a power-law,

$$\varepsilon = \varepsilon_0 \rho^\lambda T^\nu, \tag{5.39}$$

where ν is termed the *temperature exponent* and λ the *density exponent*. Equation (5.39) with constant exponents is usually valid only for a limited range of temperatures and densities. Since energy production mechanisms for stars often are important only in a very narrow temperature–density range, this can still be a useful parameterization. Temperature and density exponents for an arbitrary energy production rate function $\varepsilon(\rho, T)$ may be defined through

$$\lambda = \left(\frac{\partial \ln \varepsilon}{\partial \ln \rho}\right)_T \qquad \nu = \left(\frac{\partial \ln \varepsilon}{\partial \ln T}\right)_\rho. \tag{5.40}$$

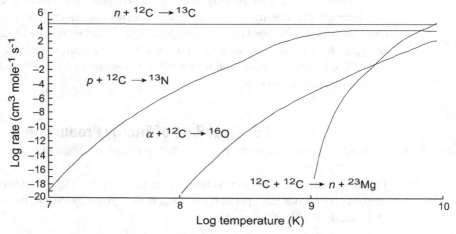

Table 5.2 Density and temperature exponents		
Stellar process	Density (λ)	Temperature (ν)
PP chains	1	~ 4
CNO cycle	1	~ 16
Triple-α	2	~ 40

Fig. 5.9 Variation of rates for some reactions on ^{12}C involving different Coulomb barriers. In the entrance channel the height of the Coulomb barrier is proportional to the product $Z_1 Z_2$, which is 0 for the neutron capture reaction, 6 for the proton capture reaction, 12 for the α capture reaction, and 36 for the ^{12}C+^{12}C reaction.

Some values of λ and ν for the PP chains, the CNO cycle, and the triple-α process (see Section 6.3), are displayed in Table 5.2. They will be discussed further below. Notice from Table 5.2 the extremely strong dependence of reaction rate on temperature for charged-particle reactions. This is illustrated dramatically in Fig. 5.9, where for increasing values of $Z_1 Z_2$ the reaction threshold is pushed to much higher temperature and the slope of the reaction rate relative to temperature steepens rapidly. Conversely, the neutron capture reaction, which has no Coulomb barrier, has very weak temperature dependence.

5.11 Neutron Reactions and Weak Interactions

Most reactions important for stars involve charged particles, for which the Coulomb barrier has a decisive influence. However, reactions involving uncharged neutrons are crucial for some aspects of stellar structure, particularly in the production of the heavy elements. This is discussed further in Box 5.2 and in Sections 13.7.2 and 20.5. Energy-producing reactions in stars tend to be mediated by strong interactions but weak interactions have

Box 5.2 Neutron Capture Reactions

Neutron reactions are unique in that there is no Coulomb barrier. They are of minor significance in the energy budgets of stars but they generally compete with β^- decays and are important for element production. No charge means no Gamow window and neutron reaction rates are generally not very sensitive to temperature. In the absence of resonances the reaction cross section $\langle \sigma v \rangle$ often is approximately constant for a given neutron capture reaction (see the neutron capture rate plotted in Fig. 5.9).

Slow Neutron Capture

When the neutron capture rate is much smaller than the rate for β-decay, stars may produce heavier elements by the slow capture of a few neutrons, followed by a β-decay:

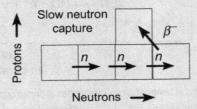

This slow neutron capture or *s-process* can produce new nuclides only along the *valley of β-stability* (the isotopes that are stable against β-decay; see Fig. 13.11). Element production by the s-process in red giant stars will be discussed further in Section 13.7.2.

Fast Neutron Capture

On the other hand, a neutron capture rate much larger than the rate for β-decay can lead to a rapid neutron capture or *r-process,* where many neutrons are captured before there is time for a β-decay to occur. The following diagram illustrates.

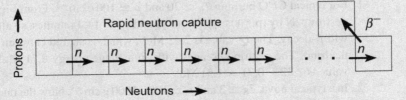

Because the r-process leads to many successive neutron captures without an intervening β-decay, it produces neutron-rich isotopes well out of the β-stability valley, and also can bypass gaps in the stability valley to produce the heavy elements beyond bismuth (Fig. 20.21). It is thought to occur primarily in neutron star mergers and core collapse supernovae, and will be discussed further in Sections 20.5 and 22.6.

significant influence. For example, as we shall discuss in Section 6.1, the rate-determining step for the reaction sequence that powers the Sun is a weak interaction. As Box 5.2 illustrates, the weak interactions may compete with neutron capture. They also may compete with charged-particle reactions, thereby influencing the energy production for a star, as illustrated in the following example.

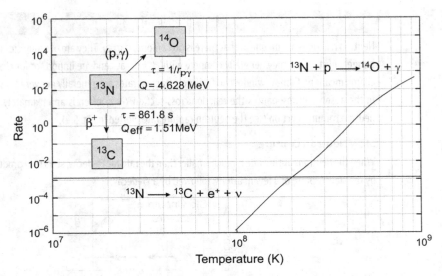

Fig. 5.10 Rates as a function of temperature for competition of proton capture and β-decay in the CNO cycle. Rate units for the 2-body reaction are cm^3 mole^{-1} s^{-1}, and for the decay s^{-1}. To compare, multiply the 2-body rate by ρY_p to convert to units of s^{-1}.

Example 5.7 Consider the competition between β^+ decay and a (p,γ) reaction for ^{13}N. The β-decay rate is independent of temperature but the (p,γ) reaction has a strong temperature dependence because of the Coulomb barrier. Figure 5.10 and 6.3 illustrate using realistic rates. Let's estimate the average time to destroy ^{13}N by these routes in two scenarios:

1. For typical CNO burning $T_6 \simeq 20$ and $\rho \simeq 100$ g cm^{-3}. From Fig. 6.3 the mean time to destroy ^{13}N by (p,γ) is $\tau_{p\gamma} \sim 10^6$ yr and $\tau_\beta \sim 14.4$ minutes, so almost all ^{13}N produced will β-decay. The Q-value is 2.22 MeV, which is shared between the β-particle and the neutrino. The neutrino escapes with an average energy 0.71 MeV, so the effective Q-value is $2.22 - 0.71 = 1.51$ MeV.
2. In a typical nova, $T_8 \simeq 3$ and $\rho \simeq 100$–1000 g cm^{-3}. Now the rate for ^{13}N$(p,\gamma)^{14}$O can exceed the rate for β-decay so the charged particle reaction dominates the β-decay. The Q-value for ^{13}N$(p,\gamma)^{14}$O is 4.628 MeV and in this case the (p,γ) reaction populating ^{14}O leads to a *breakout* from the CNO cycle into a broader set of reactions called the *hot-CNO cycle* that will be discussed in Section 19.1.1.

Thus, competition of weak interactions with other reactions can determine both energy release and element production, in a highly temperature-dependent way.

Finally, β^+ decay, β^- decay, and electron capture are the only low-energy reactions that interconvert neutrons and protons in stars. Thus the weak interactions are responsible for establishing the ratio of protons and electrons to neutrons in stellar material.

5.12 Reaction Selection Rules

Sometimes the importance of various nuclear reactions may be estimated based on *conservation laws* and *selection rules*, without invoking detailed calculations.

5.12.1 Angular Momentum Conservation

Angular momentum is conserved always. Thus the angular momentum J of a state populated in a 2-body reaction must satisfy

$$j_1 + j_2 + l = J, \tag{5.41}$$

where j_1 and j_2 are the angular momenta associated with the colliding particles and l is the angular momentum of relative orbital motion.

5.12.2 Isotopic Spin Conservation

The quantity called *isotopic spin*[8] is conserved to a high degree in strong interactions and the isotopic spins in a 2-body reaction must approximately satisfy

$$t_1 + t_2 = T, \tag{5.42}$$

where t_1 and t_2 are the isotopic spin vectors associated with the colliding particles and T is the isotopic spin vector of the state that is populated in the reaction. Unlike angular momentum symmetry, which is exact, isotopic spin symmetry is often broken at the several percent level, so isospin selection rules are only approximate.

5.12.3 Parity Conservation

Parity is the symmetry of the quantum-mechanical wavefunction under inversion of the spatial coordinate system. It is maximally broken in the weak interactions but is conserved in the strong and electromagnetic reactions. Moreover, there is a relationship between orbital angular momentum and the parity of nuclear states: in a 2-body nuclear reaction that does not involve the weak force the parities must satisfy

$$(-1)^l \pi(j_1) \pi(j_2) = \pi(J), \tag{5.43}$$

---

[8] Isotopic spin (often termed *isospin*) is a quantum number associated with an abstract approximate symmetry of the strong interactions. It has nothing whatsoever to do with ordinary spin physically, but behaves mathematically exactly as if it were a spin. Isospin symmetry implies that in a certain sense the neutron and the proton are different projections of the same particle, just as the spin up and spin down states of a spin-$\frac{1}{2}$ electron are normally viewed as two different projections of the same spin. Isospin is important in the present context because often it is approximately conserved by reactions in nuclear physics, and thus places quantum-mechanical selection rules on which reactions can take place with significant probability (similar to conservation of angular momentum imposing various quantum-mechanical selection rules on permissible transitions in atomic physics).

where $\pi = \pm$ denotes the parity of the states labeled by angular momentum quantum numbers corresponding to Eq. (5.41). Thus, compound nucleus states with angular momentum, isospin, and parity quantum numbers that do not satisfy the conditions implied by equations (5.41)–(5.43) will not be populated strongly in reactions.

For nuclei with even numbers of protons and neutrons (*even–even nuclei*) the ground states always have angular momentum and parity $J^\pi = 0^+$. Therefore, if the colliding particles in the entrance channel are even–even nuclei in their ground states, the angular momentum J and parity π of the state excited in the compound nucleus are *both* determined completely by the orbital angular momentum l of the entrance channel:

$$J = l \qquad \pi(J) = (-1)^l. \tag{5.44}$$

Resonance states satisfying this condition are said to be states of *natural parity*.

Example 5.8 Consider the reaction $\alpha + {}^{16}\text{O} \to {}^{20}\text{Ne}^*$ (where * indicates an excited state). Under normal astrophysical conditions the α-particle and ${}^{16}\text{O}$ will be in their ground states and thus will each have $J^\pi = 0^+$. Therefore, parity conservation requires that any state excited in ${}^{20}\text{Ne}$ by this reaction have parity

$$\pi({}^{20}\text{Ne}) = (-1)^l = (-1)^J.$$

Hence states in ${}^{20}\text{Ne}$ having $J^\pi = 0^+, 1^-, 2^+, 3^-, \ldots$ may be populated because they are natural parity, but population of states having (say) $J^\pi = 2^-$ or 3^+ is forbidden by parity conservation. In the ${}^{20}\text{Ne}$ spectrum there is a state at 4.97 MeV of excitation relative to the ground state having $J^\pi = 2^-$. This state cannot be excited significantly in the capture reaction ${}^{16}\text{O}(\alpha,\gamma){}^{20}\text{Ne}$ because it is not a natural parity state. As we shall explain later, this simple fact has an enormous influence on the relative abundance of ${}^{16}\text{O}$ and ${}^{20}\text{Ne}$ in the Universe (see Section 6.4.2).

Background and Further Reading

Many of the basic principles for reactions and element production in stars were laid down in seminal work by Burbidge, Burbidge, Fowler, and Hoyle (often referred to as the B²FH paper) [62] and Cameron [65] in the late 1950s. Much of the current standard formalism derives from the work of William Fowler (1911–1995) and collaborators; see for example Caughlan and Fowler [69]; and Fowler, Caughlan, and Zimmerman [88]. More concise and pedagogical textbook versions of this material may be found in Hansen, Kawaler, and Trimble [107]; Phillips [169]; Rolfs and Rodney [188]; Pagel [166]; Iliades [128]; and Ryan and Norton [192]. The proton–proton chains are discussed in Bahcall [30]. Weak interaction effects in stars were touched upon but not covered in great depth in this chapter. That will be remedied partially in later chapters, and an extensive review assuming some knowledge of nuclear physics may be found in Langanke and Martínez-Pinedo [139].

Problems

5.1 Using atomic mass tables, calculate the amount of energy released from the fusion of one gram of ^{1}H into ^{4}He. What percentage is this of the original rest mass energy of the hydrogen?***

5.2 The binding energy per nucleon for ^{4}He is 7.074 MeV and that of ^{12}C is 7.6802 MeV. What is the Q-value for the reaction ^{4}He + ^{4}He + ^{4}He $\to$ ^{12}C in MeV? How much energy is released if one gram of ^{4}He is converted to ^{12}C?

5.3 According to certain model calculations the elemental composition of the Sun at formation was $X = 0.71$, $Y = 0.27$, and $Z = 0.02$, while presently typical solar models indicate that the core has a composition $X = 0.34$, $Y = 0.64$, and $Z = 0.02$. What are the mean molecular weights at these two times?***

5.4 Show that the maximum of the Gamow factor is located at an energy

$$E_0 = 1.22(Z_\alpha^2 Z_X^2 \mu T_6^2)^{1/3} \text{ keV},$$

where Z denotes the atomic number for the relevant ion, μ is the mean molecular weight, and T_6 is the temperature in units of 10^6 K. Show that for the reaction p + ^{12}C $\to$ ^{13}N + γ, the Gamow peak is at $E_0 \simeq 31$ keV for $T_6 = 22.5$, and that the height of the peak varies as $\sim T^{17}$ for this reaction in this temperature range.***

5.5 Derive an approximate expression for the width Δ of the Gamow peak and the velocity averaged cross section $\langle \sigma v \rangle_{\alpha\beta}$ in a nonresonant reaction by approximating the Gamow peak as a gaussian of the same height and curvature at the maximum as the actual Gamow peak.***

5.6 Use the Caughlan and Fowler compilation in Ref. [88] to find a formula for the rate of the reaction ^{12}C(p,γ)^{13}N as a function of temperature in units of cm^3 s^{-1}; include resonant and nonresonant contributions. *Hint*: See Appendix D.

5.7 If the CN part of the CNO cycle is running at equilibrium under central solar conditions, what is the ratio of abundances Y_i for ^{13}C versus ^{12}C, ^{14}N versus ^{12}C, and ^{15}N versus ^{14}N? *Hint*: Use the rates in Fig. 6.3(a).

5.8 Show that for the rate-controlling step of the PP chain the temperature exponent is

$$\nu_{pp} = 11.3 T_6^{-1/3} - \tfrac{2}{3},$$

so that $\nu_{pp} \simeq 4$ for $T_6 = 15$.***

5.9 By summing net reactants and products around the cycles implied by Eqs. (5.10)–(5.11) starting at different parts of the cycle, convince yourself that any of the carbon, nitrogen, or oxygen isotopes appearing in the CNO cycle may be viewed as a catalyst promoting the effective fusion reaction 4^1H $\to$ ^{4}He.***

5.10 Suppose the Sun were capable of converting all its hydrogen to iron by a sequence of nuclear reactions. How much energy would be released? To what radius would the Sun have to shrink in order to generate the same amount of energy by gravitational contraction? What would be the Sun's density following this contraction?

5.11 Look up the mass excesses of ^{4}He and ^{20}Ne in Ref. [29] and use these to compute the corresponding masses in atomic mass units. Check your results against the tabulated values in Ref. [29].

5.12 Show by two different methods that the energy production rate per unit volume in the core of the Sun is remarkably low – only several hundred watts per cubic meter.

(a) First, show this using the tabulated rates for the rate-controlling step of the PP chain and the average energy released by the PP chain, in conjunction with expected conditions in the core of the Sun.

(b) Second, show this using the observed luminosity of the Sun, the assumption that it is in equilibrium, and the assumption that most of the energy responsible for this luminosity is generated in approximately the inner 10% of the solar radius.***

5.13 For the reaction ^{12}C(p, γ)^{13}N in Example 5.6, show that the excited state at 2.37 MeV in ^{13}N should give a resonance at a laboratory proton energy of 46 keV (as observed in Fig. 5.8), assuming validity of the Breit–Wigner formula (5.33).***

6 Stellar Burning Processes

We will now use the reaction formalism developed in Chapter 5 to address in more technical depth the most important thermonuclear reaction sequences found in stars. These sequences will entail two intrinsically linked components, each of fundamental importance: (1) the release of energy and the corresponding influence on stellar evolution, and (2) the thermonuclear transmutation of elements into new elements. This discussion will address first the PP chains and the CNO cycle that power stars on the main sequence, and the triple-α process that is the primary thermonuclear energy source for red giant stars. Then we will consider advanced burning stages for stars, including the burning of carbon, oxygen, neon, and silicon, which finally lead to the iron group of nuclei as end products. We will find that this is the end of the line for producing heavier elements by charged-particle reactions in normal stars, but we will introduce capture of (uncharged) neutrons as a possible means to produce even heavier elements.

6.1 Reactions of the Proton–Proton Chains

It is natural to begin with the PP chains that power our local star. The reaction rates of primary importance in the PP chains are displayed in Fig. 6.1 and the steps of the PP-I chain are illustrated pictorially in Fig. 6.2.

6.1.1 Reactions of PP-I

The slowest reaction in PP-I and in the overall PP chains, and therefore the one that governs the rate at which the chains produce energy, is the initial step $^1\text{H} + {}^1\text{H} \rightarrow {}^2\text{H} + e^+ + \nu_e$.[1] This reaction is nonresonant and from the tabulation of reactions rates in Caughlan and Fowler [69], using the parameterization in Eq. (D.1) of Appendix D with only the first term in the parentheses retained, the reaction rate is [107]

$$r_{pp} = \tfrac{1}{2}n_p^2 \langle \sigma v \rangle_{pp} = \frac{1.15 \times 10^9}{T_9^{2/3}} X^2 \rho^2 \exp(-3.38/T_9^{1/3}) \, \text{cm}^{-3}\,\text{s}^{-1}, \qquad (6.1)$$

---

[1] Figure 6.1 indicates that the rate for $^1\text{H} + {}^1\text{H} \rightarrow {}^2\text{H} + e^+ + \nu_e$ is very small. This is because it proceeds by the weak interaction rather than strong interaction (the appearance of a neutrino is the dead giveaway that it is a weak interaction). There is a reason why weak interactions are called weak: their reaction rates are typically many orders of magnitude smaller than corresponding strong interaction rates.

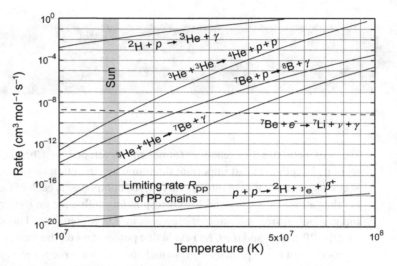

Fig. 6.1 Reaction rates important in PP-I, PP-II, and PP-III. Rates were obtained from a compilation described in Appendix D.2. To convert these rates into units of s^{-1} they should be multiplied by factors of density times abundance, as discussed in Appendix D.

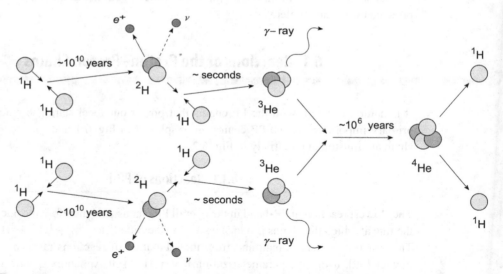

Fig. 6.2 The strongest branch PP-I of the proton–proton chains in Fig. 5.2. The mean times for each reaction in the chain are shown; the inverses of these times indicate the relative rates for each step under solar conditions. About 85% of the Sun's energy is currently being produced by this sequence of reactions.

where X is the hydrogen mass fraction and Eq. (5.19) has been used, with account taken that the reactants are identical particles [the factor of $\frac{1}{2}$, which comes from $\eta_{\alpha X}$ in Eq. (5.16)]. As shown in Problem 5.8, the temperature exponent (5.40) corresponding to the rate in Eq. (6.1) is $\nu_{pp} \simeq 4$ for $T_6 = 15$, which justifies the entry in Table 5.2 for the PP chains and the label in Fig. 5.4 indicating that the solar energy production rate is proportional to T^4. The mean lifetime of a proton against depletion by the initial step of the PP chains may be estimated from Eq. (D.12) of Appendix D,

$$\frac{dY_p}{dt} = -\lambda_p Y_p = -\frac{1}{\tau_p} Y_p, \qquad \tau_p = \frac{1}{\rho R_{pp} Y_p}, \tag{6.2}$$

where λ_p is the rate of proton depletion, $\tau_p = 1/\lambda_p$ is the corresponding mean life, and R_{pp} is defined in Eqs. (D.2)–(D.3) and is plotted versus temperature in Fig. 6.1.

Example 6.1 From Eq. (6.2) the mean life for the proton abundance as depleted by the rate-determining step of the PP chains may be estimated as $\tau_p \sim (\rho R_{pp} Y_p)^{-1}$. Taking for the center of the Sun a temperature of $T_6 = 15$, a density of $\rho = 150 \, \text{g cm}^{-3}$, a hydrogen abundance of $Y_p = 0.4$, and reading off from Fig. 6.1 that $R_{pp} \sim 9 \times 10^{-20} \, \text{cm}^3 \, \text{mol}^{-1} \, \text{s}^{-1}$ at the current central solar temperature, yields $\tau_p \simeq 6 \times 10^9$ years. This is remarkably long and sets the scale for the main sequence life of the Sun because it approximates how long it will take the Sun to burn its available hydrogen. This time is commonly termed the *nuclear burning timescale* for the star. More generally, nuclear burning timescales for other fuels burned later in stellar evolution may also be defined (see Section 13.6.1).

The fundamental reason that the initial step in the PP chains is so slow and thus that the lifetime of the Sun is so long is that the diproton (^{2}He) is not a bound system (of the three possible 2-nucleon systems, only the deuteron, ^{2}H, is bound; the diproton and dineutron are unbound). If the diproton were bound, the first step of the PP chains could be a strong interaction and the lifetime would be much shorter. Instead, the first step must wait for a highly improbable event: a weak decay of a proton from a broad p–p resonance having a very short lifetime.

In contrast, Fig. 6.1 indicates that at current solar temperatures the reaction $p + d \rightarrow {}^3\text{He} + \gamma$ corresponding to the second step of the PP chains is about 17 orders of magnitude faster than the initial rate-determining step (because it is a strong rather than weak interaction). Since the mean life of a proton in the first step was determined to be $\sim 10^{10}$ yr $\sim 10^{17}$ s, the mean life for the deuterium produced in the first step of the PP chains and consumed in the next step ($p + d \rightarrow {}^3\text{He} + \gamma$) is a few seconds under the conditions prevailing in the solar core (see Problem 10.11 for a quantitative estimate). The final fusion of two helium-3 isotopes to form helium-4 is much slower than the second step ($\tau \sim 10^6$ years), but is still a strong interaction that is orders of magnitude faster than the first step (Problem 10.11). Thus, the initial step of the PP chains governs the overall rate of the reaction and in turn sets the main sequence lifetime for stars that are running on the PP chains.

6.1.2 Branching for PP-II and PP-III

From Fig. 5.2, the relative importance of PP-I versus PP-II and PP-III depends on the competition between the reactions $^3\text{He}(^3\text{He}, 2p)^4\text{He}$ and $^3\text{He}(^4\text{He}, \gamma)^7\text{Be}$. The relevant velocity-averaged rates of Fig. 6.1 may be used in Eq. (5.19) to determine the total rate per unit volume for each reaction. Such an analysis indicates that over the range of temperatures where PP is expected to be important the first reaction is faster than the second, ensuring the dominance of PP-I over PP-II and PP-III for current conditions in the Sun.

From Fig. 5.2, the branching between PP-II and PP-III depends on the competition between electron capture and radiative proton capture on ^{7}Be. These rates are also plotted in Fig. 6.1. The electron capture changes slowly with temperature because it depends on the electron density in the plasma; in contrast, the proton capture on ^{7}Be has a strong dependence on temperature because it is a barrier penetration process. At the temperature of the Sun, electron capture dominates and PP-II is much stronger than PP-III. At somewhat higher temperatures, Fig. 6.1 indicates that PP-III would begin to make much larger relative contributions. (However, at higher temperatures the CNO cycle would quickly become more important than the PP chains in stellar energy production.)

6.1.3 Effective Q-Values

The effective Q-values for the PP chains depends on whether PP-I, PP-II, or PP-III is followed because the amount of energy carried off by neutrinos differs among these subchains. The effective Q-values are listed in Table 6.1. Utilizing this information, the average energy released per PP chain proton fusion in the Sun is

$$\overline{\Delta E}_{\rm pp} = 0.85 \left(\frac{26.2}{2} \right) + (0.15)(25.7) = 15 \, {\rm MeV},$$

where the tiny contribution of PP-III to the energy production has been ignored and where the factor of two in the denominator of the first term results from the requirement that the first two steps of PP-I must run twice to provide the two ^{3}He isotopes required for the last step (see Fig. 6.2).

Although PP-III has negligible influence on solar energy production, the values of $Q_{\rm eff}$ in Table 6.1 indicate that it produces much higher energy neutrinos than PP-I or PP-II. The PP-III chain is strongly temperature dependent because it is initiated by proton capture on a $Z = 4$ nucleus, which has a relatively high Coulomb barrier. Therefore, detection of the high-energy neutrinos emitted from the PP-III chain can provide a very sensitive probe of the central temperature of the Sun. This issue will be revisited when the production of solar neutrinos is discussed in Chapter 10.

Table 6.1 Some effective Q-values

Process	$Q_{\rm eff}$ (MeV)	% Solar energy
PP-I	26.2	83.7
PP-II	25.7	14.7
PP-III	19.1	0.02
CNO	23.8	1.6

6.2 Reactions of the CNO Cycle

The important rates in the main part of the CNO cycle are plotted as a function of temperature in Fig. 6.3(a), which indicates that the slowest reaction typically is $p + {}^{14}N \rightarrow {}^{15}O + \gamma$. Since this reaction is the slowest and the cycle is closed (neglecting small branching), it determines the rate at which the CNO cycle generates energy.

Example 6.2 The results in Fig. 6.3 may be used to estimate the rate of CNO energy generation in the Sun. The mean life for ^{14}N in the core of the Sun is approximately $\tau_{14-N} = 5 \times 10^8$ years (see Section D.2.4). The abundance of ^{14}N presently at the core of the Sun is estimated[2] from the calculation in Fig. 6.3(b) to be $Y_{14N} = 3.35 \times 10^{-4}$,

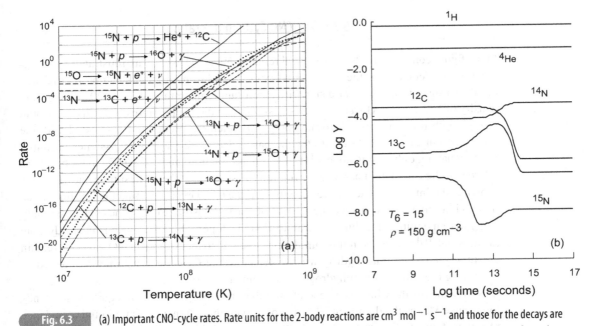

Fig. 6.3 (a) Important CNO-cycle rates. Rate units for the 2-body reactions are $cm^3\ mol^{-1}\ s^{-1}$ and those for the decays are s^{-1}. To compare in equivalent units, the 2-body rates should be multiplied by a density and abundance factor (see Appendix D). The reaction $p + {}^{13}N \rightarrow {}^{14}O + \gamma$ breaks out of the CNO cycle into the hot CNO cycle that powers nova explosions, as illustrated in Fig. 19.3. (b) Abundances in the CNO cycle at the center of the Sun as a function of time. Initial solar abundances were assumed and equations of the form described in Box 6.1 were integrated at a constant temperature $T_6 = 15$ and constant density $\rho = 150\ g\ cm^{-3}$ corresponding approximately to the present Sun.

---

[2] The CNO rate estimate will be intrinsically more uncertain than that of the PP chains, largely because of uncertainty in how much ^{14}N is in the center of the Sun. Although the *sum of abundances* for the CNO isotopes is conserved by the cycle and thus may be well represented by present surface abundances, as the cycle runs it tends to convert the other CNO isotopes into ^{14}N (see the figure in Box 6.1). Thus the present amount of ^{14}N depends on the detailed history of the Sun. A simulation to estimate the current abundance of ^{14}N in the

corresponding to a number density of 2.6×10^{22} cm^{-3}, while from the Standard Solar Model to be described in Section 10.1 the central hydrogen abundance is $Y_H \sim 0.35$, corresponding to a hydrogen concentration of 3.2×10^{25} cm^{-3}, and in Example 6.1 the mean life for consumption of a proton by PP chain fusion was estimated to be about 6×10^9 years. These numbers imply that the ratio of PP chain to CNO cycle reactions in the core of the Sun is approximately

$$\left(\frac{\text{rate for PP}}{\text{rate for }^{14}\text{N} + \text{p}}\right) = \left(\frac{\tau_{14\text{-N}}}{\tau_{\text{pp}}}\right)\left(\frac{3.2 \times 10^{25}}{2.6 \times 10^{22}}\right) \simeq 100.$$

Hence for conditions prevailing in the Sun the PP chains dominate over the CNO cycle.

More sophisticated calculations within the Standard Solar Model indicate that the Sun is producing 98.4% of its energy from the PP chains and only 1.6% from the CNO cycle, corroborating the simple estimate of Example 6.2.

6.2.1 The CNO Cycle in Operation

In the figure contained in Box 6.1 several remarks made in Box 5.1 are illustrated by implementing a calculation of the CNO abundances carried to hydrogen depletion (conversion of all hydrogen to helium) for a star with a constant temperature of $T_6 = 20$ and constant density of 100 g cm^{-3}. (In a more realistic simulation the temperatures and densities would change with time in response to the energy and element production associated with the reactions, but this example illustrates the basic idea.) The calculation assumes an initial mixture having only two isotopes: ^{1}H (0.995 mass fraction) and ^{12}C (0.005 mass fraction). (Note that the calculation has already run for 10^4 seconds at the earliest time shown.) Also shown is the integrated energy production from the CNO cycle as a function of time. Even though initially there is only a trace amount of one CNO isotope (^{12}C), the cycle eventually generates an equilibrium abundance of all CNO isotopes. Once the cycle is in steady state (after about 10^{12} seconds in this simulation), the abundances of the CNO isotopes remain essentially constant until hydrogen is depleted, so the CNO isotopes may be viewed as catalyzing the conversion of hydrogen to helium, as discussed earlier in Box 5.1.

Notice the result (which is a general one) that the CNO cycle run to equilibration tends to produce ^{14}N as the dominant CNO isotope, even though there was no initial abundance of this isotope at all in this simulation (to understand this, see Problem 6.4). Most of the ^{14}N found in the Universe probably has been produced by the CNO cycle. Finally, note that the catalytic role of the CNO isotopes does not mean that the abundances Y_i are conserved separately; for example, the abundance of ^{14}N grows at the expense of the other CNO isotopes as the cycle runs. The meaning of CNO catalysis is that the *sum of the abundances* Y_i for the CNO isotopes is conserved by the cycle, as illustrated by the dotted line labeled $\sum Y_{\text{CNO}}$. Once steady state is reached, then the CNO isotopes also become conserved individually.

Sun's core is displayed in Fig. 6.3(b). With the assumptions used there, over the age of the Sun almost all CNO abundance has been converted to ^{14}N, with a current ^{14}N abundance of $Y_{14\text{N}} = 3.35 \times 10^{-4}$, corresponding to a number density of 2.6×10^{22} cm^{-3}.

Box 6.1   Thermonuclear Burning Networks

The evolution of isotopic abundances over time in an astrophysical environment is described by a coupled set of differential equations called a *thermonuclear burning network* that describes transitions among a set of isotopes coupled by various reactions. The general form of such a network coupling N isotopic species is

$$dY_1/dt = F_+^{(1)}(\mathbf{Y}) - F_-^{(1)}(\mathbf{Y})$$
$$dY_2/dt = F_+^{(2)}(\mathbf{Y}) - F_-^{(2)}(\mathbf{Y})$$
$$\vdots$$
$$dY_N/dt = F_+^{(N)}(\mathbf{Y}) - F_-^{(N)}(\mathbf{Y}),$$

where $\mathbf{Y} = (Y_1, Y_2, \ldots Y_N)$ is a vector of isotopic abundances, $F_+^{(i)}(\mathbf{Y})$ denotes a sum of terms increasing Y_i, $F_-^{(i)}(\mathbf{Y})$ denotes a sum of terms decreasing Y_i, and the notation indicates that these terms depend on the other Y_j in the system (which couples the equations). Appendix D describes these equations in more depth.

A numerical solution of such a network for evolution of the isotopic abundances in the main CNO cycle is displayed in the following figure.

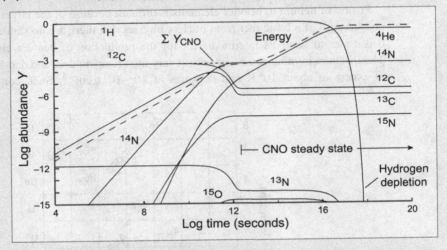

Constant $T_6 = 20$ and $\rho = 100$ g cm^{-3} were assumed, with only ^{1}H ($X_{1H} = 0.995$) and ^{12}C ($X_{12C} = 0.005$) present initially. The dashed line shows integrated energy release (arbitrary scale). The sum of CNO abundances is given by the dotted line.

6.2.2 Rate of CNO Energy Production

The effective Q-value for the CNO cycle is 23.8 MeV (Table 6.1). The rate of energy production is [107]

$$\varepsilon_{\text{CNO}} = \frac{4.4 \times 10^{25} \rho XZ}{T_9^{2/3}} \exp(-15.228/T_9^{1/3}) \,\text{erg g}^{-1}\,\text{s}^{-1}, \tag{6.3}$$

where X is the hydrogen mass fraction and Z the mass fraction of metals, which corresponds to a temperature exponent $\nu_{\rm CNO} \simeq 18$ for $T_6 = 20$. This very strong temperature dependence implies that, were the Sun only slightly hotter, the CNO cycle instead of the PP chains would be the dominant energy production mechanism (see Fig. 5.4).

6.3 The Triple-α Process

Main sequence stars produce their energy by converting hydrogen to helium using either the PP chains or the CNO cycle, which builds up a thermonuclear ash of helium in the core of the star. Nothing heavier than helium is produced in significant amounts by main-sequence burning.[3] The star continues to burn hydrogen to helium in a shell surrounding the central core of helium that is built up. This *hydrogen shell burning* adds gradually to the accumulating core of helium and the star remains on the main sequence until about 10% of its initial hydrogen has been consumed.

Fusion of helium to heavier elements is difficult because of the larger Coulomb barrier, and because of a basic fact from nuclear physics that there are no stable mass-5 or mass-8 isotopes to serve as intermediaries for the production of heavier elements (Fig. 6.4). Thus, helium burning can occur only at very high temperatures and densities: temperatures in excess of about 10^8 K and densities of 10^2–10^5 g cm^{-3}. Such conditions can result

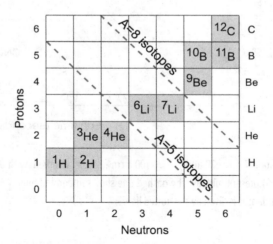

Fig. 6.4 The shaded boxes indicate stable isotopes and unshaded boxes indicate unstable isotopes. The unshaded boxes along the two dashed diagonal lines illustrate that there are no stable mass-5 or mass-8 isotopes. This basic fact of nuclear physics has large implications both for nucleosynthesis in the big bang and for production of carbon in red giant stars.

---

[3] The CNO cycle requires an initial abundance of carbon, nitrogen, and oxygen isotopes to serve as catalysts and alters their relative abundance, but the *total abundance* of these isotopes is conserved by the CNO cycle. Likewise, PP-II and PP-III involve $Z = 3, 4$, and 5 isotopes in intermediate steps, but not as final products. Thus neither of the primary hydrogen-burning sequences leads to net production of isotopes heavier than ^{4}He.

when stars exhaust their hydrogen fuel and their cores begin to contract (see Chapter 13). Because there are no stable mass-8 isotopes, the resulting fusion of helium must involve a two-step process in which two helium ions (α-particles) combine to form highly unstable ^{8}Be, and this in turn combines with another helium ion to form carbon. The resulting sequence, which is crucial to the power generated by red giant stars and to the production of most of the carbon and oxygen in the Universe, is called the *triple-α process*. Our bodies are composed of about 65% oxygen and 18% carbon, so the triple-α process is of more than academic interest! This will be discussed further in Section 6.4.2. The burning of helium to carbon by the triple-α process may be viewed as taking place in three steps:

1. *Formation of a transient ^{8}Be population*: A small transient population of ^{8}Be is built up by the reaction ^{4}He + ^{4}He $\rightleftharpoons$ ^{8}Be. (The forward reaction is the inverse of the terminating reaction in the PP-III chain; see Fig. 5.2.)
2. *Formation of a transient population of ^{12}C in an excited state*: A small population of ^{12}C in an excited state is built up by the reaction ^{4}He + ^{8}Be $\rightleftharpoons$ ^{12}C*. To produce a finite population of ^{12}C* this reaction must be resonant; otherwise it would be too slow to compete with the decay of ^{8}Be back to two α-particles.
3. *Electromagnetic decay of the excited state of ^{12}C to its ground state*: A small fraction of the ^{12}C* excited states decay electromagnetically by ^{12}C* $\rightarrow$ ^{12}C + 2γ to the ground state of carbon-12.

This (highly improbable) sequence of reactions has the net effect of converting three helium ions to ^{12}C, with an energy release of $Q = +7.275$ MeV. We now consider each of these steps of the triple-α mechanism in more detail, guided by the discussion of helium burning in Refs. [107, 169, 188].

6.3.1 The Equilibrium Population of ^{8}Be

The mass of ^{8}Be is 92 keV greater than the mass of two free α-particles and ^{8}Be is unstable against decaying back to α-particles, releasing 92 keV of energy. The width for this decay is $\Gamma_{\text{8-Be}} = \hbar\tau^{-1} = 6.8$ eV, which corresponds to a mean life for ^{8}Be of $\tau \simeq 10^{-16}$ seconds. This is a very short lifetime but it is much larger than the average time between collisions for α-particles in a hot stellar plasma. However, the capture to produce ^{8}Be will be too slow to compete with the decay back into α-particles unless the corresponding resonance peak overlaps substantially with the Gamow peak. Thus, this initial step of the triple-α process is expected to be significant only when the energy of the Gamow peak is comparable to the mass difference of 92 keV between ^{8}Be and two α-particles, and this in turn sets the required conditions for triple-α to proceed. The energy E_0 corresponding to the maximum of the Gamow peak is given by Eq. (5.30), which implies a temperature of 1.2×10^8 K for $E_0 = 92$ keV. Therefore, only for such temperatures can the initial step of the triple-α reaction produce a sufficient equilibrium concentration of ^{8}Be to allow the subsequent steps to proceed.[4]

---

[4] These simple considerations ignore details such as electron screening that must be included for a more precise estimate of the temperature for helium burning, but it sets the correct order of magnitude. This temperature estimate also prompts the question of why helium was not consumed in big bang nucleosynthesis by the triple-α mechanism. The answer is that during the short period of big bang nucleosynthesis the temperature was

The equilibrium concentration of ^{8}Be may be estimated by viewing ^{8}Be $\to \alpha + \alpha$ as the "ionization" of ^{8}Be and applying to nuclei a suitable modification of the Saha equations introduced in Section 2.1.2 to describe the dependence of this "ionization" on temperature and density. The required changes are [107]

1. Replace the number densities of ions and electrons with the number densities of α-particles, and the number density of neutral atoms with the number density of ^{8}Be.
2. Replace the statistical factors g for atoms with corresponding statistical factors associated with nuclei. This is trivial for the present case since the ground states of both ^{8}Be and ^{4}He have angular momentum zero and $g = 1$ for both.
3. Replace the ionization potentials entering the atomic Saha equations by Q-values in the nuclear case. In the present example, $Q = 91.8$ keV for ^{8}Be $\to \alpha\alpha$ ("ionization" of ^{8}Be to two α-particles).
4. Replace the electron mass entering the atomic Saha equations by the reduced mass $m_\alpha m_\alpha / 2 m_\alpha = \frac{1}{2} m_\alpha$.

The resulting *nuclear Saha equation* is (see Problem 6.1 and reference [107])

$$\frac{n_\alpha^2}{n(^8\text{Be})} = \left(\frac{\pi m_\alpha kT}{h^2} \right)^{3/2} \exp(-Q/kT). \tag{6.4}$$

The conditions under which such equations apply are termed *nuclear statistical equilibrium (NSE)*. Nuclear statistical equilibrium will figure prominently in energy and element production processes for stars once they move beyond the hydrogen-burning phase.

Example 6.3 *Helium flashes* are explosive helium-burning events in red giant stars that will be described in Section 13.5.3. Typical helium flash conditions in lower-mass red giants correspond to a temperature of $T_9 \simeq 0.1$ and a density of $\rho \simeq 10^6$ g cm^{-3}. As shown in Problem 6.2, for a helium flash in a pure helium core Eq. (6.4) yields $n(^8\text{Be})/n_\alpha = 7 \times 10^{-9}$, corresponding to an equilibrium ^{8}Be number density of 10^{21} cm^{-3} during the flash.

From the preceding example we see that under the right conditions the first step of the triple-α sequence can produce a small equilibrium abundance of ^{8}Be, which sets the stage for the second step.

6.3.2 Formation of the Excited State in ^{12}C

The second step of the triple-α process, ^{8}Be$(\alpha,\gamma)^{12}$C*, has $Q = 7.367$ MeV and proceeds through a $J^\pi = 0^+$ resonance in ^{12}C at an excitation energy relative to the ^{12}C ground state of 7.654 MeV, as illustrated in Fig. 6.5(a). This state is called the *Hoyle resonance* because its existence was *predicted* by Fred Hoyle (1915–2001) as a necessary condition for red giant stars to produce their energy. As illustrated in Fig. 6.5(b), the population of this state

high enough but not the density. Both high temperatures and high densities, which had to await the formation of stars, were required to produce significant amounts of carbon by the triple-α mechanism. A more extensive discussion of big bang nucleosynthesis may be found in Chapter 20 of Ref. [100].

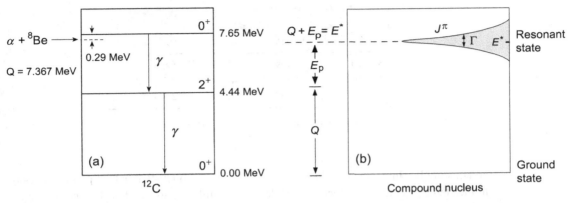

Fig. 6.5 (a) Nuclear energy levels in ^{12}C labeled by J^π and excitation energy for the final steps of the triple-α reaction. The 0^+ state at 7.65 MeV is the Hoyle resonance. (b) Relationship of Q-value, resonance energy E^*, and center of mass energy E_p when an isolated resonance is maximally excited in a reaction [see Eqs. (5.32)–(5.34)]. The resonance condition corresponds to the energy $E_p + Q$ matching the excitation energy of the resonance E^*.

is optimized if the center of mass energy plus the Q-value is equal to the resonance energy relative to the ground state of ^{12}C. Once this excited state is formed the dominant reaction is a rapid decay back to $\alpha + {}^8$Be, but a small fraction of the time the ground state of ^{12}C may instead be formed by emission of two γ-ray decays, as also illustrated in Fig. 6.5(a).[5] If nuclear statistical equilibrium is assumed, the concentration of ^{12}C* excited states is given by

$$\frac{n\left({}^{12}\text{C}^*\right)}{n_\alpha^3} = 3^{3/2}\left(\frac{h^2}{2\pi m_\alpha kT}\right)^3 \exp[(3m_\alpha - m_{12}^*)c^2/kT], \tag{6.5}$$

where m_{12}^* is the mass of ^{12}C in the excited state (see Problem 6.1).

6.3.3 Formation of the Ground State in ^{12}C

The preceding considerations determine a dynamical equilibrium

$$^{4}\text{He} + {}^{4}\text{He} + {}^{4}\text{He} \rightleftharpoons {}^{4}\text{He} + {}^{8}\text{Be} \rightleftharpoons {}^{12}\text{C}^*. \tag{6.6}$$

This produces an equilibrium population of ^{12}C*, almost all of which decays back to ^{4}He + ^{8}Be. However, the excited state of ^{12}C can decay electromagnetically to its ground state with a mean life of

$$\tau\left({}^{12}\text{C}^* \to {}^{12}\text{C(gs)}\right) = 1.8 \times 10^{-16}\,\text{s}, \tag{6.7}$$

---

[5] The energy of the excited state relative to the ground state is well above the energy threshold required to form an electron–positron pair ($2 \times 0.511 = 1.022$ MeV). Therefore, the ^{12}C excited state also decays electromagnetically to the ground state by $e^+ e^-$ pair production (but with a much smaller width than for γ-decay, so the dominant decay is by γ-rays). The energy release in pair production appears in the kinetic energy of the e^- and in the γ-rays produced when the e^+ annihilates with an electron.

Table 6.2 Parameters governing the triple-α rate*

T (K)	kT (keV)	q/kT	$\exp(-q/kT)$
5×10^7	4.309	88.08	5.6×10^{-39}
1×10^8	8.617	44.04	7.5×10^{-20}
2×10^8	17.234	22.02	2.7×10^{-10}

*The activation energy from Eq. (6.9) is $q \equiv (m_{12}^* - 3m_\alpha)c^2$.

and this implies that one in about every 2500 excited carbon nuclei that are produced decay to the stable ground state. Because this decay probability is so small, it does not influence the equilibrium appreciably in Eq. (6.6) and the entire triple-α process may be represented schematically as

$$^4\text{He} + {}^4\text{He} + {}^4\text{He} \rightleftharpoons {}^4\text{He} + {}^8\text{Be} \rightleftharpoons {}^{12}\text{C}^* \to {}^{12}\text{C}(\text{gs}), \qquad (6.8)$$

where left–right arrows indicate nuclear statistical equilibrium and the one-way arrow denotes a small leakage from the equilibrium that does not disturb it significantly. The production rate for ^{12}C in its ground state may be approximated then by the product of the equilibrium ^{12}C* population and the decay rate to the ground state [169],

$$\frac{dn\,(^{12}\text{C})}{dt} = \frac{n\,(^{12}\text{C}^*)}{\tau\,(^{12}\text{C}^* \to {}^{12}\text{C}(\text{gs}))}$$

$$= \frac{n_\alpha^3}{\tau\,(^{12}\text{C}^* \to {}^{12}\text{C}(\text{gs}))} \left(\frac{3^{1/2} h^2}{2\pi m_\alpha kT}\right)^3 \exp[-(m_{12}^* - 3m_\alpha)c^2/kT], \qquad (6.9)$$

where Eq. (6.5) and that the decay rate is the inverse of the mean life have been used. Hence the rate of carbon production is governed by temperature, the α-particle density, and:

1. An activation energy $q = (m_{12}^* - 3m_\alpha)c^2 = 379.5$ keV that must be borrowed to create the ^{12}C* intermediate state.
2. The mean life for the decay ^{12}C* $\to$ ^{12}C(gs), which is given by Eq. (6.7).

The strong temperature dependence for the triple-α reaction is a consequence of the exponential factor in Eq. (6.9) because at helium burning temperatures the average thermal energy kT is much less than the activation energy of 379.5 keV. This is illustrated dramatically in Table 6.2, where we see that doubling the temperature in the vicinity of 10^8 K changes the exponential factor by 10–20 orders of magnitude.

6.3.4 Energy Production in the Triple-α Reaction

The total energy released in each triple-α reaction is $Q = -0.0918 + 7.367 = 7.275$ MeV and the energy production rate is given by Eq. (5.38) using the reaction rate from Eq. (6.9). This may be represented conveniently as [107]

$$\varepsilon_{3\alpha} = \frac{5.1 \times 10^8 \rho^2 Y^3}{T_9^3} \exp(-4.4027/T_9)\,\text{erg}\,\text{g}^{-1}\text{s}^{-1}, \qquad (6.10)$$

where Y is the helium mass fraction. From Eq. (5.40), this corresponds to density and temperature exponents

$$\lambda_{3\alpha} = 2 \qquad \nu_{3\alpha} = 4.4 T_9^{-1} - 3, \qquad (6.11)$$

respectively. The density dependence is quadratic because the reaction is effectively 3-body, since three α-particles must be combined to form the carbon. The dependence on temperature is very strong, as was already seen in connection with Table 6.2. For $T_8 = 1$, Eq. (6.11) yields a temperature exponent $\nu_{3\alpha} \simeq 40$. This implies that a helium core constitutes a very explosive fuel,[6] which is a fact of some importance for stellar evolution beyond the main sequence (see, for example, Section 13.5.3).

Example 6.4 The triple-α reaction consumes three ^{4}He nuclei ($\sim$12 amu) to release $Q = 7.275$ MeV of energy. The energy release per unit mass is then

$$\frac{\Delta E}{\Delta M} = \frac{7.275 \text{ MeV}}{(12)(1.66 \times 10^{-24} \text{ g})} = 3.65 \times 10^{23} \text{ MeV g}^{-1} = 5.85 \times 10^{17} \text{ erg g}^{-1}.$$

Comparison with Example 5.3 indicates that the energy release per gram is about 10 times less for the burning of helium to carbon than for the burning of hydrogen to helium. This represents a first example of an important rule of diminishing returns for evolution beyond the main sequence: Each successive burning stage releases less energy per gram of fuel than the preceding stage. Ultimately this follows from the behavior of the curve of binding energy below the iron peak in Fig. 5.1.

6.4 Helium Burning to C, O, and Ne

After carbon has been created by the triple-α reaction sequence, heavier nuclei can be formed by successive α-capture reactions on the carbon. The probability for this to occur will depend on the stellar environment, and on the nuclear structure of the additional isotopes that could be produced.

6.4.1 Oxygen and Neon Production

Oxygen can be produced through the radiative capture reaction

$$^4\text{He} + {}^{12}\text{C} \rightarrow {}^{16}\text{O} + \gamma. \qquad (6.12)$$

There are no resonances near the Gamow window so the rate is low; the currently accepted value is plotted in Fig. 6.6, but it has substantial uncertainty. This is of consequence because the α-capture rate on carbon determines the ratio of C to O production, which

---

[6] An explosion is runaway burning (whether ordinary chemical burning or thermonuclear burning). If a thermonuclear fuel has a large temperature exponent, in the corresponding temperature range the rate of burning can increase enormously if the temperature increases only a little. This greatly enhances the probability that burning becomes explosive once it is initiated.

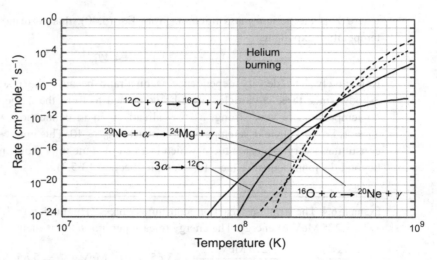

Fig. 6.6 Triple-α and radiative α-capture rates important in helium burning. The vertical gray band indicates the characteristic temperature range for helium burning.

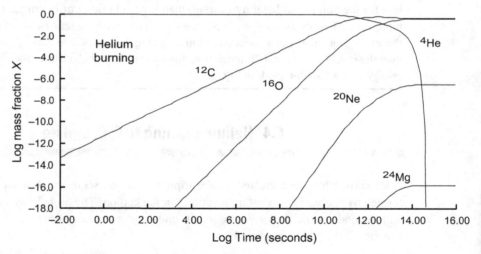

Fig. 6.7 Thermonuclear burning of pure helium using rates from Fig. 6.6 at constant $\rho = 10^4$ g cm^{-3} and $T = 1.5 \times 10^8$ K. Realistically the temperature and density would change with time in response to the burning but this simplified example illustrates the basic features expected for a more sophisticated calculation.

can impact late stellar evolution. For example, the composition of white dwarfs and of the cores of massive stars depend on this rate, so it can influence how stars die and what is left behind when they do. Once oxygen has been produced by the reaction (6.12), neon can be formed by an additional α capture,

$$^4\text{He} + {}^{16}\text{O} \rightarrow {}^{20}\text{Ne} + \gamma. \qquad (6.13)$$

The rate for this reaction also is plotted as a function of temperature in Fig. 6.6. The reaction is slow under helium burning conditions because it is nonresonant and has a larger Coulomb barrier than for (6.12). Hence little neon is produced during helium burning and the primary residue is a carbon–oxygen core. The carbon is produced by the triple-α sequence and the oxygen by radiative capture (6.12) on the carbon, with the ratio of carbon to oxygen depending strongly on the uncertain rate for the radiative capture reaction.

A simulation of helium burning at constant temperature and density is shown in Fig. 6.7. Initially only ^{4}He is present but a population of ^{12}C begins to grow because of its production in the triple-α reaction, with a corresponding depletion of ^{4}He. This population of ^{12}C produces a growing population of ^{16}O by α-capture, and by the time the ^{4}He fuel has been depleted almost equal mass fractions of ^{12}C and ^{16}O have been produced with this choice of parameters. On the other hand, because the rate to produce ^{20}Ne by α-capture on ^{16}O is small, the mass fraction of ^{20}Ne at ^{4}He depletion is 6–7 orders of magnitude below that for carbon and oxygen, and that for ^{24}Mg is suppressed by a factor $\sim 10^{-16}$. This result illustrates concisely that the normal products of helium burning are carbon and oxygen.

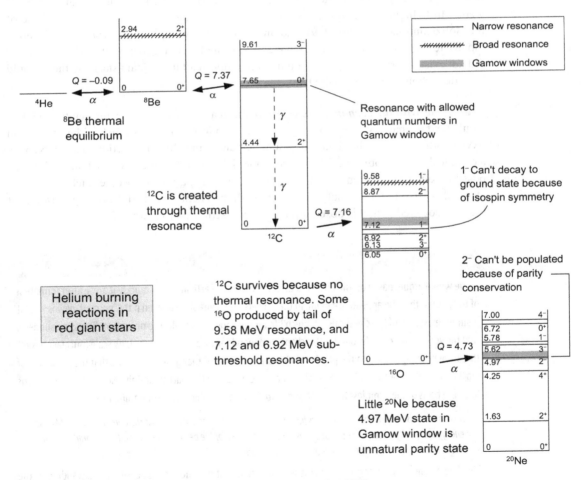

Fig. 6.8 An overview of helium burning adapted from Rolfs and Rodney [188].

6.4.2 The Outcome of Helium Burning

The outcome of helium burning, summarized in Fig. 6.8, illustrates how different the Universe would be if just a few seemingly unremarkable parameters had slightly different values (see also the discussion in Refs. [169] and [188]). The proportion of carbon to oxygen in the Universe is determined by the competition between the carbon-producing triple-α reaction and the carbon-depleting, oxygen-producing radiative capture reaction (6.12). Furthermore, that much carbon or oxygen exists at all depends crucially on the existence of the Hoyle resonance and the slowness of the neon-producing reaction (6.13).

A universe with no carbon: If – contrary to fact – a resonance existed near the energy window for the reaction (6.12), the corresponding rate would be large and almost all carbon produced by triple-α would be converted rapidly to oxygen, leaving little carbon in the Universe. A similar fate would follow if the 0^+ excited state in ^{12}C at 7.65 MeV were a little higher in energy, since this would greatly slow the triple-α rate by virtue of the Boltzmann factor in Eq. (6.9), and any carbon that was produced would be converted rapidly to oxygen through (6.12). Conversely, if the Hoyle resonance at 7.65 MeV in ^{12}C did not exist the triple-α reaction would not function at all in red giant stars and there would be little carbon or oxygen in the Universe.

A universe without carbon or oxygen: If the reaction (6.13) were resonant—which it would be if the parity of the 2^- excited state in neon were positive—much of the carbon and oxygen produced by helium burning could be transformed by this reaction to neon. Neon is a chemically inert noble gas, in contrast to the rich chemistry of carbon that makes biology possible: our very existence seems to depends on the *parity* of obscure nuclear states in atoms that have nothing whatsoever to do with the chemistry of life! Some philosophical issues associated with this observation are discussed further in Box 6.2.

Box 6.2 — **The Anthropic Principle and Helium Burning**

Some would argue, based on the observed diversity of life on Earth and how quickly it arose after formation of the planet, that the appearance of life in the Universe is inevitable. But even if this point of view is valid it assumes existence of the chemicals on which life (as we know it) is built. The preceding discussion suggests that the very existence of the building blocks of life depends on arcane facts on the MeV scale (nuclear physics) that have nothing to do with the physics of eV scales (chemistry) that governs life, and that the possibility of biochemistry may be an accident of physical parameter values in this particular Universe. Such considerations lie at the basis of the (simplest form of the) *anthropic principle,* which goes something like:

> It is not surprising that the Universe has just the right values of constants and just the detailed physics like the energy and parity of obscure nuclear states that are required for life because, if it didn't, there would be no life in the Universe and therefore no one to ask the question.

This line of thinking is intriguing but it is unclear whether it enhances our scientific understanding of the Universe (or other possible universes, provided that 'other universes' makes sense scientifically).

Nuclear fuel	Nuclear products	Ignition temperature	Minimum main sequence mass	Duration in 25 $M_\odot$ star
H	He	4×10^6 K	$0.1\, M_\odot$	7×10^6 years
He	C, O	1.2×10^8 K	$0.4\, M_\odot$	5×10^5 years
C	Ne, Na, Mg, O	6×10^8 K	$4\, M_\odot$	600 years
Ne	O, Mg	1.2×10^9 K	$\sim 8\, M_\odot$	1 years
O	Si, S, P	1.5×10^9 K	$\sim 8\, M_\odot$	~ 0.5 years
Si	Ni–Fe	2.7×10^9 K	$\sim 8\, M_\odot$	~ 1 day

Table 6.3 Burning stages in massive stars [231]

6.5 Advanced Burning Stages

If a star is massive enough, more advanced burnings are possible by virtue of the high temperature and density that results as the core contracts after exhausting successive fuels. Typical burning stages in massive stars are listed in Table 6.3, and are described briefly below (see also Fig. 14.2 and Section 20.3.2); some rates important in advanced burning stages are shown as a function of temperature in Fig. D.3 of Appendix D.

6.5.1 Carbon, Oxygen, and Neon Burning

The carbon and oxygen produced by helium burning can be converted into successively heavier elements through a sequence of reactions that can occur only at the elevated temperatures and densities possible in more-massive stars.

Carbon burning: Carbon begins to burn at a temperature of $T \sim 5 \times 10^8$ K, primarily through the reactions

$$^{12}\text{C} + {}^{12}\text{C} \rightarrow {}^{20}\text{Ne} + \alpha \qquad {}^{12}\text{C} + {}^{12}\text{C} \rightarrow {}^{23}\text{Na} + p \qquad {}^{12}\text{C} + {}^{12}\text{C} \rightarrow {}^{23}\text{Mg} + n.$$

As indicated in Table 6.3, these reactions are possible late in the lives of stars having masses of $4\,M_\odot$ or larger. Burning stages beyond that of carbon require conditions realized only for stars having $M > 8M_\odot$ or so. At the required temperatures a new feature enters because the most energetic photons can disrupt the nuclei produced in earlier burning stages.

Neon burning: At $T \sim 10^9$ K, the neon produced in carbon burning can burn by a two-step process. First, neon is *photodisintegrated* by a high-energy photon (which become more plentiful as the temperature is increased) in the reaction $\gamma + {}^{20}\text{Ne} \rightarrow {}^{16}\text{O} + \alpha$. Then the α-particle produced in this step can initiate a radiative capture reaction $\alpha + {}^{20}\text{Ne} \rightarrow {}^{24}\text{Mg} + \gamma$. This burning sequence leaves a residue of ^{16}O and ^{24}Mg.

Oxygen burning: At a temperature of 2×10^9 K, oxygen can fuse through the reaction $^{16}\text{O} + {}^{16}\text{O} \rightarrow {}^{28}\text{Si} + \alpha$. The silicon that is produced can react only at temperatures where the photon spectrum is sufficiently "hard" (has significant high-energy components) that photodissociation reactions begin to play a dominating role.

6.5.2 Silicon Burning

For $T \sim 3 \times 10^9$ K, silicon may be transformed into heavier elements by a mechanism similar to that of neon burning described above. At these temperatures the photons are quite energetic and those in the high-energy tail of the Maxwell–Boltzmann distribution can readily photodissociate nuclei. A network of photodisintegration and capture reactions in approximate nuclear statistical equilibrium develops and the population in this network evolves preferentially to those isotopes that have the largest binding energies. From Fig. 5.1 the most stable nuclei are in the iron group, so silicon burning carried to completion under equilibrium conditions tends to produce iron-group nuclei.

The silicon-burning network: The initial step in silicon burning is a photodisintegration such as $\gamma + {}^{28}\text{Si} \to {}^{24}\text{Mg} + \alpha$, which is highly endothermic ($Q = -9.98$ MeV). This implies a large activation energy that must be supplied by a photon in the gas having an energy of ~ 10 MeV or greater. Such photons are rare except at high temperatures and the astonishingly strong temperature dependence that results for the reaction is illustrated in Fig. 6.9. The α-particles liberated in the initial step can now initiate radiative capture reactions on seed isotopes in the gas; a representative sequence is

$$\alpha + {}^{28}\text{Si} \rightleftharpoons {}^{32}\text{S} + \gamma$$
$$\alpha + {}^{32}\text{S} \rightleftharpoons {}^{36}\text{Ar} + \gamma$$
$$\vdots$$
$$\alpha + {}^{52}\text{Fe} \rightleftharpoons {}^{56}\text{Ni} + \gamma. \tag{6.14}$$

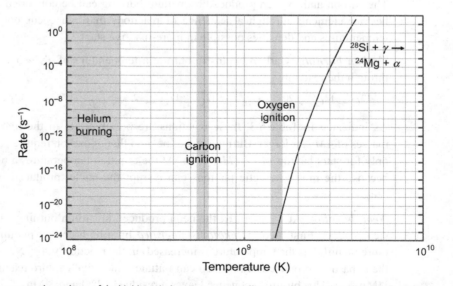

Fig. 6.9 Temperature dependence of the highly endothermic, rate-controlling initial step in silicon burning. For reference, the typical range of temperatures corresponding to helium burning and for carbon and oxygen ignition are indicated. Silicon burning requires temperatures more than an order of magnitude larger than for helium burning, and exhibits an extremely strong temperature dependence.

6.5 Advanced Burning Stages

These reactions are in quasi-equilibrium and much faster than $\gamma + {}^{28}\text{Si} \rightarrow {}^{24}\text{Mg} + \alpha$, so the initial photodisintegration of Si controls the rate for silicon burning. The rates for some of the competing capture and photodisintegration reactions are illustrated in Fig. D.4 of Appendix D. From that figure, note the steep temperature dependence of the photodisintegration reactions, and that for high enough density and α-particle abundance many photodisintegration rates become comparable to the rates for their inverse capture reactions somewhere in the temperature range 10^9–10^{10} K.

Numerical simulation of silicon burning: A calculation of silicon burning that starts with pure ^{28}Si and uses the reaction rates of Fig. D.4 is shown in Fig. 6.10. The net

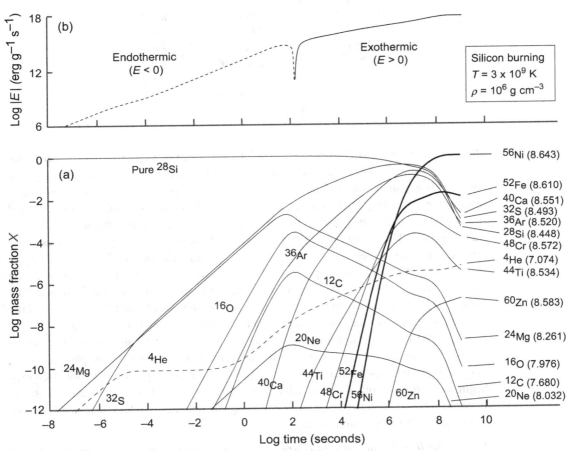

Fig. 6.10 (a) Burning of ^{28}Si at constant $T = 3 \times 10^9$ K and $\rho = 10^6$ g cm^{-3}. Binding energy per nucleon in MeV (Table 6.4) is indicated in parentheses for each isotope. The calculation used a highly simplified network having only the 15 isotopic species shown. A more realistic simulation would include several hundred isotopes and allow the temperature and density to be modified by the burning but this simplified calculation illustrates the essential features of silicon burning. (b) Absolute value of the net energy release. It is initially negative but eventually becomes positive.

Table 6.4 Some binding energies per nucleon B/A

Isotope	A	B (MeV)	B/A (MeV)
^{4}He	4	28.295	7.074
^{12}C	12	92.162	7.680
^{16}O	16	127.619	7.976
^{20}Ne	20	160.645	8.032
^{24}Mg	24	198.257	8.261
^{28}Si	28	236.537	8.448
^{32}S	32	271.781	8.493
^{36}Ar	36	306.716	8.520
^{40}Ca	40	342.052	8.551
^{44}Ti	44	375.475	8.534
^{48}Cr	48	411.462	8.572
^{52}Fe	52	447.697	8.609
^{56}Ni	56	483.988	8.643
^{60}Zn	60	514.992	8.583
^{64}Ge	64	545.954	8.531
^{68}Se	68	576.398	8.476

effect of the simulation is to convert ^{28}Si into a mixture of primarily iron-group nuclei through a sequence of reactions as in Eq. (6.14). In Fig. 6.10(a) there is initially only ^{28}Si but quickly a population of α-particles (dashed curve) and ^{24}Mg begins to accumulate, followed by a growing abundance of ^{32}S, and so on. In the quasi-equilibrium established by the silicon burning the population becomes concentrated in the most thermodynamically stable species. The binding energies per nucleon B/A for the isotopic species used for this simulation are given in Table 6.4 and in Fig. 6.10(a). The isotope ^{56}Ni has the largest B/A so it is the most stable species and ends up with the largest mass fraction as the network evolves. For the simulation in Fig. 6.10 the mass fractions for ^{56}Ni and ^{52}Fe (shown as heavier curves) sum to 0.997 after 10^9 seconds, implying that almost all of the silicon has been converted into iron-group nuclei. The integrated energy release is shown in Fig. 6.10(b) on a logarithmic scale. The log of the absolute value is plotted because initially the energy release is negative (endothermic) since 9.98 MeV of photon energy must be supplied to photodissociate ^{28}Si. Once α-particles are produced by this step they begin to capture on other nuclei, which generally corresponds to exothermic reactions. Eventually when enough α-capture is occurring the net energy release becomes positive (after $\log t \sim 2.1$ in this simulation).

End of the line for fusion and charged-particle capture: Since iron-group nuclei have the greatest stability, Si burning represents the last stage by which fusion and radiative capture reactions can build heavier elements in equilibrium. It might be thought that still heavier elements could be made by increasing the temperature to overcome the Coulomb barriers. But this is self-defeating in equilibrium because the higher temperatures lead also to increased photodissociation and iron-group nuclides are still the equilibrium product. Subsequent chapters will address other mechanisms by which stars can produce the elements heavier than iron found in the Universe.

> **Box 6.3** **The compressed timescale for advanced burning**
>
> To get a perspective on how short the advanced burning timescale is, imagine the lifetime of the 25 $M_\odot$ star of Table 6.3 to be compressed into a single year. Then the hydrogen fuel would be gone by about December 7 of that year, the helium would burn over the next 24 days, the carbon would burn in the 42 minutes before midnight, the neon and oxygen would burn in the last several seconds before midnight, and the silicon would be converted to iron in the last 1/100 second of the year (with a quite impressive New Year's Eve fireworks display in the offing – see the discussion of core collapse supernovae in Chapter 20)!

6.6 Timescales for Advanced Burning

Table 6.3 indicates that the timescales for advanced burning are greatly compressed relative to earlier burning stages. These differences are particularly striking for very massive stars, which race through all stages of their lives at breakneck speed. For example, the 25 $M_\odot$ example of Table 6.3 takes only 10 million years to advance through hydrogen and helium burning, completes its burning of oxygen in only six months, and transforms its newly minted silicon into iron group nuclei in a single day! An analogy illustrating how remarkably short these advanced-burning timescales are may be found in Box 6.3.

These timescales are set by the amount of fuel available, the energy per reaction derived from burning the fuel, and the rate of energy loss from the star, which ultimately governs the burning rate. The last factor is particularly important because energy losses are large when the reactions must run at high temperature. Each factor separately shortens the timescale for advanced burning; taken together they make the timescales for the most advanced burning stages almost instantaneous on the scale set by the hydrogen burning. To quote David Arnett [26] concerning massive stars in late burning stages: "these stars are not so much *objects* as *events!*"

Background and Further Reading

Similar coverage of the material in this chapter may be found in Hansen, Kawaler, and Trimble [107]; Phillips [169]; Rolfs and Rodney [188]; Pagel [166]; Iliades [128]; and Ryan and Norton [192].

Problems

6.1 Derive the nuclear Saha equations (6.4), (6.5), and (6.9) for helium burning, beginning from Eq. (3.55) and the assumption of nuclear statistical equilibrium. *Hint*: The condition for concentration equilibrium in the reaction a + b $\rightleftharpoons$ c is that the sum of chemical potentials on the two sides of the equation balance, $\mu_a + \mu_b = \mu_c$.***

6.2 Under helium flash conditions in low-mass red giant stars (see Section 13.5.3), $T_9 \simeq 0.1$ and the density is $\rho \simeq 10^6 \, \text{g cm}^{-3}$. Estimate the ^{8}Be concentration and the ratio of ^{8}Be to ^{4}He concentrations assuming that the flash takes place in a pure helium core.***

6.3 It was stated in Section 6.4 that radiative alpha capture on ^{16}O to produce ^{20}Ne is inhibited relative to radiative alpha capture on ^{12}C to produce ^{16}O by a higher Coulomb barrier. Estimate the difference in Coulomb barrier penetration rates for these two reactions under conditions characteristic of helium burning. Does this estimate explain the basic rate difference between the two reactions exhibited in Fig. 6.6 under helium burning conditions?

6.4 Use the rates given in Fig. 6.3(a) to estimate the mean life for each step of the main part of the CNO cycle (the CN cycle) illustrated in Fig. 5.3(a), assuming a temperature of 20 million K and a density of 100 g cm^{-3}. Based on these results, which isotope would you expect to become the most abundant after the CNO cycle has run long enough to reach equilibrium? Would you expect ^{13}N to have a high or low abundance after the CNO cycle has been running for a while in a star?***

6.5 For a $p + p$ reaction at a temperature of $T_6 = 15$, calculate the average energy of particles in the gas, the location of the Gamow peak, and its approximate width. For the second step of the triple-α process, $^8\text{Be} + \alpha \rightarrow {}^{12}\text{C}^*$, estimate the location and width of the Gamow peak for a temperature of $T_8 = 3$.

6.6 Consider an idealized closed system in which ^{12}C can undergo a (p,γ) reaction to produce ^{13}N and the ^{13}N can β-decay to ^{13}C, with no other reaction or decay channels open. Write a differential equation expressing the time dependence of the ^{13}N number density in terms of the number densities of the other species, the velocity-averaged (p,γ) reaction rate $\langle \sigma v \rangle$, and the β-decay constant λ. Rewrite this expression in terms of the abundances Y_i introduced in Eq. (3.18). Assuming that the abundances of protons and ^{12}C do not change significantly over a long period of time and that temperature and density are constant, show that the abundance of ^{13}N approaches a constant for long times and derive an expression for this equilibrium concentration of ^{13}N.

6.7 Consider the burning of pure ^{28}Si as in Fig. 6.10, with an initial step corresponding to photodisintegration of silicon: $\gamma + {}^{28}\text{Si} \rightarrow \alpha + {}^{24}\text{Mg}$. Using estimates from Figs. 6.9 and D.3 for the relevant rates, to what abundance does the α-particle population have to grow before the rate of α-capture on silicon, $\alpha + {}^{28}\text{Si} \rightarrow {}^{32}\text{S} + \gamma$, becomes equal to the rate for photodisintegration of silicon, $\gamma + {}^{28}\text{Si} \rightarrow \alpha + {}^{24}\text{Mg}$, assuming a density of 10^6 g cm^{-3} and temperature of $T_9 = 3$? *Hint*: The relevant equations are given in Section D.2.3, and note that the rates in Figs. 6.9 and D.3 are in different units.

7 Energy Transport in Stars

Most energy production in stars takes place in the deep interior where densities and temperatures are high, but most electromagnetic energy emitted from stars is radiated from the photosphere, which is a very thin layer at the surface. Thus, a fundamental issue in stellar astrophysics is how the energy produced in the interior makes its way to the surface of the star. This transport of energy occurs through four basic mechanisms: (1) *conduction* because of thermal motion of electrons and ions, (2) *radiative transport* by photons, (3) *convection* of macroscopic packets of stellar material, and (4) *neutrino emission* from the core. These modes of stellar energy transport will be introduced and discussed in this chapter.

7.1 Modes of Energy Transport

Conduction and radiative transport result from random thermal motion of constituent particles (electrons in the first case and photons in the second), while convection is a macroscopic or collective phenomenon. We shall find that in normal stars conduction is negligible but that it is important in stellar environments containing degenerate matter (for example, in the interior of white dwarfs), and that radiative transport normally is dominant unless the temperature gradient in the gravitational field exceeds a critical value, in which case convection quickly becomes the most efficient means of transporting energy.

Neutrino emission is important for core cooling late in the life of more massive stars. It differs from the other energy transport mechanisms in that it can operate only at extremely high temperatures and densities, and that the neutrinos have little interaction with the star as they carry energy out of the core at essentially the speed of light.[1] As a result, the first three modes of energy transport typically lead to thermalization of the energy (sharing of the energy among many particles), which is then eventually emitted as light of various wavelengths from the photosphere of the star, but the energy of the neutrinos almost always remains in the emitted neutrinos.

To understand stellar structure and stellar evolution, we must consider the role of all four modes of energy transport listed above, since each can be important in particular

---

[1] It is now known that neutrinos are not *completely* inert in their passage out of the star. As will be discussed in more detail in Chapters 11–12, because of flavor oscillations the type (flavor) of neutrino may be altered in this process. This has large implications both for our understanding of neutrinos and the Standard Model of elementary particle physics, and for our understanding of neutrino emission from stars, but will not be significant for the particular considerations of this chapter.

circumstances. We will find that energy transport in normal stars is dominated either by radiative transport or by convective transport. Thus our first task will be to understand radiative and convective transport, the crucial issue of how a star decides between the two in transporting energy, and how to calculate the rate of energy transport in either the radiative or convective case. Then we will address the issue of how neutrino transport becomes the dominant mode of core cooling late in the life of more massive stars. Conductive transport in degenerate matter will be largely deferred until later chapters that deal explicitly with compact objects.

7.2 Diffusion of Energy

We begin with a discussion of how energy can be transported by random thermal motion (*diffusion*), following the general discussion given by Phillips [169]. Consider the volume enclosed by the small cube illustrated in Fig. 7.1 and introduce a random velocity distribution with a small temperature gradient in the x direction. On average, it may be assumed that at any instant approximately $\frac{1}{6}$ of the particles move in the positive x direction with mean velocity $\langle v \rangle$ and mean free path λ (Box 7.1 defines the mean free path).

Let $u(x)$ be the thermal energy density. Because of the temperature gradient, particles crossing a plane at x from left to right have a different thermal energy than those crossing from right to left (Fig. 7.1(b)). Therefore, energy is transported across the surface by virtue of the temperature gradient. The rate of this transport in units of energy per unit time across a unit area is given by the energy current $j(x)$,

$$j(x) \simeq \tfrac{1}{6}\langle v \rangle u(x-\lambda) - \tfrac{1}{6}\langle v \rangle u(x+\lambda)$$
$$\simeq -\tfrac{1}{3}\langle v \rangle \lambda \frac{du}{dx} = -\tfrac{1}{3}\langle v \rangle \lambda \frac{du}{dT}\frac{dT}{dx}$$
$$\simeq -\tfrac{1}{3}\langle v \rangle \lambda C \frac{dT}{dx},$$

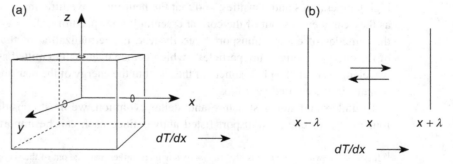

Fig. 7.1 Diffusion of energy by random thermal motion. The temperature is assumed uniform in the y and z directions, but has a gradient dT/dx along the x axis. Thus particles crossing the boundary at x from left to right have a slightly higher energy than those crossing from right to left.

7.2 Diffusion of Energy

> **Box 7.1** **Mean Free Path**
>
> The *mean free path* λ is the average distance that a particle travels before interacting with something. Qualitatively the mean free path is long if the interactions are rare and short if the interactions are frequent. A quantitative measure of the mean free path may be formulated through the following considerations [192].
>
> If a particle travels a small distance Δx through a medium containing identical particles with number density n and the cross section for interaction is σ (see Section 5.3), the probability of an interaction in the interval Δx is $n\sigma \Delta x$ and the probability for survival (no interaction) in that interval is $1 - n\sigma \Delta x$. If the particle passes through N successive intervals Δx, the probability of survival $P_s(x)$ over the distance $x = N\Delta x$ will be the product of survival probabilities in each interval Δx,
>
> $$P_s(x) = \left(1 - \frac{n\sigma x}{N}\right) \times \left(1 - \frac{n\sigma x}{N}\right) \times \cdots = \left(1 - \frac{n\sigma x}{N}\right)^N,$$
>
> and in the limit $N \to \infty$ the survival probability is
>
> $$P_s(x) = e^{-n\sigma x}.$$
>
> The mean free path λ for the particle (mean distance between interactions) is then
>
> $$\lambda = \int_0^\infty x P(x)\, dx,$$
>
> where $P(x)dx$ is the probability that the particle travels from 0 to x with no interaction and then interacts in the interval x to $x + dx$. From the considerations above, the probability to travel from 0 to x with no interaction is $P_s(x) = \exp(-n\sigma x)$ and the probability to interact in the interval dx is $n\sigma\, dx$. Thus $P(x)dx = n\sigma \exp(-n\sigma x)dx$ and
>
> $$\lambda = \int_0^\infty x P(x)\, dx = n\sigma \int_0^\infty x e^{-n\sigma x}\, dx = n\sigma \frac{1}{(n\sigma)^2} = \frac{1}{n\sigma}.$$
>
> As will be discussed further in Section 7.4.4, the mean free path λ for photons is inversely related to the *opacity* κ for a medium: $\kappa = (\lambda \rho)^{-1}$, where ρ is the mass density.

where the heat capacity per unit volume, $C = du/dT$, has been introduced. Therefore, the current across the surface may be written as

$$j(x) = -K \frac{dT}{dx}, \tag{7.1}$$

where K is termed the *coefficient of thermal conductivity*,

$$K \equiv \tfrac{1}{3} \langle v \rangle \lambda C. \tag{7.2}$$

Equation (7.1) is *Ficke's Law*; it is characteristic of diffusive transport. This result has been obtained in a carelessly heuristic way but a more careful derivation leads to essentially the same result.

7.3 Energy Transport by Conduction

Consider first heat transport by random electronic and ionic motion. As discussed in Section 3.3, for a nonrelativistic ideal gas of electrons the internal energy density u_e, heat capacity C_e, and average velocity $\langle v_e \rangle$ are given by

$$u_e = \tfrac{3}{2} n_e k T \qquad C_e = \tfrac{3}{2} n_e k \qquad \langle v_e \rangle = \sqrt{3kT/m_e}. \tag{7.3}$$

Electron–electron collisions are much less effective than electron–ion collisions in transferring energy, so the relevant mean free path is $\lambda_{ei} = 1/n_i \sigma_{ei}$, where n_i is the number density of ions and σ_{ei} is the cross section for electron–ion collisions. As a first crude estimate of the electron–ion cross section we may assume that $\sigma_{ei} \simeq \pi R^2$, where R is the separation between electron and ion at which the potential energy is equal to the average kinetic energy kT in the gas, $Ze^2/R \simeq kT$. Thus, the cross section is approximately

$$\sigma_{ei} = \pi R^2 = \pi \left(\frac{Ze^2}{kT} \right)^2,$$

and substitution in (7.2) yields

$$K_e = \frac{k}{2\pi} \left(\frac{n_e}{n_i} \right) \left(\frac{kT}{Ze^2} \right)^2 \sqrt{\frac{3kT}{m_e}}. \tag{7.4}$$

The corresponding expression for ionic conduction is obtained by the exchanges $n_e \leftrightarrow n_i$ and $m_e \leftrightarrow m_i$, giving

$$\frac{K_e}{K_i} = \frac{n_e^2}{n_i^2} \sqrt{\frac{m_i}{m_e}}. \tag{7.5}$$

As an estimate the gas may be assumed to be completely ionized, so that $n_e = Z n_i$ and

$$\frac{K_e}{K_i} = Z^2 \sqrt{\frac{m_i}{m_e}}.$$

But generally $Z \geq 1$ and $m_i \gg m_e$; therefore, $K_e \gg K_i$ and we see that conduction by electrons is much more important than conduction by the ions. This is just a mathematical statement that there are more electrons and they move faster, so electrons are more efficient than ions at transporting heat. In summary, the current produced by conduction is given approximately by

$$j(x) = K_c \frac{dT}{dx}, \tag{7.6}$$

where $K_c \simeq K_e$ is dominated by the electronic contribution.

7.4 Radiative Energy Transport

Assuming stars to be blackbody radiators, the photons may be viewed as constituting an ultrarelativistic, bosonic gas with

$$\langle v \rangle = v = c \qquad u = aT^4 \qquad C = \frac{du}{dT} = 4aT^3. \tag{7.7}$$

By analogy with Eq. (7.1), the current associated with radiative diffusion is

$$j(x) = -K_r \frac{dT}{dx}, \tag{7.8}$$

where the *coefficient of radiative diffusion* is

$$K_r = \tfrac{4}{3} c \lambda a T^3. \tag{7.9}$$

All quantities in Eqs. (7.7)–(7.9) are known except for the mean free path λ. To assess it we need to consider various contributions to the interaction of photons with the medium that are responsible for their effective mean free path in stellar environments.

7.4.1 Thomson Scattering

At high temperatures and low densities, *Thomson scattering* (the elastic scattering of electromagnetic radiation by free charged particles) from electrons dominates, with a cross section that is *independent of frequency and temperature*,

$$\sigma_T = \frac{8\pi}{3} \left(\frac{e^2}{m_e c^2} \right)^2 = 6.652 \times 10^{-25} \text{ cm}^2. \tag{7.10}$$

For a gas in thermal equilibrium this expression is valid if $kT \ll m_e c^2$, which is satisfied if $T \ll 6 \times 10^9$ K. This is true in most stars except in extreme circumstances.[2] The corresponding mean free path is

$$\lambda_T = \frac{1}{n_e \sigma_T}, \tag{7.11}$$

where n_e is the electron number density, and inserting this in (7.9) gives for the coefficient of radiative diffusion in the Thomson scattering approximation

$$K_r \simeq K_T \equiv \frac{acT^3}{2\pi n_e} \left(\frac{m_e c^2}{e^2} \right)^2. \tag{7.12}$$

As shown in Problem 7.1, assuming complete ionization and that Thomson scattering dominates then permits the ratio of coefficients for radiative and conductive transport to be approximated as [169]

---

[2] At sufficiently high energy special relativistic effects for the electron must be considered. This leads to *Compton scattering*. Thomson scattering may be viewed as the low-energy limit of Compton scattering, valid when $kT \ll m_e c^2$. Equivalently, Thomson scattering is the limit of Compton scattering when the wavelength of the photons is much larger than the Compton wavelength h/mc.

$$\frac{K_r}{K_e} \simeq \frac{K_T}{K_e} = \sqrt{3}\, Z\, \frac{P_r}{P_e} \left(\frac{m_e c^2}{kT}\right)^{5/2}, \qquad (7.13)$$

where the radiation pressure is $P_r = \frac{1}{3}aT^4$ and ideal-gas electron pressure is $P_e = n_e kT$.

Example 7.1 Equation (7.13) yields $K_r/K_e \simeq 2\times 10^5$ for the Sun, supporting the earlier assertion that radiative transport dominates over conduction in normal stars. This conclusion is based on the assumption of pure Thomson scattering but will not be altered significantly by additional photon absorption processes that will be considered in Section 7.4.3. However, it is not true if the matter in a star has a degenerate equation of state (see Section 3.7).

7.4.2 Conduction in Degenerate Matter

Electronic conduction in degenerate matter is altered in several important ways relative to that for an ideal gas:

1. Degeneracy increases the electron speed by a factor $(\varepsilon_F/kT)^{1/2}$ and decreases the heat capacity by a factor of roughly kT/ε_F, where ε_F is the Fermi energy.
2. The mean free path λ is increased because the Pauli exclusion principle allows an electron to scatter to a state only if that state is not already occupied, and in a degenerate gas there are no unoccupied states below the Fermi surface.

The net effect of these changes is that a degenerate gas behaves much like a metal and transport of energy by conduction becomes important. The issue of conduction in degenerate matter will be revisited when the structure of white dwarfs is considered in Chapter 16. There we will find that in very dense matter such as for white dwarfs or the centers of evolved stars, conduction can become a dominant mode of energy transport.

7.4.3 Absorption of Photons

In addition to undergoing simple Thomson scattering, photons may be absorbed. (Recall from Section 3.10 that the number of photons is in general not conserved.) Simultaneous conservation of energy and momentum prohibits such absorption on free electrons but it can occur for electrons in the vicinity of ions. Thus, absorption becomes more important at higher densities and lower temperatures. The two most important absorptive processes in stellar environments are[3]

1. *Bound–free absorption*, where the electron that the photon interacts with is bound initially to an ion and is ejected by the interaction. This process is also called *photoionization*.

---

[3] *Bound–bound absorption,* corresponding to transitions between two bound atomic or molecular states through absorption of photons, can occur also. These transitions involve small energies and low temperatures, and thus are significant only very near the surface for stars. They will be ignored in the present discussion but they will be important in understanding the kilonova thought to result from the merger of two neutron stars that will be discussed in Section 22.6.2.

2. *Free–free absorption*, where the electron is unbound before and after the interaction. This process is also called *inverse bremsstrahlung*.

Unlike the case for Thomson scattering, both classes of absorptive processes imply a mean free path that depends on frequency. In the frequency range ν to $\nu + d\nu$, the photon energy density u_ν and heat capacity C_ν are given by

$$u_\nu d\nu = \frac{8\pi}{c^3}\left(\frac{h\nu^3}{e^{h\nu/kT}-1}\right) d\nu \qquad C_\nu d\nu = \frac{\partial u_\nu}{\partial T} d\nu. \qquad (7.14)$$

Let λ_ν be the mean free path for photons at frequency ν. Then from Eq. (7.2) we expect the total coefficient of radiative transport to be given by an integral over frequency,

$$K_{\rm r} = \frac{1}{3}\int_0^\infty \langle v \rangle \lambda_\nu C_\nu \, d\nu = \frac{c}{3}\int_0^\infty \lambda_\nu C_\nu \, d\nu.$$

Introducing the *Rosseland mean* $\lambda_{\rm ross}$ through the definition

$$\lambda_{\rm ross} \equiv \frac{1}{4aT^3}\int_0^\infty \lambda_\nu C_\nu \, d\nu, \qquad (7.15)$$

allows the above expression for $K_{\rm r}$ to be written as

$$K_{\rm r} = \frac{4}{3}c\lambda_{\rm ross}aT^3. \qquad (7.16)$$

This is the same form as the previous expression (7.9) with the replacement $\lambda \rightarrow \lambda_{\rm ross}$. Thus we may cast equations for frequency-dependent mean free paths in the same form as those for constant ones by employing the Rosseland mean defined in Eq. (7.15).

7.4.4 Stellar Opacities

It was useful to introduce diffusion processes in terms of the intuitive idea of a mean free path λ, but it is more usual in astronomy to discuss photon transport in terms of a closely related quantity called the *opacity*. The total probability of photon interaction is a sum of contributions from electron and ion scattering, and from Box 7.1 the total cross section σ and mean free path λ are related by $\lambda \sim (n\sigma)^{-1}$, where n is a number density. Hence

$$\lambda = \frac{1}{n_e\sigma_e + n_i\sigma_i}. \qquad (7.17)$$

Now both the electron number density n_e and the ion number density n_i are proportional to the matter density ρ, which suggests that we parameterize (7.17) in the form

$$\lambda = \frac{1}{\rho\kappa}, \qquad (7.18)$$

where κ is termed the *opacity* and has units of area divided by mass. Then (7.18) may be used to rewrite the preceding formulas in terms of the opacity κ instead of the mean free path λ. For example, the current (7.8) associated with radiative diffusion may be rewritten in terms of opacity κ as

$$j(x) = -\frac{4ac}{3}\frac{T^3}{\rho\kappa}\frac{dT}{dx}, \qquad (7.19)$$

and from Eq. (7.15) we find that

$$\frac{1}{\kappa_{\text{ross}}} = \frac{1}{4aT^3} \int_0^\infty \frac{C_\nu}{\kappa_\nu} d\nu \qquad (7.20)$$

defines the *Rosseland mean opacity* κ_{ross}.

7.4.5 General Contributions to Stellar Opacity

The following qualitative remarks apply to ionization and various components of the stellar opacity.

1. The contribution of bound–free absorption is important at low temperatures where atoms are only partially ionized.
2. The contribution of free–free absorption is dominant at higher temperature where atoms become fully ionized, producing many free electrons with which to interact.
3. Thomson scattering contributes a constant background that is independent of temperature.

Let us now consider some simple opacity parameterizations that are relevant for stellar material.

Kramer's opacity law: An approximate expression for the frequency-averaged opacity deriving from the first two of these mechanisms is given by *Kramer's Law*:

$$\kappa_{\text{ab}} = \kappa_0 \rho T^{-3.5}, \qquad (7.21)$$

where the subscript denotes that this is an absorption-dominated opacity. The calculation of such opacities is a complex issue but approximate formulas for the free–free and bound–free frequency-averaged absorption opacities above $T \sim 10^4$ K may be given in the Kramer's form [107]

$$\kappa_{\text{ff}} \simeq 4 \times 10^{22}(X+Y)(1+X)\rho T^{-3.5} \text{ cm}^2 \text{ g}^{-1} \qquad (7.22)$$

$$\kappa_{\text{bf}} \simeq 4 \times 10^{25} Z(1+X)\rho T^{-3.5} \text{ cm}^2 \text{ g}^{-1} \qquad (7.23)$$

where X is the hydrogen mass fraction, Y is the helium mass fraction, and Z is the mass fraction of metals.

Thomson opacities: Thomson electron scattering gives a constant background opacity

$$\kappa_{\text{T}} = \frac{1}{\lambda_{\text{T}} \rho} = \frac{n_e \sigma_{\text{T}}}{\rho}. \qquad (7.24)$$

Introducing that for a fully ionized gas $n_e \sim (1+X)\rho/2M_u$ (see Problem 7.2), the Thomson scattering opacity may be approximated as

$$\kappa_{\text{T}} \simeq \frac{(1+X)\sigma_{\text{T}}}{2M_u} = 0.20\,(1+X)\,\text{cm}^2\,\text{g}^{-1}. \qquad (7.25)$$

Realistic solar opacities: Sophisticated calculations of solar opacities as a function of temperature and density that go beyond the considerations of the preceding simple

7.4 Radiative Energy Transport

Table 7.1 Solar opacities and mean free paths

$R/R_\odot$	$T(K)^\dagger$	$\rho\,(g\,cm^{-3})^\dagger$	$\kappa(cm^2\,g^{-1})^*$	$\lambda\,(cm)$
0	1.6×10^7	157	1	0.006
0.3	6.8×10^6	12.0	2	0.042
0.6	3.1×10^6	0.50	8	0.25
0.9	6.0×10^5	0.026	100	0.39

†Temperature and density from Standard Solar Model, Table 10.1.
*Opacities interpolated from Fig. 7.2.

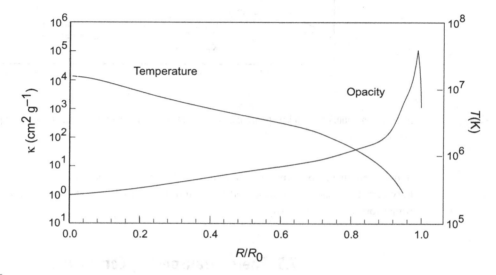

Fig. 7.2 Temperatures [32] and photon Rosseland mean opacities [190] calculated for the interior of the Sun as a function of distance from the center. The peak near the surface ($R/R_0 \sim 1$) is because of complex ionization processes in that region, primarily the photoionization of hydrogen (bound–free absorption) around $T \sim 10^4$ K. The steep falloff of opacity at even lower temperatures closer to the surface is because the average photon energies there are no longer high enough to either ionize hydrogen (bound–free absorption) or to excite its electronic states (bound–bound absorption). Over most of the solar interior the opacity lies in the range 1–100 cm^2 g^{-1}, implying mean free paths less than a centimeter.

formulas may be found in Ref. [187] and in Fig. 7.2. These opacities indicate that the interior of the Sun is extremely opaque to electromagnetic radiation, implying a very short mean free path for photons. This is illustrated for some representative values of the solar radial coordinate in Table 7.1.

Temperature and density dependence: The dominant contributions to the opacity as a function of temperature and density are illustrated in Fig. 7.3. The boundaries between regions are defined by lines where the corresponding opacities are equal. At high temperature and low density Thomson scattering dominates. At low temperature and high

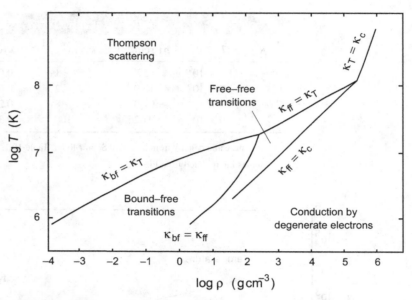

Fig. 7.3 Dominant contributions to stellar opacity as a function of temperature and density. Adapted from a figure in Ref. [107].

density electrons become degenerate and the gas becomes a very good conductor. In between, the opacity is dominated by bound–free transitions at low density and free–free transitions at higher density.

7.5 Energy Transport by Convection

In some circumstances the energy to be transported is too large to be carried efficiently by either radiative transport or conduction. In those instances the system can become unstable to macroscopic overturn of material in a process called *convection*. Because convection is a collective process that moves entire blobs of material up and down in the gravitational field, it can transport energy very efficiently when it operates. It will be important for subsequent discussion to make a distinction between two broad categories of convection. The first may be termed *microconvection*; it applies when the convective blobs are small relative to the size of the region that is unstable. The second may be termed *macroconvection*; it corresponds to convection in which the blob sizes are a substantial fraction of the size of the convective region. This distinction has an important practical implication. For microconvection it is possible to introduce approximations (*mixing-length theory* – see Section 7.9) that allow retention of spherical symmetry in the problem; thus, 1-dimensional hydrodynamics may suffice under such conditions. However, if macroconvection prevails spherical symmetry is broken strongly and multidimensional hydrodynamics is required to model the convection dynamically.

The initial discussion of convection to be presented here will be somewhat more general than is normally required for the structure of ordinary stars. This will provide foundation for later discussions of events like supernovae or neutron star mergers, in which rapid and complex convective processes may play a significant role, and for which the convection cannot be modeled adequately by simple models and must be simulated on large computers.

7.6 Conditions for Convective Instability

We begin by analyzing some conditions under which a region of fluid may be expected to become unstable against convective motion, following the discussion of Ref. [227]. Imagine a blob of matter in a gravitating fluid that moves upward a distance λ from position 1 to position 2 because of some infinitesimal stimulus, as illustrated in Fig. 7.4(a). If the region is convectively stable the displaced blob experiences a restoring force that tends to return it to its original position, as illustrated in Fig. 7.4(b). Because of overshooting, the blob executes a stable oscillation around an equilibrium height with a frequency termed the *Brunt–Väisälä frequency*. However, if the blob of material at position 2 is less dense than the surrounding medium it will be driven upward continuously by buoyancy forces and the region is then said to be convectively unstable, as illustrated in Fig. 7.4(c).[4] We

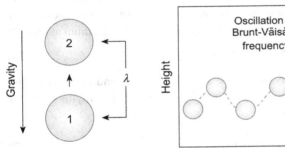
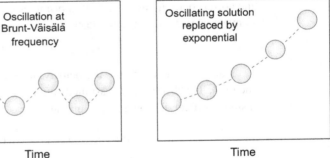

(a) Convective motion   (b) Convective stability   (c) Convective instability

Fig. 7.4 (a) Schematic illustration of convective motion. (b) Convectively stable situation: a blob displaced vertically a small amount oscillates around a stable equilibrium at the *Brunt–Väisälä frequency*. (c) Convectively unstable situation: a blob displaced vertically a small amount continues to rise as time goes on.

---

[4] Mathematically, the differential equation describing motion of the blob has an oscillating solution for a certain range of parameters and an exponential solution for a different range. Onset of the convective instability corresponds to transition from the oscillating to exponential solution. This conceptual discussion has been implemented in terms of an upwardly moving blob that continues to rise because it is less dense than the surrounding medium, but in a convectively unstable region there also will be downward-moving blobs that continue to sink because they find themselves *more dense* than the surrounding medium. Thus convection involves upward and downward plumes that transport no net mass but transport energy upward.

(a) Schwarzschild instability

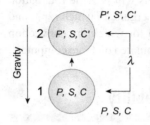

Region unstable if $\rho(P', S', C') - \rho(P', S, C') \geq 0$

(b) Schwarzschild stability condition

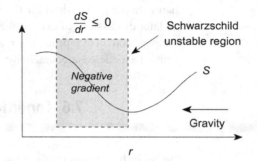

Fig. 7.5 Convective instability according to the Schwarzschild criterion.

may choose to impose particular physical conditions on how the blob of matter is moved, and these lead to three separate criteria for convective instability that we now discuss.

7.6.1 The Schwarzschild Instability

Suppose that the blob moves adiabatically (constant entropy), but in pressure and composition equilibrium with the surrounding medium. Denote the pressure, entropy, and composition of the medium at position 1 by P, S, and C, respectively, and at position 2 by the corresponding variables with a prime, as illustrated in Fig. 7.5(a).[5] The condition for convective instability is that the blob is less dense than the surrounding medium at point 2:

$$\rho(P', S', C') - \rho(P', S, C') \geq 0. \tag{7.26}$$

Let's analyze quantitatively the density difference between the blob and the surrounding medium. The difference is assumed small, which justifies expansion in a Taylor series for a small vertical displacement λ, giving

$$\rho(P', S', C') - \rho(P', S, C') = \left.\frac{\partial \rho}{\partial S}\right|_{P,C} \lambda \frac{dS}{dr}. \tag{7.27}$$

By using Eq. (3.10) to introduce the heat capacity at constant pressure C_p, the entropy S may be exchanged for the temperature T and Eq. (7.26) becomes

$$\left(\frac{T}{C_p} \left.\frac{\partial \rho}{\partial T}\right|_{P,C}\right) \lambda \frac{dS}{dr} \geq 0. \tag{7.28}$$

---

[5] Physically, these conditions can be realized if the medium has uniform composition, diffusive heat transport is slow because of high opacity, and the dynamical timescale is fast compared with the timescale for vertical displacement of the blob (the speed of the blob is small compared with the speed of sound in the medium). Then $C = C'$, so the blob has the same composition as the surrounding medium at both points 1 and 2, the pressure will equilibrate quickly due to expansion or contraction of the blob so that the blob remains in pressure equilibrium with the surroundings, but the blob at point 2 will generally have a different entropy than the surrounding medium if there is an entropy gradient ($S \neq S'$). Such conditions are not uncommon over large regions of typical stars (and in atmospheres).

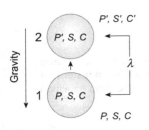 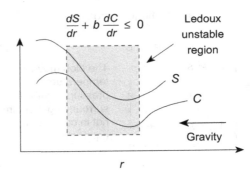

Fig. 7.6 Convective instability according to the Ledoux criterion.

For normal equations of state the partial derivative in the preceding expression is negative (for example, recall that $\rho = \mu P/kT$ for an ideal gas). Thus the quantity in parentheses is typically negative and Eq. (7.28) is equivalent to the *Schwarzschild condition* for convective instability,

$$\frac{dS}{dr} \leq 0 \quad \text{(Schwarzschild condition)}. \tag{7.29}$$

A region is unstable against Schwarzschild convection if there is a *negative entropy gradient*, as illustrated schematically in Fig. 7.5(b).

7.6.2 The Ledoux Instability

Now suppose that the blob moves adiabatically with no composition change, but in pressure equilibrium, as illustrated in Fig. 7.6(a).[6] The condition for convective instability is

$$\left(\frac{T}{C_p}\frac{\partial \rho}{\partial T}\bigg|_{P,C}\lambda\right)\frac{dS}{dr} + \left(\frac{\partial \rho}{\partial C}\bigg|_{P,S}\lambda\right)\frac{dC}{dr} \geq 0, \tag{7.30}$$

where the first term is the same as for the Schwarzschild instability and the second term arises because of the composition difference between the blob and surrounding medium at point 2. For most cases of interest both of the partial derivatives in Eq. (7.30) are negative and the Ledoux condition for instability takes the form

$$\frac{dS}{dr} + b\frac{dC}{dr} \leq 0 \quad \text{(Ledoux condition)}, \tag{7.31}$$

---

[6] Physically, this would be the case if the medium has a composition gradient over the distance λ and diffusion of composition is slow, diffusive heat transport is slow, and the dynamical timescale is fast compared with the displacement time. Then $C \neq C'$, so the blob has a different composition than the surrounding medium at point 2, it will remain in pressure equilibrium with the surroundings, and the blob at point 2 will generally have a different temperature and entropy than the surrounding medium if there is a temperature gradient. Although less common than the conditions leading to the Schwarzschild instability, such conditions can also be realized in stellar evolution; see Section 20.3.6.

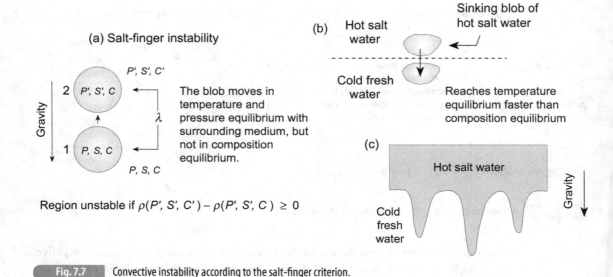

Fig. 7.7 Convective instability according to the salt-finger criterion.

where b is positive. Therefore, a region is unstable against Ledoux convection if both the entropy and the concentration variables have negative gradients, as illustrated schematically in Fig. 7.6(b). If the entropy gradient and concentration gradient have opposite signs, the stability of the region is dependent on the relative size of the two terms in Eq. (7.31). For example, a region of a star could be Schwarzschild-stable but Ledoux-unstable.

7.6.3 Salt-Finger Instability

Finally, consider a situation where the blob is in temperature and pressure equilibrium with the surrounding medium, but not in composition equilibrium, as illustrated in Fig. 7.7(a). The condition for convective instability now takes the form,

$$\rho(P', T', C') - \rho(P', T', C) = \left(\frac{\partial \rho}{\partial C}\bigg|_{P,T} \lambda\right) \frac{dC}{dr} \geq 0. \tag{7.32}$$

Consider a layer of hot salt water that lies over a layer of cold fresh water, and imagine a blob of the hot salt water that is perturbed and begins to sink into the underlying cold fresh water [Fig. 7.7(b)]. This blob of sinking material will be able to come into heat equilibrium with its surroundings faster than it will be able to come into composition equilibrium because transfer of heat by molecular collisions generally is faster than the motion of the sodium and chlorine ions that causes the composition to equilibrate. It follows that such a blob may be in approximate temperature equilibrium but remain out of composition equilibrium. The heat diffusion will cool the blob of salt water and, since salt water is more dense than fresh water at the same temperature, the blob continues to sink in the surrounding fresh water. As this motion continues, the medium develops "fingers" of salt water reaching down into the fresh water, as illustrated in Fig. 7.7(c).

> **Box 7.2** **Doubly Diffusive Instabilities**
>
> The salt-finger instability is a particular example of a general class of instabilities that are termed *doubly diffusive instabilities*. They may occur when
>
> 1. Two diffusing substances are present (heat and salt in this example).
> 2. One of the substances diffuses more rapidly than the other (heat in this example).
> 3. The substance diffusing more rapidly has a stabilizing gradient and the substance diffusing more slowly has a destabilizing gradient (in this example, cold salt water is more dense than cold fresh water).
>
> For an informative discussion of the effects of doubly diffusive instabilities in lakes and oceans on Earth, see Ref. [67].

As discussed in Box 7.2, this *salt-finger instability* is a particular example of what are more generally termed *doubly diffusive instabilities*. This type of instability is observed in oceans and lakes on Earth but whether it is important for stellar structure and stellar evolution is uncertain. For an example of a possible salt-finger instability associated with off-center helium burning in stars, see Section 6.5 of reference [134]. It also has been suggested that an analog of the salt-finger instability discussed above may occur in a core collapse supernova when a neutron-rich region surrounds a central region that is less neutron rich, but the claim is controversial.

7.7 Critical Temperature Gradient for Convection

The most important convective instability for the structure and evolution of stars is that set by the Schwarzschild condition (7.29), which is driven by entropy gradients. The instability criterion for Schwarzschild convection may be expressed also in terms of a critical temperature gradient (see Problem 7.4 and Fig. 7.8)

$$\frac{dT}{dr} < \left(\frac{dT}{dr}\right)_{ad}, \qquad (7.33)$$

with the *adiabatic temperature gradient* defined by (Problems 7.4 and 7.5)

$$\left(\frac{dT}{dr}\right)_{ad} = \left(\frac{\gamma-1}{\gamma}\right)\frac{T}{P}\frac{dP}{dr} = -\frac{g}{c_P}, \qquad (7.34)$$

where c_P is the specific heat (heat capacity per unit mass) at constant pressure, and where both derivatives generally are negative in Eq. (7.33). Thus, by the Schwarzschild criterion:

> A region is convectively unstable if the actual temperature gradient is steeper than the local adiabatic temperature gradient.

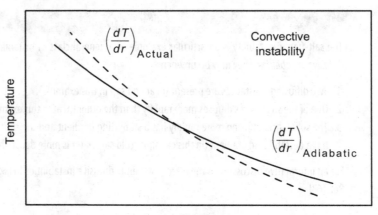

Fig. 7.8 Schematic illustration (solid line) of the critical temperature gradient for convection implied by Eq. (7.33). In this example the actual gradient (dashed line) is steeper than the adiabatic gradient, so the region is convectively unstable.

The difference between the actual temperature gradient dT/dr and the adiabatic gradient (7.34) is sometimes termed the *superadiabatic temperature gradient* $\delta(dT/dr)$,

$$\delta\left(\frac{dT}{dr}\right) \equiv \frac{dT}{dr} - \left(\frac{dT}{dr}\right)_{\text{ad}}. \quad (7.35)$$

Conditions for which this quantity is negative (so that $|dT/dt| > |(dT/dr)_{\text{ad}}|$, since both derivatives are negative) are said to be *superadiabatic*. If Eq. (7.34) is substituted into Eq. (7.33) and both sides divided by dT/dr, the instability condition may be expressed in the alternative form

$$\frac{d\ln P}{d\ln T} < \frac{\gamma}{\gamma - 1}. \quad (7.36)$$

Thus, convective instability according to the Schwarzschild criterion requires the temperature to fall off sufficiently fast with radius that the actual temperature gradient satisfies Eq. (7.33) or (7.36). Equivalently, convective instability is heralded by the appearance of a negative superadiabatic gradient (7.35). We won't prove it here but these temperature-gradient conditions for convection are equivalent to the entropy-gradient condition (7.29).

The two most important quantities governing the behavior of the right side of (7.34) are the adiabatic index γ and the pressure gradient dP/dr. Let's now examine in more depth how these can influence the critical temperature gradient that marks the boundary of convective instability.

7.7.1 Convection and the Adiabatic Index

For an ideal gas the adiabatic index defined in Eq. (3.12) may be expressed also as

$$\gamma = \frac{C_{\text{P}}}{C_{\text{V}}} = \frac{1 + s/2}{s/2} = \frac{s + 2}{s}, \quad (7.37)$$

where s is the number of classical degrees of freedom per particle,[7] each carrying average thermal energy $E = \frac{1}{2}kT$. Therefore, for a monatomic gas with only three translational degrees of freedom for each atom the adiabatic index is

$$\gamma = \frac{1+3/2}{3/2} = \frac{5}{3},$$

and the condition (7.36) for convective instability is that

$$\frac{d \ln P}{d \ln T} < \frac{5}{2}. \tag{7.38}$$

But if the gas has additional degrees of freedom the adiabatic index will decrease and for sufficiently many degrees of freedom it will approach unity:

$$\lim_{s \to \infty} \gamma = \lim_{s \to \infty} \left(\frac{1+s/2}{s/2} \right) = 1. \tag{7.39}$$

As $\gamma \to 1$ the factor $(\gamma - 1)/\gamma$ tends to zero and the adiabatic temperature gradient of Eq. (7.34) becomes less steep. Thus, an increase in the effective number of degrees of freedom for a gas will decrease the adiabatic index γ and thereby enhance the possibility of convective instability. Three processes common in stellar astrophysics illustrate how an increase in the number of degrees of freedom may occur:

1. Energy may be absorbed by exciting vibrations and rotations of molecules, and emitted by the corresponding deexcitations.
2. Energy may be absorbed by the dissociation of molecules and emitted in their recombination.
3. Energy may be absorbed by ionization of atoms or molecules and emitted in the corresponding recombination.

In each of these cases the associated physical process can contribute to convective instability by increasing the effective number of degrees of freedom in the gas. This decreases the adiabatic index toward the critical value of unity, making the condition (7.36) easier to fulfill. As we shall see in later examples, the physical reason for this decreased convective stability typically is that these processes permit rising blobs of gas to remain buoyant longer, thereby enhancing convection.

7.7.2 Convection and the Pressure Gradient

Under conditions of hydrostatic equilibrium the pressure gradient is given by Eq. (4.10),

$$\frac{dP}{dr} = -\frac{Gm(r)}{r^2}\rho(r) = -g(r)\rho(r), \tag{7.40}$$

---

[7] Physically it is clear that the heat capacity (change in heat divided by the change in temperature) should depend on the number of degrees of freedom for the particles in the gas. Internal degrees of freedom such as molecular vibrations allow energy to be stored without increasing the (translational) kinetic energy of the gas particles, so the total number of degrees of freedom (translation plus internal) for the gas will affect how the temperature changes with added heat.

where $g(r)$ is the local gravitational acceleration. Thus, pressure falls off more gradually where $g(r)$ is small and a smaller value of dP/dr makes the condition implied by (7.33) and (7.34) easier to fulfill, thereby favoring convective instability. It follows that regions in which the local gravity is weak will be more susceptible to convective instabilities than those with stronger gravity.

Example 7.2 In close binary systems (see Chapter 18), a star may be tidally distorted by its companion. The decreased gravity in tidally distended regions (which occurs because a gas particle there feels a gravitational attraction from one star that is partially canceled by the gravitational attraction of the other star) may initiate convective instability.

7.8 Stellar Temperature Gradients

The condition (7.33) defines a (theoretical) critical temperature gradient for convective instability in terms of the adiabatic temperature gradient (7.34). Therefore, we must investigate the *actual* temperature gradients dT/dr of stars to assess their stability against convection.

7.8.1 Choice between Radiative or Convective Transport

It may be assumed that the temperature gradients of normal stars that are not convective and not cooled by neutrino emission are determined by the rate of radiative energy transport, since conductive transport is negligible in nondegenerate stars. This suggests the following approach for determining the mode of energy transport in stars, if neutrino transport is ignored:

1. Calculate the temperature gradient for radiative transport according to Eq. (7.43) below.
2. If this gradient is sub-critical according to the criterion of Eq. (7.33), assume no convection and that energy transport is by radiative diffusion.
3. If the resulting gradient is critical or supercritical, assume that convection – because it is very efficient in transporting energy – prevails as the means of energy transport as long as the temperature gradient remains critical.

Notice that these considerations could lead to different conclusions in different regions of a star; thus, we may expect that stars could be convective in some regions and radiative in others. In regions of a star that are convectively unstable it may be expected that the actual temperature gradient remains very near the adiabatic gradient because of a natural process of self-regulation: (1) A small fluctuation that steepens the temperature gradient will tend to increase the outward heat flow, which will work to make the temperature gradient more shallow. (2) Conversely, a fluctuation that makes the temperature gradient more shallow will suppress convection, which will reduce the outward heat flow and steepen the temperature gradient.

7.8.2 Radiative Temperature Gradients

To use the transport algorithm outlined above we require an estimate of the radiative temperature gradient. Let $L(r)$ denote the rate of energy flow through a shell of thickness dr at a radius r, and let $\varepsilon(r)$ denote the nuclear power per unit volume generated at radius r. Then the power generated in the shell of thickness dr at radius r is given by $4\pi r^2 \varepsilon(r) dr$. This adds to the outward power flow from interior shells, so the energy flow is governed by the differential equation

$$\frac{dL}{dr} = 4\pi r^2 \varepsilon(r). \tag{7.41}$$

Outside the central power-generating regions for a star it may be expected that $L(r)$ approaches a constant equal to the surface luminosity of the star. If this energy flow is assumed to be caused by radiative diffusion,

$$L(r) = 4\pi r^2 j(r), \tag{7.42}$$

where Eq. (7.19) gives the current for radiative transport. Inserting (7.19) in (7.42) and solving for the temperature gradient associated with *transport by radiative diffusion*,

$$\left(\frac{dT}{dr}\right)_{\text{rad}} = -\frac{3\rho(r)\kappa(r)}{4acT^3(r)} \frac{L(r)}{4\pi r^2}. \tag{7.43}$$

If the gradient (7.43) becomes steeper than the critical (adiabatic) gradient defined by Eq. (7.34), the system switches to convective transport *with a temperature gradient approximated by Eq. (7.34)*.

Example 7.3 According to the Standard Solar Model (see Table 10.1), for a shell at a radius $R/R_\odot = 0.30$ (corresponding to an enclosed mass $M/M_\odot = 0.61$), the luminosity is essentially the surface luminosity of $3.8 \times 10^{33}\,\text{erg}\,\text{s}^{-1}$ (meaning that almost all power production takes place inside this radius), the density is $\rho \sim 12\,\text{g}\,\text{cm}^{-3}$, and the temperature is $T \sim 6.8 \times 10^6$ K. From solar opacity tables (for example, Fig. 7.2 or Ref. [127]), the opacity may be estimated as $\kappa \sim 2\,\text{cm}^2\,\text{g}^{-1}$, and from Eq. (7.43) the corresponding radiative temperature gradient is $dT/dr \simeq -10^{-4}\,\text{K}\,\text{cm}^{-1}$, the average mean free path is $\lambda = 1/\rho\kappa \simeq 0.04$ cm, and from these numbers the fractional change in temperature over a characteristic distance of one mean free path is of order 10^{-12}. Thus, the solar interior is highly opaque, the temperature changes slowly over a characteristic diffusion distance, and the preceding radiative diffusion approximations are seen to be amply justified.

7.9 Mixing-Length Treatment of Convection

A proper treatment of convection in stars is a difficult subject because it requires the solution of 3-dimensional hydrodynamics for a turbulent, compressible fluid that is strongly coupled to other complex aspects of the problem such as radiation transport and thermonuclear energy generation. Modern high-performance computers are able to handle this with increasing levels of sophistication but historically much of the understanding of convection has derived from simple models based on *mixing-length approximations*. These models have rather murky theoretical foundation but they appear to work well as phenomenological descriptions of the most important aspects of convection in normal stars.

Part of that success is because of the empirical nature of mixing-length models; part is because: (1) Convection is such an efficient source of energy transport that it often dominates all other modes, so that we do not have to give undue thought to partitioning energy transport between radiation and convection. (2) Convection often operates with convective velocities that are well below sound speed, so that no shockwaves are produced. (3) Convection often operates on a timescale that is well-separated from other relevant timescales such as the hydrodynamic response time in the star. However, mixing-length models are basically empirical, with the most essential parameter (the mixing length) not specified by any fundamental theory. As a consequence, they break down in a variety of cases. For example, mixing-length models are generally not very appropriate for situations where

1. Radiative transport competes strongly with convection, such as in the surface of a convective star.
2. Convective transport is supersonic and thus produces shockwaves, as in supernovae and neutron star mergers.
3. Convection may violate spherical symmetry strongly, as in supernova explosions or in mergers of compact objects like neutron stars.
4. Convective timescales are comparable to other dynamical timescales, as may happen for some pulsating variable stars and for supernova explosions and neutron star mergers.

With this as introduction, and taking due note of the caveats, we now consider a basic mixing-length model of convection (see Refs. [55, 68, 107, 134]).

7.9.1 Pressure Scale Height

For a gravitating gas the *pressure scale height* H_p is defined by

$$H_p \equiv -\frac{dr}{d \ln P} = -P\frac{dr}{dP}. \qquad (7.44)$$

If we assume H_p to be constant the solution of this differential equation is

$$P = P_0 \, e^{-r/H_p}. \qquad (7.45)$$

Therefore, the scale height has the dimension of length and is the characteristic vertical scale for variation of the pressure in a star or in an atmosphere, since H_p is the vertical distance over which the pressure changes by a factor of e. Using the equation for hydrostatic equilibrium (4.10) and the ideal gas law, the scale height may be expressed as

$$H_p = \frac{P}{g\rho} = \frac{kT}{g\mu M_u}, \qquad (7.46)$$

where g is the local gravitational acceleration, k is Boltzmann's constant, μ is the mean molecular weight, and $M_u = 1/N_A$ is the atomic mass unit.

7.9.2 The Mixing-Length Philosophy

The mixing-length idea assumes that the stellar fluid in a certain region is composed of identifiable blobs that can move vertically in the gravitational field between regions of differing heat content, as illustrated in Fig. 7.9. For example, some blobs may move upward because of buoyancy forces, carrying warm fluid outward, while other blobs may move downward because of negative buoyancy, carrying cooler fluid inward (it is assumed that there is no net vertical mass flow, so that hydrostatic equilibrium is preserved). This leads to an outward transport of energy. The average distance over which blobs move before dissipating in the surrounding medium is termed the *mixing length, ℓ*. Mixing-length approaches then analyze the motion of these blobs over a characteristic scale defined by the mixing length with the following assumptions [107].

1. Characteristic dimensions of the rising and falling blobs are of the same order as the mixing length ℓ.
2. The mixing length is much shorter than other length scales of physical significance in the star. (This assumption may be violated in practice, which is one reason for the somewhat dubious theoretical underpinnings of the approach.)
3. The temperature, density, pressure, and composition of a blob differ only slightly from that of the surrounding medium. The requirement that the internal pressure of blobs remains approximately the same as the surrounding fluid means that the timescales associated with convective processes are long enough that pressure equilibrium is maintained. This implies that the vertical speeds of blobs are small compared with the local speed of sound, so that acoustic and shock phenomena may be neglected.

We shall omit the details here but in Appendix E these assumptions and guidelines are used to construct a simple mixing-length model of convection.

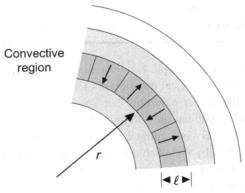

Fig. 7.9 Schematic illustration of a mixing-length approximation for convective motion. The mixing length ℓ determines the vertical distance scale over which rising and falling blobs move before merging with the surrounding medium.

7.9.3 Analysis of Solar Convection

The following example tests the mixing-length model developed in Appendix E by applying it to some quantitative estimates for subsurface convection in the Sun.

Example 7.4 The equations of Appendix E and parameters from the Standard Solar Model (see Section 10.1) may be used to investigate a mixing-length picture of the Sun's surface convection layer with a base lying at $r \simeq 0.7 R_\odot$. As you are invited to demonstrate in Problem 7.7, with the assumption $\alpha = 1$ the adiabatic temperature gradient is found to be

$$(dT/dr)_{\rm ad} \simeq -1.6 \times 10^{-4} \, {\rm K \, cm^{-1}},$$

the superadiabatic gradient is

$$\delta \, (dT/dr) = -9.5 \times 10^{-11} \, {\rm K \, cm^{-1}},$$

their ratio indicates that the connective gradient is only slightly steeper than adiabatic,

$$\frac{\delta(dT/dr)}{(dT/dr)_{\rm ad}} = 5.9 \times 10^{-7},$$

and the average convective velocity is $\bar{v} = 1 \times 10^4 \, {\rm cm \, s^{-1}} = 0.1 \, {\rm km \, s^{-1}}$, which is much less than the local speed of sound, $v_{\rm s} = 2.3 \times 10^7 \, {\rm cm \, s^{-1}}$. Since $\alpha = 1$, the corresponding mixing length is the pressure scale height,

$$\ell = \alpha H_{\rm p} = H_{\rm p} = 5.8 \times 10^4 \, {\rm km},$$

and the timescale for the blob to travel the mixing-length distance $\alpha H_{\rm p} = H_{\rm p}$ is

$$t = H_{\rm p}/\bar{v} = 5.8 \times 10^5 \, {\rm s} \simeq 6.7 \, {\rm days}.$$

For comparison, observations and detailed calculations suggest that the solar convective zone is about 200,000 km thick, with the characteristic convection cell size being about 10^4 km at the base of the convection zone and about 10^3 km at its top.

The results of Example 7.4 indicate that a mixing-length model can give a reasonable phenomenological description of convection. More generally, mixing-length models support the earlier assertion that convection often is such an efficient process that a temperature gradient only slightly steeper than the adiabatic one is sufficient to carry all flux convectively, indicating that the temperature gradient in convective regions can be well approximated by the adiabatic gradient. The earlier reservations should be kept in mind, however. For example, very near the surface of a star temperature gradients in convective regions may differ substantially from the adiabatic gradient and a slightly superadiabatic gradient may be a very poor approximation for the actual temperature gradient in that regime.

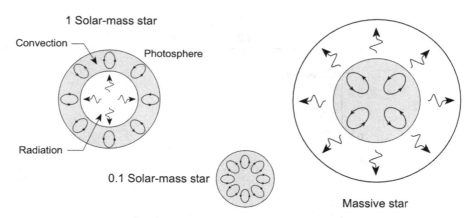

Fig. 7.10 Characteristic regions of convective and radiative transport for main sequence stars in different mass ranges. Ovals and gray shading indicate convective transport; wiggly arrows and no shading indicate radiative transport.

7.10 Examples of Stellar Convective Regions

We may expect that convection will dominate radiative transport as soon as the critical temperature gradient (7.33) is reached in a particular region of a star. Generally, it is believed that the most massive stars are centrally convective and radiative in their outer envelopes, that stars of a solar mass or so have subsurface convection zones but that the central region is not convective, and that in the least-massive stars essentially the entire star is convective. Figure 7.10 illustrates schematically and Fig. 7.11 displays simulations of the radial extent of convection in main sequence stars as a function of total mass [134]. These calculations support the schematic overview of Fig. 7.10, indicating that the lightest stars are convective through their entire volumes, that with increasing mass this convection is pushed outward to the surface layers until it disappears completely above about 1 $M_\odot$,[8] and that stars of more than about 1 $M_\odot$ have no outer convection zones but their cores are convective. Let us now examine two of these convective regions in more detail: (1) the cores of massive main sequence stars, and (2) ionization zones in the sub-surface layers of intermediate-mass stars.

7.10.1 Convection in Stellar Cores

Convection in the cores of stars is favored if the power is generated in a compact central region for two basic reasons:

(i) In the core there is a large energy flow through a small region.

---

[8] Although less-massive stars are often convective over significant portions of their outer layers, the amount of mass involved in convection is relatively small because of low densities near stellar surfaces. For example, the Sun is convective from about 70% to 90% of its radius but from Table 10.1 this convective region contains less than 3% of the Sun's mass (see also Fig. 7.11).

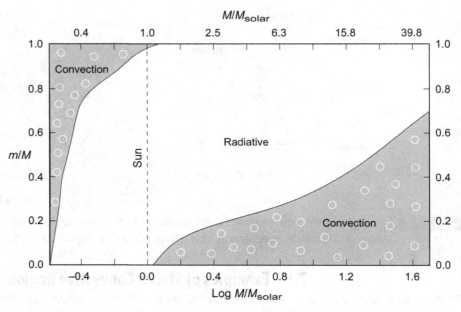

Fig. 7.11 Radial extent of convection in main sequence stars as a function of stellar mass. Convective zones are shaded gray. The vertical axis is in Lagrangian units of enclosed mass and the mass of the Sun is indicated. Figure adapted from Ref. [134]. Adapted by permission from Springer: *The Main Sequence*, Kippenhahn R. and Weigert A. COPYRIGHT (1990).

(ii) Gravity is weak in the core because at a small radius there is little enclosed mass. Hence pressure falls off gradually and rising packets of gas remain buoyant longer because they do not need to expand much to equilibrate in pressure.

Setting the radiative temperature gradient (7.43) equal to the critical temperature gradient (7.34), utilizing (7.40), and rearranging the resulting expression gives [169]

$$\frac{L(r)}{m(r)} = \frac{16\pi aGc}{3\kappa} \left(\frac{\gamma - 1}{\gamma}\right) \left(\frac{T^4}{P}\right). \qquad (7.47)$$

This defines a critical value of $L(r)/m(r)$ favoring convection over radiative diffusion. Generally, we may expect convection to develop for any regions of a star in which the luminosity reaches the critical value (which depends on location in the star). Some possibilities are indicated schematically in Fig. 7.12. Of immediate interest for the present discussion is the suggestion in Fig. 7.12(c) that convective cores of radius r and enclosed mass $m(r)$ can develop in stars if the critical value of $L(r)/m(r)$ defined by Eq. (7.47) is exceeded inside the radius r.

Example 7.5 The critical luminosity and actual luminosity for the Sun inside a radius $R = 0.1\,R_\odot$ are estimated in Problem 7.14. There it is concluded that the actual luminosity of the solar core is about a factor of two less than the critical luminosity. Therefore, we

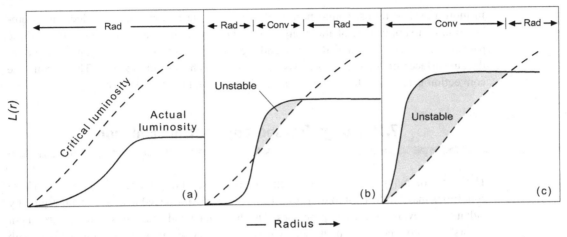

Fig. 7.12 Schematic illustration of the competition between radiative and convective energy transport for three qualitatively different situations. The critical luminosities computed from (7.47) are illustrated by the dashed curves and the actual luminosities by the solid curves. For each case the star is convectively unstable for those regions in which the actual luminosity exceeds the critical luminosity (denoted by gray shading), and radiative in the other regions.

may expect that the core of the Sun is not convective but that the centers of main sequence stars only a little more massive than the Sun are likely to be convectively unstable. The simulation displayed in Fig. 7.11 supports this conclusion.

As suggested by Example 7.5 and shown explicitly in Fig. 7.11, convective cores tend to develop in main sequence stars more massive than the Sun. The physical reason is that in these stars the CNO cycle dominates energy production and the strong temperature dependence (energy production varying as $\sim T^{16}$–T^{18}) confines the energy source to a small central region. Conversely, for less massive stars where the PP chains are the dominant power source the temperature dependence is much weaker (energy production varying as $\sim T^4$); then the power source is spread over a larger central region and core convection becomes less likely.

7.10.2 Surface Ionization Zones

Convection is favored in the surface layers of less-massive stars where constant ionization and recombination transitions are taking place. There are two basic reasons for this [169]:

1. The opacity is large, making the temperature gradient for radiative transport steep [see Eq. (7.43)].
2. The critical temperature gradient Eq. (7.34) required for convection is not very steep because there are many degrees of freedom associated with the ionization–recombination transitions and the adiabatic index [see Eq. (7.37)] is decreased toward unity.

In more physical terms, convection is favored in these regions because electron recombination can supply part of the energy to expand the rising packets of gas. Thus, these packets do not cool much as they rise and are more likely to remain buoyant. In the Sun, the subsurface convective layer is associated with such ionization zones. This subsurface convection is responsible for the *granules* observed on the solar surface.

7.11 Energy Transport by Neutrino Emission

The cores of massive stars become extremely dense and hot late in their lives. These conditions make it difficult to transport the large energy produced in a very small region by radiative or even convective processes. On the other hand, this dense, hot environment favors the production of neutrinos which—by virtue of their weak interactions with matter—can leave the star essentially unimpeded. As a result, neutrino emission is thought to be the dominant mechanism for cooling stellar cores that proceed beyond carbon burning.

Neutrino emission typically is not related to the temperature gradient and the energy outflow from neutrino cooling is directly proportional to the local rate at which the neutrinos are produced in the core of the star. (David Clayton has stated it nicely: as far as stellar structure and evolution are concerned, neutrinos function mostly as a local refrigerator [71].) The interaction of electron neutrinos with matter scales quadratically with the neutrino energy and has an average cross section that may be approximated by

$$\sigma_\nu \simeq 10^{-44} \left(\frac{E_\nu}{m_e c^2}\right)^2 \text{ cm}^2, \qquad (7.48)$$

where E_ν is the neutrino energy and $m_e c^2 = 511$ keV is the electron rest mass energy.

Example 7.6 For most processes important in stars $E_\nu/m_e c^2$ differs by less than a factor of 10 from unity and a (very) crude estimate results from approximating Eq. (7.48) for the interaction of neutrinos with stellar matter by $\sigma_\nu \simeq 10^{-44}$ cm^2. Assuming an average density 1 g cm^{-3} (corresponding roughly to a number density of $n \simeq 10^{24}$ cm^{-3}), the mean free path for an electron neutrino in average stellar matter is

$$\lambda = \frac{1}{\sigma_\nu n} \simeq \frac{1}{(10^{-44} \text{ cm}^2) \times (10^{24} \text{ cm}^{-3})} \simeq 10^{20} \text{ cm} = 1.4 \times 10^9 \, R_\odot.$$

Obviously there is little chance that the neutrino scatters from the matter on its way out of a normal star (but see Problem 14.5 and the MSW effect discussed in Chapter 12).

7.11.1 Neutrino Production Mechanisms

Several neutrino production mechanisms influence the evolution of stars. Let us summarize them briefly [55, 71, 194].

7.11 Energy Transport by Neutrino Emission

Neutrinos from β-decay: The most familiar stellar neutrino sources are β-decay and electron capture occurring in stellar energy production and nucleosynthesis. The neutrino energy losses from these sources are typically relatively small. For example, in the CNO cycle the average energy carried off by neutrinos in each traversal of the cycle is only a few percent of the total energy generated by the cycle.

Pair annihilation neutrinos: Neutrino–antineutrino pairs can be produced by the reaction $e^- + e^+ \to \nu + \bar{\nu}$. The requisite positrons (denoted by e^+ or β^+) can be produced in abundance by $\gamma + \gamma \to e^+ + e^-$, provided that $kT \geq 2m_e c^2 \sim 1$ MeV, which requires a temperature of $\sim 10^9$ K or greater. The electromagnetic interaction is *much* stronger than the weak interaction and the overwhelming majority of the positrons that are produced annihilate with electrons to give two photons by $e^- + e^+ \to \gamma + \gamma$, but about once every 10^{19} times a $\nu\bar{\nu}$ pair is produced instead. For temperatures exceeding $\sim 10^9$ K this is the dominant neutrino production mechanism, except at very high densities where the plasmon process described below dominates.

Photoneutrinos: When the energy is too low to produce significant numbers of neutrinos by pair production, neutrinos can still be produced by the reaction $e^- + \gamma \to e^- + \nu + \bar{\nu}$, which may be viewed as a Compton scattering process (Section 7.4.1) between a photon and an electron in which the exit-channel photon is replaced by a $\nu\bar{\nu}$ pair. These are called *photoneutrinos*. Except at low temperatures, photoneutrinos typically are a relatively minor component of the neutrino flux.

Plasma neutrinos: In dense ionized gases a photon can interact with free electrons to form a collective excitation called a *plasmon*.[9] Direct free-space decay of a photon to a neutrino–antineutrino pair is forbidden because it cannot satisfy both energy and momentum conservation, but a plasmon γ_{pl} can decay directly to neutrino–antineutrino pairs: $\gamma_{pl} \to \nu + \bar{\nu}$. As discussed in Box 7.3, the plasma frequency ω_0, which can be expressed in terms of the electron number density n_e and effective mass m_f as $\omega_0^2 = 4\pi n_e e^2 c^2 / m_f$, acts as an effective mass for the plasmon that increases with higher electron density.[10] Thus plasmons behave like "heavy photons" that can decay to neutrino–antineutrino pairs while still conserving energy and momentum. Plasmon decay becomes important when the relation $\hbar\omega_0 \geq kT$ is satisfied, so plasma neutrino emission is particularly important at high densities. It becomes the dominant neutrino production mechanism at all temperatures in very dense and electron-degenerate environments. For example, neutrino emission from plasmons is the dominant neutrino decay mode for cooling of hot white dwarfs.

Bremsstrahlung neutrinos: Deceleration of electrons in the Coulomb field of a nucleus can cause *bremsstrahlung* ("braking radiation") to be emitted in the form of photons (the free–free absorption discussed in Section 7.4.3 is the inverse process). At high temperature and low density there is a finite probability that bremsstrahlung leads to emission of a $\nu\bar{\nu}$

---

[9] Plasmons are not unique to stellar plasmas. For example, they are important in a variety of condensed matter systems where (among other things) they influence optical properties of materials.

[10] This formula is valid for nondegenerate matter; if electrons are degenerate ω_0 is modified by a factor that depends on the electron density (Box 7.3).

Box 7.3 — Plasmons and Neutrino–Antineutrino Pairs

A photon in free space cannot decay into a $\nu\bar{\nu}$ pair because the reaction would not simultaneously conserve angular momentum, momentum, and energy. This follows from the peculiar spin structure associated with neutrinos and antineutrinos.

Selection Rules for Neutrino Pair Production

Neutrinos have a helicity (projection of spin on the direction of motion) of $-\frac{1}{2}$ and antineutrinos a helicity of $+\frac{1}{2}$ (see Box 7.4), and the photon has a spin of 1. To add up to a helicity of 1 to conserve the angular momentum carried by the initial photon, the ν and $\bar{\nu}$ must be produced moving in exactly opposite directions. This makes it impossible to preserve both the momentum and energy of the original photon since the photon, ν, and $\bar{\nu}$ are all ultrarelativistic particles with energy pc, where p is the momentum [from Eq. (3.46) with $m = 0$].

Heavy Photons

A plasmon is a kind of "heavy photon" that acquires an effective mass through interactions with the medium, much as will be discussed in Box 12.1. The classical dispersion relation for a photon propagating through a plasma is given by [71]

$$\omega^2 = k^2 c^2 + \omega_0^2,$$

where ω is the frequency of the electromagnetic wave, k is the photon wavenumber, and ω_0 is a characteristic *plasma frequency*. Comparing with the relativistic expression for energy $E^2 = p^2 c^2 + m^2 c^4$ given in Eq. (3.46) suggests that the plasma frequency ω_0 endows the plasmon with an effective mass, which takes the value $\hbar\omega_0/c^2$ when the plasmon is quantized. The plasma frequency is given by [71]

$$\omega_0^2 = \frac{4\pi e^2 n_e}{m_e}\left[1 + \left(\frac{\hbar}{m_e c}\right)^2 (3\pi^2 n_e)^{2/3}\right]^{-1/2},$$

where n_e is the electron number density and m_e the electron mass, and the factor in square brackets is absent if the gas is nondegenerate. Because it has an effective mass, a plasmon can decay directly to a neutrino–antineutrino pair while still conserving energy and momentum. As a corollary, the minimum energy of a propagating photon is $\hbar\omega_0$, so electromagnetic waves with ω less than the plasma frequency ω_0 are damped out and do not propagate in the medium.

---

pair instead of a photon. This process depends strongly on the charge of the nucleus and is important only if heavy ions are present. It will be of minor significance for the present discussion.

Recombination and Urca neutrinos: Neutrinos can be emitted if a free electron is captured in the atomic K-shell of a completely ionized nucleus according to the reaction $e^- + (Z, A) \rightarrow (Z - 1, A) + \nu$, where (Z, A) denotes an ion with atomic number Z and atomic mass A. If electron capture on a stable nucleus (Z, A) produces a daughter nucleus $(Z - 1, A)$ that is β-unstable, the reaction sequence $e^- + (Z, A) \rightarrow (Z - 1, A) + \nu$ followed

Box 7.4   Chirality and Helicity for Neutrinos

Neutrinos and antineutrinos possess a property called *handedness* that can be expressed in terms of a *chirality quantum number* taking one of two values, denoted by the labels L ("left-handed") or R ("right-handed"). It is a fundamental but unexplained fact (see Problem 11.10 for one possible explanation) that all neutrinos of any flavor are found to be left-handed and all antineutrinos of any flavor are found to be right-handed. This asymmetry has deep significance because it leads to *parity non-conservation* for the weak interactions. If neutrinos and antineutrinos were each equal mixtures of R and L chiralities, the weak interactions would conserve parity (as do all other basic interactions).

Neutrinos and antineutrinos have an intrinsic spin **s** of magnitude $\frac{1}{2}$ (in units of $\hbar$), which by the rules of quantum mechanics can take two projections, $\pm\frac{1}{2}$, on some reference "quantization" axis. It is often convenient to take that reference axis in the direction of the momentum **p** of the particle. The projection of the spin on the direction of motion for a particle is called its *helicity*, h. If ν or $\bar{\nu}$ *are massless*, there is a one-to-one mapping between the two possible chirality states and the two possible helicity states, and it is common to refer to neutrinos as being left-handed with negative helicity and to antineutrinos as being right-handed with positive helicity. The following figure illustrates schematically,

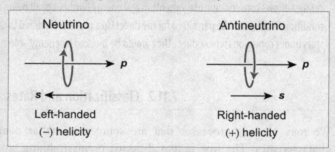

with the direction of the spin vector related to the cartoon spin motion by the usual "right hand rule" for angular velocity. If ν and $\bar{\nu}$ are massless, this labeling by helicity is a relativistically invariant characterization.[a]

If neutrinos have mass (now known to be the case), chirality and helicity are not in one-to-one correspondence and helicity is not a relativistic invariant (because $v < c$ if $m \neq 0$, so there *do exist* reference frames moving faster than the neutrino). However, neutrino masses are very small so helicity and chirality are almost the same, and the common blurring of the distinction between them for neutrinos is conceptually in error but not so serious practically since neutrino masses are tiny.

[a] This figure is a cartoon of the schematic relationship among spin, momentum, chirality, and helicity for massless ν or $\bar{\nu}$, but spin is an *internal quantum variable* that is not associated with angular motion in space. Helicity is relativistically (Lorentz) invariant if the particle is massless because to change the sign of h requires boosting to a frame moving faster than the particle (which reverses the direction of **p** but not **s** for the observer). This is impossible for massless particles, which move at lightspeed.

by $(Z - 1, A) \rightarrow (Z, A) + e^- + \bar{\nu}$ can occur. This is an example of an *Urca process* (see Box 7.5 for elaboration), which leads to neutrinos being emitted and carrying off energy with no net change in Z and A. The Urca process or a variant called the *modified area process*, are important sources of cooling for neutron stars and core collapse supernovae.

> **Box 7.5** — **Urca Processes**
>
> The label "Urca" derives from the name of a once-famous but long-closed gambling establishment – *Cassino da Urca* – in Rio de Janeiro. Allegedly the Brazilian physicist Mário Schönberg (1914–1990) remarked to George Gamow (1904–1968) that the rapid disappearance of energy from a cooling supernova was much like the rapid disappearance of a gambler's money at the Urca roulette wheel. "Urca" has since become standard terminology for a class of reactions that cool dense matter through neutrino emission with no net change in Z/A.
>
> **Direct Urca Processes**
>
> The reaction $n \rightarrow p + e^- + \bar{\nu}$ followed by $p + e^- \rightarrow n + \nu$ is an example of a *direct Urca process*. It leads to loss of energy through emission of neutrinos without a net change in the neutron and proton number. (This example starts with a neutron and ends up with a neutron and a ν and $\bar{\nu}$ that escape, carrying away energy.) The direct Urca mechanism is important in the cooling of neutron stars, for example.
>
> **Modified Urca Processes**
>
> The *modified Urca process* $n + N \rightarrow p + N + e^- + \bar{\nu}$ followed by $p + N + e^- \rightarrow n + N + \nu$ (where N is either a proton or a neutron) is like the direct Urca process except that a second nucleon (N) acts as a catalyst. The catalyst allows conservation of momentum and energy (without itself being consumed) for a broader range of conditions than would be permitted for the direct Urca process. Modified Urca is important for cooling neutron stars under conditions where direct Urca would be blocked by energy–momentum conservation.

7.11.2 Classification and Rates

We may classify processes that are sources of stellar neutrinos according to several characteristics. The first is that β-decay, bremsstrahlung, and Urca neutrino production all involve nuclei, while the other neutrino sources listed above are purely leptonic and governed by the Standard Model for electroweak interactions (see Section 11.2). The second is that because of lepton family number conservation in the Standard Model the processes that create neutrino–antineutrino pairs can produce any flavor pair of neutrino and antineutrino, but the other processes deal only with electron neutrinos or antineutrinos. For example, the Sun produces neutrinos predominantly by nuclear processes and so (in the absence of flavor oscillations) the flux of solar neutrinos is essentially all of electron flavor (Chapter 11), but in the high-density, high-temperature environment of a core collapse supernova neutrino pair production is rampant and most of the supernova energy is carried off by neutrinos and antineutrinos of *all flavors* (Chapter 20).

The preceding neutrino reactions also may be classified according to their importance for neutrino cooling of stars. Under conditions encountered in normal stellar evolution, cooling by neutrino emission is typically dominated by (1) pair production neutrinos, (2) photoneutrinos, and (3) plasma neutrinos. Approximate formulas for energy production rates are given in Ref. [38] and Figs. 7.13 and 7.14 show energy production rates from these dominant processes, the schematic dependence on temperature and density, and total neutrino energy production rates summed over pair production, photoneutrino production, and plasma neutrino production. Since the neutrinos are assumed to leave the

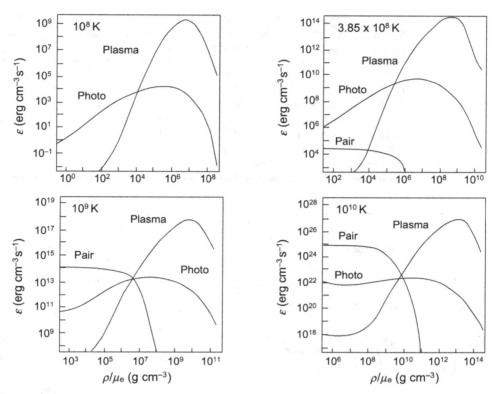

Fig. 7.13 Neutrino energy emission rates at four different temperatures (adapted from Ref. [38]) Adapted from G. Beaudet, V. Petrosian, and E. E. Salpeter, *Astrophys. J.*, **150**, 979 (1967). 10.1086/149398.

star unimpeded, the neutrino energy production rates also are the neutrino energy loss rates. These results may be used to analyze the importance of neutrino emission for later stages of stellar evolution, as illustrated in Example 7.7.

Example 7.7 Typical conditions for silicon burning correspond to $T = 3$–5×10^9 K and $\rho = 10^5$–10^7 g cm^{-3}. From Figs. 7.13 and 7.14(a), neutrinos are produced mostly by pair annihilation in this temperature and density range. From Fig. 7.14(b) the total neutrino energy production rate under these conditions is $\varepsilon_\nu/\rho \simeq 10^{12}$–$10^{15}$ erg g^{-1} s^{-1}. This is comparable to the Si-burning energy generation rate. For example, in the simulation of silicon burning shown in Fig. 6.10 the peak energy generation rate was 2.8×10^{12} erg g^{-1} s^{-1}. This suggests that most of the energy released by silicon burning is being transported from the star by neutrino emission rather than by radiative diffusion or convection. See Problem 7.13 and Table 7.2 for further elaboration on neutrino cooling in late burning stages.

An example of neutrino cooling dominating photon emission in late burning stages is given in Table 7.2 for evolution of a 20 $M_\odot$ star. It is seen that neutrino luminosity L_ν exceeds photon luminosity L_γ by orders of magnitude for carbon burning and beyond in this case.

Table 7.2 Photon and neutrino luminosities for a 20 solar mass star [27]

Fuel	ρ_c (g cm^{-3})[†]	T_c (10^9 K)[†]	Duration (yr)	L_γ (erg s^{-1})	L_ν (erg s^{-1})
Hydrogen	5.6	0.040	1.0×10^7	2.7×10^{38}	—
Helium	9.4×10^2	0.19	9.5×10^5	5.3×10^{38}	$< 1.0 \times 10^{36}$
Carbon	2.7×10^5	0.81	300	4.3×10^{38}	7.4×10^{39}
Neon	4.0×10^6	1.7	0.38	4.4×10^{38}	1.2×10^{43}
Oxygen	6.0×10^6	2.1	0.50	4.4×10^{38}	7.4×10^{43}
Silicon	4.9×10^7	3.7	0.0055	4.4×10^{38}	3.1×10^{45}

[†] The columns ρ_c and T_c give the critical density and critical temperature, respectively, to ignite a fuel.

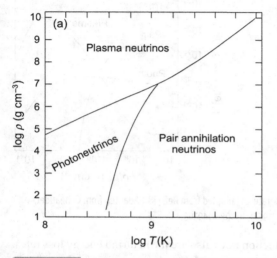
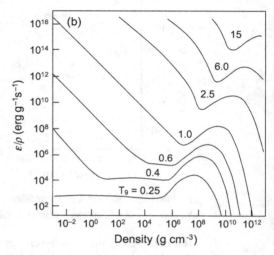

Fig. 7.14 (a) Dependence of dominant mechanisms for neutrino energy losses on density and temperature (adapted from Ref. [38]). Adapted from G. Beaudet, V. Petrosian, and E. E. Salpeter, *Astrophys. J.*, **150**, 979 (1967). 10.1086/149398. (b) Energy production rates by neutrino processes (adapted from Ref. [38]). Adapted from G. Beaudet, V. Petrosian, and E. E. Salpeter, *Astrophys. J.*, **150**, 979 (1967). 10.1086/149398. Curves are labeled by temperatures in units of 10^9 K.

7.11.3 Coherent Neutrino Scattering

The Standard Electroweak Theory of elementary particle physics predicts that neutral weak currents (those mediated by the Z^0 gauge boson) can scatter coherently off the A nucleons of a composite nucleus rather than off individual nucleons. The neutrino–nucleon scattering cross section is of the form [compare Eq. (7.48)] $\sigma_{\text{nucleon}} \propto E_\nu^2$, where E_ν is the neutrino energy, but the coherent cross section on a nucleus of nucleon number A takes the form

$$\sigma_{\text{coherent}} \propto A^2 E_\nu^2. \tag{7.49}$$

For the most massive stars, Si burning produces iron-group nuclei in the core and the coherent cross section is enhanced by a factor $A^2 \sim (56)^2 \sim 3000$ relative to normal nucleonic weak interactions. Enhancements of this magnitude, coupled with the increase in the basic weak interactions strength because of the very high temperature and density of the core, can lead to very large neutrino interactions in hot, dense matter. Because of the large mass difference between neutrinos and heavy nuclei, coherent scattering transfers momentum but not much energy, so it is nearly elastic. However, the shorter mean free path implied by the enhanced neutrino cross sections causes neutrinos to stay in the core longer before escaping, increasing the probability that they undergo an inelastic scattering that does change the energy. It will be argued in Section 20.3.3 that coherent elastic scattering of neutrinos from composite nuclei through the neutral weak current can have a large influence on core collapse in a supernova explosion.

Background and Further Reading

The presentation in this chapter has adapted often from the discussion of energy transport in Phillips [169]; see also Kippenhahn, Weigert, and Weiss [134]; Carrol and Ostlie [68]; and Hansen, Kawaler, and Trimble [107]. For a summary of neutrino reactions in stars, see Clayton [71]; Bowers and Deeming [55]; and Salaris and Cassisi [194].

Problems

7.1 Derive an approximate expression for the ratio of coefficients for radiative and electronic diffusion, assuming the dominant photon scattering process to be Thomson scattering and complete ionization. Estimate this ratio for the Sun.***

7.2 Prove that the electron number density n_e and ion number density n_i are given approximately by

$$n_e \simeq \frac{(1+X)\rho}{2M_u} \qquad n_i \simeq \frac{(2X + \frac{1}{2}Y)\rho}{2M_u},$$

where X is the hydrogen and Y the helium mass fractions.***

7.3 Use data from Table 10.1 and Fig. 7.2 to confirm the solar mean free path entries in Table 7.1.

7.4 By analyzing rising blobs of fluid in a gravitational field, show that the Schwarzschild condition for convective instability may be expressed in the form (7.33). *Hint:* For an adiabatic process $P \propto \rho^\gamma$.***

7.5 Demonstrate that the second form of the adiabatic temperature gradient in Eq. (7.34), $(dT/dr)_{ad} = -g/c_P$, follows from the first. *Hint:* See Section 3.3.***

7.6 Assume the photosphere of the Sun to be an ideal gas of hydrogen and helium with the typical ZAMS abundance, and to have a density of $1.8 \times 10^{-7}\,\mathrm{g\,cm^{-3}}$ and a temperature of 5800 K. What is the scale height in the solar photosphere?

7.7 Apply the mixing-length model developed in Section 7.9 and Appendix E to the analysis of solar convection, assuming a Standard Solar Model (Section 10.1) and ideal gas behavior. Calculate the mixing length, the adiabatic temperature gradient, the superadiabatic gradient, the percentage by which the superadiabatic gradient exceeds the adiabatic one assuming all transport in this region is convective, the average convective velocity relative to the local sound speed, and the timescale for convection set by the average time for a blob to travel one mixing length. Make your estimates for the convective region with a base around $r = 0.7 R_\odot$ and assume a mixing-length parameter $\alpha = 1$ and a mean molecular weight of 0.6.***

7.8 Stars are commonly treated as if they were in thermodynamic equilibrium but the whole discussion of energy transport in stars indicates that this cannot generally be a correct assumption. Real stars radiate net energy into space and have temperature gradients and thus are not blackbodies. However, if the characteristic distance over which the temperature changes is large compared with the mean free paths of particles and photons, the idea of local thermodynamic equilibrium (LTE) may still be retained: particles and photons cannot quickly escape from their local "box" and within that local box an approximately constant temperature exists. It is common (particularly in discussions of stellar atmospheres) to define a *temperature scale height* H_T that is analogous to the pressure scale height introduced in Eq. (7.44),

$$H_\text{T} = \frac{T}{|dT/dr|},$$

and to use this scale as the characteristic one for measuring how rapidly the temperature is changing. Show that in the deep interior of stars the LTE assumption is valid for both matter and photons, but that in the photosphere LTE is typically expected for matter but is problematic for photons. (Note: this result notwithstanding, the assumption of LTE is often used for stellar atmospheres. Notice also that this result is an inevitable consequence of the *definition* of the photosphere, since the photosphere is precisely where photon mean free paths must become long.) You may use the Standard Solar Model (see Table 10.1) for representative conditions in the solar interior, and you may use that for a typical atmospheric model of the Sun (for example, see Table 14.9 of Ref. [75]) the temperature changes from 5980 K to 6180 K over a distance of 20 km, in a surface region having density $2.5 \times 10^{-7}\,\text{g cm}^{-3}$ and opacity $\sim 0.25\,\text{cm}^{-2}\,\text{g}^{-1}$ at the dominant photon wavelengths.

7.9 Show that Eq. (7.36) follows from Eq. (7.33). *Hint*: This problem is simple, except that you must be careful about the signs of factors multiplying both sides of the inequality.

7.10 Estimate the adiabatic index for Earth's atmosphere at standard temperature and pressure. *Hint*: Is the gas ideal and will molecular vibrational modes play a role under those conditions? What is the corresponding speed of sound?

7.11 The formulas that have been employed for convective instability and related issues also may be applied to planetary atmospheres. For Mars, assume an atmosphere composed entirely of carbon dioxide at a temperature of 220 K, and that the density of the atmosphere at the surface is $0.02\,\text{kg m}^{-3}$. Calculate the gravitational acceleration at the surface and the scale height of the atmosphere. What is the corresponding atmospheric pressure at the surface?

7.12 (a) Use Archimedes' principle to derive a formula for the buoyancy acceleration a_b of a packet of gas having density ρ immersed in a larger volume of the same gas having density ρ'. (b) Consider a layer of air over a parking lot at a temperature of 28 °C, with the temperature of the surrounding air being 27 °C. What is the buoyancy acceleration of this layer of air?

7.13 For a density of 10^6 g cm^{-3} and temperatures of $T_9 = 3$ and $T_9 = 1$, respectively,

(a) What is the radiation energy density of the photon gas for each case?

(b) Clayton [71] gives approximate formulas for the neutrino energy loss rate as

$$\frac{df_\nu}{dt} \simeq 4.9 \times 10^{18} T_9^3 \exp(-11.86/T_9) \,\mathrm{erg\,cm^{-3}\,s^{-1}} \quad (T_9 \lesssim 1),$$

$$\frac{df_\nu}{dt} \simeq 4.6 \times 10^{15} T_9^9 \,\mathrm{erg\,cm^{-3}\,s^{-1}} \quad (T_9 \gtrsim 3),$$

assuming the neutrinos to originate from the pair production mechanism. How good is this assumption for the two temperatures? Compare eyeball interpolation of the fluxes in Fig. 7.14 with the results of the above formulas for $T_9 = 3$ and $T_9 = 1$.

(c) Estimate the timescale for the thermal energy of the photons to be radiated as neutrinos using the neutrino energy loss rates eyeball-interpolated from Fig. 7.14 for the two temperatures. How does this compare with the thermal adjustment (Kelvin–Helmholtz) timescale for the star, assuming it to have a mass of 18 $M_\odot$?

What conclusions do you draw from these results about cooling of the star?***

7.14 Use Eq. (7.47), parameters for the Sun from the Standard Solar Model in Table 10.1, and solar opacities estimated from Fig. 7.2 to show that for the volume contained in the inner 10% of the solar radius the luminosity is less than critical. Thus, we may expect the Sun to not be convective in its central regions.***

8 Summary of Stellar Equations

The preceding chapters have introduced and explained the essential physical concepts that govern stars. These will be employed in succeeding chapters to understand stellar evolution: how stars are born, how they live their normal lives, how they die, and what this stellar life cycle entails for our overall understanding of the Universe. As preparation for that task, this brief chapter pulls together the material introduced in Chapters 1–7 to summarize in a single place the essential equations that govern the physics of stars, and introduces some methods of finding solutions for these equations. As will be seen in the subsequent discussion, the equations governing the structure and evolution of stars often can be expressed in remarkably compact form but their actual application and solution in realistic systems can entail substantial complication.

8.1 The Basic Equations Governing Stars

Let us now summarize the minimal set of equations that might be used to describe stellar structure and stellar evolution. These may be broken down into equations governing (1) hydrostatic equilibrium, (2) luminosity, (3) temperature gradients, (4) composition changes caused by nuclear reactions, and (5) equations of state.

8.1.1 Hydrostatic Equilibrium

From Chapter 4 one requires equations that describe the conditions for hydrostatic equilibrium, expressed either in Eulerian or Lagrangian form. In Eulerian form, these equations may be written as

$$\frac{dm}{dr} = 4\pi r^2 \rho(r) \quad \text{[mass conservation (4.4)]}$$

$$\frac{dP}{dr} = -\frac{Gm(r)}{r^2} \rho \quad \text{[hydrostatic equilibrium (4.10)]},$$

where P is pressure, $m(r)$ is total mass contained within the radius r, the mass density is ρ, and the labels indicate the physical interpretation of the equation and the original equation number where it was introduced. (The corresponding Lagrangian equations are given in Table 4.1.) The first of these ensures that no mass is gained or lost, while the second establishes an equilibrium between inwardly directed gravitational forces and outwardly directed forces that arise from the radial pressure gradient.

8.1.2 Luminosity

Next, we require an equation for the luminosity L, which was given in differential form in Chapter 7 as

$$\frac{dL}{dr} = 4\pi r^2 \varepsilon(r) \qquad \text{[luminosity (7.41)]},$$

where $\varepsilon(r)$ is the power per unit volume produced in a concentric shell at r with thickness dr. The total luminosity of a concentric shell at any radius r is then the integral of the differential luminosity from the center to the radius r.

8.1.3 Temperature Gradient

The radial temperature gradient for a star in hydrostatic equilibrium may be expressed by one of two equations,

$$\frac{dT}{dr} = -\frac{3\rho(r)\kappa(r)}{4acT^3(r)}\frac{L(r)}{4\pi r^2} \qquad \text{[radiative } T \text{ gradient (7.43)]},$$

$$\frac{dT}{dr} = \left(\frac{\gamma-1}{\gamma}\right)\frac{T}{P}\frac{dP}{dr} \qquad \text{[convective } T \text{ gradient (7.34)]},$$

where T is the temperature, P is the pressure, κ is the opacity, L is the luminosity, and γ is the adiabatic coefficient. These two equations for the temperature gradient are to be employed in the complementary fashion suggested in Section 7.8: the radiative gradient (7.43) should be used unless the condition (7.33) for convective instability

$$\left(\frac{dT}{dr}\right)_{\text{actual}} < \left(\frac{dT}{dr}\right)_{\text{ad}} \equiv \left(\frac{\gamma-1}{\gamma}\right)\frac{T}{P}\frac{dP}{dr},$$

is satisfied (meaning that the actual temperature gradient is steeper than the adiabatic gradient), in which case the adiabatic gradient (7.34) should be used.

8.1.4 Changes in Isotopic Composition

The nuclear reactions powering stars lead to elemental composition changes in addition to releasing energy. The abundances of isotopes are described by the coupled set of ordinary differential equations called a *thermonuclear reaction network* that is discussed in Box 6.1 (see also Appendix D), which takes the general form

$$\begin{aligned} dY_1/dt &= F_+^{(1)}(Y) - F_-^{(1)}(Y) \\ dY_2/dt &= F_+^{(2)}(Y) - F_-^{(2)}(Y) \\ &\vdots \\ dY_N/dt &= F_+^{(N)}(Y) - F_-^{(N)}(Y), \end{aligned} \qquad (8.1)$$

with one equation for each of the N species, where $Y = (Y_1, Y_2, \ldots Y_N)$ is a vector of isotopic abundances, $F_+^{(i)}(Y)$ denotes a sum of terms that increase the value of Y_i over time, $F_-^{(i)}(Y)$ denotes a sum of terms that decrease the value of Y_i over time, and the argument Y indicates that in general these terms depend on the other Y_j in the system. For elementary examples the number of species (and differential equations) N may be only a few, but more realistic simulations may require hundreds or thousands of isotopic species with a differential equation for each in the reaction network.

8.1.5 Equation of State

Generally, the preceding equations are not closed without a further constraint imposed by the equation of state describing the star, which takes the schematic form

$$P = P(T, \rho, X_i, \ldots) \qquad \text{[equation of state (3.1)]},$$

giving a relationship among thermodynamic variables for the problem. The equation of state for stars will depend strongly on the microscopic physics of the gas and radiation. If the density is not too high an ideal gas equation of state is often a good initial approximation for stars, but at high density or in regions of partial ionization the equation of state can be more complex. In realistic simulations the equation of state often does not have a completely analytical form and must be computed numerically.

8.2 Solution of the Stellar Equations

The equations listed above represent a considerably simplified description of a star, but even in this simplified form they typically must be solved on large computers and the solution for realistic cases presents formidable numerical problems. Relatively specialized techniques must be used for some aspects of the solution because of the boundary conditions required for a star, and because these equations couple processes having characteristic time and length scales that may differ by many orders of magnitude.

Example 8.1 In the large-scale numerical simulation of a Type Ia supernova explosion (see Chapter 20) the thermonuclear burning front that passes through the white dwarf matter and consumes it in about a second can have a width as small as millimeters, while the scale characteristic of the white dwarf that is incinerated is of order 10,000 km, and in the burn front simulation temperatures are observed to change at a rate as high as $10^{17}\,\text{K}\,\text{s}^{-1}$. The numerical challenges of dealing with such extremes are among the most serious encountered for computer simulations of any known physical system.

On the other hand, under certain approximations the equations of stellar structure decouple from each other in such a way that they can be solved more simply. Such an approach yields solutions that are only approximate, but may capture enough of the essential physics to be useful in particular contexts. As an example, in Section 8.4 we will show that under the approximation that the equation of state is of polytropic form it is possible to decouple the equations describing hydrostatic equilibrium from the rest of the equations and to obtain solutions to these equations that can be expressed in terms of simple formulas. Then in Section 8.5 a brief overview will be presented of the issues associated with solving the full set of stellar equations numerically on a computer.

8.3 Important Stellar Timescales

Before discussing solution of the stellar structure equations it is appropriate to review the timescales important for stellar evolution that have been introduced in preceding chapters. This is useful to set solutions in context, but also because for many problems an understanding of relevant timescales can yield important physical insight without having to solve the full set of equations governing stellar structure and evolution. Timescales have been introduced earlier through a variety of physical arguments but the discussion can be systematized by noting that a timescale τ_s characteristic of some important physical process represented by a quantity s may be defined as $\tau_s = s/\dot{s}$.[1] The three most important stellar timescales are summarized below and in Table 8.1.

1. *Dynamical timescale*: In Section 4.4 we saw that a dynamical timescale can be introduced in terms of a characteristic time to either expand or contract if there is a deviation from hydrostatic equilibrium. Taking the former as representative, a characteristic dynamical time can be estimated as the ratio of the radius R to the escape velocity $v_{\rm esc}$,

$$\tau_{\rm dyn} = \frac{R}{v_{\rm esc}} = \sqrt{\frac{R^3}{2GM}} \sim \sqrt{\frac{1}{G\bar{\rho}}} \qquad (8.2)$$

where $v_{\rm esc} = (2GM/R)^{1/2}$ was used and an average density $\bar{\rho} = 3M/4\pi R^3$ has been introduced in the last step. The time-dependent differential equation characterized by this timescale is the hydrodynamical equation (4.14). Applying Eq. (8.2) to the Sun gives a dynamical timescale of about 55 minutes (see Example 4.1).

Table 8.1 Some important stellar timescales

Timescale	Characteristic value	Value for Sun	Discussion
Dynamical	$\tau_{\rm dyn} \sim \sqrt{\dfrac{R^3}{2GM}}$	55 min	Section 4.4
Thermal adjustment	$\tau_{\rm therm} \sim \dfrac{GM_0^2}{RL}$	3×10^7 yr	Section 4.9
Nuclear burning	$\tau_{\rm nuc} \sim \eta \dfrac{Mc^2}{L}$	10^{10} yr	Example 6.1

---

[1] This is just a generalization of the standard example from introductory physics of estimating a time to travel some distance x as $t = x/\dot{x} = x/v$, where v is the average velocity.

2. *Thermal adjustment timescale*: The thermal adjustment (Kelvin–Helmholtz) timescale is associated with time for a star to shed thermal energy. It may be approximated by

$$\tau_{\text{therm}} = \frac{U}{L} = \frac{GM^2}{LR}, \tag{8.3}$$

where U is the internal energy and L the luminosity, and the virial theorem has been used to estimate $U \sim GM^2/R$. The time-dependent differential equation governed by this timescale is given in Eq. (4.35). Applying Eq. (8.3) to the Sun gives a thermal adjustment timescale of about 3×10^7 yr (see Section 4.9).

3. *Nuclear burning timescale*: The time to burn the star's hydrogen fuel may be approximated by the timescale

$$\tau_{\text{nuc}} = \frac{\eta M_0 c^2}{L}, \tag{8.4}$$

where η is the efficiency for conversion of mass into energy in hydrogen fusion (about 0.007; see Example 5.3), M_0 is the mass of hydrogen available to burn in the star, and it has been assumed that $L = L_{\text{nuc}}$ in thermal equilibrium (where L_{nuc} is the luminosity associated with the nuclear reactions). A representative differential equation governed by the timescale (8.4) is given by Eq. (6.2). It will be argued in later chapters that only about 10% of a star's mass can be accessed under conditions suitable for main sequence hydrogen burning, so realistically the mass entering Eq. (8.4) should be $M_0 \sim 0.1 M$, where M is the mass of the star. For the Sun this gives $\tau_{\text{nuc}} \sim 10^{10}$ yr. In Example 6.1 more realistic estimates give $\tau_{\text{nuc}} \sim 6 \times 10^9$ yr for the Sun.[2]

In all these cases a more careful solution of the problem utilizing the differential equation responsible for the timescale yields values comparable to those deduced from simple timescale analysis, differing by numerical factors that are often of order one. This shows clearly the potential of timescale analysis to give correct order of magnitude estimates in stellar structure problems.

8.4 Hydrostatic Equilibrium for Polytropes

It is useful to have a solvable formalism that describes a star in hydrostatic equilibrium. One way to accomplish that is to approximate the equation of state so that the mechanical (hydrostatic) properties may be decoupled from the thermal and transport properties, allowing the equations for hydrostatic equilibrium to be solved separately from the other equations governing the star.

---

[2] Stars also burn heavier fuels after they leave the main sequence and we can associate additional nuclear burning timescales with each fuel. But these timescales typically are small compared with that for hydrogen burning on the main sequence. Thus the time to consume available hydrogen on the main sequence is a rough approximation to the total nuclear burning timescale for a star.

8.4.1 Lane–Emden Equation and Solutions

The two first-order differential equations of hydrostatic equilibrium, (4.4) and (4.10), may be combined to give the second-order differential equation

$$\frac{1}{r^2}\frac{d}{dr}\left(\frac{r^2}{\rho}\frac{dP}{dr}\right) = -4\pi G\rho. \tag{8.5}$$

Let us solve this equation using a polytropic equation of state

$$P = K\rho^{(n+1)/n} \qquad K = \frac{P_c}{\rho_c^{(n+1)/n}}, \tag{8.6}$$

where n is the polytropic index, $\rho_c \equiv \rho(r=0)$ is the central density, and $P_c \equiv P(r=0)$ is the central pressure.

Dimensionless variables: To solve Eq. (8.5) with the equation of state (8.6) it is convenient to introduce dimensionless variables ξ and θ through

$$\rho = \rho_c \theta^n \qquad r = a\xi \qquad a = \sqrt{\frac{(n+1)K\rho_c^{(1-n)/n}}{4\pi G}} = \sqrt{\frac{(n+1)P_c}{4\pi G\rho_c^2}}, \tag{8.7}$$

where a has the dimension of length and θ parameterizes the density in units of the central density and therefore ranges between 0 and 1. Then the differential equation embodying hydrostatic equilibrium for a polytropic equation of state may be expressed in terms of the new independent variable ξ and new dependent variable $\theta(\xi)$ as

$$\frac{d^2\theta}{d\xi^2} + \frac{2}{\xi}\frac{d\theta}{d\xi} + \theta^n = 0, \tag{8.8}$$

which is known as the *Lane–Emden equation*.[3] The boundary conditions in these new variables are

$$\theta(0) = 1 \qquad \theta'(0) \equiv \left.\frac{d\theta}{d\xi}\right|_{\xi=0} = 0, \tag{8.9}$$

where the first equation follows from the requirement that the correct central density $\rho_c = \rho(0)$ be reproduced and the second from requiring that the pressure gradient dP/dr vanish at the origin (a necessary condition for hydrostatic equilibrium). This equation may be integrated outward from the origin ($\xi = 0$) until the point $\xi \equiv \xi_1$ where θ first vanishes, which defines the surface of the star [since at this point $\rho = P = 0$; see Eq. (8.7)].

Lane–Emden solutions: Equation (8.8) has analytical solutions for the special cases $n = 0, 1$, and 5, but in the physically most-interesting cases the equations must be integrated numerically to define the Lane–Emden constants ξ_1 and $\xi_1^2|\theta'(\xi_1)|$ for given n. These are tabulated for various values of n and γ in Table 8.2, and the corresponding

---

[3] Often the Lane–Emden equation is stated in the equivalent form

$$\frac{1}{\xi^2}\frac{d}{d\xi}\left(\xi^2\frac{d\theta}{d\xi}\right) = -\theta^n.$$

This is more compact but not as useful computationally as the form (8.8).

Table 8.2 Lane–Emden constants

n	$\gamma \equiv \frac{n+1}{n}$	ξ_1	$\xi_1^2 \lvert \theta'(\xi_1) \rvert$
0	∞	2.4494	4.8988
0.5	3	2.7528	3.7871
1.0	2	3.14159	3.14159
1.5	5/3	3.65375	2.71406
2.0	3/2	4.35287	2.41105
2.5	1.4	5.35528	2.18720
3.0	4/3	6.89685	2.01824
4.0	5/4	14.97155	1.79723
4.5	1.22	31.83646	1.73780
5.0	1.2	∞	1.73205

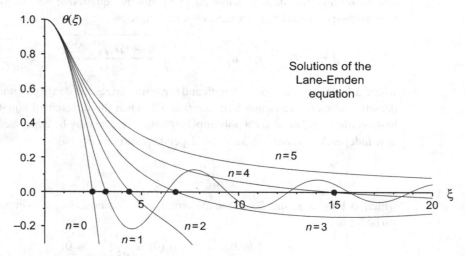

Fig. 8.1 Lane–Emden solutions $\theta(\xi)$ obtained by integrating Eq. (8.8) numerically for integer n from 0 to 5. For each curve a dot indicates the value ξ_1 given in Table 8.2. The solutions for $n \geq 5$ do not intersect the ξ axis; they are not suitable for describing stars because they correspond to configurations without a finite radius.

solutions are plotted in Fig. 8.1 for integer n from 0 to 5. Not all Lane–Emden solutions are of physical interest in stellar applications. In hydrostatic equilibrium the density must decrease monotonically from the center and vanish at a finite radius (surface of the star). The scaled density $\theta(\xi)$ for a Lane–Emden solution decreases monotonically and has a zero at finite $\xi \equiv \xi_1$ only if $n < 5$, so only these solutions are candidates for describing stars in hydrostatic equilibrium. Of those only polytropic indices around $n = \frac{3}{2}$ and 3 lead to physical properties such as density distributions that have some resemblance to those of actual stars. Pressure profiles computed for polytropic equations of state with several values of n are shown in Fig. 8.2. The $n = 3$ polytrope is seen to bear some resemblance to the actual pressure profile of the Sun, which was inferred from the Standard Solar Model (Section 10.1).

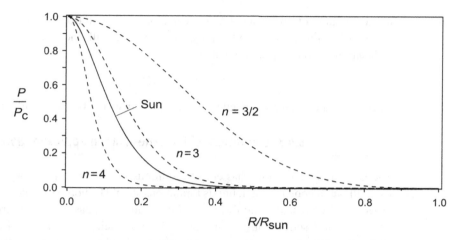

Fig. 8.2 Pressure versus radius for polytropes of index n. Also shown is the pressure profile of the Sun calculated in the Standard Solar Model that will be described in Chapter 10. The $n = 3$ polytrope yields a profile not too different from the actual pressure profile of the Sun.

8.4.2 Computing Physical Quantities

The Lane–Emden solution and the transformation equations (8.7) may be used to express quantities of physical interest in terms of the Lane–Emden constants tabulated in Table 8.2 for given values of the polytropic index n. For example, from Eq. (8.7) the stellar radius R is

$$R = a\xi_1 = \sqrt{\frac{(n+1)K}{4\pi G}}\, \rho_c^{(1-n)/2n}\xi_1, \qquad (8.10)$$

the central density ρ_c is given by (Problem 8.6)

$$\rho_c = -\left(\frac{\xi_1}{3\theta'(\xi_1)}\right)\frac{M}{\frac{4}{3}\pi R^3}, \qquad (8.11)$$

and the mass M is given by (Problem 8.3)

$$M \equiv 4\pi a^3 \rho_c \left[-\xi^2 \frac{d\theta}{d\xi}\right]_{\xi=\xi_1}$$

$$= 4\pi \left[\frac{(n+1)K}{4\pi G}\right]^{3/2} \rho_c^{(3-n)/2n}\xi_1^2|\theta'(\xi_1)|. \qquad (8.12)$$

Eliminating ρ_c between Eqs. (8.10) and (8.12) gives a general relationship between the mass and the radius,

$$M = 4\pi R^{(3-n)/(1-n)}\left(\frac{(n+1)K}{4\pi G}\right)^{n/(n-1)}\xi_1^{(3-n)/(n-1)}\xi_1^2|\theta'(\xi_1)|, \qquad (8.13)$$

for a solution of the Lane–Emden equation with polytropic index n. As shown in Problem 8.9, for a Lane–Emden polytropic solution with index n, the central density ρ_c and central pressure P_c are related by

$$P_c = \beta_n M^{2/3} \rho_c^{4/3}, \tag{8.14}$$

where β_n is a constant independent of M and ρ_c that slowly varies with m [176].

8.4.3 Limitations of the Lane–Emden Approximation

The Lane–Emden equation has elegant solutions with a direct physical interpretation that we shall put to good use later, but its limitations should be borne in mind. It reflects only the property of hydrostatic equilibrium, and then only for a gas with a polytropic equation of state. Thus it describes only the mechanical part of stellar structure and has nothing to say about temperature gradients and energy transport, and their coupling to the full problem.

There are two general situations where a polytropic equation of state for a star may be a reasonable approximation (and thus the Lane–Emden solutions very useful) [134]. The first is when the realistic equation of state contains a temperature dependence in addition to a density dependence (for example, an ideal gas), but additional physical constraints between T and P lead to an equation of state having a polytropic form. For example, the adiabatic constraint applied to an ideal gas leads to a polytropic equation of state [see the first of Eqs. (3.40) with $V \propto \rho^{-1}$]. In such a case the temperature is effectively fixed by a constraint $T = T(P)$ and not by coupling to the full set of equations describing stellar structure. The second is when the realistic equation of state actually is at least roughly approximated by the polytropic form (8.6). This is often the case in very dense matter such as white dwarfs and neutron stars, as will be discussed further in Chapter 16.

8.5 Numerical Solution of the Stellar Equations

Numerical solution of the full set of equations describing stellar structure and stellar evolution is a specialized topic that would take us too far afield for the present discussion. As noted previously, in addition to features shared with many other large-scale scientific computations, the stellar structure and evolution problem has some specific features that complicate obtaining numerical solutions. These issues fall primarily into two categories: boundary conditions and extreme scale differences in the equations.

1. The boundary conditions are unusual in that for the structure of a star in hydrostatic equilibrium some of the boundary conditions must be imposed at the center and some at the surface. This requires specialized techniques to ensure compatibility of the solutions with the two sets of boundary conditions.
2. The scale issue arises because in both spatial and time dimensions quantities of relevance may differ in size by many orders of magnitude. This leads to stability issues that also require the use of specialized numerical methods. Large-scale scientific computer simulations in other disciplines have similar problems of scale, but few are as

severe as in the stellar problem. To give one representative example, solution of even the relatively simple case of equations (8.1) governing changes in isotopic composition and energy release for the proton–proton chains with only the eight nuclear species appearing in Fig. 5.2 involve timescales that can differ by 10–20 orders of magnitude. (In mathematical parlance, equations governed by more than one timescale that differ by large amounts are said to be *stiff*.) Such equations can be solved only with numerical methods customized to deal with this issue.

A more extensive discussion of the particular issues associated with solving the stellar equations and methods for obtaining numerical solutions may be found in Ref. [134].

Background and Further Reading

Discussions of the basic equations governing stellar structure and evolution may be found in Phillips [169]; Kippenhahn, Wiegert, and Weiss [134]; Prialnik [176]; Carrol and Ostlie [68]; Bowers and Deeming [55]; and Hansen, Kawaler, and Trimble [107]. Kippenhahn, Wiegert, and Weiss [134] summarize how these equations are solved in practical applications and Bodenheimer et al. [51] give an overview of the numerical methods used for such solutions on computers. Press et al. [175] address general issues of scientific computing, with algorithms and sample code. Prialnik [176] gives a particularly clear overview of stellar timescales that has influenced our presentation.

Problems

8.1 Express the basic equations of stellar structure (4.4), (4.10), (7.41), (7.43), and (7.34) in Lagrangian form.

8.2 Apply pure scaling (dimensional) arguments to the equation for hydrostatic equilibrium to obtain directly the Lane–Emden result

$$M \propto R^{(3-n)/(1-n)}$$

relating the mass M and radius R for a star with a polytropic equation of state

$$P = K\rho^{1+1/n} \equiv K\rho^{\gamma},$$

where K is a constant, P is the pressure, ρ is the mass density, and n is the polytropic index.

8.3 Use Eqs. (8.7) and (8.8) to demonstrate that the mass for a Lane–Emden solution is given by Eq. (8.12).***

8.4 (a) Show that Eq. (8.5),

$$\frac{1}{r^2}\frac{d}{dr}\left(\frac{r^2}{\rho}\frac{dP}{dr}\right) = -4\pi G\rho,$$

with appropriate boundary conditions is equivalent to the usual formulation of the hydrostatic equilibrium problem for Newtonian gravity.

(b) Prove that Eq. (8.8) with the assumption (8.6) is equivalent to this result and therefore describes hydrostatic equilibrium for a star governed by a polytropic equation of state.

(c) Show formally that the parameter a defined in Eq. (8.7) has the dimension of length.

(d) Show that the coefficient K in Eq. (8.6) and the parameter a in Eq. (8.7) can be written as

$$K = \frac{P_c}{\rho_c^{(n+1)/n}} \qquad a = \sqrt{\frac{(n+1)P_c}{4\pi G \rho_c^2}},$$

respectively. Thus, for a given polytropic index n the parameters K and a are determined completely by the central pressure P_c and the central density ρ_c.

8.5 Using the form of the Lane–Emden equation given in footnote 3 after Eq. (8.8), find an analytical solution for $n = 0$ and evaluate the corresponding Lane–Emden constants ξ_1 and $\xi_1^2 |\theta'(\xi_1)|$.

8.6 Show that for the Lane–Emden solution the central density ρ_c can be expressed as

$$\rho_c = -\left(\frac{\xi_1}{3\theta'(\xi_1)}\right) \frac{M}{\frac{4}{3}\pi R^3},$$

where $\theta'(\xi_1) \equiv (d\theta/d\xi)_{\xi_1}$.***

8.7 Verify the polytropic curves in Figs. 8.1 and 8.2, and the entries for ξ_1 in Table 8.2, by integrating the Lane–Emden equation (8.8) numerically. *Hint*: There are various ways to do this but the simplest is to use a combined numerical/graphics package like MatLab, Maple, or Mathematica to integrate and visualize results.

8.8 Prove that the relationship (8.13) between the mass and radius for a Lane–Emden solution follows from Eqs. (8.10) and (8.12).

8.9 Use Eqs. (8.13), (8.6), and (8.11) to show that for a Lane–Emden solution with polytropic index n the relationship between the central density and central pressure is given by Eq. (8.14).***

PART II

STELLAR EVOLUTION

9 The Formation of Stars

Substantial direct and indirect observational evidence and theoretical understanding indicate that stars are born in the clouds of gas and dust called nebulae. This chapter uses the understanding of stellar processes developed in the preceding chapters to address some specifics of how regions of such nebulae become unstable and collapse gravitationally to form stars. The basics are well understood but we will find that many details are not. Thus, there must be considerable subtlety to the detailed steps in the formation of stars that will be glossed over at our level of presentation. This will be justified by the observation that stars exist and, therefore, something like these assumptions are likely to be correct.

9.1 Evidence for Starbirth in Nebulae

The observation of many hot O and B spectral class stars in and near nebulae is a rather strong indicator that stars are being born there. These stars are so luminous that they must consume their nuclear fuel at a prodigious rate. Thus, their time on the main sequence is limited to millions of years and it follows that they cannot have wandered far from their birthplace. Also observed, usually in association with complexes of hot O and B stars embedded in dust clouds, is a class of stars called *T Tauri variables*. These are red irregular variables with a number of unusual characteristics. They exhibit emission lines and spectral lines with P Cygni profiles, as illustrated in Fig. 9.1, which indicate the presence of expanding shells of low-density gas around the stars. T Tauri stars typically are more luminous than corresponding main sequence stars, implying that they are larger. They exhibit strong winds (*T Tauri winds*), often with bipolar jet outflows, and *Herbig–Haro Objects* are sometimes found in the directions of these jets. An example of outflow from a young star and the associated Herbig–Haro objects is shown in Fig. 9.2.

These considerations indicate that T Tauri stars still are in the process of contracting to the main sequence. They are less massive than the O and B stars that often accompany them, indicating that they have contracted more slowly and many have not yet had time to reach the main sequence by the time that the more rapidly evolving O and B stars have done so. The HR diagram for a young cluster is illustrated in Fig. 9.3, where there is strong evidence for many young stars that are more luminous than the main sequence and so are still contracting. The stars marked with horizontal and vertical bars in this figure exhibit observational characteristics typical of T Tauri stars. The observed bipolar outflows could be explained by an accretion disk around the young stars formed as a result of conservation

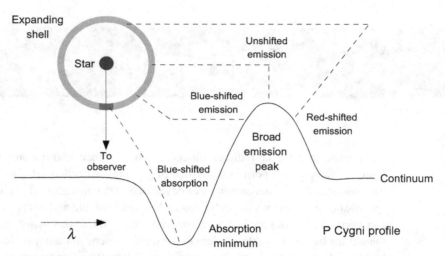

Fig. 9.1 Origin of P Cygni profiles in Doppler shifts associated with expanding gas shells. The emission peak is both red-shifted and blue-shifted for the observer because of gas moving radially away from the star. The blue-shifted dip is caused by absorption in gas moving toward the observer on the near side.

Fig. 9.2 Jets and Herbig–Haro objects produced by outflow from a young star hidden in dust at the center of the image. The Herbig–Haro Objects HH-1 and HH-2 correspond to the nebulosity at the ends of the jets. The entire width of this image is about one lightyear.

of angular momentum for the infalling matter. Then, if there are strong winds emanating from the star they would tend to be directed in bipolar flows perpendicular to the plane of the disk (see Fig. 9.14). However, it is difficult to explain the tight collimation of the jets by such a mechanism, and the source of the energy driving the winds is also not explained by such a simple model. The Herbig–Haro objects are likely the result of shocks formed when the matter flowing out of the T Tauri star interacts with low-density gas clouds.

These observations suggest that stars are born in nebulae. They also suggest that the life of protostars contracting to the main sequence may be more complex (and more violent) than the following simple considerations might indicate. (References [84, 110, 206] give a more detailed analysis of the processes involved in star formation than that presented here.) Taking note of those warnings, we turn now to a more quantitative discussion

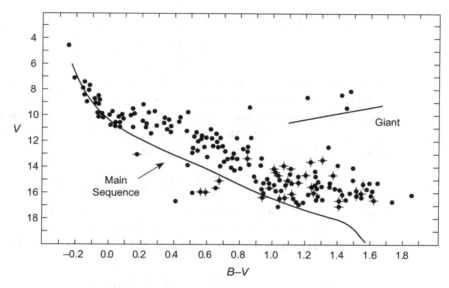

Fig. 9.3 HR diagram for the young open cluster NGC2264. Horizontal bars denote stars with H_α line emission and vertical bars denote variable stars, both characteristic of T Tauri stars. Adapted from R. L. Bowers and T. Deemings, *Astrophysics Volumes I and II*, Jones and Bartiett Publishers, Inc. 1984.

of gravitational stability as a foundation for understanding how regions of nebulae can collapse to form stars.

9.2 Jeans Criterion for Gravitational Collapse

We may investigate the general question of gravitational collapse by considering a spherical cloud composed primarily of hydrogen that has radius R, mass M, and uniform temperature T, consisting of N particles having average mass μ. The question of stability will be assumed to be one of competition between gravitation, which would collapse the cloud if left to its devices, and gas pressure, which attempts to expand the cloud. The gravitational energy is of the form

$$\Omega = -f \frac{GM^2}{R}, \tag{9.1}$$

where the factor f is $\tfrac{3}{5}$ if the cloud is spherical and of uniform density, and larger (but still of order one) if the density increases toward the center (see Problem 9.6). The thermal energy is assumed to be that of an ideal gas, $U = \tfrac{3}{2} NkT$. From the virial theorem (4.21) describing a gravitating gas in equilibrium, the static condition for gravitational instability is $2U < |\Omega|$, implying that the system is unstable if it has a mass M with

$$M > M_J \equiv \frac{3kT}{fG\mu M_u} R = \left(\frac{3kT}{fG\mu M_u}\right)^{3/2} \left(\frac{3}{4\pi\rho}\right)^{1/2}, \tag{9.2}$$

where $N = M/\mu M_u$ and $R = (3M/4\pi\rho)^{1/3}$ have been employed. The quantity

$$M_J = \left(\frac{3kT}{fG\mu M_u}\right)^{3/2}\left(\frac{3}{4\pi\rho}\right)^{1/2} \qquad (9.3)$$

is termed the *Jeans mass*. It defines a critical mass beyond which the system becomes unstable to gravitational contraction. Since $M_J \propto T^{3/2}\rho^{-1/2}$, the Jeans mass will be smaller for colder, more dense clouds. This makes physical sense: it is easier to collapse a cloud of a given mass gravitationally if the cloud is cold and dense than if it warm and diffuse. Equation (9.2) may be solved for the *Jeans length*,

$$R_J = \frac{fG\mu M_u}{3kT}M_J = \left(\frac{9kT}{4\pi fG\mu M_u \rho}\right)^{1/2}, \qquad (9.4)$$

which defines the distance scale associated with the Jeans mass and thus characterizes the size of gravitationally unstable regions. Finally, often it is more useful to solve the condition (9.2) for a critical density for gravitational collapse called the *Jeans density*,

$$\rho_J = \frac{3}{4\pi M^2}\left(\frac{3kT}{f\mu M_u G}\right)^3. \qquad (9.5)$$

Notice that the Jeans critical density is lowest (and thus is more easily achieved) if the mass is large and the temperature low, as would be expected on intuitive grounds.

Example 9.1 From Eq. (9.5), a uniform cloud of molecular hydrogen at $T = 20\,\text{K}$ with $M = 1000\,M_\odot$ has a Jeans density $\rho_J = 1.2 \times 10^{-22}\,\text{g}\,\text{cm}^{-3}$ if $f = \frac{3}{5}$ and $\mu = 1$. A molecular hydrogen cloud at the same temperature containing only $1\,M_\odot$ has a Jeans density 6 orders of magnitude larger.

The Jeans criterion is a static condition that says nothing about gas dynamics, and it neglects potentially important factors influencing stability such as magnetic fields, dust formation and vaporization, turbulence, jets, and radiation transport.[1] Nevertheless, it is a useful starting point for understanding how stars form from clouds of gas and dust.

9.3 Fragmentation of Collapsing Clouds

The foregoing considerations indicate that the collapse of more massive clouds is favored, but this alone is at odds with the observation that most stars contain less than a solar mass of material. The solution to this dilemma is thought to lie in fragmentation of the collapsing

---

[1] Shortcomings of the Jeans criterion are suggested by nothing that our galaxy contains $\sim 10^9 M_\odot$ of gas suitable to form stars, yet it produces only $\sim 1 M_\odot$ of new stars each year. This implies a star formation rate one to two orders of magnitude smaller than that expected from gravitational considerations alone [84]. The formal basis of the Jeans condition has been questioned by some because it is doubtful whether the conditions assumed in its derivation are realized in star-formation regions. Specifically, the implicit assumption of constant background density is generally inconsistent with conditions for hydrostatic equilibrium because there are then no pressure gradients to balance the gravitational forces. See the discussion in Chapter 5 of Ref. [46] and Appendix 2 of Ref. [110].

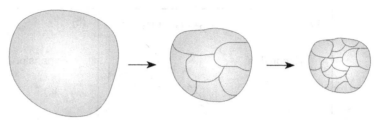

Fig. 9.4 Fragmentation of a collapsing cloud into gravitationally unstable subclouds, each of which can pursue its own independent collapse. However, actual fragmentation may be much more complex than this simple, orderly cartoon would suggest.

Box 9.1 **Dependence of Collapse on Composition**

The globular clusters in our galaxy, which typically each contain $\sim 10^5$ stars, are very old and were formed early in the galaxy's history. However, present-day star formation in the galaxy appears to be producing much less populous open clusters (having 100–1000 stars), not clusters of 10^5 stars. Thus, as the Milky Way Galaxy has evolved, something has changed in the way clouds collapse to form clusters of stars.

A possible clue is supplied by observations in the Large and Small Magellanic Clouds, which are satellites of our own galaxy where larger clusters are still forming (see Böhm-Vitense [52], Vol. 3). The Magellanic clouds have a composition somewhat richer in hydrogen with less dust and metals than the present Milky Way. These conditions are closer to those thought to have been prevalent in the Milky Way when the globular clusters formed. Thus, the size of gravitationally unstable clouds, and the conjectured fragmentation mechanism operating in those clouds, may be sensitive to the elemental composition of the galaxy.

---

clouds, as illustrated schematically in Fig. 9.4. As will be seen shortly, the initial collapse is expected to occur at almost constant temperature. Therefore, from Eq. (9.3) we may expect the Jeans mass to decrease in the initial phases of the collapse (speaking loosely, since the Jeans criterion was constructed for a cloud in equilibrium, not one that is already collapsing). Hence as large clouds, which have the smallest Jeans density, begin to collapse their average density increases and at some point subregions of the original cloud may exceed the critical density and become unstable in their own right to gravitational collapse.

If there are sufficient perturbations on the right length scales in the cloud (caused by turbulence, magnetic fields, and jets, for example), these subregions may separate and begin to pursue independent collapse. Within these collapsing subclouds this scenario may be repeated: as the density increases, subregions may themselves become gravitationally unstable and begin to pursue an independent collapse. By such a hierarchy of fragmentations it is plausible that clusters of protostars might be formed that have individual masses more comparable to those of observed stars (see also Section 9.7), or are sufficiently close that mass loss in the collapse to the main sequence would produce the observed spectrum of stellar masses (recall the strong outflows associated with T Tauri stars). Box 9.1 gives an example of additional complexity that might influence the way star clusters form.

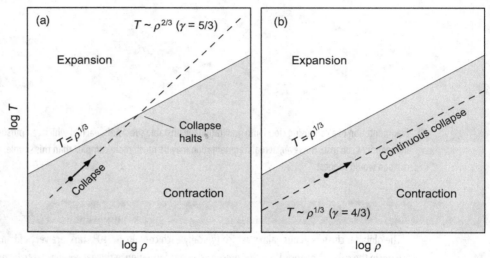

Fig. 9.5 Gravitational equilibrium in temperature–density space assuming adiabatic conditions. The solid line $T \propto \rho^{1/3}$ corresponds to hydrostatic equilibrium by the Jean's condition (9.5). The dashed lines correspond to adiabatic behavior assuming (a) $\gamma = \frac{5}{3}$ and (b) $\gamma = \frac{4}{3}$.

9.4 Stability in Adiabatic Approximation

Consider the adiabatic contraction (or expansion) of a homogeneous cloud. Real clouds will exchange energy with their surroundings and so are not completely adiabatic, and they may not be completely homogeneous, but the results obtained in this limit will be instructive in understanding more realistic situations. From Eq. (9.5), equilibration of gravity and pressure forces requires that the temperature T and density ρ be related by $T \propto \rho^{1/3}$. In Fig. 9.5, this relation has been used to divide the T–ρ plane into regions above and below the line $T = \rho^{1/3}$. For points above the stability line (unshaded area), pressure-gradient forces are larger than gravitational forces and the system is unstable to expansion. For points below the stability line (shaded area), pressure-gradient forces are weaker than gravitational forces and the system is unstable with respect to contraction.

9.4.1 Dependence on Adiabatic Exponents

From Eq. (3.40), for ideal-gas adiabats (curves of adiabatic heating or cooling) the temperature T and density $\rho \propto V^{-1}$ are related by

$$T \propto \rho^{\gamma - 1}. \tag{9.6}$$

First consider a monatomic ideal gas, for which the adiabatic exponent is $\gamma = \frac{5}{3}$. This corresponds to the dashed line $T \sim \rho^{2/3}$ in Fig. 9.5(a), which is steeper than the equilibrium line and therefore crosses it at some point. A cloud that is unstable to gravitational contraction (corresponding to a point on the dashed line in the shaded area) will follow the dashed line to the right as it collapses (direction of increasing density), as indicated by the arrow. But in this case the collapse will be halted where the dashed line reaches the stability line. Likewise, a cloud unstable to expansion (corresponding to a point on the dashed line lying in the unshaded area) will follow the dashed line to the left as it expands (direction of decreasing density), until it reaches the stability line and the expansion halts.

Now consider Fig. 9.5(b), where it is assumed that the cloud has an adiabatic exponent $\gamma = \frac{4}{3}$. In this case, the contraction (or expansion) follows an adiabat for which $T \propto \rho^{1/3}$. Since this adiabat is *parallel to the stability line,* the two lines never cross and a system lying on the dashed line collapses and continues to collapse indefinitely. This also will be the case if $\gamma < \frac{4}{3}$. Likewise, a system with $\gamma \leq \frac{4}{3}$ that is above the stability line will continue to expand adiabatically as long as $\gamma \leq \frac{4}{3}$.

9.4.2 Physical Interpretation

The physical meaning of the preceding discussion is that a gas is less able to generate the pressure differences required to resist gravity if the energy released by gravitational contraction can be absorbed into internal degrees of freedom and thus is not available to increase the kinetic energy of the gas particles. The parameter γ is relevant because it is related to the heat capacity for the gas. Typical sinks of energy that can siphon off energy internally are rotations and vibrations of molecules, ionization, and molecular dissociation. In the large clouds of gas and dust that are candidates for stellar birthplaces, γ can be reduced to a value of $\frac{4}{3}$ or less by[2]

- polyatomic molecules with more than five degrees of freedom [see Eq. (7.37)],
- the ionization of hydrogen at $T \sim 10,000$ K, or
- the dissociation of hydrogen molecules at $T \sim 4,000$ K.

The large molecules required for the first situation are rare in the interstellar medium but their presence enhances the chance of gravitational collapse for a cloud. In hydrogen ionization or molecular dissociation zones, typically $\gamma \leq \frac{4}{3}$ and this causes an instability until the ionization or dissociation is complete, at which point γ will typically return to normal values ($\gamma \simeq \frac{5}{3}$) and collapse on the corresponding adiabat will reach the equilibrium line and stabilize [see Fig. 9.5(a)].

9.5 The Collapse of a Protostar

The preceding introduction sweeps much under the rug but the existence of stars implies that protostars form by some mechanism similar to the one outlined above, so let us follow

---

[2] To simplify the initial discussion the gas will be assumed to consist only of molecular and atomic hydrogen, with trace concentrations of polyatomic molecules. In reality about 25% of the mass would be in helium gas, which has first and second ionization energies of 24.6 eV and 54.4 eV, respectively, and there will likely be dust. These more realistic features are important for a quantitative description but would not change the conceptual picture that will be painted here.

the consequences of the gravitational collapse of such a protostar [169]. Assuming that $f\mu \sim 1$, for a one solar mass protostar the Jeans criterion gives $\rho_J \simeq 3 \times 10^{-16} \,\mathrm{g\,cm^{-3}}$ for $T = 20$ K and $M = 1\,M_\odot$. Thus, we may expect that a $1\,M_\odot$ cloud can collapse if this average density is exceeded. The size of this initial cloud may be estimated by assuming the density to be constant and distributed spherically, implying that $R_1 \sim 3 \times 10^{16}$ cm ~ 2000 AU $\sim 4 \times 10^5\,R_\odot$. Thus, the initial protostar has a radius approximately 50 times that of the present Solar System.

9.5.1 Initial Free-Fall Collapse

The initial collapse may be assumed to be free fall and isothermal, as long as the gravitational energy released is not converted into thermal motion and thereby into pressure. This will be the case as long as the energy not radiated away is largely taken up by

- dissociation of hydrogen molecules into hydrogen atoms
- ionization of the hydrogen atoms.

The dissociation energy for hydrogen molecules is $\varepsilon_d = 4.5$ eV and the ionization energy for hydrogen atoms is $\varepsilon_{\mathrm{ion}} = 13.6$ eV. The energy required to dissociate and ionize all the hydrogen in the original cloud is then

$$E = N(H_2)\varepsilon_d + N(H)\varepsilon_{\mathrm{ion}} = \frac{M}{2m_H}\varepsilon_d + \frac{M}{m_H}\varepsilon_{\mathrm{ion}}, \qquad (9.7)$$

where N denotes the number of the corresponding species and m_H is the mass of a hydrogen atom. For a protostar of one solar mass the requisite energy is approximately 3×10^{46} erg. If this energy is supplied by gravitational contraction from an initial radius R_1 to a final radius R_2, energy conservation gives

$$\frac{GM^2}{R_2} - \frac{GM^2}{R_1} = \frac{M}{2m_H}\varepsilon_d + \frac{M}{m_H}\varepsilon_{\mathrm{ion}}. \qquad (9.8)$$

Solving this for R_2 assuming a mass of $1\,M_\odot$ and an initial radius of 3×10^{16} cm gives $R_2 = 9 \times 10^{12}$ cm $\simeq 130\,R_\odot \simeq 0.6$ AU. The corresponding time for collapse is set by the free-fall timescale (see Problem 4.1), which is $t_{\mathrm{ff}} = (3\pi/32G\bar{\rho})^{1/2} \simeq 13{,}000$ yr. Therefore, in this very simple model a protostar of one solar mass collapses from a radius of about 50 times the radius of the Solar System to a radius about half the radius of the Earth's orbit in near free fall on a timescale of order 10^4 years.[3]

At this point we may expect the collapse to slow because all hydrogen molecules have been dissociated and all hydrogen atoms have been ionized, the photon mean free path

---

[3] The characteristic timescale for free fall $t_{\mathrm{ff}} \sim (G\bar{\rho})^{-1/2}$ is independent of the radius of the collapsing mass distribution. This behavior is termed *homologous collapse*. One consequence of homologous collapse is that if the initial density is uniform it will remain uniform for the entire collapse. Because configurations in homologous systems are self-similar (related by a scale transformation), they are particularly easy to deal with mathematically. Therefore, simple but reasonable approximate treatments of the initial phases for gravitational collapse are often possible by making the homology assumption. Another situation where homology can be employed to simplify the description is in a core collapse supernova, where part of the core is expected to collapse in approximately homologous fashion (see Chapter 20).

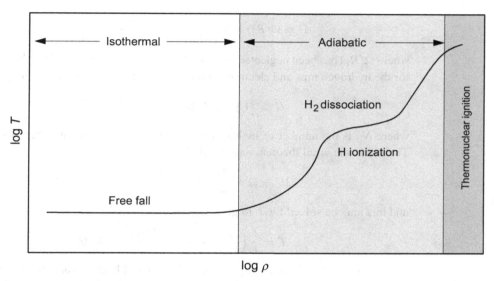

Fig. 9.6 Schematic track in density and temperature for the collapse of a gas cloud to form a star. Flat regions of the curve correspond to isothermal collapse.

becomes short because of interactions with the free electrons so that the cloud becomes opaque to its own radiation, the temperature increases as heat begins to be trapped, and the resulting pressure gradients counteract the gravitational force and bring the system nearly into hydrostatic equilibrium. Thus, the virial theorem (4.21) in near-adiabatic conditions may be applied henceforth.

9.5.2 A Little More Realism

The picture presented above is oversimplified in that a realistic cloud does not wait until all hydrogen has been dissociated and ionized before it begins trapping heat, but its essential features are supported by more sophisticated considerations. A more correct variation of temperature and density for star formation is illustrated in Fig. 9.6. In this more realistic picture the temperature of the cloud increases and it begins to deviate from free-fall behavior once significant heat has been trapped. After all hydrogen has been dissociated and ionized, the collapse is governed by approximately adiabatic conditions in near hydrostatic equilibrium.

9.6 Onset of Hydrostatic Equilibrium

The temperature at which approximate hydrostatic equilibrium sets in may be estimated in our simple model as follows. The virial theorem gives $2U + \Omega = 0$ and the gravitational energy Ω is

$$\Omega \equiv \Omega(R_2) = -\frac{GM^2}{R_2} = -\left(\frac{M}{2m_H}\varepsilon_d + \frac{M}{m_H}\varepsilon_{ion}\right), \quad (9.9)$$

where $\Omega(R_1)$ has been neglected compared with $\Omega(R_2)$. From Eq. (3.9), the internal energy for the hydrogen ions and electrons in the fully ionized gas is approximately

$$U \simeq \tfrac{3}{2}(N_H + N_e)kT = 3N_H kT = \frac{3M}{m_H}kT, \quad (9.10)$$

where N_H is the number of hydrogen ions and $N_e \sim N_H$ is the number of free electrons. Therefore, the virial theorem requires that

$$2U + \Omega = \frac{6M}{m_H}kT - \frac{M}{2m_H}\varepsilon_d - \frac{M}{m_H}\varepsilon_{ion} = 0,$$

and this may be solved for T to give

$$T = \frac{1}{k}\left(\frac{\varepsilon_d}{12} + \frac{\varepsilon_{ion}}{6}\right) \simeq \frac{2.6\text{ eV}}{k} \simeq 30{,}000\text{ K},$$

as a rough estimate for the temperature at the onset of hydrostatic equilibrium.

Subsequent contraction of the protostar is in near hydrostatic equilibrium and is controlled by the opacities, which govern how fast energy can be brought to the surface and radiated. As was discussed in Section 4.9, this too is a consequence of the virial theorem and leads to the Kelvin–Helmholtz timescale of about 10^7 years to form a star of one solar mass. Although the energy production of the protostar is assumed to derive primarily from gravity, some energy also may be supplied by thermonuclear burning of primordial deuterium, as discussed further in Box 9.2. Theoretical evolutionary tracks for protostars of various masses to collapse to the main sequence are shown in Fig. 9.7.

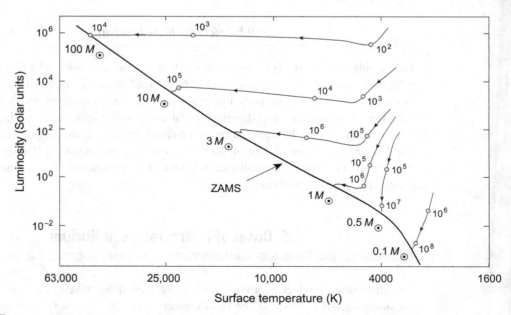

Fig. 9.7 Evolutionary tracks for collapse to the main sequence. Numbers on tracks are times in years since the onset of near-adiabatic collapse. Notice the large difference in timescales for collapse as a function of mass.

Box 9.2 **Deuterium Burning in Protostars**

In the present discussion protostars have been assumed to derive their energy from gravitational contraction and not from nuclear reactions because the temperatures and densities are too low for the PP chains or the first steps of the CNO cycle until the protostar nears the main sequence. However, this is a small oversimplification because from Fig. 6.1 the rate for the second step of PP-I, $^2\text{H} + {}^1\text{H} \rightarrow {}^3\text{He} + \gamma$, is many orders of magnitude larger than the rate for the first step of the PP chains, $^1\text{H} + {}^1\text{H} \rightarrow {}^2\text{H} + \beta^+ + \nu_e$. Thus, if deuterium were already present in the protostar, PP-I could proceed directly from its second step at a much lower temperature.

Primordial Deuterium Production

Trace amounts of deuterium are expected in the contracting protostar because a small amount is made in the big bang. The following figure illustrates light-element production in big bang nucleosynthesis (the calculation is described in more detail in Chapter 20 of [100] and uses the methods outlined in Appendix D).

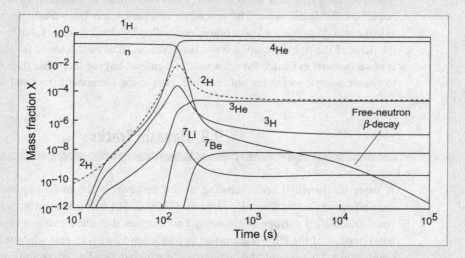

Big bang nucleosynthesis leaves the Universe dominated by ^1H and ^4He, but traces of deuterium (dotted ^2H curve) and a few other light isotopes are produced.

Burning Primordial Deuterium

If a protostar collapses from material never processed in stars it will contain the abundance of deuterium (mass fraction $X_d \sim 10^{-5}$) produced in the big bang, and the deuterium will burn when the temperature exceeds about 6×10^5 K (see also the discussion of lithium burning in Section 9.10.1). This makes some contribution beyond the gravitational energy to the internal heating of the protostar, which can in turn affect the contraction and can add to the initial abundance of ^3He in the forming star. A more extensive discussion of deuterium burning prior to main sequence evolution is given in Vol. I of Iben [130] and in Iliades [128].

9.7 Termination of Fragmentation

The discussion in Section 9.3 indicated that gravitational collapse of large clouds is likely to fragment into a hierarchy of sub-collapses, and this fragmentation was invoked to explain why so many low-mass stars are produced. However, this argument is incomplete (even leaving aside that little quantitative justification was given for it) because it fails to provide a mechanism for *stopping* the fragmentation in the vicinity of 0.1–1 solar masses (since no stars would form otherwise; see Section 9.9). As will now be explained, one possible resolution of this dilemma is that the transition from isothermal to adiabatic collapse for the protostar implies a modification of the Jeans criterion, and that this modification dictates a lower limit for the mass of the fragments produced by a hierarchical collapse.

Substitution of Eq. (9.6) for adiabats in Eq. (9.2) implies that $M_J \simeq \rho^{(3\gamma-4)/2}$ for adiabatic clouds. For $\gamma = 5/3$ then, $M_J \propto \rho^{1/2}$ and the Jeans mass *increases* as the collapse proceeds, implying that the transition from isothermal to adiabatic collapse sets a *lower bound* on the Jeans mass. The preceding discussion has been qualitative. More realistic calculations do suggest a lower bound controlled by cloud opacities, but it is often less than the mass of the lightest known stars. However, as shall be discussed in Section 9.10, there is observational evidence for collapse of fragments having less mass than the lightest stars to *brown dwarfs*, which are not stars but share some features of stars and some of planets.

9.8 Hayashi Tracks

A more fundamental understanding of the collapse to the main sequence follows from the demonstration by Chushiro Hayashi (1920–2010) that *a star cannot reach hydrostatic equilibrium if its surface is too cool*. This implies that there exists a region in the right-hand portion of the HR diagram that is forbidden to a given star while it is in hydrostatic equilibrium. This region, the boundaries of which depend on mass and composition of the star, is called the *Hayashi forbidden zone;* it is illustrated in Fig. 9.8 for a star of mass M and composition c.

9.8.1 Fully Convective Stars

Protostars contracting to the main sequence must have large surface areas and (relatively) high surface temperatures, so they have *high luminosities*. Furthermore, once the hydrogen is ionized they will have *high opacities*. The combination of high opacity and high luminosity ensures that the temperature gradients necessary to transport the energy radiantly are steeper than the adiabatic gradient and, according to the discussion in Section 7.7, the contracting protostar is expected to be *almost completely convective*. As noted in Box 3.4, such objects are described approximately by a $\gamma = \frac{5}{3}$ polytrope.

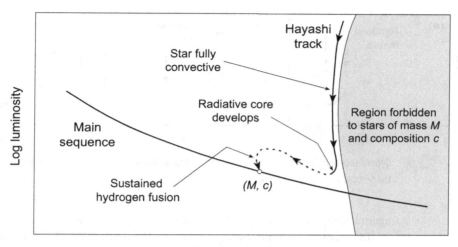

Fig. 9.8 Evolution of a protostar to the main sequence. While the protostar is fully convective it follows an almost vertical path in the HR diagram called the *Hayashi track*.

By examining fully convective stars with a thin radiative envelope, Hayashi demonstrated that these contracting protostars follow an almost vertical path in the HR diagram that is now called the *Hayashi track* for the star. (The T Tauri stars of Fig. 9.3 are thought to be on their Hayashi tracks.) Numerical simulations and simplified models indicate that objects to the right of the Hayashi track in the HR diagram cannot achieve hydrostatic equilibrium, so no stable stars of protostars can exist in this region. Thus the Hayashi track marks the left boundary of the Hayashi forbidden zone, as illustrated in Fig. 9.8.

9.8.2 Development of a Radiative Core

As the shrinking star descends the Hayashi track its central temperature is increased by the gravitational contraction and this decreases the central opacity (recall from Section 7.4.5 that for the representative Kramers opacity, $\kappa \propto T^{-3.5}$). Eventually this lowers the temperature gradient in the central region sufficiently that it drops below the critical value for convective stability and a radiative core develops. As contraction proceeds the radiative core begins to grow at the expense of the convective regions, which are eventually pushed out to the final subsurface zones characteristic of stars like the Sun (see Section 7.10.2).[4]

While the fully convective star is on the Hayashi track its luminosity decreases rapidly because of the shrinking surface area. However, as the opacity decreases over more and more of the interior because of the increasing temperature the luminosity begins to rise

---

[4] In more massive stars the subsurface convective zones are eliminated completely but the core may become convective after the star enters the main sequence if the power generation is sufficiently large; see Section 7.10.1. In the least massive stars, a radiative core never develops and they remain completely convective (see below).

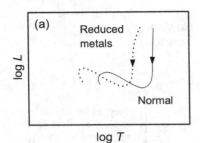 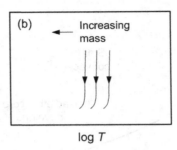 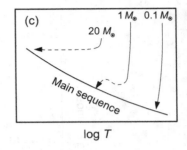

Fig. 9.9 Dependence of Hayashi tracks on (a) composition and (b) mass. The solid portions of each curve in (c) represent the descent on the Hayashi track.

again because more energy can flow out radiantly. Since at this point the star is shrinking while its luminosity is increasing, the surface temperature must increase and the star begins to follow a track to the left and somewhat upward in the HR diagram (Fig. 9.8). Finally, the onset of hydrogen burning causes the track to bend over and enter the main sequence, as also illustrated in Fig. 9.8. Thus, the contraction to the vicinity of the main sequence for the protostar is composed of two basic periods: a vertical descent in the HR diagram for fully convective protostars, followed by a drift up and to the left as the interior of the protostar becomes increasingly radiative at the expense of the convective envelope.

9.8.3 Dependence on Composition and Mass

Hayashi tracks depend weakly on the mass and composition, as illustrated in Fig. 9.9. Thus, for more massive stars of fixed composition the Hayashi tracks are almost parallel to each other but increasingly shifted to the left in the HR diagram [Fig. 9.9(b)]. The Hayashi tracks also depend on stellar composition, because this influences the opacity. For example, a metal-poor star of a given mass will generally have a Hayashi track to the left of an equivalent metal-rich star because of lower opacity [Fig. 9.9(a)]. Finally, the transition from convective to radiative interiors, and the corresponding transition from downward motion to more horizontal leftward drift on the HR diagram, is generally faster in more massive stars because of more rapid heating of the interior. Therefore, as illustrated in Fig. 9.9(c) and Fig. 9.7, more massive stars leave the Hayashi track quickly and approach the main sequence almost horizontally for much of their protostar evolution. Conversely, the least massive stars never leave the Hayashi track as they collapse and drop almost vertically to the main sequence [Fig. 9.9(c)]. They are expected to remain completely convective, even after entering the main sequence.

9.9 Limiting Lower Mass for Stars

A contracting protostar becomes a star only if its core temperature rises enough to initiate thermonuclear reactions, which requires $T \sim 4 \times 10^6$ K. The results of Problem 4.8 show that for a star composed of a monatomic ideal gas having uniform temperature and density, the temperature varies approximately with the cube root of the density:

$$T \simeq 4.09 \times 10^6 \mu \left(\frac{M}{M_\odot}\right)^{2/3} \left(\frac{\rho}{\text{g cm}^{-3}}\right)^{1/3} \text{ K.} \qquad (9.11)$$

However, this behavior assumes an ideal classical gas; the temperature will no longer increase with contraction if the equation of state becomes that of a degenerate gas. In Problem 4.8, the critical temperature and density for onset of electron degeneracy were estimated by setting kT equal to the Fermi energy, giving a critical density defined by $\rho \simeq 6 \times 10^{-9} \mu_e T^{3/2}$ g cm^{-3}. Inserting this into Eq. (9.11) gives for the temperature at which the critical density is reached in the contracting protostar,

$$T \simeq 5.6 \times 10^7 \mu \mu_e^{1/3} \left(\frac{M}{M_\odot}\right)^{2/3} \text{ K.} \qquad (9.12)$$

Assuming representative values $M \sim 1 \, M_\odot$ and $\mu \mu_e^{1/3} \sim 1$ gives $T \sim 10^7$ K. This is more than enough to ignite hydrogen fusion before the core electrons become degenerate, so the Sun (you will be reassured to know) is a star.

On the other hand, decreasing the mass of the protostar eventually leads to a situation where the core will become degenerate before the temperature rises to the value required to sustain hydrogen fusion. Detailed calculations indicate that this limiting mass is approximately $0.08 \, M_\odot$–$0.10 \, M_\odot$. But the Jeans criterion knows nothing of the above considerations and it is expected that clouds of mass smaller than this could become unstable to gravitational collapse. What is the fate of collapsing clouds having less than the critical mass required to form stars? For them the growth in temperature is halted by electron degeneracy pressure before fusion reactions can begin. It is speculated that many such *brown dwarfs* may exist in the Universe, supported hydrostatically by electron degeneracy pressure; they have no thermonuclear furnaces but are radiating energy left over from earlier gravitational contraction.

9.10 Brown Dwarfs

Brown dwarfs collapse out of hydrogen clouds, not out of protoplanetary disks (like stars), but produce energy only by gravitational contraction, not from hydrogen fusion (like planets).[5] Their masses are expected to range from several times the mass of Jupiter to a few percent of the Sun's mass. The cooler brown dwarfs may resemble gas giant planets in chemical composition, while hotter ones may look chemically more like stars. Brown dwarfs are intrinsically difficult to detect because they are small and of very low luminosity, and once detected they are difficult to distinguish from low-mass stars and from planets.

---

[5] However, just as for protostars, the formation of brown dwarfs could involve a small amount of deuterium burning; see Box 9.2.

9.10.1 Spectroscopic Signatures

The first brown dwarf discovered was Gliese 229B, which appears to be too hot and massive to be a planet but too small and cool to be a star. The IR spectrum of GL229B looks like the spectrum of a gas giant planet. Most telling is evidence of methane gas, common in gas giants but not found in stars because methane can survive only in atmospheres having temperatures lower than about 1500 K.

In addition to searching for gases like methane that should not be present in stars, observational quests for brown dwarfs look for evidence of the element lithium. Hydrogen fusion destroys lithium in stars: at temperatures above about 2×10^6 K, a proton encountering a lithium nucleus has a high probability to react with it, converting the lithium to helium. The amount of lithium that can survive is a function of how strongly the material of the star is mixed down to the core fusion region by convection. Protostars are convective, so stars start off with a strongly mixed interior, but the initial core temperature in the protostar is not high enough to burn lithium. The lightest stars (red dwarfs) remain convective once on the main sequence, so lithium is mixed down to the fusion region and destroyed in red dwarfs. Because these stars are cool, it takes some time to burn the lithium, but calculations indicate that lithium could survive no longer than about 2×10^8 years in the lightest true star.[6] The basic interior structures expected for stars, brown dwarfs, and gas giant planets are summarized schematically in Fig. 9.10.

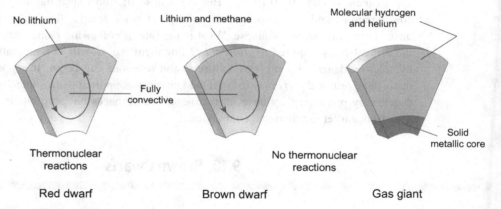

Fig. 9.10 Contrasting interiors of a red dwarf star, a brown dwarf, and a gas giant planet. Stars initiate thermonuclear reactions but brown dwarfs and planets do not; thus, primordial lithium is destroyed in stars but not in planets or brown dwarfs. The presence of methane is an indication that temperatures are too low for the object to be a star. Gas giant planets can contain lithium and methane, like brown dwarfs, but their upper interiors tend to be dominated by molecular hydrogen and helium.

---

[6] For stars more massive than red dwarfs, as the protostar contracts the center becomes radiative and the region of convective mixing begins to retreat toward the surface at about the same time that fusion is initiated in the core. Whether all the lithium is destroyed depends on how fast this happens. If the retreat is fast enough to separate a well-mixed surface from the interior, a small amount of the lithium will survive. Observations of the Sun indicate that about 1% of its original lithium was preserved in this manner.

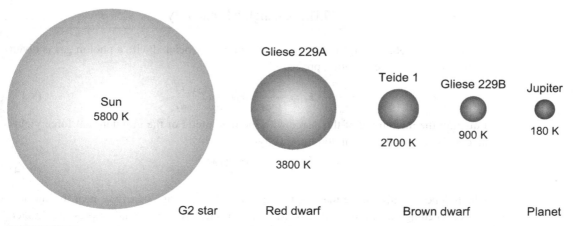

Fig. 9.11 Relative size (to scale) and surface temperature trends (darker is cooler) for examples of stars, brown dwarfs, and gas giant planets.

9.10.2 Stars, Brown Dwarfs, and Planets

Figure 9.11 summarizes the size and surface temperature trend from middle main sequence stars like the Sun, through the lowest mass stars (red dwarfs), through brown dwarfs, and finally to planets. Brown dwarfs can have surface temperatures comparable to that of the lowest-mass stars, but atmospheric compositions similar to large planets. The challenge is to distinguish them from these other two kinds of objects at interstellar distances. A number of brown dwarf candidates have now been identified but in many cases there is uncertainty about whether they are brown dwarf companions of stars, or giant planets orbiting stars.

9.11 Limiting Upper Mass for Stars

As we discussed in the preceding section, a limiting lower mass for stars is set by the requirement that sufficient temperature be generated by gravitational collapse to initiate the burning of hydrogen to helium in the core. An upper limiting mass for stars is thought to exist because of the opposite extreme: if the star is too massive, the intensity of the thermonuclear energy production makes the star unstable to disruption by the radiation pressure. Pressure associated with radiation grows as the fourth power of the temperature and thus will be most important for hot stars. It is instructive to ask what photon luminosity is required such that the magnitude of the force associated with the radiation field is equivalent to the magnitude of the gravitational force. This critical luminosity, which defines limiting configurations that are marginally stable gravitationally with respect to the pressure of the photon flux, is termed the *Eddington luminosity*.

9.11.1 Eddington Luminosity

As shown in Problem 9.1, the force per unit volume associated with a photon gas is given by the gradient of the radiation pressure,

$$\frac{1}{V} F_r = -\frac{dP_r}{dr} = -\frac{4}{3} a T^3 \frac{dT}{dr}. \tag{9.13}$$

Equating the magnitude of this force with the magnitude of the gravitational force yields an expression for the Eddington luminosity

$$L_{\text{Edd}} = \frac{4\pi c G M}{\kappa}, \tag{9.14}$$

which depends only on the mass of the star M and an average opacity κ near the surface. Stars exceeding the luminosity (9.14) can expel surface layers by radiation pressure. In fact, since the total pressure will always exceed the radiation pressure because of contributions from the gas, the stability limit typically will be lower than the Eddington luminosity.[7]

9.11.2 Estimate of Upper Limiting Mass

Stars for which radiation pressure is important are hot and from Fig. 7.3 the opacities near their surfaces are expected to be dominated by Thomson scattering. Taking the Thomson opacity (7.25) for pure, fully ionized hydrogen as a rough estimate of κ, Eq. (9.14) may be expressed as

$$L_{\text{Edd}} = \frac{4\pi c G M}{\kappa_{\text{T}}} \simeq 3.3 \times 10^4 \left(\frac{M}{M_\odot}\right) L_\odot. \tag{9.15}$$

Equation (9.15) may be used to estimate a radiation-pressure mass limit by assuming that the most luminous stars observed (with luminosities $\sim$ several $\times\, 10^6\, L_\odot$) are radiating at the Eddington limit.[8] As shown in Problem 9.1, this suggests a maximum stable mass for stars of order $100\, M_\odot$. This is a very crude estimate but detailed calculations, and observations, suggest that the most massive stars indeed have masses of this order.

For example, Fig. 9.12 illustrates the frequency of stellar masses inferred from near-infrared Hubble Space Telescope observations of the Arches Cluster, which is near the galactic center and has a greater density than any other young cluster observed in the galaxy. No star more massive than about $130\, M_\odot$ is observed in this cluster. The dashed

---

[7] The ejection of material may also be abetted by stellar pulsations that result from pressure instabilities at high luminosity, and may be influenced by the presence of stellar rotation and magnetic fields. When stars are near the Eddington limit, strong winds driven by continuum absorption blow from their surfaces. Most stars are not near the Eddington limit. In those cases, any winds blowing from their surfaces are typically weaker and driven by absorption in discrete spectral lines. These are termed *line-driven winds*. In the ionized gas near the surface the radiation pressure acts primarily on the electrons but the mass resides in the protons (assuming pure hydrogen for simplicity). The radiation pressure on the electrons lifts them on average relative to the protons, producing a radially directed electric field that causes the protons and electrons to eventually be expelled together. More precise treatments of the Eddington luminosity account for ions with higher masses such as alpha particles in the gas, and the contribution of free–free and bound–free absorption to the opacity.

[8] For example, the highly unstable star η Carinae (see Section 14.3.2) has a total luminosity of about $5 \times 10^6\, L_\odot$ and is presently emitting mass at a rate of about 0.001 solar masses per year [5].

9.11 Limiting Upper Mass for Stars

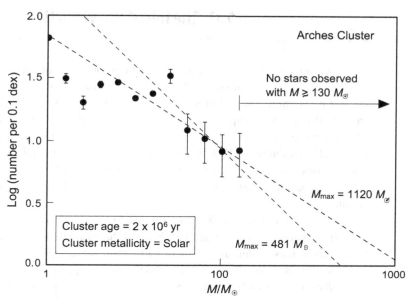

Fig. 9.12 Observed distribution of star masses in the Arches Cluster (adapted from Ref. [85]).

lines indicate two different fitted theoretical estimates for the initial mass distribution (proportional to the probability that a star formed with a given mass in the cluster; see Section 9.12). The one giving a maximum mass of 481 $M_\odot$ predicts that 18 stars should have been observed with masses greater than 130 $M_\odot$, and the one giving a maximum mass of 1120 $M_\odot$ predicts that 33 stars should have been observed with a mass greater than 130 $M_\odot$. Since no stars are observed in the cluster with a mass greater than 130 $M_\odot$, it was concluded that there is an upper limit to the mass of stars and that from the mass distribution in the Arches Cluster this upper limit is no larger than 150 $M_\odot$ though this value depends strongly on the rate of IR extinction by intervening dust [85].

This conclusion rests on the assumption that the most massive stars in the cluster have not yet been removed by supernova explosions. The absence of observed supernova remnants in the cluster is used to justify this assumption but there is evidence from core collapse supernova simulations that stars with initial masses of 150 $M_\odot$ or larger may collapse directly to black holes, producing short bursts of neutrinos, gravitational waves, and possibly gamma-rays, but ejecting little in the way of traditional supernova remnants (see Section 14.6.2). This may call into question whether the data of Fig. 9.12 really indicate a maximum star formation mass, or whether instead they are telling us something about the masses of stars that may collapse directly to black holes rather than eject a core collapse supernova remnant. At any rate, observation of stars with mass greater than about 150 $M_\odot$ is extremely rare.

As will be discussed in Section 14.3, there also is strong observational evidence that very massive stars eject large amounts of material from their envelopes. This tends to corroborate the idea of an upper limiting mass for stars set by stability against radiation, because radiation pressure and pulsational instabilities are thought to play a leading role in these observed mass-loss events.

9.12 The Initial Mass Function

Since for stars mass is destiny, a topic of great practical importance is the distribution in initial mass for a population of stars, which is called the *initial mass function (IMF)* for that population. The initial mass function $\xi(M)$ may be defined by requiring that the amount of mass bound up in stars in the mass interval M to $M + dM$ in a given volume of space be given by $MdN = \xi(M)dM$, where N denotes the number of stars in the volume. Determining the IMF requires an indirect, semiempirical chain of reasoning since observations give us apparent magnitudes, not masses, and stellar populations evolve with time so a mass distribution observed today differs from the initial mass distribution of the population.

Edwin Salpeter (1924–2008) first estimated $\xi(M)$ in 1955 by examining the luminosities of main sequence stars in the neighborhood of the Sun, relating the luminosity to the mass by empirical mass–luminosity relations like those illustrated in Eq. (1.27) and Fig. 1.11, and assuming that stars evolve away from the main sequence when about 10% of their initial hydrogen has been burned (see Section 10.6) [195]. Salpeter found a simple power law,

$$\xi(M) = \xi_0 M^{-1.35}. \tag{9.16}$$

The initial mass function for stars in the solar neighborhood determined in more recent work [180] is illustrated as points in Fig. 9.13, along with a line showing the original Salpeter estimate (9.16).

Although both observational data and methods of analysis have improved greatly since the time of Salpeter's pioneering work, it is remarkable that the original power law (9.16)

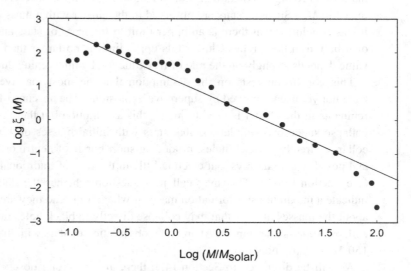

Fig. 9.13 Initial mass function (IMF). Points are from Ref. [180] for stars near the Sun and the line represents a Saltpeter power law, $\log \xi(M) = -1.35 \log M + 1.2$.

continues to work well for stellar masses in the ~ 0.2–$80\ M_\odot$ range. From Fig. 9.13 there are indications that it begins to fail for $M \sim 100\ M_\odot$ and larger, which presumably reflects the upper stability limit for stellar mass discussed in Section 9.11.2, and for very low masses. The region below $\sim 0.2\ M_\odot$ is complex, both observationally and theoretically, and requires special techniques that go beyond the assumptions of the Salpeter analysis. However, these uncertainties should not obscure the clear message of Fig. 9.13 that massive stars are rare, and that the majority of stars produced in the galaxy have masses well below $1\ M_\odot$. In fact, from Fig. 9.13 we may conclude that the most likely event in a star-forming episode is the production of a main sequence star with a mass of a few tenths of a solar mass.

9.13 Protoplanetary Disks

During the final stages of protostar collapse to the main sequence matter from the solar nebula may be expected to continue to accrete onto the star, most likely from an equatorial accretion disk because of angular momentum conservation. There is strong observational evidence that young stars tend to produce very strong winds that are focused perpendicular to the equatorial accretion disk. Thus, accretion disks and strong bipolar outflow may be a rather common feature of stars that are still collapsing to the main sequence. Figure 9.14 illustrates.

The strong winds blowing from young stars are not completely understood. One idea is that they are caused by the matter drawing a magnetic field inward as it falls into the accretion disk. The outwardly flowing wind is partially blocked by the accretion disk around the equator of the forming star and so escapes along the polar axis, thus producing bipolar outflows from the young star. However, this simple picture cannot explain the rather narrow width of some outflows. Presumably the full mechanism is more complex, perhaps involving the effect of magnetic fields to focus the ejected material. There also is strong observational evidence for dust disks around these young stars. It is likely that angular momentum plays a large role in the collapse of protostars and the formation of these circumstellar disks; this is discussed further in Box 9.3.

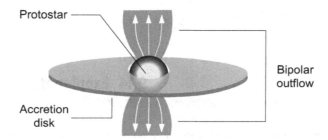

Fig. 9.14 Schematic model of an accretion disk and bipolar outflow for a young star.

> ### Box 9.3 — Stars, Disks, and Angular Momentum
>
> The preceding discussion of collapsing protostars has mentioned the role of angular momentum only in passing. The interstellar clouds from which protostars collapse will in general be rotating slowly, since it would take very special conditions to produce a cloud with no initial angular momentum at all. Doppler shifts of radio waves from opposite sides of these clouds may be used to infer their rotational velocities. Typically rotations for dense regions in stellar-mass clouds are found to have line-of-sight velocities of order 0.1 km s^{-1}.
>
> **Angular Momentum**
>
> If such a cloud collapses decoupled from the rest of the Universe its angular momentum must be preserved, so $v_0 r_0 = v_f r_f$, where v_0 and r_0 denote an initial tangential velocity and radius, respectively, and v_f and r_f denote the corresponding quantities after the collapse. In Section 9.5.1 it was estimated that a 1 $M_\odot$ cloud collapses to a star from an initial radius of order 10^{16} cm. Taking the Sun as representative of the final star, this corresponds to a decrease in radius by 5–6 orders of magnitude. Invoking conservation of angular momentum, if the 1 $M_\odot$ cloud collapsed directly to the Sun from an initial radius of 10^{16} cm and tangential velocity 0.1 km s^{-1}, the surface of the Sun should have been spun up to a velocity $v = (r_0/R_\odot)v_0 \simeq 14{,}000$ km s^{-1}. No normal star is spinning at anywhere near this rate.
>
> **Disk Formation**
>
> A basic fallacy in the preceding argument is that for finite angular momentum the collapse will not proceed directly to a star, but will terminate when the rotating cloud forms a stable disk for which the gravitational acceleration exactly keeps the particles in a circular orbit [193]. This requires that $v_f^2/r_f = GM/r_f^2$, which may be combined with $v = (r_0/R_\odot)v_0$ to give a disk radius
>
> $$r_{\text{disk}} = r_f \simeq \frac{v_0^2 r_0^2}{GM}.$$
>
> These disk radii typically are of order 100 AU (Problem 9.8).
>
> **Transfer of Angular Momentum**
>
> Thus, the initial collapse is likely to a rotating disk much larger than a star, and the final star is produced by an object that condenses at the center of this disk having much of the disk's mass but only a fraction of its angular momentum. The mechanism by which this takes place is not well understood but involves transfer of angular momentum outward in the disk. Thus, for example, in our Solar System the outer planets like Jupiter that formed from the solar disk carry much more angular momentum in their orbital motion than the Sun carries in its rotation.

9.14 Exoplanets

The dust disks observed around a number of young stars suggest that planetary formation may be taking place in these systems. Indeed, in recent decades impressive evidence has accumulated for thousands of *extrasolar planets* or *exoplanets*. Because the field of exoplanets is changing rapidly[9] and is a specialized topic only peripherally related to the main subject matter here, we shall say only a few words about exoplanets and their methods of detection. It will hardly come as a surprise that exoplanets are difficult to observe directly at their great distance. They are detected using a variety of methods but primarily through their gravitational influence on the parent star measured using precise Doppler spectroscopy, and by eclipses with the parent star, which cause variations in the light output of the system. Until about 2014 most exoplanets were detected using the Doppler spectroscopy method but more recently the majority of new discoveries have used transit methods.

9.14.1 The Doppler Spectroscopy Method

The *Doppler spectroscopy method* is illustrated in Fig. 9.15. The semiamplitude K of the stellar radial velocity caused by an orbiting planet is given by [145]

$$K = \left(\frac{2\pi G}{P}\right)^{1/3} \frac{m_p \sin i}{(M_* + m_p)^{2/3}(1 - e^2)^{1/2}}, \quad (9.17)$$

where i is the tilt angle of the orbit [see Fig. 1.9(a)], m_p is the mass of the unseen companion, M_* is the mass of the star, the orbital eccentricity is e, and the orbital period P is given by Kepler's third law,

$$P^2 = \frac{4\pi^2 a^3}{G(M_* + m_p)}, \quad (9.18)$$

where a is the semimajor axis of the relative orbit. Generally, the tilt angle i is unknown (but see the discussion of transits below), so masses are uncertain by a factor $\sin i$ in the absence of further information (see Section 1.5.2). To be useful the Doppler spectroscopy method requires that changes in radial velocity for the parent star of order 10 m s^{-1} or

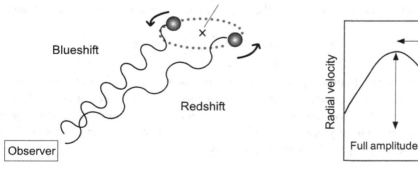

Fig. 9.15 The Doppler spectroscopy method for detecting extrasolar planets.

---

[9] The first confirmed exoplanet was discovered in 1992 and as of mid 2017 there are more than 3600 confirmed exoplanets, with thousands of additional candidates having been cataloged.

9.14.2 Transits of Extrasolar Planets

When the geometry permits an eclipse as seen from Earth, the transit of extrasolar planets across the face of their parent star may be observed, and in favorable cases the secondary eclipse of the exoplanet by the parent star can be seen. Such data allow the tilt angle i of the orbit to be tightly constrained to near $\frac{\pi}{2}$, and from eclipse timing the radius of the planet can be estimated. The IR flux associated with the planet may be deduced from the total flux decrease during the secondary eclipse, and by fitting such data to models the properties of the planet's atmosphere may be inferred from the eclipse data. Transit information such as this, coupled with data from Doppler analysis of the system when it is available, allows a rather full picture to be constructed: a detailed orbit of the planet, its mass, its size, and its density, and information about its atmosphere.

Background and Further Reading

For an introduction to star formation see Böhm-Vitense [52], Harpaz [108], Hartmann [110], Phillips [169], Iben [130], and Salaris and Cassisi [194]. A more extensive discussion of star formation may be found in Stahler and Palla [206]. Our discussion of mass loss from stars has been qualitative. An accessible introduction to a more quantitative analysis may be found in Chapter 8 of Prialnik [176]. For a more extensive introduction to the initial mass function, see Prialnik [176], and Stahler and Palla [206].

Problems

9.1 Show that in radiative equilibrium the force per unit volume associated with a photon gas is

$$\frac{1}{V} F_{\rm r} = -\frac{dP_{\rm r}}{dr} = -\frac{4}{3} a T^3 \frac{dT}{dr} = \frac{\kappa \rho L}{4\pi r^2 c},$$

where L is the luminosity, κ is the opacity, and ρ is the density. (See also Problem 9.9.) By equating the magnitude of this force per unit volume and the magnitude of the gravitational force per unit volume at the same radius, show that the maximal luminosity before radiation pressure would make the star gravitationally unstable is given by the Eddington luminosity

$$L_{\rm Edd} = \frac{4\pi c G M}{\kappa}.$$

Use this result, along with an estimate for the opacity from Thomson scattering, to make a rough estimate of the upper limiting mass for stars, assuming that the most luminous stars observed ($L \sim$ several $\times \, 10^6 \, L_\odot$) are the most massive and are radiating at the Eddington limit.***

9.2 For the main sequence, estimate and plot versus the spectral class the ratio of radiation pressure to gas pressure, and the ratio of the photon luminosity to the Eddington luminosity. *Hint*: Use Table 2.2 for main sequence properties and the results of Problem 4.6 to estimate the average internal temperature of main sequence stars, and estimate opacity near the surface assuming Thomson scattering.

9.3 From the virial theorem (4.21), show that the total energy of an ideal gas can be expressed as $E = -(3\gamma - 4)U$, where U is the internal energy and γ is the adiabatic exponent. Show that if $\gamma = \frac{4}{3}$, hydrostatic equilibrium can be attained only if the total binding energy of the gas tends to zero, implying that a $\gamma = \frac{4}{3}$ gas has precarious gravitational stability.

9.4 Estimate the maximum magnitude of the variation in the Sun's radial velocity caused by the influence of Jupiter if you were observing the Solar System from a distance of 30 lightyears. What is the spectroscopic precision required to measure this amount of Doppler shift? What is the maximum corresponding shift in angular position of the Sun on the celestial sphere caused by the influence of Jupiter as viewed from this distance? How far away would a 1-cm radius coin need to be to subtend the same angle?***

9.5 What is the Jeans density for a 1000 $M_\odot$ spherical cloud of atomic hydrogen at temperature 20 K? Suppose that in a dense region of a molecular cloud the number density is 1×10^8 cm^{-3} and the temperature is 100 K. What is the Jeans mass?

9.6 Evaluate the constant f in the gravitational potential equation (9.1), $\Omega = fGM^2/R$, for the density profile

$$\rho(r) = \rho_0 \left(1 - \left(\frac{r}{R}\right)^2\right),$$

that was used in Problem 4.15. (*Hint:* This problem will be easier to solve if you do Problem 4.15 first.)***

9.7 (a) Derive a Jeans length for instability against gravitational collapse by the following considerations. Assume a spherical cold cloud of molecular hydrogen having a uniform density ρ and radius r, governed by an ideal gas equation of state. Suppose the cloud is perturbed and begins to collapse gravitationally. Require for stability that the characteristic timescale for gravitational collapse be comparable to the characteristic timescale for the collapsing cloud to respond and produce a radial pressure gradient to counteract the collapse. Show that the radius for a cloud satisfying this condition is, up to factors of order one, equal to the Jean's length defined in Eq. (9.4).

(b) Consider a spherical cloud of pure molecular hydrogen at a temperature of 10 K with a uniform density of 10^{-18} g cm^{-3}. What is the adiabatic coefficient γ? Use the formula derived in part (a) to estimate the Jeans length for this cloud. How long would it take this cloud to collapse if there were no opposition from pressure gradients?

9.8 From angular momentum conservation, estimate typical radii for the disks that form and halt the gravitational collapse of clouds with angular momentum, as discussed in Box 9.3. Assume parameters characteristic of a 1 $M_\odot$ cloud with initial tangential velocity of 0.1 km s^{-1} and an initial radius of 2000 AU for the cloud.***

9.9 Use the equation of state for a photon gas to show that for radiative heat transport the radiation pressure $P_{\rm rad}$ has a gradient proportional to dT/dr; thus prove that for a star in hydrostatic equilibrium with heat transport by radiative diffusion

$$\frac{dP_{\rm rad}}{dP} = \frac{\kappa L(r)}{4\pi G m c},$$

where P is the total hydrostatic pressure, κ is the radiative opacity, L is the luminosity, and m is the enclosed mass at r.

9.10 Show that if gas and radiation pressure decrease outward in a star, then $\kappa L(r) < 4\pi c G m$, where $L(r)$ is the luminosity, κ is the opacity, and m is the enclosed mass at radius r. *Hint:* See the result of Problem 9.9. Thus argue that hydrostatic equilibrium and the equation for radiative energy transport cannot both be satisfied if either κ or $L(r)$ are too large. (If the inequality is violated and hydrostatic equilibrium is assumed, then the radiative transport equation must be invalid; indeed, heat transport was found earlier to be convective, not radiative, if either the luminosity or the opacity are too high.)

9.11 Consider a pair of gravitationally bound objects (two stars, or a star and a planet, for example) having masses m_1 and m_2, and separated from their common center of mass by a_1 and a_2, respectively. Assuming the validity of Kepler's third law, show that the apparent orbits as viewed from Earth satisfy

$$\frac{m_2^3}{(m_1+m_2)^2}\sin^3 i = \frac{4\pi^2}{G}\frac{(a_1 \sin i)^3}{P^2},$$

where i is the tilt angle of the orbital plane with respect to the line of sight

9.12 A spherical gravitating object of radius R and mass M consists of a monatomic ideal gas in hydrostatic equilibrium that has average density $\langle \rho \rangle$ and average temperature $\langle T \rangle$. Show that $\langle T \rangle = {\rm constant} \times \langle \rho \rangle^{1/3} M^{2/3}$, so that for two such objects of the same composition at the same average temperature the more massive object must have lower average density. *Hint:* Use the virial theorem.

9.13 Consider a gravitating sphere of uniform density ρ in hydrostatic equilibrium from which (in a thought experiment) all pressure is suddenly removed. Derive a formula giving the time for a test particle at radius r to fall to the center (zero radial coordinate) by assuming the particle to obey Kepler's third law for an orbit with eccentricity $\varepsilon = 1$ (that is, a straight line). You should find that the resulting free-fall time is independent of r, so all mass points in the sphere collapse to zero radial coordinate in the same (finite) time interval.

9.14 Assuming the validity of the virial theorem, write a differential equation governing the rate of energy emission with time (luminosity L), for a star that is contracting gravitationally assuming all the luminosity to come from release of gravitational energy. Show that the resulting timescale t_{KH} (the Kelvin–Helmholtz timescale of Section 4.9) can be interpreted as the time for the radius to decrease by a factor e^{-1} if the timescale is assumed constant, and that this timescale precludes gravitational contraction as the source of the main sequence luminosity of the Sun. Argue that because the Kelvin–Helmholtz timescale is not constant, t_{KH} estimated using main sequence values of the mass, radius, and luminosity of a star is a reasonable estimate of the time that it took the protostar to collapse to the main sequence.***

9.15 The initial collapse phase of star formation is difficult to observe because the collapsing protostar is usually shrouded in dust, and because the time for collapse is very short compared with the lifetime of the star. Verify this last statement by estimating the ratio of the time of collapse to the main sequence to the lifetime on the main sequence for stars with main sequence masses $1\,M_\odot$, $6.45\,M_\odot$, and $17.8\,M_\odot$. *Hint*: See Table 2.2 for data.

9.16 Use the Salpeter form of the initial mass function given in Eq. (9.16) to make a rough estimate of the fraction of stars that form having masses (a) greater than $1\,M_\odot$ and (b) greater than $10\,M_\odot$. Assume for purposes of this problem that the lowest-mass star has $M = 0.1\,M_\odot$ and the highest-mass star has $M = 100\,M_\odot$.

10 Life and Times on the Main Sequence

In Chapter 9 the collapse of a protostar to a Zero Age Main Sequence (ZAMS) star was considered. In this chapter the nature of life on the main sequence for such a star is examined. Since a ZAMS star achieves hydrostatic and thermal equilibrium quickly, the ZAMS state may be viewed as the initial condition for subsequent stellar evolution. This is fortunate, because the discussion of Chapter 9 indicates that many uncertainties remain in the detailed understanding of protostar collapse to the main sequence. The essence of main sequence life is stable burning of core hydrogen into helium under conditions of hydrostatic equilibrium, primarily by PP chains for stars of a solar mass or less, and by the CNO cycle for more massive stars. Because we have discussed hydrostatic equilibrium in Chapter 4 and energy production by the PP chains and CNO cycle in Chapter 6, the essential features of life on the main sequence have been introduced already.

10.1 The Standard Solar Model

Since the Sun is a main sequence star, it is appropriate to begin by examining this local star that we know the best. A large amount of relevant data have been amassed and considerable understanding exists of how the Sun functions. This has allowed the construction of a *Standard Solar Model*: a mathematical model of the Sun that uses basic understanding from fields such as nuclear and atomic physics, measured key quantities, and a few assumptions to describe all solar observations. Standard Solar Models are important because they fix the Sun's helium abundance and the convection length scale in the solar surface, and they provide a benchmark for measuring improved solar modeling and a starting point for more general stellar modeling. The essence of the Standard Solar Model is that a 1 $M_\odot$ ZAMS star is evolved to the present age of the Sun subject to a small set of assumptions:

1. The Sun was formed from a homogeneous mixture of gases.
2. The Sun is powered by nuclear reactions in its core.
3. The Sun is approximately in hydrostatic equilibrium. Some deviations from equilibrium are permitted as the Sun evolves, but these are assumed to be small and to occur slowly.
4. Energy is transported from the core of the Sun, where it is produced, to the surface, where it is radiated into space, by photons (radiative transport) and by large-scale vertical motion of packets of hot gas (convection). Conduction is considered negligible for heat transport in the Sun.

Let us now discuss each of these assumptions of the Standard Solar Model in a little more depth.

10.1.1 Composition of the Sun

The assumption that the Sun was formed from a homogeneous mixture of gases is motivated by the strong convection expected in the protostar during collapse to the main sequence that was discussed in Chapter 9. The surface abundances are then assumed to have been undisturbed in the subsequent evolution, so that present surface abundances are an accurate reflection of the composition of the original ZAMS star. The abundance of most elements in the surface can be inferred by spectroscopy, except for the noble gases He, Ne, and Ar. They are not excited significantly by the blackbody emission spectrum of the photosphere, so their abundances cannot be fixed well by spectroscopic information. Because evolution of the Sun's luminosity depends strongly on the mean molecular weight, which is in turn strongly influenced by the helium abundance, the H/He ratio is normally taken as an adjustable parameter in solar models. This parameter is determined by requiring that the luminosity of the Sun at the present age of the Solar System (4.6 billion years, as determined by dating of meteorites) be accurately reproduced by the model.

10.1.2 Energy Generation and Composition Changes

The Sun is assumed to derive its power and associated composition changes from the proton–proton chains PP-1, PP-2, and PP-III defined in Fig. 5.2, and the CNO reactions defined in Eqs. (5.10)–(5.11) and Fig. 5.3. The nuclear reaction networks (see Box 6.1) describing this energy and element production are solved at a given time by dividing the Sun into concentric shells, calculating the nuclear reactions in each shell as a function of the current temperature and density there, and using the updated composition and the energy production as input to the partial differential equations describing the solar equilibrium structure and its time evolution.

10.1.3 Hydrostatic Equilibrium

Since the dynamical timescale of the Sun defined in Eq. (4.18) and Table 4.2 is about an hour, the Sun may be expected to have reached hydrostatic equilibrium quickly. Although the Sun is assumed to be in approximate hydrostatic equilibrium, a Standard Solar Model allows small expansions and contractions in response to time evolution of the star, and it may be expected that re-equilibration of the system after such excursions is fast compared with the timescale for evolution. The pressure responsible for hydrostatic equilibrium is composed of both gas pressure and radiation pressure, but the radiation pressure even at the center of the Sun is only about 0.05% of the total pressure. A Standard Solar Model typically ignores the effects of both rotation and the part of the pressure deriving from magnetic fields on hydrostatic equilibrium. Likewise, any stellar pulsations are assumed to have negligible effect on hydrostatic equilibrium.

10.1.4 Energy Transport

It is assumed that energy transport in the Sun by acoustic or gravitationally driven waves is negligible, and that the energy produced internally in the Sun is transported to the surface by radiative diffusion and convection. In the interior, the transport is assumed to be by radiative diffusion unless the critical gradient for convective instability is exceeded, in which case the Sun transports energy in that region convectively with an adiabatic temperature gradient (see Section 7.7). In the subsurface region the actual gradient is steeper than the critical gradient and mixing length theory is used to model convection. Because convection in the subsurface region is difficult to calculate reliably, the mixing length in units of the scale height [the parameter α of Eq. (E.7)] is taken as an adjustable parameter, to be fixed by requiring the model to yield the observed radius of the Sun. (The definition of the solar radius is discussed in Box 10.1, in terms of the *optical depth* τ for the Sun.) The opacities required for radiative diffusion of energy are Rosseland mean opacities that often must be calculated numerically. They are among the least well-determined quantities entering the Standard Solar Model.

10.1.5 Constraints and Solution

Solution of the problem corresponds to evolving in time four partial differential equations in five unknowns [pressure P, temperature T, density ρ, enclosed mass $m(r)$, and luminosity L], supplemented by an equation of state that defines a relationship among T, P, and ρ, and subject to constraints that the radius, luminosity, and mass calculated within the model agree with current observations of the Sun. The network of equations required to describe nuclear energy and element production is solved separately for each timestep in each zone, as described above. The equation of state is assumed to be given by the ideal gas law for regions that are completely ionized. Otherwise, a numerical equation of state is typically used.

The Standard Solar Model solution is constructed iteratively. Starting values for the helium abundance and the mixing length parameter are used to evolve an initial zero-age model to the current age of the Sun. The model's luminosity and radius are then compared with observations, the helium abundance and mixing length parameters adjusted accordingly, and the model is evolved again. This cycle is repeated until convergence is obtained. Table 10.1 gives the temperature, density, pressure, and luminosity of a Standard Solar Model as a function of radius and enclosed mass at that radius [32]. Figure 10.1 illustrates graphically some of the parameters of this model plotted versus the radius and Fig. 10.2 plots the same quantities versus the Lagrangian enclosed mass coordinate.

The Standard Solar Model may be tested by comparing with observations. These tests range from general ones, such as accounting for the existence, age, and energy output of the Sun, to specific ones such as accounting for detailed results of solar seismology. The Standard Solar Model has passed these tests very well. Let us discuss two specific tests that probe the solar interior: helioseismology and neutrino emission.

Box 10.1 — Optical Depth and the Solar Surface

Let a radiant flux F be incident on a thin slab of stellar material characterized by a thickness dr, density ρ, and opacity κ, as illustrated in the following figure.

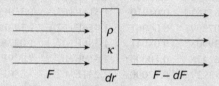

The difference between incident and transmitted flux dF is given by, $dF = -\kappa \rho F dr$. Assuming $\kappa \rho$ to be constant over dr, this has a solution

$$F = F_0 e^{-\kappa \rho r} = F_0 e^{-r/\lambda} \qquad \lambda \equiv (\kappa \rho)^{-1},$$

where λ is the mean free path (Box 7.1). Introduce a dimensionless quantity τ called the *optical depth* through the differential equation $d\tau = -\kappa \rho dr$. The optical depth at r is then defined by integrating

$$\tau = \int_r^\infty \kappa \rho \, dr.$$

Optical depth measures the probability that a photon at r will interact before leaving the star, so it characterizes the transparency of a medium. An opaque medium has a large τ, which can be because of some combination of high opacity, large physical depth, or high density. A region is *optically thin* if $\tau \ll 1$ and *optically thick* if $\tau \gg 1$.

By definition, the *photosphere* lies at the radius where $\tau = \frac{2}{3}$, so a photon emitted from the photosphere will suffer on average less than one scattering before reaching a distant observer. (Optical depth and location of the photosphere depend on frequency through the opacity, but this will be ignored for the present discussion.) The equation for hydrostatic equilibrium becomes particularly simple if τ is used to parameterize distance. As shown in Problem 10.12,

$$\frac{dP}{d\tau} = \frac{g}{\kappa} \qquad \text{(Hydrostatic equilibrium)}.$$

The region with $\tau \lesssim 1$ is called the *stellar atmosphere*. The diffusion approximation for radiative transport fails when τ is less than about 1–10 because the mean free path for photons then becomes very long (in the solar surface, λ is of order 10^7 cm or more, compared with fractions of a cm in the interior). Thus, methods used to deal with radiative transport in stellar atmospheres are much more complicated than those adequate for the interior. It is essential to model this complex region adequately because it defines the outer boundary conditions for integration of the equations describing the interior structure, and because the atmosphere mediates the emission and absorption of radiation and therefore has a large influence on the photon spectrum (our primary source of stellar information).

Table 10.1 A Standard Solar Model [32]

$M/M_\odot$	$R/R_\odot$	T(K)	ρ (g cm^{-3})	P (dyn cm^{-2})	$L/L_\odot$
0.0000298	0.00650	1.568E+07	1.524E+02	2.336E+17	0.00027
0.0008590	0.02005	1.556E+07	1.483E+02	2.280E+17	0.00753
0.0065163	0.04010	1.516E+07	1.359E+02	2.111E+17	0.05389
0.0207399	0.06061	1.456E+07	1.193E+02	1.868E+17	0.15638
0.0439908	0.08041	1.386E+07	1.027E+02	1.606E+17	0.29634
0.0762478	0.10006	1.310E+07	8.729E+01	1.349E+17	0.45135
0.1173929	0.12000	1.231E+07	7.350E+01	1.108E+17	0.60142
0.1672004	0.14056	1.150E+07	6.123E+01	8.892E+16	0.73152
0.2203236	0.16027	1.076E+07	5.114E+01	7.094E+16	0.82657
0.2800107	0.18104	1.002E+07	4.205E+01	5.517E+16	0.89658
0.3393826	0.20107	9.353E+06	3.459E+01	4.279E+16	0.94011
0.3966733	0.22038	8.762E+06	2.847E+01	3.319E+16	0.96616
0.4559683	0.24084	8.188E+06	2.301E+01	2.516E+16	0.98259
0.5114049	0.26085	7.676E+06	1.857E+01	1.907E+16	0.99183
0.5627338	0.28058	7.214E+06	1.496E+01	1.446E+16	0.99669
0.6099028	0.30016	6.794E+06	1.203E+01	1.096E+16	0.99860
0.6564038	0.32132	6.379E+06	9.484E+00	8.119E+15	0.99941
0.6952616	0.34091	6.028E+06	7.605E+00	6.156E+15	0.99976
0.7304369	0.36063	5.703E+06	6.092E+00	4.667E+15	0.99993
0.7621708	0.38053	5.400E+06	4.876E+00	3.539E+15	1.00002
0.7907148	0.40067	5.117E+06	3.900E+00	2.683E+15	1.00005
0.8163208	0.42109	4.851E+06	3.118E+00	2.034E+15	1.00007
0.8374222	0.44008	4.621E+06	2.539E+00	1.578E+15	1.00007
0.8580756	0.46112	4.383E+06	2.029E+00	1.197E+15	1.00006
0.8750244	0.48072	4.176E+06	1.651E+00	9.287E+14	1.00006
0.8902432	0.50063	3.978E+06	1.345E+00	7.206E+14	1.00005
0.9038831	0.52086	3.789E+06	1.095E+00	5.591E+14	1.00004
0.9160850	0.54139	3.606E+06	8.924E-01	4.339E+14	1.00004
0.9260393	0.56033	3.445E+06	7.413E-01	3.445E+14	1.00003
0.9358483	0.58142	3.273E+06	6.052E-01	2.673E+14	1.00003
0.9438189	0.60081	3.120E+06	5.040E-01	2.123E+14	1.00002
0.9509668	0.62036	2.969E+06	4.205E-01	1.686E+14	1.00002
0.9573622	0.64001	2.818E+06	3.517E-01	1.339E+14	1.00002
0.9636045	0.66168	2.648E+06	2.900E-01	1.039E+14	1.00001
0.9686223	0.68129	2.485E+06	2.445E-01	8.249E+13	1.00001
0.9730081	0.70042	2.315E+06	2.081E-01	6.572E+13	1.00001
0.9771199	0.72033	2.115E+06	1.780E-01	5.161E+13	1.00001
0.9811002	0.74162	1.899E+06	1.513E-01	3.936E+13	1.00000
0.9842836	0.76050	1.718E+06	1.299E-01	3.055E+13	1.00000
0.9874435	0.78148	1.526E+06	1.085E-01	2.264E+13	1.00000
0.9900343	0.80103	1.355E+06	9.066E-02	1.678E+13	1.00000
0.9922832	0.82051	1.193E+06	7.470E-02	1.215E+13	1.00000

(Continued next page ...)

$M/M_\odot$	$R/R_\odot$	T(K)	ρ (g cm^{-3})	P (dyn cm^{-2})	$L/L_\odot$
0.9942853	0.84082	1.031E+06	5.987E-02	8.406E+12	1.00000
0.9958822	0.86022	8.826E+05	4.733E-02	5.682E+12	1.00000
0.9972278	0.88035	7.356E+05	3.590E-02	3.585E+12	1.00000
0.9982619	0.90020	5.966E+05	2.613E-02	2.110E+12	1.00000
0.9990296	0.92017	4.627E+05	1.775E-02	1.107E+12	1.00000
0.9995498	0.94015	3.343E+05	1.080E-02	4.833E+11	1.00000

Table 10.1 (Continued) Standard Solar Model

$M_\odot = 1.989 \times 10^{33}$ g $R_\odot = 6.96 \times 10^{10}$ cm $L_\odot = 3.828 \times 10^{33}$ erg s^{-1}

10.2 Helioseismology

One way to study the Sun's interior is to study the propagation of waves in its body, similarly to the way geologists learn about the interior of the Earth by studying seismic waves or how the composition of a bell may be inferred by studying the sound frequencies that it produces. The corresponding field of study is called *helioseismology*. The concepts of helioseismology introduced here to study the Sun can be extended (with suitable generalization) to study the interior of other stars (see Section 16.5); this is termed *asteroseismology*.

10.2.1 Solar p-Modes and g-Modes

Solar oscillations were discovered by studying Doppler shifts of surface absorption lines [143], which showed that patches of the solar surface oscillated with a period of five minutes and velocity amplitude of 0.5 km s^{-1}. These *5-minute oscillations* represent pressure waves (*p-modes*) trapped between the surface and the lower boundary of the convective zone at about 70% of the solar radius. They are reflected from the surface by density gradients and refracted near the base of the convection zone because of changing sound speed in that region. In addition to p-modes associated with acoustical waves trapped near the solar surface, the Sun may exhibit *gravity waves* or *g-modes*, which correspond to oscillations with gravity as the restoring force. The g-modes carry information about much deeper regions of the Sun than that carried by the p-modes. As will be discussed in Sections 15.1 and 16.5, some white dwarfs are observed to be pulsating variables and the dominant oscillation modes responsible for pulsations in variable white dwarfs are expected to be g-modes.

10.2.2 Surface Vibrations and the Solar Interior

The solar interior can be studied by observing vertical motion of the Sun's surface in localized regions. The Michelson Doppler Imager (MDI) instrument on the SOHO

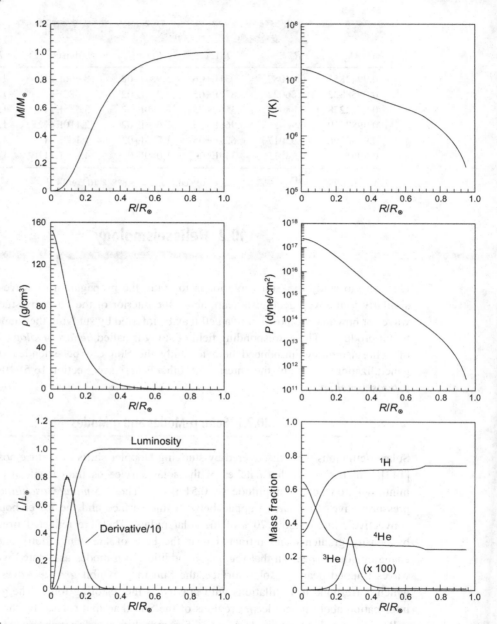

Fig. 10.1 Parameters from a Standard Solar Model (Table 10.1) plotted versus the radial coordinate R. Data are from Ref. [32].

observatory orbits the Sun 1.5 million kilometers sunward from the Earth,[1] and is capable of measuring vertical displacement of the solar surface at a million points per minute.

---

[1] SOHO orbits slowly around the Earth–Sun L1 Lagrange point (see Section 18.2.2), which lies on the Earth–Sun line about 1.5 million kilometers toward the Sun from the Earth. More recently MDI has been superceded by the Helioseismic and Magnetic Imager (HMI) on the Solar Dynamics Observatory (SDO), launched into a geosynchronous orbit in 2010. The HMI instrument scans the full solar disk at 6173 Å with arcsecond resolution.

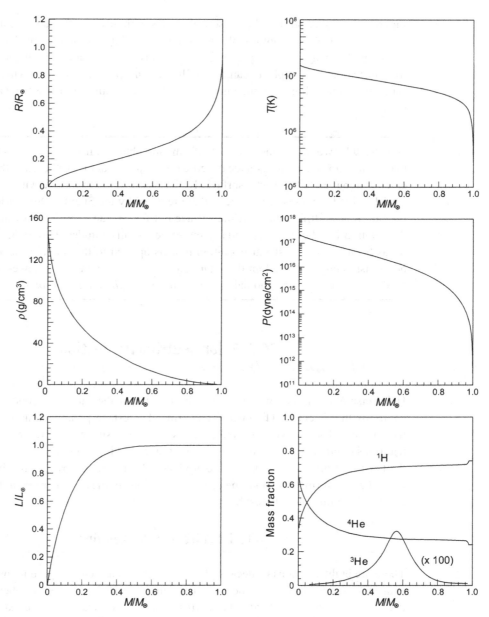

Fig. 10.2 Parameters from a Standard Solar Model (Table 10.1) plotted versus the enclosed mass $M(R)$. Data are from Ref. [32].

The Sun vibrates at a complex set of frequencies. By decomposing the observed vibrations of the Sun into a superposition of standing acoustic waves, it is possible to learn about the interior because the structure of the interior affects the wave patterns that appear on the surface. Such decompositions indicate that the observed motion of the surface is a superposition of several million resonant modes with different frequencies and horizontal wavelengths. Helioseismology places strong constraints on theories of the solar interior. Although the analysis is complex, the basic idea is simple: changes in the properties of the solar interior (for example, the amount of helium in a particular region) will affect the way sound waves travel through the interior, and this will in turn influence the way the surface vibrates.

Example 10.1 Two important pieces of information obtained from early helioseismology are that the helium abundance outside the core is the same as at the surface, and that convection extends about 30% below the surface. Helioseismology also has shown that the speed of sound inside the Sun is very close to that predicted by the Standard Solar Model, and that the Sun rotates differentially at the surface but that inside about 65% of the solar radius the rotation rate becomes essentially the same for all latitudes. However, more recently it has been found that stellar evolution models applied to the Sun are incompatible with helioseismology if they adopt the solar composition obtained by the newest spectroscopic models. This unresolved anomaly is called the *Solar abundance problem*.

10.3 Solar Neutrino Production

Helioseismology is one way to probe the interior of the Sun. A second is to study the neutrinos that it emits. The energy powering the surface photon luminosity must make its way on a 100,000-year or greater timescale to the surface before being radiated, but neutrinos emitted from the core are largely unimpeded in their exit from the Sun, reaching the Earth $8 1/2$ minutes after they were produced. Therefore, neutrinos carry immediate and more direct information about the current conditions in the solar core than do the photons emitted from the solar photosphere.

10.3.1 Sources of Solar Neutrinos

There are eight reactions or decays playing a role in solar energy production that produce neutrinos (see Section 5.2). They are listed in Table 10.2, along with labels that will be used to refer to them, and with their corresponding Q-values. Six of the reactions produce spectra with a range of Q-values and two are line sources (the neutrinos are emitted at discrete energies). Neutrinos from the CNO reactions are difficult to observe because they are weak (less than 2% of the Sun's energy comes from the CNO cycle) and the energies are low. Our primary concern will be with the reactions labeled pp, ^{7}Be, and ^{8}B, which correspond to branches of the PP chains.

The solar neutrino spectrum that is predicted by the Standard Solar Model is shown in Fig. 10.3(a), while Fig. 10.3(b) illustrates the radial regions of the Sun responsible for producing neutrinos from each of the PP reactions. From Fig. 10.3(b), the ^{8}B and ^{7}Be

10.3 Solar Neutrino Production

Table 10.2 Important neutrino-production reactions in the Sun

Label	Reaction	Q-value
pp	$p + p \to {}^2\text{H} + e^+ + \nu_e$	$Q \leq 0.420$ MeV
pep	$p + e^- + p \to {}^2\text{H} + \nu_e$	$Q = 1.442$ MeV
hep	${}^3\text{He} + p \to {}^4\text{He} + e^+ + \nu_e$	$Q \leq 18.773$ MeV
${}^7\text{Be}$	${}^7\text{Be} + e^- \to {}^7\text{Li} + \nu_e$	$Q = 0.862$ MeV (89.7%)
		$Q = 0.384$ MeV (10.3%)
${}^8\text{B}$	${}^8\text{B} \to {}^8\text{Be}^* + e^+ + \nu_e$	$Q \leq 15$ MeV
CNO	${}^{13}\text{N} \to {}^{13}\text{C} + e^+ + \nu_e$	$Q \leq 1.199$ MeV
CNO	${}^{15}\text{O} \to {}^{15}\text{N} + e^+ + \nu_e$	$Q \leq 1.732$ MeV
CNO	${}^{17}\text{F} \to {}^{17}\text{O} + e^+ + \nu_e$	$Q \leq 1.740$ MeV

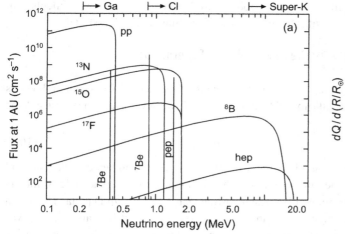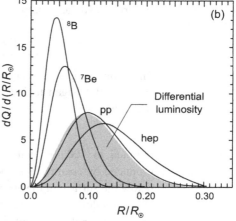

Fig. 10.3 (a) The solar neutrino spectrum. The sensitive regions for various experiments described in Section 10.4 are indicated above the graph. (b) Differential neutrino production as a function of solar radius. Labels are described in Table 10.2. The shaded area indicates the differential photon luminosity.

neutrinos probe much smaller radii than the photons or neutrinos from PP-I (labeled pp), because they are produced at higher temperatures and therefore at greater depth.

10.3.2 Testing the Standard Solar Model with Neutrinos

From Fig. 5.2 and Table 10.2, each of the three PP chains may be tagged by a particular combination of neutrinos [113]. The overall rate of energy production in all three chains is governed by the initial step $p + p \to {}^2\text{H} + e^+ + \nu_e$, which produces an electron neutrino, with a maximum neutrino energy (endpoint) of 0.42 MeV. The PP-II chain is tagged in addition by the neutrinos from electron capture on ${}^7\text{Be}$, which come at the discrete energies 0.38 and 0.86 MeV. Finally, PP-III is tagged uniquely by higher-energy neutrinos from β-decay of ${}^8\text{B}$, which has an endpoint of about 15 MeV. Thus, detection of neutrinos from the Sun can provide a direct probe of the relative rates for PP-I, PP-II, and PP-III, and a stringent test of Standard Solar Models. This is a difficult but not impossible task, as will be shown in the next section.

10.4 The Solar Electron-Neutrino Deficit

By counting the number of neutrinos produced and the average energy released for each $4H \rightarrow {}^4He$ in the PP chains, it may be estimated that the Sun should be emitting $\sim 10^{38}$ electron neutrinos per second, if it is powered by the PP chains (see Problem 10.3). However, detectors on Earth see only a fraction of the corresponding number of electron neutrinos that would be expected to reach Earth. Historically this was termed the *solar neutrino problem*. Let us introduce this problem by summarizing the experiments that confirm this deficit compared with the predictions of the Standard Solar Model.

10.4.1 The Davis Chlorine Experiment

The pioneering solar neutrino detection experiment implemented by Raymond Davis (1914–2006) began taking data in 1970 and continued for several decades. It used the reaction

$$\nu_e + {}^{37}Cl \rightarrow {}^{37}Ar + e^- \tag{10.1}$$

initiated in 100,000 gallons of cleaning fluid (C_2Cl_4). To shield against background produced by cosmic rays, the tank containing the cleaning fluid was placed 1500 meters below the surface in the abandoned Homestake Mine near Lead, South Dakota. The argon atoms produced by the reaction (10.1) are tiny in number – about one argon atom was produced per day by neutrino interactions in the tank – but they are radioactive, so their decays can be counted after separation of the argon from the cleaning fluid by bubbling helium through the liquid.[2]

The reaction (10.1) has a threshold (minimum energy for the reaction to occur) of 0.81 MeV, which is higher than the maximum energy of 0.42 MeV for neutrinos in the PP-I chain [see Fig. 10.3(a)]. Therefore, the Davis experiment was sensitive primarily to the ^8B neutrinos (and weakly to the ^7Be neutrinos) from the reactions in Table 10.2. The Davis experiment counted neutrinos at a rate that was approximately *three times smaller* than the rate predicted by the Standard Solar Model. Because this detector was based on chemical separation conducted well after the neutrino reactions had taken place, it had no directional sensitivity. It was assumed that the neutrinos were coming from the Sun because there was no other plausible source for neutrinos of that intensity; however, this could not be proved directly by the experiment.

---

[2] The chlorine neutrino-detection experiment was a brilliant scientific and technical feat, for which Davis shared the Nobel Prize for Physics in 2002. An early popular-level account giving more of the technical details and the historical context may be found in the article by John Bahcall (1934–2005) [33], who contributed much of the theory behind the Davis experiment.

10.4.2 The Gallium Experiments

The chlorine experiment was sensitive primarily to the ^{8}B neutrinos. These neutrinos probe the deepest regions of the Sun [see Fig. 10.3(b)], but they come from a very minor side branch of the solar energy production cycle (see Fig. 5.2); thus they are not related very closely to the Sun's photon luminosity. Because of the threshold for the reaction (10.1), the neutrinos from the primary energy production process in the Sun (PP-I) are not detected at all. Another chemistry-based detection system can be constructed using the reaction

$$\nu_e + {}^{71}\text{Ga} \rightarrow {}^{71}\text{Ge} + e^-. \tag{10.2}$$

The ^{71}Ge produced is radioactive so the Ge atoms can be separated chemically and their decays observed to count the number of neutrino reactions. The reaction (10.2) has a threshold of only 0.23 MeV, so it can detect neutrinos coming from the PP-I chain that produces most of the solar energy.

Two large experiments, SAGE (operated by a Russian–American collaboration underground in the Caucasus) and GALLEX (operated by a largely European collaboration in the Gran Sasso underground laboratory in Italy), were implemented based on the gallium reaction (10.2). These experiments, for which more than half of the neutrinos are expected to come from the pp reaction in Table 10.2, also measured a neutrino deficit compared with the Standard Solar Model. However, the deficit was not as large as in the chlorine experiment. They found that the electron neutrino flux is reduced by a factor of about two relative to that expected. Like the chlorine experiment, the gallium experiments are chemistry-based and have no directional sensitivity, so they could not demonstrate conclusively that the detected neutrinos originated in the Sun.

10.4.3 Super Kamiokande

The Super Kamiokande detector operates 1000 meters underground in the Mozumi Mine of the Kamioka Mining and Smelting Company in Japan, and uses a different (non-chemical) approach to detecting neutrinos.[3] A large tank containing 50,000 cubic meters of ultrapure water is monitored by more than 11,000 photodetectors. When the neutrino–electron elastic scattering reaction $\nu + e^- \rightarrow \nu + e^-$ occurs in the water, the recoiling electrons may exceed the speed of light in the medium and produce Cherenkov radiation that is then detected by the phototubes (see Box 10.2). The threshold for reliable detection in Super-K is about 7 MeV, so it is sensitive only to the more energetic ^{8}B neutrinos produced in PP-III [see Fig. 10.3(b)]. Because the detector has directional sensitivity, Super Kamiokande (unlike the Davis experiment, SAGE, or GALLEX) was able to demonstrate that the neutrinos being detected come from the direction of the Sun. The Super-K results again indicated

---

[3] This detector is commonly known as 'Super-K' in the neutrino physics community. It is a successor to an earlier, smaller version called Kamiokande. The name derives from 'Kamioka' plus 'nde', which stands for "nucleon decay experiment", because the original Kamiokande detector was built to search for radioactive decay of the proton. It did not find evidence that the proton is unstable, but its successor Super-K used similar detection technology to do Nobel-prize-winning neutrino research.

Box 10.2 — Cherenkov Radiation

Light in a medium travels at a velocity $v = c/n$, where c is the speed of light in vacuum and n is the *refractive index* of the medium. For example, the speed of light in a vacuum is 3×10^5 km s^{-1} ($n = 1$) but in water it is about 2.3×10^5 km s^{-1} ($n \sim 1.3$) and in glass about 2.0×10^5 km s^{-1} ($n \sim 1.5$). Hence, a charged particle moving through a medium can acquire a velocity that exceeds the speed of light in the medium (but does not exceed the speed of light in a vacuum, which is forbidden by special relativity).

Sonic Booms

This is similar to an airplane exceeding the speed of sound in air, which generates a shockwave heard as a "sonic boom." When the airplane breaks the sound barrier it is "outrunning its own sound." The result is a shockwave that focuses an intense burst of sound in a specific direction, as illustrated in Fig. 10.4(a). Likewise, when a charged particle exceeds the speed of light in a medium an electromagnetic 'sonic boom' produces a directed burst of light called *Cherenkov radiation* (the blue glow seen in water surrounding a nuclear reactor core is one common manifestation). This is illustrated in Fig. 10.4(b), where the uncharged neutrino interacts with a charged electron in water and the recoiling electron exceeds the speed of light in the water and emits a cone of Cherenkov radiation.

Detecting Cherenkov Radiation

Figure 10.4(c) illustrates detection of Cherenkov light by phototubes arrayed around the water. The energy, direction, point of interaction, and charged-particle type may be inferred by analyzing data from the phototubes, which permits the direction of the incoming neutrino to be determined.

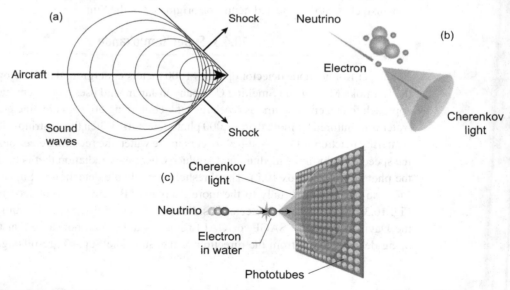

Fig. 10.4 (a) Shockwaves produced by exceeding the speed of sound in a medium. (b) Production of Cherenkov radiation by neutrinos. (c) Detection of Cherenkov radiation.

a solar neutrino deficit, with the detector seeing fewer than 40% of the electron neutrinos expected based on fluxes predicted by the Standard Solar Model.

10.4.4 Astrophysics and Particle Physics Explanations

The experiments described above were not all sensitive to the same neutrinos from the Sun and found somewhat different magnitudes for the solar neutrino deficit. However, the Davis chlorine experiment, the two gallium experiments, and the water Cherenkov experiment all found reproducibly that significantly fewer neutrinos are being detected than predicted by the Standard Solar Model. Table 10.3 summarizes solar neutrino fluxes measured by these detectors and compares them with the predictions of a Standard Solar Model. These results indicate a deficit of solar neutrinos in all detectors by an amount that depends on which neutrinos are being detected, with the size of the deficit well outside the range of experimental uncertainties.

Confirmation in the more recent experiments of the solar neutrino problem uncovered by Davis implies some combination of two alternatives: (1) How the Sun works is not understood (failure of the Standard Solar Model), or (2) How neutrinos work is not understood (failure of the Standard Model of elementary particle physics). Thus, a debate ensued over whether the solution to the solar neutrino problem lay in a modification of our astrophysics understanding or of our particle physics understanding. The apparently anomalous flux of solar neutrinos has implications far broader than would be suggested by some (difficult to measure) number being a factor of 2–3 smaller than expected. If the Standard Solar Model were wrong about the predicted neutrino flux it would call into question its description of the central region of the Sun. Since the Sun is the best-studied star, this would in turn raise issues about whether our general understanding of stellar structure and stellar evolution were on firm ground. Conversely, the Standard Model of elementary particle physics is arguably the most successful scientific theory of all time, given the breadth and accuracy with which it correlates and explains data. But despite these past successes, a failed prediction in its weak-interaction sector would indicate unequivocally that there is physics beyond that described by the Standard Model.

Table 10.3 Solar neutrino fluxes compared with Standard Solar Model			
Experiment	Observed flux	SSM	Observed/SSM
Homestake	$2.54 \pm 0.14 \pm 0.14$ SNU	$9.3^{+1.2}_{-1.4}$	0.273 ± 0.021
SAGE	$72^{+12}_{-10}{}^{+5}_{-7}$ SNU	137^{+8}_{-7}	0.526 ± 0.089
GALLEX	$69.7 \pm 6.7^{+3.9}_{-4.5}$ SNU	137^{+8}_{-7}	0.509 ± 0.089
Super Kamiokande	$2.51^{+0.14}_{-0.13}$ (10^6 cm^{-2}s^{-1})	$6.62^{+0.93}_{-1.12}$	0.379 ± 0.034

All fluxes in solar neutrino units (SNU; 1 SNU is the neutrino flux that would produce 10^{-36} reactions per target atom per second), except for Super Kamiokande results. Experimental uncertainties include systematic and statistical contributions. Comparisons with a Standard Solar Model (SSM) from [31].

As will be discussed in depth in Chapters 11–12, experiments and observations have shown rather conclusively that the "solar neutrino problem" is now resolved, and that the resolution lies in new properties for neutrinos that imply physics beyond the Standard Model of elementary particle physics. Specifically, there is now strong evidence that at least some neutrinos have a non-zero mass, and that this permits neutrinos to change their types (*flavors*) from electron neutrinos (to which the above detectors are sensitive) into other types that the detectors described above cannot see.[4] As a consequence, this will provide strong support for the validity of the Standard Solar Model. But before describing the resolution of the solar neutrino problem, let us conclude this chapter by considering broader issues of main sequence systematics, evolution of stars on the main sequence, and the evolutionary processes that eventually cause a star like the Sun to leave the main sequence.

10.5 Evolution of Stars on the Main Sequence

The main sequence is the longest and most stable period of a star's life but stars do evolve while they are on the main sequence, primarily in response to concentration changes in their core as they burn hydrogen to helium in hydrostatic equilibrium. This lowers the central pressure because it increases the mean molecular weight (distributes the mass over fewer particles). This in turn increases the core density and releases gravitational energy, half of which is radiated away and half of which raises the core temperature (the virial theorem). Energy outflow due to higher core temperature causes the outer layers to expand slightly and the star becomes more luminous. The surface temperature during this process may either increase or decrease, depending on the mass of the star. For stars below about 1.25 $M_\odot$ the surface temperature tends to increase, while for more massive stars it tends to decrease as the star evolves on the main sequence. This is illustrated in Fig. 10.5(b).

Therefore, the primary externally visible effect of a star's evolution on the main sequence is to cause a drift slightly upward and to the left on the HR diagram from the ZAMS position for lighter stars, and slightly upward and to the right for heavier stars. Internal changes are more substantial but their effects are not very visible while the star continues to burn hydrogen in its core. Significant modification of elemental abundances is taking place, but these changes are limited initially to the central regions.

Example 10.2 The Standard Solar Model [30] indicates that over the 4.6 billion year time that the Sun has spent on the main sequence the radius has increased by about 12%, the core temperature has increased by about 16%, the luminosity has increased by about 40%, the effective surface temperature has increased by about 3%, and the flux of ^{8}B neutrinos has increased by more than a factor of 40. Near the center the mass fraction of hydrogen

---

[4] Chemical experiments (Davis, SAGE, GALLEX) cannot see any flavors other than ν_e. Water Cherenkov detectors like Super-K can see muon and tau neutrinos also, but with greatly reduced efficiency because the cross section for scattering ν_μ or ν_τ from electrons is much smaller than that for scattering ν_e from electrons. Hence water Cherenkov detectors record some ν_μ and ν_τ, but cannot distinguish them easily from ν_e.

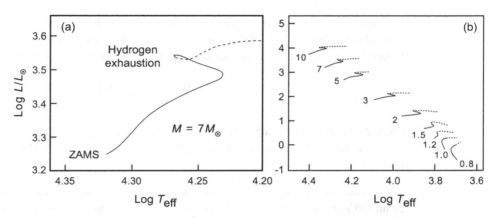

Fig. 10.5 Evolution of stars leaving the main sequence (note the expanded scale) [134]. (a) Evolution of a 7 $M_\odot$ star from the ZAMS through exhaustion of central hydrogen. Dashed part of curve indicates evolution after central hydrogen has been exhausted. (b) As for (a) but for a range of masses from 0.8 $M_\odot$ to 10 $M_\odot$. Adapted by permission from Springer Nature: *Evolution on the Main Sequence*, Kippenhahn, R., Weigert, A., and Weiss, A. COPYRIGHT (2012).

has decreased and the mass fraction of helium has increased by about a factor of 2 from their initial values, but outside of about 20% of the solar radius hydrogen and helium retain their ZAMS abundances. Although the mass fraction of hydrogen fuel has decreased substantially in the solar core over its lifetime, from Sections 6.1 and 5.9 the rate of energy production by the proton–proton chain is $dE/dt \propto \rho^2 X^2 T^4$, where ρ is the density, X is the hydrogen mass fraction, and T is the temperature. The increase in ρ and T more than offsets the effect of diminishing X as the Sun evolves, explaining why its luminosity is rising even as its hydrogen fuel is being depleted.

The internal changes discussed in the preceding example set the stage for rapid evolution away from the main sequence that will be the topic of subsequent chapters.

10.6 Timescale for Main Sequence Lifetimes

A question of basic importance for the Sun and for main sequence stars in general is how long the star will remain on the main sequence. Since the main sequence is defined by stable burning of core hydrogen, the rate of hydrogen burning relative to the amount of burnable hydrogen determines this timescale. Comparison of stellar evolution simulations with observations suggests a rule of thumb that stars leave the main sequence when about 10% of their total original hydrogen has been burned to helium.[5] We may define a *nuclear*

---

[5] This refers to the total mass of hydrogen contained in the star at formation, not just to the hydrogen in the central burning regions. The physical origin of this rule of thumb is that most stars are not strongly mixed vertically, which means that only the fraction of the total hydrogen found in the central regions is available for main sequence burning. Simulations indicate that the hydrogen found in regions conducive to burning is typically $\sim$ 10% of the total hydrogen in the star.

(hydrogen) burning or main sequence timescale τ_{nuc} by forming the ratio of the energy released from burning 10% of the hydrogen and the luminosity. The energy available from the burning of one gram of hydrogen to helium is $\sim 6 \times 10^{18}$ ergs. Therefore,

$$\tau_{\text{nuc}} = \frac{E_{\text{H}}/10}{L} = 6 \times 10^{17} \frac{XM}{L} \text{ s}, \tag{10.3}$$

where E_{H} is the energy available from burning all the hydrogen in the star, L is the present luminosity in erg s^{-1}, X is the original hydrogen mass fraction, and M is the mass of the star in grams.

Example 10.3 Inserting values characteristic of the Sun into Eq. (10.3) gives for the solar main sequence timescale

$$\tau_{\text{nuc}}^{\odot} \sim 2.2 \times 10^{17} \text{ s} \sim 10^{10} \text{ yr},$$

which, as has already been discussed, is set essentially by the rate-determining step of the PP chains being a weak-interaction process.

Expressing this main sequence timescale in solar units, for any star

$$\tau_{\text{nuc}} = 10^{10} \left(\frac{M}{M_{\odot}}\right) \left(\frac{L_{\odot}}{L}\right) \text{ yr}, \tag{10.4}$$

and utilizing the mass–luminosity relation (1.27), which is expected to be valid for $M \geq 1\,M_{\odot}$,

$$\frac{L}{L_{\odot}} \simeq \left(\frac{M}{M_{\odot}}\right)^{3.5},$$

the main sequence timescale may be expressed for $M \geq 1\,M_{\odot}$ as

$$\tau_{\text{nuc}} \simeq 10^{10} \left(\frac{M}{M_{\odot}}\right)^{-2.5} \text{ yr}, \tag{10.5}$$

implying that mass has a large influence on main sequence lifetimes.

Main sequence lifetimes as a function of star mass are illustrated in Fig. 10.6, where we see that there is indeed a strong correlation between time spent on the main sequence and mass of the star. The Sun has a main sequence lifetime of about 10 billion years but a $20\,M_{\odot}$ star remains on the main sequence for about 5.5 million years and a $100\,M_{\odot}$ star lives on the main sequence for only about 100,000 years [though $L/L_{\odot} = (M/M_{\odot})^{3.5}$ may be questionable in this case]. Conversely, for main sequence stars with $M \ll 1\,M_{\odot}$ simulations indicate that the main sequence lifetime greatly exceeds the present age of the Universe.

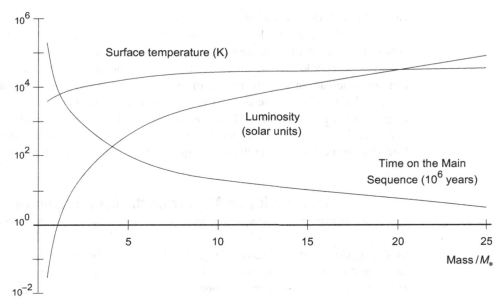

Fig. 10.6 Main sequence lifetimes, temperatures, and luminosities as a function of ZAMS mass.

10.7 Evolutionary Timescales

Evolution prior to the main sequence is governed by two primary timescales: (1) the dynamical (free-fall) timescale and (2) the Kelvin–Helmholtz (thermal adjustment) timescale. Evolution on the main sequence and beyond is governed in addition by a third set of timescales, the *nuclear burning timescales*, one for each nuclear fuel consumed in the main sequence stage and beyond. The dynamical timescale is hours to days for most stars and the thermal adjustment timescale is typically hundreds of thousands to hundreds of millions of years. The nuclear burning timescale depends on the fuel and the mass of the star (among other factors), but is typically much longer than the dynamical and Kelvin–Helmholtz timescales. Thus, stars spend much more time on the main sequence than in their formation phase because time spent on the main sequence is governed by the hydrogen burning timescale, which is much longer than the hydrodynamical and Kelvin–Helmholtz timescales.

Once stars exhaust their core hydrogen and leave the main sequence they can undergo successive burnings of heavier fuels, which introduce new nuclear burning timescales. In the periods between exhaustion of one fuel and ignition of another, thermal adjustment timescales will also be important, and in certain cases (such as stellar pulsations or gravitational core collapse) dynamical timescales will be relevant. The nuclear burning timescales that become important after the main sequence are typically longer than the corresponding Kelvin–Helmholtz and dynamical timescales, just as for the main sequence. However, post main sequence burning timescales are much shorter than those for main sequence hydrogen burning because they necessarily occur at much higher temperatures and densities, and the fuels being burned produce less energy per unit mass burned; thus, a star generally spends more time on the main sequence than in its post main sequence evolution.

We may conclude that the nuclear burning timescale on the main sequence is longer than any timescale in any other stage of the star's life, explaining why at any one time in a population of stars one expects to see the majority on the main sequence (unless the population is sufficiently old that many stars have had time to evolve off the main sequence). This accounts neatly for the existence of the main sequence, and for the age-dependent turn off point from the main sequence found in the HR diagram of clusters.

10.8 Evolution Away from the Main Sequence

As already discussed qualitatively in Section 10.5, while the star burns hydrogen in near-perfect hydrostatic equilibrium on the main sequence a series of internal changes sets the stage for the star's subsequent evolution.

1. As hydrogen burns to helium the mass in the core is concentrated in fewer particles. For the Sun, models indicate that the mean molecular weight in the core has increased from about 0.61 amu at ZAMS to about 0.85 amu presently (see Problem 5.3 and Fig. 10.7).
2. This reduction of the number of particles in the core lowers the pressure in the energy-generating zone, causing the core to contract and to become hotter and more dense.
3. The rise in core temperature increases the energy production and steepens the temperature gradient dT/dr, leading to an increased flow of energy from the star. The outer layers expand in response, increasing the luminosity and radius, and decreasing surface temperature.
4. These changes cause the star to develop a high-density core surrounded by an extended, diffuse envelope with low surface temperature – it becomes a *red giant*.

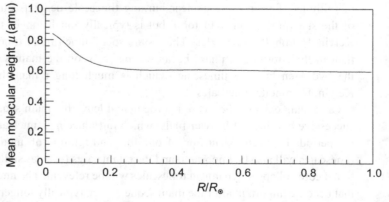

Fig. 10.7 Mean molecular weight as a function of radius for the present Sun according to the Standard Solar Model of Table 10.1.

10.8 Evolution Away from the Main Sequence

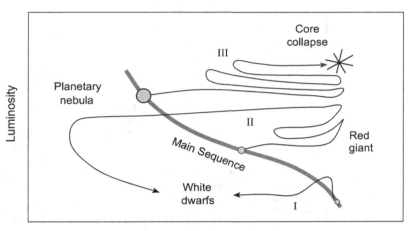

Fig. 10.8 Categories of stellar evolution after the main sequence. Very low-mass stars ($M \lesssim 0.5\,M_\odot$) in category I evolve to helium white dwarfs. Stars with ($0.5\,M_\odot \lesssim M \lesssim 8\,M_\odot$) in category II evolve to C–O or Ne–Mg white dwarfs. The fate of the highest-mass stars ($M \gtrsim 8\,M_\odot$) in category III is somewhat uncertain. In most cases these stars undergo a core collapse that produces a supernova explosion and leaves behind a neutron star. However, there is growing evidence that the most massive such stars may collapse directly to black holes, with little ejection of matter and no neutron star.

5. If the star is massive enough, rising internal temperatures and densities can ignite successively higher-mass fuels in a series of advanced burning stages (see Section 6.5).

Depending on the mass of the star, these steps can lead to the three qualitatively different scenarios that are illustrated in Fig. 10.8.

10.8.1 Three Categories of Post Main Sequence Evolution

The details of evolution after the main sequence sketched in Fig. 10.8 depend very strongly on the mass of the star. (1) For stars with $M \lesssim 0.5\,M_\odot$ the core temperatures never rise high enough to ignite the helium produced by PP-chain proton burning and the star evolves to a helium white dwarf. (2) For stars with $0.5\,M_\odot \lesssim M \lesssim 8\,M_\odot$, this evolution produces a red giant that eventually sheds much of its outer envelope as a planetary nebula and becomes a carbon–oxygen or neon–magnesium white dwarf. (3) For the most massive stars ($M \gtrsim 8\,M_\odot$), evolution from the main sequence leads to a sequence of burning episodes involving successively heavier fuels until the core of the star becomes gravitationally unstable and collapses, producing in most cases a supernova with a remnant neutron star, but for the most massive cases the collapse may drag the entire star into a black hole.

10.8.2 Examples of Post Main Sequence Evolution

Figure 10.9 summarizes the final stages of collapse to the main sequence (beginning with a brief episode of deuterium burning), evolution on the main sequence, and evolution off

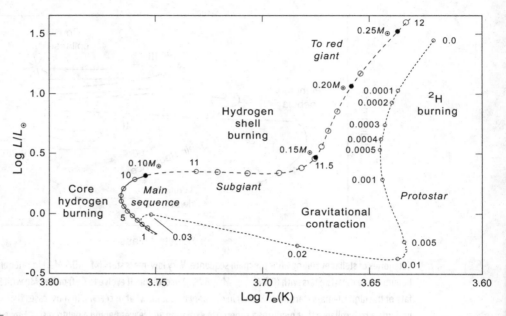

Fig. 10.9 Simulated evolution of a 1 $M_\odot$ star from the final stages of protostar collapse through main sequence core hydrogen burning and on to hydrogen shell burning as the star leaves the main sequence. See text for explanation of symbols and number labels. Initial composition was $Y = 0.275$ and $Z = 0.015$. Evolution beyond the last point shown in this figure will be shown later in Fig. 13.16. Figure adapted from Iben [130]. Adapted from *Volume I: Physical Processes in Stellar Interiors*, Icko Iben, Published by Cambridge University Press, 2012.

the main sequence with the development of a shell hydrogen source for a 1 $M_\odot$ star [130]. Small open circles denote protostar evolution, medium open circles denote main sequence evolution, and large open circles denote evolution away from the main sequence toward the red giant region. Numbers beside open circles indicate times in units of 10^9 yr, with zero time chosen near the onset of deuterium (^{2}H) burning when the protostar has shrunk to a radius 10 times the present solar radius (see Fig. 13.16). The numbers beside solid circles indicate the mass coordinate for the hydrogen shell source (discussed in Section 13.2) at that time in units of solar masses. In this example hydrogen is burning in a shell enclosing 25% of the star's mass for the last point shown, at which time the star has expanded to about 10 times its main sequence radius (see Fig. 13.16).

Evolution after the main sequence for stars of various masses is summarized in Fig. 10.10. Detailed features are seen to depend strongly on the mass of the star. Chapter 13 will address the post main sequence fate of less-massive stars. The fate of more massive stars will be examined in Chapters 14 and 20. As already noted, our discussion won't have much to say about the lightest stars because their main sequence lifetimes are so long that there is little observational evidence concerning their evolution after the main sequence.

Table 10.4 Duration of the numbered intervals displayed in Fig. 10.10 [125]

Mass:	1.0	1.25	1.5	2.25	3	5	9	15
1→2	7×10^9	2.8×10^9	1.6×10^9	4.8×10^8	2.2×10^8	6.5×10^7	2.1×10^7	1.0×10^7
2→3	2×10^9	1.8×10^8	8.1×10^7	1.6×10^7	1.0×10^7	2.2×10^6	6.1×10^5	2.3×10^5
3→4	1.2×10^9	1.0×10^9	3.5×10^8	3.7×10^7	1.0×10^7	1.4×10^6	9.1×10^4	–
4→5	1.6×10^8	1.5×10^8	1.1×10^8	1.3×10^7	4.5×10^6	7.5×10^5	1.5×10^5	7.6×10^4
5→6	$\geq 1 \times 10^9$	$\geq 4 \times 10^8$	$\geq 2 \times 10^8$	3.8×10^7	4.2×10^6	4.9×10^5	6.6×10^4	–
6→7	–	–	–	–	2.5×10^7	6.1×10^6	4.9×10^5	7.2×10^5
7→8	–	–	–	–	–	1.0×10^6	9.5×10^4	6.2×10^5
8→9	–	–	–	–	4.1×10^7	9.0×10^6	3.3×10^6	1.9×10^5
9→10	–	–	–	–	6.0×10^6	9.3×10^5	1.6×10^5	3.5×10^4

Masses in units of solar masses; times are in units of years and have been rounded to one decimal place.

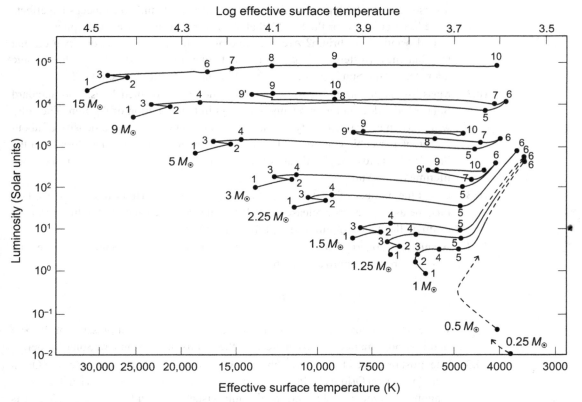

Fig. 10.10 Evolution off the main sequence for stars of different initial main sequence mass [125]. Dashed lines are estimates. Times to evolve through each numbered interval are given in Table 10.4. Adapted from Icko Iben, Jr., Annual Review of Astronomy and Astrophysics, **5**, 571–626 (1967), https://doi.org/10.1146/annurev.aa.05.040167.003035.

Background and Further Reading

Evolution on the main sequence and leaving the main sequence are discussed in all books about stellar structure and evolution. The book by Bahcall [30] is a good introduction both to the Standard Solar Model and to solar neutrinos.

Problems

10.1 Spherical symmetry has been assumed for our basic discussion of hydrostatic equilibrium and for the Standard Solar Model. Rotation could flatten stars, thus invalidating this assumption. Show that rotation of a star should cause negligible deviation from spherical symmetry if the rotational frequency is much less than the inverse of the hydrodynamical timescale for the star. Would you expect such effects to be large or small for the Sun? What about for Saturn, viewed as a sphere of gas in hydrostatic equilibrium? Are your findings consistent with the observation that the oblateness (difference in equatorial and polar diameters divided by the average diameter) of the Sun is of order 10^{-5}, while that of Saturn is about 0.1?

10.2 Mass loss is important for some stars in some phases of their lives. The Standard Solar Model does not assume any mass loss in the evolution of the Sun. The solar wind has an average velocity at Earth of about 400 km s^{-1} and a density of about 7 protons per cubic centimeter. Use this to estimate the rate of mass loss from the Sun. Will this loss be significant over the approximately 10^{10} year main sequence lifetime of the Sun?

10.3 Count the number of neutrinos emitted for each 4H $\rightarrow$ ^{4}He in the PP chains. Use this, the average energy release in the PP chains, and the solar photon luminosity to estimate the neutrino luminosity of the Sun.***

10.4 Assuming a spherical star of uniform density $\bar{\rho}$, show that the central pressure may be (very crudely) approximated by

$$P \simeq \text{constant} \times \frac{M^2}{R^4},$$

where M is the mass and R is the radius. Calculate the central pressure of the Sun in this approximation and compare with the results of the Standard Solar Model.

10.5 The most massive of the four bright stars in the Trapezium open cluster of the Orion Nebula is Theta Orionis C. Its spectrum–luminosity class is O6V, its visual apparent magnitude is $m_v = 5.13$, and its absolute visual magnitude is $M_v = -3.2$. It is estimated to have a radius of 8 $R_\odot$, a mass of 40 $M_\odot$, and its luminosity is 251,000 $L_\odot$ (much of this at UV wavelengths, since Theta Orionis C has the distinction of having the hottest surface temperature of any star visible to the naked eye). What is the parallax of Theta Orionis C? What is its surface temperature? Given that Theta Orionis C is thought presently to be about 1 million years old, how much longer will it spend on the main sequence?

10.6 Assume a completely ionized ZAMS star from Pop I and neglect the contribution of metals to the mean molecular weight relative to that of hydrogen and helium. Show that the mean molecular weight μ defined in Eq. (3.19) can then be approximated as

$$\mu \simeq \frac{4}{5X+3} \simeq 0.6,$$

where Eq. (3.28) has been used to estimate X. Compare with the results of Fig. 10.7 for the mean molecular weight of the Sun computed in the Standard Solar Model.

10.7 Estimate the main sequence lifetimes for stars of spectral class O5, A0, K5, and M5. *Hint:* See Table 2.2 for relevant data.

10.8 (a) Suppose that the central energy-producing regions of the Sun were unchanged but that (magically) the layers outside this region were made transparent so that they did not impede transport of photons. Assuming that it were still a blackbody radiator, in what region of the spectrum would the observed photons from the Sun peak?

(b) In reality the photon energy diffuses outward through the outer layers of the Sun. Estimate a diffusion time to the surface assuming a random walk with opacities characteristic of Thomson scattering. Compare this diffusion timescale with the Kelvin–Helmholtz timescale, which is a measure of how rapidly a star can convert gravitational energy into emitted photons. You should find that the diffusion timescale and the Kelvin–Helmholtz timescale are in fact very different. Can you explain what the fallacy is in assuming that the diffusion timescale estimated in this way and the Kelvin–Helmholtz timescales should be comparable?***

10.9 Use data from Fig. 7.2 and Table 10.1 to estimate the mean free path for photons in the center of the Sun and at 90% of the solar radius. From Fig. 7.2 estimate an average internal opacity (excluding the surface region) and from that opacity calculate a time for photons to diffuse to the surface by a random walk.***

10.10 Derive a formula for the ratio of energy densities carried by the photons and by the gas in a star, assuming an ideal gas and photons with a Planck distribution. Estimate this ratio for the Sun using a representative interior temperature. From Table 2.2 most main sequence stars have an average density $\rho \sim 1 \, \text{g cm}^{-3}$. Assuming this density, and that one can continue to treat the radiation and gas independently as above, at what temperature would the energy density of the radiation become equal to that of the gas?

10.11 In Example 6.1 the mean life for loss of protons in the first step of the PP chains was estimated to be about 6×10^9 yr. Make a similar estimate for the mean life of the deuterium formed in the first step and consumed in the second step in PP-I, and for the mean life of the ^3He formed in the second step and consumed in the third step of PP-I. Assume for the center of the Sun that

$$T_6 = 15 \qquad \rho = 150 \, \text{g cm}^{-3} \qquad Y_p = 0.4 \qquad Y_{3\text{He}} \sim 4 \times 10^{-6}$$

(the ^3He abundance is rather uncertain). Use Fig. 6.1 to estimate rates that you will need.***

10.12 Show that if the radial coordinate dr is parameterized in terms of the optical depth $d\tau$ (see Box 10.1), the equation for hydrostatic equilibrium can be expressed as $dP/d\tau = g/\kappa$, where g is gravitational acceleration and κ is radiative opacity.***

10.13 Stars have magnetic fields but the energy of those fields has generally been neglected in discussing the equilibrium configuration for stars. Justify this for the Sun by estimating an upper limit for the average magnetic field to be given by that in sunspots, which you may take to be about 0.1 T (the average field is certainly less than this), and computing the ratio of the energy density of the magnetic field to that of the gravitational field.***

10.14 The electron number density is about 6.4×10^{25} cm^{-3} at the center of the Sun and 5×10^{24} cm^{-3} at 30% of the solar radius. Use results from the Standard Solar Model to calculate the ratio of the actual electron density to the critical density given by Eq. (3.51) at these two radii.***

11 Neutrino Flavor Oscillations

In Section 10.4 we reviewed the observational evidence that fewer solar neutrinos are detected on Earth than should be the case if the Sun were emitting neutrinos at the rate expected from the Standard Solar Model using the neutrino physics of the Standard Model of elementary particle physics.[1] This chapter and the next will elaborate on the physics of solar neutrinos and show that this solar neutrino deficit is not really a deficit at all, but rather a failure to count *all* the neutrinos coming from the Sun because the physics of actual neutrinos is much richer than is envisioned in the Standard Model. Hence the reconciliation of solar neutrino observations with our understanding of elementary particle physics to be described here and in Chapter 12 will have fundamental implications both for astrophysics and for elementary particle physics. Although the discussion will involve directly only the Sun because it is the only normal star near enough to allow its emitted neutrinos to be detected on Earth with present technology, presumably the physics described in this chapter operates in other stars as well.

11.1 Overview of the Solar Neutrino Problem

An understanding of solar neutrinos requires basic knowledge of weak interactions in the Standard Model of elementary particle physics and in conjectured extensions of that model. This will require the language and some elementary results from relativistic quantum field theory. For readers lacking a background in quantum mechanics and quantum field theory, a concise introduction to the mathematical concepts essential to the present discussion may be found in Appendix F. Although the origin of various equations will require an acquaintance with quantum field theory, or at least with quantum mechanics, it is possible to understand the basic ideas of this chapter and the next if the reader is cognizant of the concepts reviewed in Appendix F.

Our strategy will be to introduce first in this chapter the physics of neutrinos in the Standard Model, and then to discuss possible modifications of that picture if – contrary to a fundamental assumption of the Standard Model – neutrinos are not identically massless. We will then review the evidence that neutrinos undergo *flavor oscillations* in which

---

[1] Two "standard models" will be on prominent display in this chapter. For convenience I will often refer to the Standard Model of elementary particle physics as just "the Standard Model," which should not be confused with "the Standard Solar Model."

neutrinos of one flavor can interconvert with neutrinos of other flavors, which necessarily implies that at least some neutrino flavors have a small but non-zero mass. Then in the remainder of this chapter and in Chapter 12 we shall develop the formalism required to understand the flux of neutrinos coming from the Sun if—in contradiction to the assumptions of the Standard Model—they undergo flavor oscillations. Finally, it will be shown in Chapter 12 that flavor oscillations in the Sun and in vacuum spread the flux of electron neutrinos produced in the Sun out into a mix of electron, muon, and tau neutrino flavors before they are detected on Earth. Thus, there is no solar neutrino deficit if all three flavors are detected, but there is an *apparent deficit* if (as in all the pioneering neutrino experiments discussed in Chapter 10) the detectors can identify only electron neutrinos. As a consequence, the results to be presented in this chapter and the next will (1) resolve the solar neutrino anomaly, (2) support strongly the essential correctness of the Standard Solar Model, and (3) indicate that the Standard Model of elementary particle physics is incomplete, and give hints about possible physics beyond the Standard Model.

11.2 Weak Interactions and Neutrino Physics

The Standard Model of elementary particle physics assumes that the electromagnetic and weak interactions are unified in a *local gauge theory* (also termed a *Yang–Mills field theory*) in which the leptons and quarks are grouped into *generations* or *families* and interact through the exchange of *gauge bosons*: the photon γ, W^+, W^-, and Z^0, where the superscripts denote electrical charge.[2] (See Box 11.1 for a review of some basic terminology in elementary particle physics.) This unification is not apparent at low energies, however, because the weak gauge bosons W^+, W^-, and Z^0 have finite mass by virtue of coupling to a background scalar field called the Higgs field, while the photon remains massless. This acquisition of mass through the Higgs coupling is called *spontaneous symmetry breaking by the Higgs mechanism,* or just the *Higgs mechanism.*

11.2.1 Matter and Force Fields of the Standard Model

In the Standard Model matter fields are fermion fields, the forces are mediated by gauge bosons, and the masses of the matter fields and the weak gauge bosons come from coupling to the Higgs scalar field. The matter fields are divided into three "generations" or "families," I, II, and III, as summarized in Fig. 11.1. In the Standard Model transitions across family lines are forbidden. For the leptons this is implemented formally by assigning a lepton family number to each particle and requiring interactions to conserve this number.

---

[2] The Standard Model also includes the strong interactions, which are mediated by exchange of *gluons* between quarks and between gluons. However, our concern here is primarily with the weak interaction sector, so quarks and gluons will be mentioned only in passing. The weak and electromagnetic sector of the Standard Model is sometimes called the *Standard Electroweak Model.*

Box 11.1 Some Standard Terminology in Elementary Particle Physics

The starting point for understanding elementary particle physics is the classification of fundamental particles in various ways.

The Classification Zoo

One distinction is whether particles are *fermions* or *bosons* (Box 3.6), which is a question of intrinsic spin: fermions have half-integer spins and bosons have integer spins. Thus quarks are fermions but photons are bosons. Another distinction is between particles and antiparticles. Thus, the antielectron (positron) is the antiparticle of the electron, having the same mass and spin, but opposite charge. Elementary particles also may be divided into *leptons*, which don't undergo the strong interactions (example: electrons), and *hadrons*, which do undergo strong interactions (example: quarks). Hadrons may be subdivided into the half-integer spin *baryons*, such as the proton, and the integer-spin *mesons*, such as the pion. (Protons and pions are composite particles, but are often classified as if they were elementary.)

Particles, Fields, and Symmetries

Particles are associated with fields described by (special) relativistic quantum mechanics. Some common terminology is associated with spin: spin-0 particles (or fields) are said to be *scalar*, spin-1 particles are termed *vector*, and spin-2 particles are termed *tensor*. Example: the photon is a spin-1 particle associated with a vector field. Matter fields are always fermionic but forces are mediated by the exchange of (virtual) bosons. For all non-gravitational forces the force-mediating particles are spin-1 (vector) particles exhibiting a powerful type of symmetry called *local gauge invariance*, which generalizes electric charge conservation in electromagnetism. Mediators of the strong, weak, and electromagnetic interactions – *gluons, intermediate vector bosons*, and *photons*, respectively – are collectively termed *gauge or vector bosons*. The Higgs boson, which gives mass to particles, corresponds to a spin-0 or scalar field.

Broken Symmetries

Mass differences between weak gauge bosons and the photon imply that electromagnetic interactions and weak interactions at low energy have very different properties. For example, the electromagnetic interaction has a long range but the weak interaction has a very short range. Only at high energies ($\gtrsim$ 100 GeV) do their properties merge, leading to the *unified electroweak theory*. This essential difference between electromagnetism and the weak interactions at low energy is termed *spontaneous symmetry breaking (by the Higgs mechanism)*. The convergence of the properties of electromagnetism and weak interactions at high energy leading to the unified electroweak theory is termed *symmetry restoration*.

Example 11.1 In generation I an electron family number of $+1$ may be assigned to the electron and electron neutrino, -1 to the antielectron and the electron antineutrino, and zero for all other particles. Then $\nu_e + n \rightarrow p + e^-$ conserves electron family number $(1 + 0 = 0 + 1)$ and is observed, but $\nu_e + p \rightarrow n + e^+$ violates electron family number $[1 + 0 \neq 0 + (-1)]$ and has never been seen experimentally.

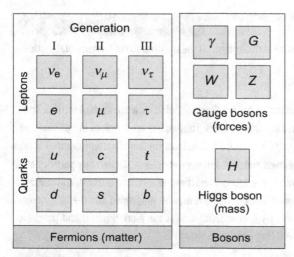

Fig. 11.1 Particles of the Standard Model and characteristic mass scales in the quark and lepton sectors for each generation [6]. Photons are labeled by γ and gluons by G.

11.2.2 Masses for Particles of the Standard Model

Also displayed in Fig. 11.1 are characteristic mass scales for quarks and neutrinos within each generation. The neutrino masses listed represent upper limits, since no neutrino mass has been measured directly thus far. These limits imply that the neutrino masses are either zero or very tiny on a mass scale set by the quarks of a generation. The explanation of this is a major unresolved issue in the theory of elementary particles. The Standard Model contains particles (photons γ and gluons G) for which the mass is zero but in those cases a fundamental principle, local gauge invariance, requires that the particles must be *identically massless*. In contrast, there is no known reason why the neutrino mass should be identically zero and, if it is not *identically* zero, why does it have such a small value when the other members of the same generation are much more massive?

The Standard Model gives no fundamental reason why, but agreement with the phenomenology of the weak interactions – in particular, with the observation that the weak interactions violate parity symmetry to the maximum extent possible – requires that *the masses of all neutrinos must be assumed to be identically zero.* The full justification of this statement would require an excursion into fundamental weak-interaction theory but the central point is that there are potentially two kinds of neutrino mass terms that could appear in the field theory: a *Dirac mass* and a *Majorana mass*. The first is appropriate if the neutrino and antineutrino are distinct particles; the second is appropriate if a neutrino acts as its own antiparticle, which is not ruled out by present experiments because the neutrino carries no charge that would distinguish the particle from the antiparticle; see Box 11.2. Both types of mass terms must vanish for Standard Model neutrinos but at least one could be non-zero in various extensions of the Standard Model.

Box 11.2 Neutrinoless Double β Decay and Majorana Neutrinos

A clear signature of a Majorana mass for neutrinos (see Section 11.2.2) would be *neutrinoless double β decay*. Normal double β decay can be observed if the mass of the isotope one neutron removed from some isotope is larger than the mass of the parent (so β^- decay is energetically forbidden), but the mass of the next isotope two neutrons removed is less than the parent. Then decay emitting two β^- particles (electrons) is possible, but with small probability: two neutrons in the nucleus are converted to protons, with the emission of two e^- and two $\bar{\nu}_e$,

$$(A, Z) \rightarrow (A, Z+2) + 2e^- + 2\bar{\nu}_e,$$

for either Dirac or Majorana neutrinos. If the neutrino is its own antiparticle, a variant of this decay in which no neutrinos or antineutrinos are emitted is possible:

$$(A, Z) \rightarrow (A, Z+2) + 2e^-.$$

This is *neutrinoless double-β decay*. These processes are illustrated in terms of Feynman diagrams (described in Box 11.3) in the following figure, where now β^- decay of the neutron is viewed to occur at the quark level, converting a down quark into an up quark, and thus transforming a neutron (udd) into a proton (uud).

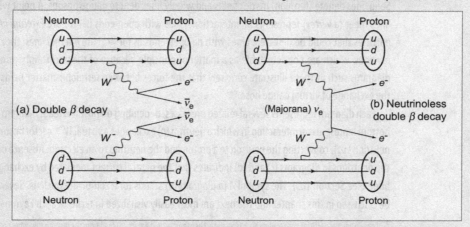

A Majorana neutrino is its own antiparticle, so in (b) the virtual neutrino emitted from one vertex can be absorbed at the other and no external neutrino is emitted. Normal double β decay is improbable but it has been observed for a few isotopes. Neutrinoless double β decay has not been observed to date.

11.2.3 Charged and Neutral Currents

The Standard Model describes two basic categories of weak interactions. In the *charged weak currents* electrical charge is transferred in the interaction (total charge is conserved, of course). Because charge is transferred the boson mediating the force must be charged. Thus, charged weak currents involve the W^+ or W^- weak gauge bosons. Diagrams (a) and (b) in Box 11.3 are examples of charged weak current interactions. Since the Standard

> **Box 11.3** **Feynman Diagrams**
>
> In quantum field theory it is common to use pictorial representations of interaction matrix elements (probability amplitudes) called *Feynman diagrams*. They are highly intuitive: given a diagram one can (with practice) write the corresponding matrix element, and given the matrix element one can sketch the corresponding diagram. Here are some important weak-interaction Feynman diagrams:
>
> [Diagrams (a)–(d) showing: (a) $n \to p$ via W^- with $\nu_e \to e^-$; (b) $e^- \to \nu_e$ via W^+ with $\nu_e \to e^-$; (c) e–ν scattering via Z^0; (d) n,p,A – ν scattering via Z^0.]
>
> Solid lines denote (fermion) matter fields and wiggly lines denote gauge bosons. A point where two or more lines meet (a *vertex*) represents an interaction. Lines with open ends like ν_e are *external lines* denoting real particles that could be detected. Lines with no open ends as for $W^\pm$ are *internal lines*; they represent *virtual particles*, which are not detectable as a matter of principle because of the uncertainty principle.[a] Feynman diagrams such as these illustrate concisely that the forces between fermionic matter fields are mediated by the exchange of virtual gauge bosons.
>
> Each diagram represents several related processes, depending on how it is read. Diagram (a) read from the bottom represents an interaction in which a neutron (n) exchanges a virtual W^- vector boson with an electron neutrino (ν_e), converting the neutron to a proton and the neutrino to an electron. Absence of flavor indices on the neutrinos in diagrams (c) and (d) indicates that the neutral current mediated by exchange of a Z^0 is flavor blind (see Section 11.3). The symbol A in diagram (d) stands for a composite nucleus. Several phenomena to be discussed in this chapter and the next are most easily visualized in terms of such Feynman diagrams.
>
> [a] Loosely, virtual particles violate energy conservation. For example, in diagram (a) read from the bottom the initial energy is the sum of the n and ν_e rest masses multiplied by c^2, plus kinetic energy. The intermediate state has in addition the rest mass energy of a virtual W^+ ($\sim$ 80 GeV), which violates energy conservation by an amount $\Delta E \geq 80$ GeV. But this violation is not observable because the time for existence of the virtual particle Δt is restricted by the uncertainty principle relation $\Delta t \sim \hbar/\Delta E$.

Model represents a partial unification of weak interactions and electromagnetism, and electromagnetism is mediated by an uncharged gauge boson (the photon), the weak sector also must have an uncharged gauge boson that is mixed with the photon to provide the full electroweak unified description of electromagnetic and weak interactions. The uncharged weak gauge boson is the Z^0, and it can mediate *neutral weak currents* in which there is no transfer of charge in the weak-interaction matrix elements. An example of a weak neutral current interaction is diagram (c) in Box 11.3. A ν_e can interact with an electron through the charged weak current or the neutral weak current, but a ν_μ or ν_τ can interact with electrons only through the neutral current.

11.3 Flavor Mixing

The Standard Model as described to this point is simple conceptually. It is made richer and more complex by the experimental observation of *flavor mixing*.

11.3.1 Flavor Mixing in the Quark Sector

The term *flavor* is used to distinguish among different species of quarks and leptons. Thus, (ν_e, ν_μ, ν_τ) are different flavors of neutrinos, and (u, d, s, ...) are different flavors of quarks. In the Standard Model, experiments require that for quarks *the mass eigenstates and the weak eigenstates are not equivalent.* This quantum-mechanical jargon means that the quark states that enter the weak interactions (the *flavor eigenstates*) are generally linear combinations of the states for propagating quarks (*mass eigenstates*). For example, restricting to the first two generations, it is found that the d and s quarks enter the weak interactions in the "rotated" linear combinations d_c and s_c defined by the matrix equation

$$\begin{pmatrix} d_c \\ s_c \end{pmatrix} = \begin{pmatrix} \cos\theta_c & \sin\theta_c \\ -\sin\theta_c & \cos\theta_c \end{pmatrix} \begin{pmatrix} d \\ s \end{pmatrix} \qquad (11.1)$$

where d and s are the mass eigenstate quark fields and θ_c is termed the *mixing angle* (in this particular context it is also called the *Cabibbo angle*). Comparison with data determines the Cabibbo mixing angle to be rather small: $\theta_c \sim 13.0°$. In the more general case of three generations of quarks, weak eigenstates are described by a 3×3 mixing matrix called the *Cabibbo–Kobayashi–Maskawa or CKM matrix* that has three real mixing angles and one complex phase. There is little fundamental understanding of this quark flavor mixing but it is clear that the data require it.

11.3.2 Flavor Mixing in the Leptonic Sector

Since in the Standard Model the quarks entering the weak interactions are known to be mixtures of different mass eigenstates, it might be expected that the corresponding leptons in these generations could also enter the weak interactions as mixed-mass eigenstates. For example, in the weak decay $p^+ \to n + e^+ + \nu_e$ the electron neutrino ν_e is produced in an eigenstate of the weak interactions $|\nu_e\rangle$ (that is, a state of definite flavor), but by virtue of Eq. (F.3) it then propagates in time as a mass eigenstate $|\nu_i\rangle$ (a state of definite mass), according to $|\nu_i(t)\rangle = \exp(-i\hbar E_i t)|\nu_i(0)\rangle$, where E_i is the energy of the state. Since if there is flavor mixing the mass eigenstate will be a linear combination of flavor eigenstates, this implies that the flavor of the propagating neutrino is mixed and that the components of this mixture can oscillate with time by virtue of the $\exp(-i\hbar E_i t)$ factor.[3]

---

[3] The perceptive reader might ask whether it is possible (leaving some difficult technical issues aside) to measure the mass of a neutrino in flight and determine it to be in a definite mass eigenstate. Kayser [132] has shown that this is a textbook Heisenberg uncertainty principle issue. Such a measurement done precisely enough would indeed determine a unique mass eigenstate. However, because of the uncertainty principle the neutrino oscillation would be destroyed by the measurement because the precision in the momentum determinations

It is easy to show that if all flavor states of neutrinos are identically massless and thus have the same mass (recall that this is an *essential hypothesis* of the the Standard Model), flavor-mixing has no observable consequences. Hence, for neutrino flavor mixing to be measurable at least one flavor of neutrino must have a non-vanishing mass. Conversely, observation of neutrino flavor mixing would be conclusive evidence that at least one neutrino flavor has a non-zero mass. Thus, either the observation of neutrino flavor oscillations, or a direct measurement of finite neutrino mass, would be indisputable evidence of physics beyond the Standard Model.

11.4 Implications of a Finite Neutrino Mass

A finite neutrino mass, implying the possibility of neutrino flavor mixing, would be of fundamental importance for elementary particle physics but it could be of similar importance for astrophysics because it suggests a possible solution of the solar neutrino problem. If neutrino flavor eigenstates are mixtures of mass eigenstates, neutrinos propagating in time will oscillate in flavor. If neutrino flavors can oscillate, then when some of the ν_e emitted by the Sun reach Earth they could have oscillated into another flavor (for example, ν_μ). But since the experiments described in Section 10.4 can identify only electron neutrinos, they would miss neutrinos that had oscillated into other flavors, thus potentially explaining the observed neutrino deficit and reconciling it with the Standard Solar Model.

Standard Model neutrinos *must* be massless. However, there are many reasons to believe that the Standard Model – despite its remarkable success – is incomplete and represents a low-energy approximation to a more complete theory. For example, there are 20 or so adjustable parameters of the Standard Model that have no convincing fundamental constraint, the origin of mass through the Higgs mechanism is purely phenomenological, the generational structure assumed for the particle multiplets and the choice of three generations is based entirely on phenomenology, the violations of symmetries such as parity that are observed in the weak interactions are put by hand in the Standard Model, and so on. Various extensions such as *Grand Unified Theories* (GUTs) have been proposed that go beyond the Standard Model. Often for these theories the reasons that mass terms are forbidden in the Standard Model are not operative and finite neutrino masses may occur naturally.

11.5 Neutrino Vacuum Oscillations

Motivated by the preceding discussion, let's address the possibility that neutrinos might undergo flavor oscillations that could account for the observed solar neutrino deficit.

necessary to resolve the mass sufficiently well would necessarily lead to an uncertainty in the position of the neutrino source or detector comparable to or larger than the characteristic spatial scale of the oscillation pattern (the *oscillation length* defined below). Thus, the uncertainty principle forbids determining simultaneously the mass and flavor of a neutrino as a matter of principle if there is flavor mixing.

11.5.1 Mixing for Two Neutrino Flavors

To get the lay of the land we consider first neutrino oscillations in a simple 2-flavor model, in the absence of matter.[4] To be definite the two flavors will be assumed to correspond to the electron neutrino ν_e and the muon neutrino ν_μ, but the formalism could be applied to the mixing of any two flavors. By analogy with Eq. (11.1) for quarks, the flavor eigenstates ν_e and ν_μ are assumed to be related to the mass eigenstates ν_1 and ν_2 through the matrix transformation

$$\begin{pmatrix} \nu_e \\ \nu_\mu \end{pmatrix} = U(\theta) \begin{pmatrix} \nu_1 \\ \nu_2 \end{pmatrix} = \begin{pmatrix} \cos\theta & \sin\theta \\ -\sin\theta & \cos\theta \end{pmatrix} \begin{pmatrix} \nu_1 \\ \nu_2 \end{pmatrix}, \qquad (11.2)$$

where θ is the (entirely phenomenological) vacuum mixing angle, chosen to lie in the range 0–45°, and the unitary matrix U is parameterized by the single real angle θ (unitary transformations are described in Appendix F). By inverting this equation using the unitarity condition $UU^\dagger = 1$ of Eq. (F.7), the mass eigenstates may in turn be expressed as a linear combination of the flavor eigenstates (see Problem 11.3):

$$\begin{pmatrix} \nu_1 \\ \nu_2 \end{pmatrix} = U(\theta)^\dagger \begin{pmatrix} \nu_e \\ \nu_\mu \end{pmatrix} = \begin{pmatrix} \cos\theta & -\sin\theta \\ \sin\theta & \cos\theta \end{pmatrix} \begin{pmatrix} \nu_e \\ \nu_\mu \end{pmatrix}. \qquad (11.3)$$

Assuming that the respective masses are different, the different mass eigenstates will travel with slightly different energies as neutrinos propagate in time. Thus, the probability of detecting a particular flavor of neutrino will oscillate with time, or equivalently, with the distance traveled. From basic quantum field theory (see Section 12.2.1) the mass eigenstates evolve with time t according to

$$|\nu_i(t)\rangle = e^{-iE_i t}|\nu_i(0)\rangle, \qquad (11.4)$$

where the index i labels mass eigenstates of energy E_i, so the time evolution of the ν_e state in Eq. (11.2) will be given by

$$|\nu(t)\rangle = \cos\theta\, e^{-iE_1 t}|\nu_1(0)\rangle + \sin\theta\, e^{-iE_2 t}|\nu_2(0)\rangle, \qquad (11.5)$$

and this may be written as the mixed-flavor state (see Problem 11.3)

$$|\nu(t)\rangle = (\cos^2\theta\, e^{-iE_1 t} + \sin^2\theta\, e^{-iE_2 t})|\nu_e\rangle \\ + \sin\theta\cos\theta(-e^{-iE_1 t} + e^{-iE_2 t})|\nu_\mu\rangle. \qquad (11.6)$$

Squaring the overlap of Eq. (11.6) with the flavor eigenstates gives the probabilities that a neutrino initially in an electron neutrino flavor state will remain an electron neutrino, or instead be converted to a muon neutrino after a time t [see Eqs. (F.4)–(F.5)],

---

[4] A realistic model would consider three neutrino flavors but the physics relevant here is reasonably well described by a simpler model with only two neutrino flavors. The full 3-flavor case will be considered briefly in section 11.6. The discussion in much of this chapter will employ $\hbar = c = 1$ units, which are standard in elementary particle physics and are explained in Appendix B.2 and illustrated in Example 11.2.

$$P(\nu_e \to \nu_e, t) = |\langle \nu_e | \nu(t) \rangle|^2 = 1 - \tfrac{1}{2}\sin^2(2\theta)[1 - \cos(E_2 - E_1)t],$$

$$P(\nu_e \to \nu_\mu, t) = |\langle \nu_\mu | \nu(t) \rangle|^2 = \tfrac{1}{2}\sin^2(2\theta)[1 - \cos(E_2 - E_1)t], \tag{11.7}$$

with the sum of these probabilities equal to unity by construction.

11.5.2 The Vacuum Oscillation Length

Neutrinos have at most a tiny mass and the mass eigenstates are assumed to have the same momentum p but slightly different masses m_i and energy E_i. Since $E_i \gg m_i c^2$ we may approximate $E_i = (p^2 + m_i^2)^{1/2} \simeq p + m_i^2/2p$, where $P = |\mathbf{P}|$. Thus, assuming that $E_1 \sim E_2 \equiv E$,

$$E_2 - E_1 \simeq \frac{\Delta m^2}{2E} \qquad \Delta m^2 \equiv m_2^2 - m_1^2. \tag{11.8}$$

The probabilities for flavor survival and flavor conversion as a function of distance traveled $r \sim ct$ are then (see Problem 11.2)

$$P(\nu_e \to \nu_e, r) = 1 - \sin^2 2\theta \sin^2\left(\frac{\pi r}{L}\right) \qquad P(\nu_e \to \nu_\mu, r) = \sin^2 2\theta \sin^2\left(\frac{\pi r}{L}\right), \tag{11.9}$$

where θ is the mixing angle and the *oscillation length L*, defined by

$$L \equiv \frac{4\pi E}{\Delta m^2}, \tag{11.10}$$

is the distance required for one complete flavor oscillation (for example, $\nu_e \to \nu_\mu \to \nu_e$), assuming that $v \simeq c$. Neutrino oscillations for a 2-flavor model using these formulas are illustrated in Fig. 11.2.

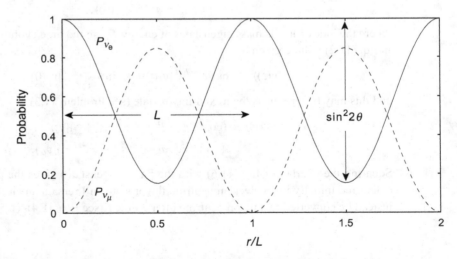

Fig. 11.2 Neutrino vacuum oscillations in a 2-flavor model as a function of distance traveled r in units of the vacuum oscillation length L. The probability as a function of r to be an electron neutrino is denoted by P_{ν_e} and that to be a muon neutrino by P_{ν_μ}. The period of the oscillation is L and its amplitude is $\sin^2 2\theta$, where θ is the vacuum mixing angle. In this calculation $\theta = 33.5°$, the neutrino energy is $E = 5$ MeV, the difference in squared masses for the two flavors is $\Delta m^2 c^4 = 7.5 \times 10^{-5}$ eV^2, and the corresponding oscillation length L is 165.3 km.

Example 11.2 Let's practice converting from the $\hbar = c = 1$ units used in this chapter to standard or "engineering units" by expressing L in units of kilometers. As discussed more extensively in Appendix B, this is just a dimensional analysis problem. First multiply Eq. (11.10) by $c^4/c^4 = 1$ to give

$$L = \frac{4\pi E c^4}{\Delta m^2 c^4} = \frac{4\pi E c^4}{\Delta E^2},$$

where $\Delta E^2 \equiv \Delta m^2 c^4$ has units of energy squared. Let $[x]$ denote the units of x and define our standard length unit as $\mathscr{L}$, our standard energy unit as $\mathscr{E}$, and our standard time unit as $\mathscr{T}$. Then dimensionally,

$$[L] = \left[\frac{4\pi E c^4}{\Delta E^2}\right] = \frac{[E][c^4]}{[\Delta E^2]} = \frac{\mathscr{E}\mathscr{L}^4\mathscr{T}^{-4}}{\mathscr{E}^2} = \mathscr{L}\frac{\mathscr{L}^3}{\mathscr{E}\mathscr{T}^4}.$$

The final expression should have units of length $\mathscr{L}$, so this result must be multiplied by a combination of $\hbar$ and c having units of $\mathscr{E}\mathscr{T}^4/\mathscr{L}^3$. Since $[\hbar] = \mathscr{E}\mathscr{T}$ and $[c] = \mathscr{L}/\mathscr{T}$, the required factor is $\hbar/c^3$ and

$$L = \frac{\hbar}{c^3} \times \frac{4\pi E c^4}{\Delta E^2} = \frac{4\pi \hbar c E}{\Delta E^2} = 2.48 \times 10^{-3} \left(\frac{E}{\text{MeV}}\right)\left(\frac{\text{eV}^2}{\Delta E^2}\right) \text{ km}, \qquad (11.11)$$

for neutrino energy E specified in MeV and energy squared difference ΔE^2 corresponding to the mass squared difference Δm^2 specified in eV^2. A variation on this approach is illustrated in Problem 11.9(a).

Going the opposite direction (converting from normal units to $\hbar = c = 1$ units) is of course trivial: write the equations and then remove all factors of $\hbar$ and c, effectively setting them to one.

11.5.3 Time-Averaged or Classical Probabilities

Practically the oscillation wavelength may be smaller than the uncertainties in position for emission and detection of neutrinos. For example, there are thousands of kilometers variation in the distance traversed between production and detection of solar neutrinos caused by varying production location in the Sun, varying detection location since the Earth is rotating, and varying Earth–Sun orbital separation.

Time or distance averaging: If the oscillation length is less than the averaging introduced by the preceding considerations, the detectors will see a distance (or time) average of Eqs. (11.9). Denoting the averaged detection probability by a bar gives (Problem 11.7)

$$\bar{P}(\nu_e \to \nu_e) = 1 - \tfrac{1}{2}\sin^2 2\theta \qquad \bar{P}(\nu_e \to \nu_\mu) = \tfrac{1}{2}\sin^2 2\theta, \qquad (11.12)$$

for the probability (11.9) of detecting the two flavors distance-averaged over the oscillating factor. For two flavors the instantaneous probability to remain a ν_e can approach zero if the mixing angle is large (see Fig. 11.2), but Eq. (11.12) indicates that the *average* survival probability has a lower limit of $\frac{1}{2}$ for two flavors. For n flavors the lower limit is n^{-1}, but that limit can be realized only for a precisely tuned flavor mixture (maximal mixing).

Classical probabilities: The results of Eq. (11.12) have an alternative interpretation if we define the quantum probability

$$P(\nu_e \to \nu_e) \equiv |\langle \nu_e(t)|\nu_e(0)\rangle|^2, \tag{11.13}$$

where $\langle \nu_e(t)|\nu_e(0)\rangle$ may be interpreted as the probability amplitude to find the neutrino as an electron neutrino at time t if it was created as an electron neutrino at time $t = 0$. Since the propagating states are the mass eigenstates $|\nu_i\rangle$, it is useful to expand this expression in terms of those states. In Dirac notation the unit operator may be written as $\hat{1} = \sum_i |i\rangle\langle i|$, where i labels a complete set of states. Thus, for a 2-flavor basis

$$\begin{aligned}P(\nu_e \to \nu_e) &= |\langle \nu_e(t)|\hat{1}|\nu_e(0)\rangle|^2 = \left|\sum_i \langle \nu_e(t)|\nu_i\rangle\langle \nu_i|\nu_e(0)\rangle\right|^2 \\ &= |\langle \nu_e(t)|\nu_1\rangle\langle \nu_1|\nu_e(0)\rangle + \langle \nu_e(t)|\nu_2\rangle\langle \nu_2|\nu_e(0)\rangle|^2 \\ &= |\langle \nu_e(t)|\nu_1\rangle\langle \nu_1|\nu_e(0)\rangle|^2 + |\langle \nu_e(t)|\nu_2\rangle\langle \nu_2|\nu_e(0)\rangle|^2 \\ &\quad + 2\langle \nu_e(t)|\nu_1\rangle\langle \nu_1|\nu_e(0)\rangle\langle \nu_e(t)|\nu_2\rangle\langle \nu_2|\nu_e(0)\rangle,\end{aligned} \tag{11.14}$$

where in the second line the summation has been written out explicitly for a 2-flavor model with basis mass eigenstates $|\nu_1\rangle$ and $|\nu_2\rangle$. The final term in the last line is the interference between amplitudes that arises because in quantum mechanics the probability is not the sum of individual probabilities (as it would be in classical mechanics), but rather the square of the sum of probability amplitudes. Dropping the interference term gives the *classical average*

$$P_{\text{class}}(\nu_e \to \nu_e) = P(\nu_e \to \nu_1) P(\nu_1 \to \nu_e) + P(\nu_e \to \nu_2) P(\nu_2 \to \nu_e),$$

where $P(\nu_e \to \nu_1) \equiv |\langle \nu_e(t)|\nu_1\rangle|^2$ and so on. Evaluating this expression for the 2-flavor model (Problem 11.8) and comparing with Eq. (11.12) gives

$$\begin{aligned}P_{\text{class}}(\nu_e \to \nu_e) &= \bar{P}(\nu_e \to \nu_e) = 1 - \tfrac{1}{2}\sin^2 2\theta, \\ P_{\text{class}}(\nu_e \to \nu_\mu) &= \bar{P}(\nu_e \to \nu_\mu) = \tfrac{1}{2}\sin^2 2\theta.\end{aligned} \tag{11.15}$$

Thus $\bar{P}(\nu_e \to \nu_e)$ and $\bar{P}(\nu_e \to \nu_\mu)$ may be viewed either as time or distance averages, or as classical probabilities resulting from discarding quantum-mechanical interference effects. Because of the inevitable averaging associated with practical observations, these are the probabilities that would be measured in a typical solar neutrino experiment. Notice for later use that these equations also may be written in matrix form. For example, letting the row vector (1 0) and its corresponding column vector denote pure ν_e flavor states,

$$\begin{aligned}\bar{P}(\nu_e \to \nu_e) &= \sum_{i=1}^{2} P(\nu_e \to \nu_i) P(\nu_i \to \nu_e) \\ &= (1\ 0) \begin{pmatrix} \cos^2\theta & \sin^2\theta \\ \sin^2\theta & \cos^2\theta \end{pmatrix} \begin{pmatrix} 1 & 0 \\ 0 & 1 \end{pmatrix} \begin{pmatrix} \cos^2\theta & \sin^2\theta \\ \sin^2\theta & \cos^2\theta \end{pmatrix} \begin{pmatrix} 1 \\ 0 \end{pmatrix},\end{aligned} \tag{11.16}$$

which you can verify by matrix multiplication and trigonometric identities is equivalent to the first of Eqs. (11.15), $\bar{P}(\nu_e \to \nu_e) = 1 - \tfrac{1}{2}\sin^2 2\theta$ (see also Problems 12.11 and 12.12).

11.6 Neutrino Oscillations with Three Flavors

We will demonstrate in the next chapter that solar neutrinos may be understood well in the simple 2-flavor formalism developed above. Thus at fixed energy a single vacuum mixing angle and one mass-squared difference characterizes the theory. However, in the general case there are three known flavors of neutrinos (and their corresponding antineutrinos), so the correct treatment of neutrino oscillations requires additional parameters associated with a 3×3 mixing matrix. This matrix relates the flavor eigenstates $\{\nu_e, \nu_\tau, \nu_\tau\}$ to the mass eigenstates $\{\nu_1, \nu_2, \nu_3\}$ through [115][5]

$$\begin{pmatrix} |\nu_e\rangle \\ |\nu_\mu\rangle \\ |\nu_\tau\rangle \end{pmatrix} = \mathcal{U}(\theta_{12}, \theta_{13}, \theta_{23}, \delta) \begin{pmatrix} e^{\frac{i}{2}\alpha_1}|\nu_1\rangle \\ e^{\frac{i}{2}\alpha_2}|\nu_2\rangle \\ |\nu_3\rangle \end{pmatrix}, \qquad (11.17)$$

where the *Pontecorvo–Maki–Nakagawa–Sakata (PMNS) matrix*

$$\mathcal{U}(\theta_{12}, \theta_{13}, \theta_{23}, \delta) \equiv \begin{pmatrix} c_{12}c_{13} & s_{12}c_{13} & s_{13}e^{-i\delta} \\ -s_{12}c_{23} - c_{12}s_{23}s_{13}e^{i\delta} & c_{12}c_{23} - s_{12}s_{23}s_{13}e^{i\delta} & s_{23}c_{13} \\ s_{12}s_{23} - c_{12}c_{23}s_{13}e^{i\delta} & -c_{12}s_{23} - s_{12}c_{23}s_{13}e^{i\delta} & c_{23}c_{13} \end{pmatrix}$$

with $c_{ij} \equiv \cos\theta_{ij}$ and $s_{ij} \equiv \sin\theta_{ij}$ is a function of four parameters: three mixing angles θ_{12}, θ_{13}, and θ_{23}, and a phase δ. In addition, there are two independent mass-squared differences,

$$\Delta m_{21}^2 \equiv m_2^2 - m_1^2 \qquad \Delta m_{31}^2 \equiv m_3^2 - m_1^2.$$

Experimental values for neutrino oscillation parameters are displayed in Table 11.1 (assuming the normal mass hierarchy; see Section 11.6.2).

Table 11.1 Neutrino oscillation parameters [115]

Parameter	Value (normal hierarchy)
$\Delta m_{21}^2 = m_2^2 - m_1^2$	$7.54^{+0.26}_{-0.22} \times 10^{-5}$ eV^2
$\Delta m_{31}^2 = m_3^2 - m_1^2$	$2.47^{+0.06}_{-0.10} \times 10^{-3}$ eV^2
$\sin^2\theta_{12}$	$0.307^{+0.018}_{-0.016}$
$\sin^2\theta_{13}$	0.0241 ± 0.0025
$\sin^2\theta_{23}$	$0.386^{+0.024}_{-0.021}$
δ	0–2π (unknown)

---

[5] The phases α_1 and α_2 for the mass eigenstates appearing in Eq. (11.17) are relevant only if it should turn out that neutrinos are Majorana particles (see Section 11.2.2 and Box 11.2). If instead they are Dirac particles, α_1 and α_2 can be ignored by setting $e^{i\alpha_1/2} = e^{i\alpha_2/2} = 1$. These phases play no role in flavor oscillations in any case, but they could be relevant for neutrinoless double β-decay.

> **Box 11.4** **C, P, and T Symmetries**
>
> In elementary particle physics C refers to the symmetry of *charge conjugation* (what happens if particles are exchanged for their antiparticles), P refers to symmetry under *parity* (what happens if the spatial coordinate system is inverted), and T refers to symmetry under *time reversal* (what happens if the time axis is inverted).
>
> **CPT Invariance**
> Very general theorems suggest that the product of all three symmetries, CPT, should always be conserved (that is, all physical observables should be unchanged under the combined operations of reversing time, inverting the spatial coordinates, and replacing particles with antiparticles), but combinations of only some of these symmetries are known to be violated in some experiments. For example, P is violated by all weak interactions and CP (simultaneously exchange particles for antiparticles and invert the coordinate system) is violated at a very small level by weak interactions in the quark sector.
>
> **CP-Violating Processes**
> The presence of a complex phase in the flavor-mixing matrix allows for processes having different rates for particles and their corresponding antiparticles, and thus to CP violation. This CP violation requires at least three flavors, because for fewer flavors any complex phase can be absorbed into a redefinition of the wavefunctions and thus is not observable.

11.6.1 CP Violation in Neutrino Oscillations

Were an inequivalence to be observed between flavor oscillations for neutrinos and the corresponding antineutrinos, it would indicate CP violation in the neutrino sector (C, P, and T symmetries are discussed in Box 11.4). If neutrinos are Dirac particles (meaning that neutrinos and antineutrinos are distinct particles; see Section 11.2.2 and Box 11.2), the source of any CP violation would be the phase factor $e^{\pm i\delta}$ appearing in the matrix $\mathcal{U}$ of Eq. (11.17), which is often termed the *Dirac phase*. If neutrinos are Majorana particles instead (neutrino and antineutrino are the same particle) *Majorana phases* $e^{i\alpha_1/2}$ and $e^{i\alpha_2/2}$ appearing in the column vector defining the mass eigenstates in Eq. (11.17) also can violate CP. Majorana phases could influence double β-decay but not neutrino oscillations [78].

The phases δ, α_1, and α_2 all are presently unknown, so it is unclear if CP is violated in Eq. (11.17). If it occurs, CP violation in the neutrino sector might be important for cosmology because to generate the large preponderance of matter over antimatter densities observed in the present Universe a CP-violating process operating in the early Universe is necessary (see Section 21.4 of Ref. [100]). The experimentally observed CP violation in the quark sector is too small to account for this. If there is CP violation in the leptonic (neutrino) sector, this might help explain the dominance of matter over antimatter in the Universe, but presently there is no evidence for leptonic CP violation. The idea that (as yet unobserved) CP violation in neutrino flavor mixing could account for the dominance of matter over antimatter is called *leptogenesis*.

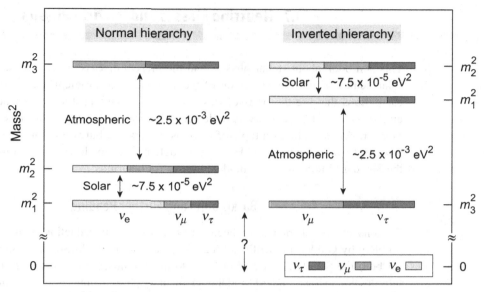

Fig. 11.3 The neutrino mass hierarchy in a 3-flavor model [133]. Approximate mass-square differences that have been inferred from present atmospheric (see Section 12.9.1) and solar neutrino data are indicated. Since only values of Δm^2 and not absolute masses are known, two orderings of the known mass square differences are consistent with data: the *normal hierarchy* and the *inverted hierarchy*. Shading indicates the relative contribution of the three neutrino flavors to each mass eigenstate in the two possible orderings. Adapted from Stephen F. King and Chrisopher Luhn, *Reports on Progress in Physics*, **76**(5), 2013, with permission of IOP Publishing Ltd.

11.6.2 The Neutrino Mass Hierarchy

Direct mass measurements have succeeded only in placing the upper limits on neutrino masses indicated in Fig. 11.1. Oscillation measurements (1) indicate that at least some neutrino flavors have non-zero mass, and (2) constrain the mixing angle and the mass squared difference between flavors participating in neutrino oscillation, but cannot give the actual masses. This leads to the hierarchy ambiguity displayed in Fig. 11.3. In principle the correct hierarchy can be inferred from matter oscillations but evidence for these has been seen thus far only for solar neutrinos and not for atmospheric neutrinos (decribed in Section 12.9.1).

11.6.3 Recovering 2-Flavor Mixing

For three flavors the dominant mixing angle for atmospheric neutrinos is θ_{23} and that for solar neutrinos is θ_{12}. Thus the parameters of most importance for solar neutrinos are the mixing angle θ_{12} and the mass-squared difference Δm_{21}^2, and by restricting to ν_e and ν_μ one may (with suitable redefinitions) recover the 2-flavor formalism for solar neutrino oscillations described earlier. The possibility of approximating most neutrino physics in terms of a 2-flavor mixing matrix, with the effect of the third flavor added using perturbation theory if necessary, rests on two empirical facts: (1) the characteristic splitting in the mass-squared hierarchies of Fig. 11.3, and (2) the small value of θ_{13} (see Table 11.1).

11.7 Neutrino Masses and Particle Physics

This is a book about stellar physics and the present discussion has focused primarily on solar neutrinos and the resolution of the solar neutrino problem. However, the neutrino oscillations that are the basis of this solution also have profound consequences for the understanding of elementary particle physics because they point to physics beyond the Standard Model. This is a topic of obvious importance but would take us too far off our primary track to pursue here. However, Problem 11.10 explores one idea for extension of the Standard Model to accommodate masses for neutrinos.

Background and Further Reading

Portions of this chapter have been adapted from an introduction to the solar nuetrino problem by Guidry and Billings [102]. The book by Bahcall [30] is a good starting point for both the Standard Solar Model and to the significance of solar neutrinos. A general introduction to the Standard Model of elementary particle physics may be found in Collins, Martin, and Squires [74] or in Guidry [97]. An introduction to the general topic of neutrinos is presented in two articles by Haxton and Holstein [113, 114]. A conceptual, non-mathematical approach to understanding neutrino oscillations is given by Waltham [222]. The idea that neutrinos might undergo flavor oscillations that would have physical implications is generally attributed to conjectures by Bruno Pontecorvo (1913–1993), long before it was possible to measure such effects; an historical account is given in Bonolis [54].

Problems

11.1 Show that if a neutrino has a mass m_ν that is very small, a spread in energy ΔE_ν for a set of detected neutrinos having an average flight time t from source to detector will lead to a spread in arrival times given by

$$\frac{\Delta t}{t} = \frac{\Delta v}{v} \simeq \frac{m_\nu^2}{E_\nu^2} \frac{\Delta E_\nu}{E_\nu},$$

where E_ν is the average energy and ΔE_ν is the spread in energy. Some 19 $\bar{\nu}_e$ events were detected on Earth from Supernova 1987A. These neutrinos arrived over about a 10-second time span with an average energy of about 25 MeV and a spread in energy of about 10 MeV. Use these results to place an upper limit on the $\bar{\nu}_e$ (and thus ν_e) mass, assuming the supernova to have occurred about 170,000 lightyears away.

11.2 Show that in a 2-flavor (ν_e and ν_μ) model of neutrino oscillations, if the two energies E_1 and E_2 are approximately equal and much larger than the neutrino masses, then ($\hbar = c = 1$ units)

$$E_2 - E_1 \simeq \frac{\Delta m^2}{2E},$$

where $E_1 \sim E_2 \equiv E$ and $\Delta m^2 \equiv m_2^2 - m_1^2$. Thus, show that the probability for flavor survival and conversion as a function of distance traveled $r \sim ct$ may be expressed as

$$P(\nu_e \to \nu_e, r) = 1 - \sin^2(2\theta)\sin^2\left(\frac{\pi r}{L}\right),$$

$$P(\nu_e \to \nu_\mu, r) = \sin^2(2\theta)\sin^2\left(\frac{\pi r}{L}\right),$$

where θ is the mixing angle and the oscillation length $L \equiv 4\pi E/\Delta m^2$ is the distance (assuming $v \sim c$) over which one complete flavor oscillation occurs.***

11.3 Starting from the neutrino flavor eigenstates in a 2-flavor model with vacuum mixing angle θ,

$$|\nu_e\rangle = \cos\theta|\nu_1\rangle + \sin\theta|\nu_2\rangle$$
$$|\nu_\mu\rangle = -\sin\theta|\nu_1\rangle + \cos\theta|\nu_2\rangle,$$

show that the mass eigenstates may be expressed as

$$|\nu_1\rangle = \cos\theta|\nu_e\rangle - \sin\theta|\nu_\mu\rangle$$
$$|\nu_2\rangle = \sin\theta|\nu_e\rangle + \cos\theta|\nu_\mu\rangle,$$

so that the time-evolved neutrino state

$$|\nu(t)\rangle = \cos\theta e^{-iE_1 t}|\nu_1(0)\rangle + \sin\theta e^{-iE_2 t}|\nu_2(0)\rangle$$

may be written as the mixed-flavor state

$$|\nu(t)\rangle = (\cos^2\theta e^{-iE_1 t} + \sin^2\theta e^{-iE_2 t})|\nu_e\rangle$$
$$+ \sin\theta\cos\theta(-e^{-iE_1 t} + e^{-iE_2 t})|\nu_\mu\rangle.$$

Use this result to show that the probabilities to detect a ν_e or ν_μ after a time t are given by

$$P(\nu_e \to \nu_e, t) = 1 - \tfrac{1}{2}\sin^2(2\theta)[1 - \cos(E_2 - E_1)t],$$
$$P(\nu_e \to \nu_\mu, t) = \tfrac{1}{2}\sin^2(2\theta)[1 - \cos(E_2 - E_1)t],$$

respectively.***

11.4 A smoking-gun signature of a Majorana mass for neutrinos (see Section 11.2.2) would be the observation of *neutrinoless double β-decay*, as discussed in Box 11.2. Which of the following isotopes are potentially candidates for double β-decay experiments: ^{76}Ge, ^{78}Ge, ^{82}Se, ^{80}Se, ^{238}U, and ^{236}U? *Hint*: Compare the masses relative to those for β-decay daughters.

11.5 Show that for a general 2×2 matrix

$$A = \begin{pmatrix} \alpha & \beta \\ \gamma & \delta \end{pmatrix}$$

the eigenvalues are given by

$$\lambda_\pm = \tfrac{1}{2}T \pm \sqrt{\tfrac{1}{4}T^2 - D},$$

where $T \equiv \operatorname{Tr} A$ is the trace of A and $D \equiv \det A$ is the determinant of A.

11.6 From basic quantum mechanics we may expect the transformation between mass and flavor basis states in a 2-flavor neutrino model to be unitary and to have unit determinant.

(a) Prove that a unitary transformation preserves matrix elements. *Hint*: This is trivial if you use the Dirac notation for matrix elements.

(b) Prove that the most general form of a 2×2 unitary matrix U with $\det U = 1$ can be parameterized as

$$U = \begin{pmatrix} a & b \\ -b^* & a^* \end{pmatrix},$$

subject to the constraint $aa^* + bb^* = 1$. Thus justify the parameterization of the transformation implemented by Eq. (11.2). Verify that the transformation (11.2) is both unitary and orthogonal.

11.7 Show that the averaged neutrino probabilities are given by Eq. (11.12) in a 2-flavor model described by Eq. (11.9). If for neutrinos in vacuum $\theta = 20°$, $E = 5$ MeV, and $\Delta m^2 = 10^{-5}$ eV^2, what is the minimum distance over which the probabilities must be averaged for Eq. (11.12) to be valid, and what is the averaged probability to detect an electron neutrino if the neutrino was produced as an electron neutrino?***

11.8 Show that the classical probability $P_{\text{class}}(\nu_e \to \nu_e)$ of Section 11.5.3 is given by Eq. (11.15) for a 2-flavor model.***

11.9 (a) Repeat the derivation of Eq. (11.11) starting from Eq. (11.10) and the observation that $\hbar c = 197.3 \times 10^{-13}$ MeV cm $= 1$ in $\hbar = c = 1$ units. (b) In $\hbar = c = 1$ units the weak coupling constant takes the value $G_F \equiv G_F/(\hbar c)^3 = 1.16637 \times 10^{-5}$ GeV^{-2} [6]. Express G_F in more conventional units of MeV cm^3.***

11.10 Observation of neutrino oscillations implies a finite mass for neutrinos and thus physics beyond the Standard Model. One idea for extending the Standard Model is to add the missing right-handed component of the neutrino, which is assumed to be very massive since it is not seen. By some elementary arguments, it is speculated that the 2-flavor neutrino mass matrix $\mathcal{M}$ (see Section 12.2) for such a minimal extension of the Standard Model to include a heavy right-handed neutrino may be written

$$\mathcal{M} \simeq \begin{pmatrix} 0 & m \\ m & M \end{pmatrix},$$

where it is expected that m is of order 100 GeV and M is of order 10^{15} GeV.

(a) Find the eigenvalues of this matrix and show that this gives one very light and one very massive eigenvalue for the neutrino masses. *Hint*: See Problem 11.5.

(b) Find the corresponding eigenvectors and argue that they can be interpreted approximately as a left-handed Standard Model neutrino but with a small but non-zero mass, and a right-handed neutrino with mass much too large to be detected in present experiments.***

12 Solar Neutrinos and the MSW Effect

The vacuum neutrino flavor oscillations described in Chapter 11 might contribute to depressing the flux of solar neutrinos measured on Earth, but a large value of the mixing angle would be required to suppress the electron neutrino flux by significant amounts. Because quark-sector flavor mixing angles are small, initial theoretical prejudice—but no substantial evidence—favored a small mixing angle in the neutrino sector also. We will return to the appropriate value for the vacuum mixing angle below, but first there is another issue to be addressed: neutrinos must transit out of the Sun. Neutrinos interact extremely weakly with matter so the effect of solar matter on neutrino propagation might seem unimportant. But even a feeble interaction can have significant consequences if the wavefunction has interfering components and the coupling strength to the different components is unequal. Indeed, electron neutrinos couple more strongly to normal matter than do other flavor neutrinos because electron neutrinos and the particles making up normal matter all reside in the first generation of the Standard Model. In this chapter we shall demonstrate that the flavor-dependent interaction with the medium alters the effective mass of a propagating electron neutrino and influences the flavor oscillation in a highly non-trivial way.

12.1 Propagation of Neutrinos in Matter

Following the insight of Mikheyev, Smirnov, and Wolfenstein (MSW), we now consider the effect of weak matter coupling on the neutrino flavor oscillations that were discussed in Chapter 11 [30, 43, 138, 151, 229]. We will find that the interaction of neutrinos with the solar medium causes an unanticipated resonance condition called the *MSW effect* that can lead to resonant flavor conversion of the electron neutrinos produced by the Sun.

12.1.1 Matrix Elements for Interaction with Matter

For the low neutrino energies relevant in the Sun inelastic scattering is negligible and electron neutrinos interact only through *elastic forward scattering*, mediated by both charged and neutral weak currents. The relevant Feynman diagrams are displayed in Fig. 12.1(a). For mu or tau neutrinos only the neutral current can contribute, as illustrated in Fig. 12.1(b). Since the neutral current interaction contributes to both electron and muon neutrino scattering, it provides an overall shift that can be neglected for the

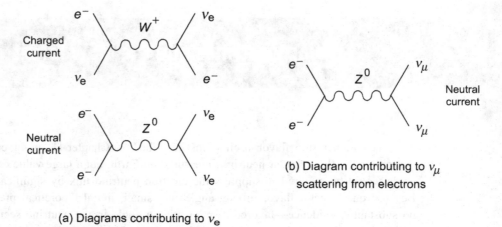

Fig. 12.1 Feynman diagrams (see Box 11.3) responsible for neutrino–electron scattering in a 2-flavor model with ν_e and ν_μ.

present analysis. Therefore, the vacuum neutrino oscillations will be modified in the presence of matter because the charged-current diagram in Fig. 12.1(a) contributes to ν_e elastic scattering but not to ν_μ scattering in the 2-flavor model. By evaluating the Feynman diagrams using standard methods of quantum field theory, the charged-current diagram in Fig. 12.1 is found to contribute an additional medium-dependent interaction that has the form of an effective potential energy V seen *only by the electron neutrinos or antineutrinos*,

$$V = \pm\sqrt{2}G_F n_e, \qquad (12.1)$$

where the positive sign is for neutrinos, the negative sign is for antineutrinos, n_e is the local electron number density, and G_F is the weak (Fermi) coupling constant.

12.1.2 The Effective Neutrino Mass in Medium

For the electron neutrino subject to the additional potential V [see Eq. (12.4) below] we have $E - V = (p^2 + m^2)^{1/2}$ and hence

$$p^2 + m^2 = (E - V)^2 = E^2\left(1 - \frac{V}{E}\right)^2 \simeq E^2 - 2EV, \qquad (12.2)$$

where the last step is justified by the assumption that $V \ll E$. Thus $E^2 \sim p^2 + \tilde{m}^2$, where $\tilde{m} = (m^2 + 2EV)^{1/2}$ may be interpreted as an *effective mass* that has been modified from its value in vacuum by interaction with the medium. Since V is positive, an electron neutrino behaves effectively as if it is slightly heavier when propagating through matter than in vacuum, with the amount of increase governed by the electron number density of the matter. Figure 12.2 illustrates this. From this figure, an electron neutrino that is less

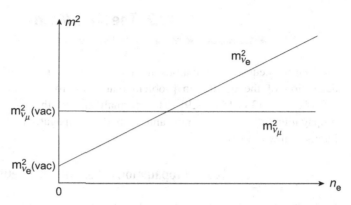

Fig. 12.2 The effective mass-squared of an electron neutrino and a muon neutrino as a function of electron number density n_e, neglecting flavor mixing. Because the ν_μ does not couple to the charged weak current its m^2 does not depend on n_e but the effective m^2 of ν_e increases linearly with the electron density. Thus the order of states in the m^2 spectrum in vacuum (left side) can become *inverted in matter* at high electron density (right side).

Box 12.1   **Effective Masses**

There is nothing unusual about the effective mass of a particle being modified from its vacuum value by interaction with a medium; it is an everyday occurrence in the quantum mechanics of interacting many-body systems.

Classical Interpretation

The effective mass may be *defined* through $m = F/a$. If for a fixed force applied to a particle the acceleration is large, one deduces that the particle has a small mass, and if the corresponding acceleration is small, one deduces that the particle has a large mass. It is obvious intuitively that the mass inferred from this thought experiment will be different for a particle propagating in the vacuum and one interacting with a medium. Two other well-known examples of particles (photons in these cases) acquiring an effective mass through interactions with a medium are superconductivity, described below, and plasmons, described in Box 7.3.

Example: Photons in Superconductors

One way to understand the properties of a superconductor is that the photon, which is identically massless when propagating in the vacuum, acquires an effective non-zero mass in the superconductor through interaction with the medium. It may then be shown that the basic properties of a superconductor such as the Meissner effect (the bulk expulsion of a magnetic field) follow from the photon having acquired this non-zero effective mass through interactions in the superconductor. A more extensive discussion may be found in Chapter 8 of Ref. [97].

---

massive than a muon neutrino in vacuum will become effectively *more massive* than its oscillation partner in matter if the electron density is sufficiently high (Problem 12.4). As discussed in Box 12.1, acquiring an effective mass through interaction with a medium is common for particles in many contexts.

12.2 The Mass Matrix

To address neutrino oscillations in matter it is desirable to introduce a more formal derivation of the oscillation problem that will prove useful in subsequent discussion [30, 43, 74, 138, 154, 229]. That formalism will then be used to consider first the propagation of free neutrinos and then the propagation of neutrinos with an added interaction with matter.

12.2.1 Propagation of Left-Handed Neutrinos

In Box 7.4 it was remarked that only left-handed neutrinos and right-handed antineutrinos couple to the weak interactions. This implies that the full spin structure does not influence the propagation of ultrarelativistic neutrinos (in the absence of magnetic fields). Only the left-handed component of the neutrino couples to the weak interactions and chirality is conserved to order m/E, where m is the neutrino mass and E its energy. Thus for $E \gg m$ only the propagation of left-handed neutrinos is relevant. The Schrödinger equation of ordinary quantum mechanics is not relativistically invariant and so is not appropriate for ultrarelativistic particles. For fermions the wave equation must be generalized to the *Dirac equation*, while for spinless particles the corresponding relativistic wave equation is the *Klein–Gordon equation*. Neutrinos are ultrarelativistic fermions but if the spin structure is eliminated from consideration the propagation of the (left-handed component of the) free neutrino may be described by the simple free-particle Klein–Gordon equation

$$(\Box + m^2)|\nu\rangle = 0 \qquad \Box \equiv -\frac{\partial^2}{\partial t^2} + \frac{\partial^2}{\partial x^2} + \frac{\partial^2}{\partial y^2} + \frac{\partial^2}{\partial z^2}, \qquad (12.3)$$

where $\Box$ is termed the *d'Alembertian operator* and for n neutrino flavors $|\nu\rangle$ is an n-component column vector in the mass-eigenstate basis and m^2 is an $n \times n$ matrix (for either Dirac or Majorana masses; see Section 11.2.2).

Because of oscillations, the solutions to Eq. (12.3) of interest correspond to the propagation of a linear combination of mass eigenstates. For ultrarelativistic neutrinos we make only small errors by assuming neutrinos of tiny mass and slightly different energies to propagate with the same 3-momentum p. In that approximation a solution of Eq. (12.3) for definite momentum is given by

$$|\nu_i\rangle = e^{-iE_i t} \cdot e^{-ip \cdot x} \qquad E_i = \sqrt{p^2 + m_i^2}. \qquad (12.4)$$

As you are asked to show in Problem 12.13, for ultrarelativistic particles this may be approximated as

$$|\nu_i(t)\rangle \simeq e^{-i(m_i^2/2E)t}. \qquad (12.5)$$

Differentiating Eq. (12.5) with respect to time gives an equation of motion for a single mass eigenstate

$$i\frac{d}{dt}|\nu_i(t)\rangle = \frac{m_i^2}{2E}|\nu_i(t)\rangle, \qquad (12.6)$$

which may be generalized for a 2-flavor model to the matrix equation[1]

$$i\frac{d}{dt}\begin{pmatrix}\nu_1\\\nu_2\end{pmatrix} = M\begin{pmatrix}\nu_1\\\nu_2\end{pmatrix} = \begin{pmatrix}m_1^2/2E & 0\\0 & m_2^2/2E\end{pmatrix}\begin{pmatrix}\nu_1\\\nu_2\end{pmatrix}, \quad (12.7)$$

where M is termed the *mass matrix*. These results are valid for both Dirac and Majorana masses (with a corollary that oscillation measurements alone cannot distinguish whether neutrinos are Dirac or Majorana particles).

12.2.2 Evolution in the Flavor Basis

Neutrinos propagate in mass eigenstates but they are produced and detected in flavor eigenstates, so it is useful to express the preceding equation in the flavor basis. The required transformations matrices are given in Eqs. (11.2) and (11.3), permitting Eq. (12.7) to be written in the form

$$i\frac{d}{dt}\begin{pmatrix}\nu_1\\\nu_2\end{pmatrix} = i\frac{d}{dt}U^\dagger\begin{pmatrix}\nu_e\\\nu_\mu\end{pmatrix} = \begin{pmatrix}m_1^2/2E & 0\\0 & m_2^2/2E\end{pmatrix}U^\dagger\begin{pmatrix}\nu_e\\\nu_\mu\end{pmatrix}.$$

Multiplying from the left by U and using the unitarity condition $UU^\dagger = 1$ gives

$$i\frac{d}{dt}\begin{pmatrix}\nu_e\\\nu_\mu\end{pmatrix} = U\begin{pmatrix}m_1^2/2E & 0\\0 & m_2^2/2E\end{pmatrix}U^\dagger\begin{pmatrix}\nu_e\\\nu_\mu\end{pmatrix}, \quad (12.8)$$

where the flavor states and transformation matrices are explicitly

$$\begin{pmatrix}\nu_e\\\nu_\mu\end{pmatrix} = U\begin{pmatrix}\nu_1\\\nu_2\end{pmatrix} \quad U = \begin{pmatrix}\cos\theta & \sin\theta\\-\sin\theta & \cos\theta\end{pmatrix} \quad U^\dagger = \begin{pmatrix}\cos\theta & -\sin\theta\\\sin\theta & \cos\theta\end{pmatrix}.$$

As is clear by substitution, Eq. (12.8) has a solution

$$\begin{pmatrix}\nu_e(t)\\\nu_\mu(t)\end{pmatrix} = U\begin{pmatrix}e^{-i(m_1^2/2E)t} & 0\\0 & e^{-i(m_2^2/2E)t}\end{pmatrix}U^\dagger\begin{pmatrix}\nu_e(0)\\\nu_\mu(0)\end{pmatrix}. \quad (12.9)$$

Since the masses m_1 and m_2 are presently unknown it is convenient to rewrite Eq. (12.8) in terms of Δm^2, which is measurable. Adding a multiple of the unit matrix to the matrix in Eq. (12.8) will not modify observables (a trick that will be employed several times in what follows), so we may subtract $m_1^2/2E$ times the unit 2×2 matrix and use

$$\begin{pmatrix}0 & 0\\0 & \Delta m^2/2E\end{pmatrix} = \begin{pmatrix}m_1^2/2E & 0\\0 & m_2^2/2E\end{pmatrix} - \begin{pmatrix}m_1^2/2E & 0\\0 & m_1^2/2E\end{pmatrix}$$

to replace Eq. (12.8) by the equivalent form

$$i\frac{d}{dt}\begin{pmatrix}\nu_e\\\nu_\mu\end{pmatrix} = U\begin{pmatrix}0 & 0\\0 & \Delta m^2/2E\end{pmatrix}U^\dagger\begin{pmatrix}\nu_e\\\nu_\mu\end{pmatrix}, \quad (12.10)$$

where $\Delta m^2 \equiv m_2^2 - m_1^2$ was introduced in Eq. (11.8).

---

[1] Equations of motion typically will use time t as the independent variable but the distance traveled r may be substituted as the independent variable using $r \sim ct$ for ultrarelativistic neutrinos. Those familiar with elementary quantum mechanics will note that Eq. (12.6) is mathematically similar to the time-dependent Schrödinger equation for nonrelativistic particles. This similarity arises from ignoring the spin structure of the ultrarelativistic neutrinos and our approximation of the energy.

12.2.3 Propagation in Matter

The evolution equation (12.10) is just a reformulation of our previous treatment of neutrinos propagating in vacuum as described in Chapter 11. Let us now add a charged-current interaction with matter. By previous arguments the charged current couples only elastically and only to electron neutrinos, so we add to the Klein–Gordon equation (12.3) an interaction potential given by Eq. (12.1), which modifies the equation of motion (12.10) to

$$i\frac{d}{dt}\begin{pmatrix}\nu_e\\ \nu_\mu\end{pmatrix} = \left[U\begin{pmatrix}0 & 0\\ 0 & \Delta m^2/2E\end{pmatrix}U^\dagger + \begin{pmatrix}V(t) & 0\\ 0 & 0\end{pmatrix}\right]\begin{pmatrix}\nu_e\\ \nu_\mu\end{pmatrix}, \qquad (12.11)$$

where $V(t) = \sqrt{2}G_F n_e(t)$ depends on time (or position) because it depends on the electron density. As shown in Problem 12.14, this may be expressed as

$$i\frac{d}{dt}\begin{pmatrix}\nu_e\\ \nu_\mu\end{pmatrix} = \begin{pmatrix} V & \frac{\Delta m^2}{4E}\sin 2\theta \\ \frac{\Delta m^2}{4E}\sin 2\theta & \frac{\Delta m^2}{2E}\cos 2\theta \end{pmatrix}\begin{pmatrix}\nu_e\\ \nu_\mu\end{pmatrix} \equiv M\begin{pmatrix}\nu_e\\ \nu_\mu\end{pmatrix}, \qquad (12.12)$$

where M is the mass matrix in the flavor basis. Equation (12.12) is the required result but it is conventional to write the mass matrix M appearing in it in a more symmetric form. First define

$$A \equiv 2EV = 2\sqrt{2}EG_F n_e, \qquad (12.13)$$

(which has units of mass squared) and then subtract $A/4E + (\Delta m^2/4E)\cos 2\theta$ multiplied by the unit matrix to give the mass matrix in traceless form

$$M = \frac{\pi}{L}\begin{pmatrix}\chi - \cos 2\theta & \sin 2\theta\\ \sin 2\theta & \cos 2\theta - \chi\end{pmatrix} \qquad (12.14)$$

(see Problem 12.15), where the dimensionless charged-current coupling strength χ is defined by

$$\chi \equiv \frac{L}{\ell_m} = \frac{2EV}{\Delta m^2} \qquad \ell_m \equiv \frac{\sqrt{2}\pi}{G_F n_e} \qquad L \equiv \frac{4\pi E}{\Delta m^2}, \qquad (12.15)$$

with L the vacuum oscillation length defined in Eq. (11.10) and ℓ_m an additional contribution to the oscillation length caused by the matter interaction.[2]

12.3 Solutions in Matter

For a fixed density the mass eigenstates in matter – which generally will differ from the mass eigenstates in vacuum because of the interaction V – may be found by diagonalizing

---

[2] Electron neutrinos in the Sun interact only through elastic forward scattering and the effect of the medium on ν_e propagation can be described as a *refraction* characterized by an *index of refraction* $n_{\text{ref}} = 1 + V/p$, where $p = |\boldsymbol{p}|$. This is analogous to refraction of light in a medium, except that the neutrino index of refraction depends on flavor. The quantity ℓ_m is termed the *refraction length*, because it is the distance over which an additional phase of 2π is acquired through refraction in the matter [203]. Notice that in vacuum $n_e \to 0$ so that $\ell_m \to \infty$ and the coupling term $\chi \equiv L/\ell_m$ in Eq. (12.14) vanishes.

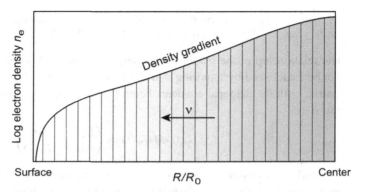

Fig. 12.3 Solar density gradient. Neutrinos are produced near the center at high density and propagate out through regions of decreasing density. In a given concentric layer, the density may be assumed nearly constant.

(finding the eigenvalues) of the mass matrix M at that density. However, since the interaction V depends on the density, in a medium with varying density such as the Sun the mass eigenstates in matter at one time (or position) will generally not be eigenstates at another time or position. We may imagine dividing the Sun up into concentric layers and that within each the density may be assumed to be approximately constant, as illustrated in Fig. 12.3. Our strategy will be to calculate the mass eigenstates within a single layer assuming it to have a constant density, and then consider how to determine the evolution of neutrino states as they propagate through successive layers of decreasing density on the way out of the Sun.

12.3.1 Mass Eigenvalues for Constant Density

At constant density the problem resembles vacuum oscillations but with a different potential. The time-evolved mass states in matter, $|\nu_1^m\rangle$ and $|\nu_2^m\rangle$, may be obtained by diagonalizing the mass matrix (12.14), giving two eigenvalues $\lambda_\pm$ (Problem 12.8)

$$\lambda_\pm = \left(\frac{m_1^2 + m_2^2}{2} + \frac{\Delta m^2}{2}\chi\right) \pm \frac{\Delta m^2}{2}\sqrt{(\cos 2\theta - \chi)^2 + \sin^2 2\theta}. \qquad (12.16)$$

The splitting between the two eigenstates is given by the second term and reaches a minimum at the density where $\chi = \cos 2\theta$. As for the vacuum case, the mass eigenstates in matter, $|\nu_1^m\rangle$ and $|\nu_2^m\rangle$ at fixed time t, are assumed to be related to the flavor eigenstates by a unitary transformation

$$\begin{pmatrix} \nu_e \\ \nu_\mu \end{pmatrix} = U_m(t) \begin{pmatrix} \nu_1^m \\ \nu_2^m \end{pmatrix}, \qquad (12.17)$$

where $U_m(t)$ is a unitary matrix yet to be determined.

12.3.2 The Matter Mixing Angle θ_m

The matrix $U_m(t)$ depends on time and can be parameterized as for the vacuum mixing angle, but now in terms of a time-dependent matter mixing angle $\theta_m(t)$ with

$$U_{\mathrm{m}} = \begin{pmatrix} \cos\theta_{\mathrm{m}} & \sin\theta_{\mathrm{m}} \\ -\sin\theta_{\mathrm{m}} & \cos\theta_{\mathrm{m}} \end{pmatrix} \qquad U_{\mathrm{m}}^{\dagger} = \begin{pmatrix} \cos\theta_{\mathrm{m}} & -\sin\theta_{\mathrm{m}} \\ \sin\theta_{\mathrm{m}} & \cos\theta_{\mathrm{m}} \end{pmatrix}. \qquad (12.18)$$

The relationship of the matter mixing angle θ_{m} and the vacuum mixing angle θ at time t can be established by requiring that a similarity transform by $U_{\mathrm{m}}(t)$ diagonalize the mass matrix, with the diagonal elements being the time-dependent eigenvalues in matter $E_1(t)$ and $E_2(t)$,

$$U_{\mathrm{m}}^{\dagger}(t) M U_{\mathrm{m}}(t) = \begin{pmatrix} E_1(t) & 0 \\ 0 & E_2(t) \end{pmatrix}. \qquad (12.19)$$

(A similarity transformation generally is implemented by $U^{-1}MU$ but $U^{\dagger} = U^{-1}$ for a unitary matrix.) As you are asked to demonstrate in Problem 12.9, inserting the explicit values of the matrices $U, U^{\dagger}$, and M from Eqs. (12.14) and (12.18) in Eq. (12.19) gives an equation that can be satisfied only if the matter and vacuum mixing angles are related by

$$\tan 2\theta_{\mathrm{m}} = \frac{\sin 2\theta}{\cos 2\theta \pm \chi} = \frac{\tan 2\theta}{1 \pm \chi/\cos 2\theta}, \qquad (12.20)$$

where the plus sign is for $m_1 > m_2$ and the negative sign for $m_1 < m_2$. From Eq. (12.20), in vacuum $\theta_{\mathrm{m}} = \theta$ because for vanishing electron density $\ell_{\mathrm{m}} \to \infty$ and $\chi = L/\ell_{\mathrm{m}} \to 0$, but in matter the mixing angle will be modified from its vacuum value by an amount that depends on density.

12.3.3 The Matter Oscillation Length L_{m}

From Eq. (11.10) the oscillation length in vacuum $L = 4\pi E/\Delta m^2$ is proportional to the inverse of the mass-squared difference Δm^2 between the states participating in the oscillation. In matter the neutrino effective mass is altered by interaction with the medium and the vacuum mass-squared difference is rescaled, $\Delta m^2 \to f(\chi)\Delta m^2$, where from the splitting of the two eigenvalues in Eq. (12.16)

$$f(\chi) = \sqrt{(\cos 2\theta - \chi)^2 + \sin^2 2\theta} = \sqrt{1 - 2\chi \cos 2\theta + \chi^2}. \qquad (12.21)$$

Hence the oscillation length in matter L_{m} is given by

$$L_{\mathrm{m}} = \frac{4\pi E}{f(\chi)\Delta m^2} = \frac{L}{f(\chi)} = \frac{L}{\sqrt{(\cos 2\theta - \chi)^2 + \sin^2 2\theta}}, \qquad (12.22)$$

which reduces to the vacuum oscillation length L if the interaction χ vanishes. The variations of θ_{m}, L_{m}, and f with the dimensionless coupling χ are illustrated in Fig. 12.4 for several values of the vacuum mixing angle θ.

From Fig. 12.4(a) the matter mixing angle θ_{m} reduces to the vacuum mixing angle θ for vanishing coupling, but $\theta_{\mathrm{m}} \to \frac{\pi}{2}$ at large coupling for *any value of the vacuum mixing angle*. From Fig. 12.4(b) the matter oscillation length is equal to the vacuum oscillation length at zero coupling, but increases to a maximum at the coupling strength where $\theta_{\mathrm{m}} = \frac{\pi}{4}$ [compare Fig. 12.4(a)], and then decreases again. Notice that the coupling strength at which L_{m} is maximal coincides with highest rate of change in θ_{m}, and with the

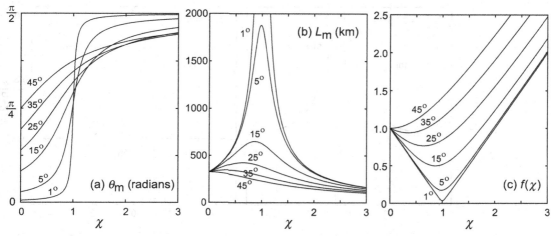

Fig. 12.4 (a) Mixing angle in matter $\theta_m(\chi)$, (b) oscillation length in matter $L_m(\chi)$, and (c) the scaling factor $f(\chi)$ of Eq. (12.21) as a function of the dimensionless matter coupling parameter χ. All calculations assumed $E = 10$ MeV and $\Delta m^2 = 7.6 \times 10^{-5}$ eV^2, and curves are marked with the assumed vacuum mixing angle θ.

minimum of the scaling function $f(\chi)$ governing the separation between the mass states in matter displayed in Fig. 12.4(c), suggesting something special about the density where $\theta_m = \frac{\pi}{4}$. In particular, the rapid change of θ_m and the strong peaking of L_m near this density are suggestive of *resonant behavior*. We will address the implications of this observation shortly.

12.3.4 Flavor Conversion in Constant-Density Matter

In matter of *constant density* the electron neutrino state after a time t becomes

$$|\nu(t)\rangle = (\cos^2\theta_m e^{-iE_1 t} + \sin^2\theta_m e^{-iE_2 t})|\nu_e\rangle \\ + \sin\theta_m \cos\theta_m(-e^{-iE_1 t} + e^{-iE_2 t})|\nu_\mu\rangle, \quad (12.23)$$

which is analogous to the vacuum equation Eq. (11.6) but with the vacuum mixing angle θ replaced by the matter mixing angle θ_m defined through Eq. (12.20). Hence for a constant density n_e the flavor conservation and flavor retention probabilities are given by Eq. (11.9) with the replacements $\theta \to \theta_m$ and $L \to L_m$,

$$P(\nu_e \to \nu_e, r) = 1 - \sin^2 2\theta_m \sin^2\left(\frac{\pi r}{L_m}\right), \quad (12.24)$$

with $P(\nu_e \to \nu_\mu, r) = 1 - P(\nu_e \to \nu_e, r)$. The corresponding classical averages are

$$\bar{P}(\nu_e \to \nu_e) = 1 - \tfrac{1}{2}\sin^2 2\theta_m \qquad \bar{P}(\nu_e \to \nu_\mu) = \tfrac{1}{2}\sin^2 2\theta_m, \quad (12.25)$$

which are valid when the uncertainty in distance between source and detection exceeds the oscillation length (see Section 11.5.3).

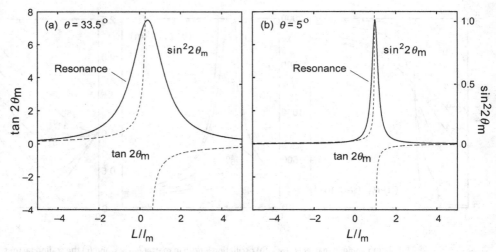

Fig. 12.5 The MSW resonance condition for two values of the vacuum mixing angle θ. When $L/\ell_m \to \cos 2\theta$ the denominator of Eq. (12.20) goes to zero, $\tan 2\theta_m$ goes to $\pm\infty$ so that $|\theta_m| \to \frac{\pi}{4}$, and the flavor conversion probability $\sin^2 2\theta_m$ attains its maximum value. Thus, at the resonance Eq. (12.24) indicates that significant flavor conversion can be obtained for any non-vanishing vacuum oscillation angle θ.

12.4 The MSW Resonance Condition

From Eq. (12.24), optimal flavor mixing occurs whenever $\sin^2 2\theta_m$ achieves its maximum value of unity, which occurs when $|\theta_m| = \frac{\pi}{4}$. The most significant property of Eq. (12.20) is that if Δm^2 and L are positive [which requires that $m_1 < m_2$ and selects the negative sign in Eq. (12.20)],[3] then $\tan 2\theta_m \to \pm\infty$ and $\theta_m \to \frac{\pi}{4}$ whenever the coupling strength satisfies

$$\chi = \frac{L}{\ell_m} = \cos 2\theta, \qquad (12.26)$$

as illustrated in Fig. 12.5. From (12.15), this occurs when the electron density satisfies

$$n_e = \frac{\cos 2\theta \, \Delta m^2}{2\sqrt{2} G_F E} \equiv n_e^R. \qquad (12.27)$$

---

[3] This sign criterion depends on the definition $\Delta m^2 \equiv m_2^2 - m_1^2$ in Eq. (11.8). If the opposite convention, $\Delta m^2 \equiv m_1^2 - m_2^2$ is used, the resonance sign criteria are opposite that used here. That is just convention but there is an interesting history associated with the implications of choosing the physically correct sign for the neutrino–electron interaction itself [30, 43]. Our discussion assumes (correctly) that the effective charged-current interaction V of Eq. (12.1) is positive for neutrinos. In Wolfenstein's original paper [229] the sign was correctly given as positive but in a later Wolfenstein paper it was given as negative. When Mikheyev and Smirnov elaborated on Wolfenstein's work several years later they followed his second paper containing the erroneous sign. Thus they deduced, incorrectly, that the resonance for neutrinos would occur if the electron neutrino were more massive than the muon neutrino in vacuum. Only when it was realized that the sign should be opposite that assumed originally by Mikheyev and Smirnov (Paul Langacker pointed out the error to Hans Bethe) did it become clear that the MSW resonance could occur in the Sun if the electron neutrino is *less massive* than the muon neutrino in vacuum. As Bethe has noted [43], the correct sign is a technically crucial but conceptually minor issue. The seminal insight that changed neutrino physics fundamentally was the realization by Mikheyev and Smirnov that there could be a *resonance* in the interaction of neutrinos with matter, and that this resonance could lead to large flavor conversion for either large or small vacuum mixing angles.

12.4 The MSW Resonance Condition

From Eq. (12.24), this corresponds to a resonance condition leading to maximal mixing between electron neutrinos and muon neutrinos, with a ν_e survival probability

$$P(\nu_e \to \nu_e, r) = 1 - \sin^2\left(\frac{\pi r}{L_m}\right) \quad \text{(at resonance)}, \tag{12.28}$$

and, from Eqs. (12.22) and (12.26), an oscillation length at resonance L_m^R given by

$$L_m^R = L_m(\chi = \cos 2\theta) = \frac{L}{\sin 2\theta}. \tag{12.29}$$

This is the *Mikheyev–Smirnov–Wolfenstein* or *MSW resonance* [151, 229]. It implies that, no matter how small the vacuum mixing angle θ, as long as it is not zero there is some critical value n_e^R of the electron density defined by Eq. (12.27) where the resonance condition is satisfied and *maximal flavor mixing* ensues. The important resonance parameters are plotted in Fig. 12.6 as a function of electron density for two values of θ.

The effects of the MSW resonance on variation of the matter mixing angle θ_m and the oscillation length in matter L_m are illustrated for a small and large angle solution in Fig. 12.7. The values of θ_m and L_m will vary with the solar depth since they depend on the number density n_e through χ. From Eq. (12.20), $\theta_m \to \theta$ as the electron density tends to zero, while in the opposite limit of very large electron density $\theta_m \to \frac{\pi}{2}$. From Fig. 12.7, the manner in which the two limits are approached depends on whether the vacuum mixing angle is large or small. Figure 12.7(a) corresponds to parameters valid for solar neutrinos. At the solar center ($\chi \sim 2.13$) the matter mixing angle is $\theta_m \sim 76°$, compared with a vacuum mixing angle $33.5°$ at the solar surface. Conversely, for Fig. 12.7(b) with $\theta = 5°$ at the solar surface, the matter mixing angle at a density corresponding to the solar center is $\theta \sim 86°$.

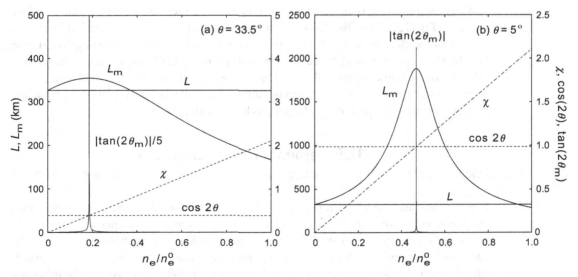

Fig. 12.6 Resonance parameters versus the electron number density n_e in units of the central solar value $n_e^0 \sim 6.3 \times 10^{25}$ cm^{-3} for $\theta = 33.5°$ and $5°$, with $\Delta m^2 = 7.6 \times 10^{-5}$ eV^2 and $E = 10$ MeV. The coupling strength $\chi = L/\ell_m$ is linear in the density. Intersection of the dashed lines specifies the electron density giving the resonance condition.

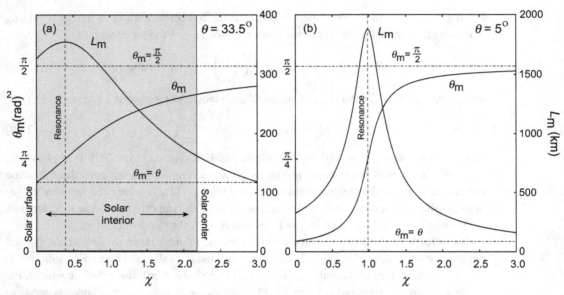

Fig. 12.7 The matter mixing angle θ_m as a function of the dimensionless coupling strength $\chi \equiv L/\ell_m$ for vacuum mixing angles of (a) $\theta = 33.5°$ and (b) $\theta = 5°$. Also shown is the oscillation length in matter L_m, which has a maximum at the position of the MSW resonance (see Section 12.4), marked by the dashed vertical line. The oscillation length was computed from Eqs. (12.22) and (11.11) assuming $E = 10$ MeV and $\Delta m^2 = 7.6 \times 10^{-5}$ eV^2. Case (a) is realistic for solar neutrinos and the density at the center of the Sun corresponds to $\chi \sim 2.13$ (Problem 12.2). Hence the shaded region on the left side of (a) indicates the range of coupling strengths available to electron neutrinos in the interior of the Sun.

If $m_1 > m_2$ there is no resonance for ν_e because then the positive sign would be chosen in Eq. (12.20) and the denominator doesn't vanish. In that case there is a resonance instead for the electron antineutrino $\bar{\nu}_e$ [which traces back to choosing the negative sign in Eq. (12.1) for antineutrinos]. Since the Sun emits primarily neutrinos and not antineutrinos, a discussion of antineutrinos will be omitted in the present context. However, antineutrino oscillations could occur in environments such as core collapse supernovae or neutron star mergers, where all flavors of ν and $\bar{\nu}$ are produced in abundance.

12.5 Resonant Flavor Conversion

If $m_1 < m_2$ and the electron density in the central region of the Sun where neutrinos are produced satisfies $n_e > n_e^R$, a solar neutrino will inevitably encounter the MSW resonance on its way out of the Sun. If the change in density is sufficiently slow that the additional phase mismatch between the ν_e and ν_μ components produced by the charged-current elastic scattering from electrons changes very slowly with density (the *adiabatic condition* discussed further in Section 12.7), the ν_e flux produced in the core can be almost entirely converted to ν_μ by the MSW resonance near the radius where the condition (12.27)

| Box 12.2 | **Resonant Flavor Conversion and Adiabatic Level Crossings** |

MSW flavor conversion can be viewed as an *adiabatic quantum level crossing* [43, 112],[a] where adiabatic means that the *density scale height* $H_{n_e} \equiv -n_e/(dn_e/dr)$ near n_e^R is large relative to a neutrino oscillation length (see Section 7.9.1). This is equivalent to requiring that the oscillation length be less than the width of the resonance peak in Fig. 12.5. The situation is illustrated schematically in the following figure,

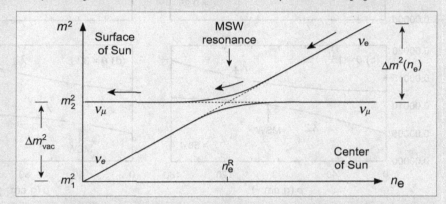

and more realistically in Fig. 12.8. A neutrino produced near the center of the Sun (high density on right side) will be in a ν_e flavor eigenstate that coincides with the *higher-mass eigenstate,* since V representing interaction with the medium increases the mass of the electron neutrino but not the muon neutrino. As the neutrino propagates out of the Sun (right to left in this diagram) the density decreases so V and the effective mass of the neutrino decrease. Conversely, the lower-mass eigenstate (primarily ν_μ flavor) remains constant in mass as n_e decreases (no coupling to the charged current). Thus the two levels cross at the critical density where the effect of V exactly cancels the vacuum mass-squared difference $\Delta m_{\text{vac}}^2 \equiv \Delta m^2(n_e = 0)$ between the eigenstates, and the neutrino remains in the high-mass eigenstate and changes adiabatically into a ν_μ flavor state by the time it exits the Sun (left side), because *in vacuum the high-mass eigenstate approximately coincides with* ν_μ.

In summary, *if the crossing is adiabatic* the neutrino remains in the high-mass eigenstate in which it was created and follows the upper curved trajectory through the resonance in the level-crossing region, as indicated by the arrows. It emerges from the Sun in a different flavor state than the one in which it was created in the core of the Sun because *the high-mass eigenstate is primarily ν_e in the dense medium but is primarily ν_μ in vacuum.* Adiabatic level crossing is common in a variety of quantum systems and is often termed *avoided level crossing.*

[a] In quantum physics "adiabatic" has a different meaning than in classical thermal physics, where it means *thermally isolated*. In quantum mechanics adiabatic means that a process is *slow on some timescale*. The two definitions can converge since the slow process could be heat exchange, but the quantum definition encompasses a broader range of phenomena than the classical thermal one.

is satisfied. As Box 12.2 illustrates, the MSW effect can be understood intuitively as an energy-level crossing for a quantum system. Solutions of the MSW eigenvalue problem illustrating this level crossing for various choices of the vacuum mixing angle are displayed in Fig. 12.8. At zero density on the left side of these plots the eigenvalues $\lambda_\pm$ converge to

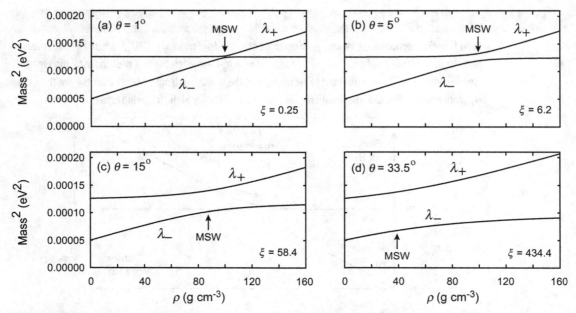

Fig. 12.8 Solutions $\lambda_\pm$ of the MSW eigenvalue problem as a function of mass density according to Eq. (12.16). Each case corresponds to the choices $\Delta m^2 = 7.6 \times 10^{-5}$ eV^2 and $E = 10$ MeV, but to different values of the vacuum mixing angle θ. The individual neutrino masses are presently unknown but for purposes of illustration $m_1^2 = 5 \times 10^{-5}$ eV^2 has been assumed in vacuum, so that $m_2^2 = m_1^2 + \Delta m^2 = 1.26 \times 10^{-4}$ eV^2. The critical density leading to the MSW resonance (corresponding to minimum splitting between the eigenvalues) and the value of the adiabaticity parameter $\xi = \delta r_R / L_m^R$ defined in Eq. (12.33) are indicated for each case (see Problem 12.5). As will be discussed in Section 12.9, realistic conditions in the Sun are expected to imply the very adiabatic crossing exhibited in case (d).

the vacuum m^2 values, but for non-zero density the masses are altered by the interaction of the electron neutrino with the medium and the mixing of the solutions by the neutrino oscillation. (Compare with the unmixed case in Fig. 12.2.)

The number density of electrons in the Sun computed in the Standard Solar Model is illustrated in Fig. 12.9(a), along with an approximation that is rather good in the region where the MSW effect is most important. In Fig. 12.9(b) the approximate locations where electron neutrinos of various energies would encounter the MSW resonance condition [obtained by solving Eq. (12.27) for each energy] are illustrated. As explored in Problem 12.3, only neutrinos having an energy larger than some minimum energy $E_{\min}$ can experience the MSW resonance in the Sun because the neutrino must be produced at a density higher than the critical resonance density. The conditions used to obtain Fig. 12.9(b) give $E_{\min} \sim 1.6$ MeV, so the MSW effect should be more efficient at converting higher-energy neutrinos. Flavor conversion is found to be preferentially suppressed for lower-energy polar neutrinos, which will prove to be important evidence favoring the MSW mechanism over vacuum oscillations for the flavor conversion of solar neutrinos.

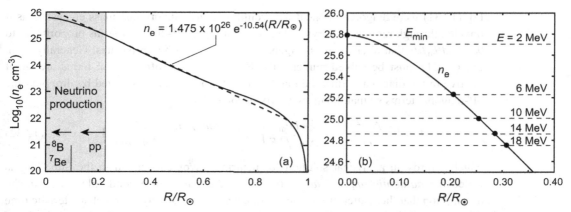

Fig. 12.9 (a) Electron number density as a function of fractional solar radius from the Standard Solar Model [32]. The dashed line is an exponential approximation that will be employed in discussing the MSW effect. Regions of primary neutrino production in the PP chains are indicated. (b) Radius where the MSW critical density for a 2-flavor model is realized (dots at intersection of dashed lines with the curve for n_e) for neutrinos of energies ranging from 2 to 18 MeV. A vacuum mixing angle $\theta = 35°$ and $\Delta m^2 c^4 = 7.5 \times 10^{-5}$ eV^2 have been assumed. The minimum energy of an electron neutrino $E_{\text{min}} \sim 1.6$ MeV that could be produced in the Sun and still encounter the MSW resonance is indicated (see Problem 12.3). Thus Fig. 10.3 suggests that the MSW resonance will affect mostly the ^{8}B neutrinos

12.6 Propagation in Matter of Varying Density

We are now ready to consider realistic neutrino propagation in the Sun. A neutrino produced in the center will encounter decreasing density as it travels toward the solar surface, as illustrated in Fig. 12.3. The neutrino flavor evolution will be governed by the analog of the differential equations for vacuum propagation, but with $U \to U_m(t)$ since the flavor–mass basis transformation now depends on time. From Eq. (12.8) with this replacement

$$i \frac{d}{dt} \left[U_m(t) \begin{pmatrix} \nu_1(t) \\ \nu_2(t) \end{pmatrix} \right] = \frac{1}{2E} U_m(t) \begin{pmatrix} m_1^2 & 0 \\ 0 & m_2^2 \end{pmatrix} \begin{pmatrix} \nu_1(t) \\ \nu_2(t) \end{pmatrix}, \quad (12.30)$$

where both the wavefunctions and the transformation matrix U_m are indicated explicitly to depend on the time. Taking the derivative of the product in brackets on the left side and multiplying the equation from the left by $U_m^\dagger$ gives (Problem 12.10)

$$i \frac{d}{dt} \begin{pmatrix} \nu_1 \\ \nu_2 \end{pmatrix} = \begin{pmatrix} -\Delta m^2/4E & -i\dot\theta_m \\ i\dot\theta_m & \Delta m^2/4E \end{pmatrix} \begin{pmatrix} \nu_1 \\ \nu_2 \end{pmatrix}, \quad (12.31)$$

where $\dot\theta_m \equiv d\theta_m/dt$, Eq. (12.18) was used, and the constant $(m_1^2 + m_2^2)/4E$ times the unit matrix has been subtracted from the matrix on the right side (which does not affect observables).

The earlier statement that mass eigenstates at some density in the Sun generally will not be eigenstates at a different density may now be quantified. If the mass matrix in

Eq. (12.31) were diagonal the neutrino would remain in its original mass eigenstate as it traveled through regions of varying density, so it is the off-diagonal terms proportional to $\dot{\theta}_m = d\theta_m/dt$ that alter the mass eigenstates as the neutrino propagates. Generally then, Eq. (12.31) must be solved numerically. However, if the off-diagonal terms are small relative to the diagonal terms, the mass matrix M may be approximated by dropping the off-diagonal terms so that the mass matrix becomes

$$M = \begin{pmatrix} -\Delta m^2/4E & -i\dot{\theta}_m \\ i\dot{\theta}_m & \Delta m^2/4E \end{pmatrix} \simeq \begin{pmatrix} -\Delta m^2/4E & 0 \\ 0 & \Delta m^2/4E \end{pmatrix}, \qquad (12.32)$$

which affords an analytical solution for neutrino flavor conversion in the Sun. This is called the *adiabatic approximation* (see Box 12.2), and corresponds physically to the assumption that the matter mixing angle θ_m changes only slowly over a characteristic time for motion of the neutrino. A neutrino travels at nearly the speed of light so $r \sim ct$ and the adiabatic condition also may be interpreted as a limit on the spatial gradient of θ_m. These observations may be used to quantify the conditions appropriate for the adiabatic approximation.

12.7 The Adiabatic Criterion

The adiabatic condition for resonant flavor conversion can be expressed as a requirement that the spatial width of the resonance layer δr_R (defined by the radial distance over which the resonance condition is approximately satisfied) be much greater than the oscillation wavelength in matter evaluated at the resonance, L_m^R. This can be characterized by an *adiabaticity parameter* ξ defined by [see Eq. (12.29) and the discussion in Box 12.3]

$$\xi \equiv \frac{\delta r_R}{L_m^R} \qquad \delta r_R = \frac{n_e^R}{(dn_e/dr)_R} \tan 2\theta \qquad L_m^R = \frac{L}{\sin 2\theta}, \qquad (12.33)$$

where the label R denotes quantities evaluated at the resonance, L is the vacuum oscillation length, and θ is the vacuum oscillation angle [43, 203]. The adiabatic condition corresponds to requiring that $\xi \gg 1$, implying physically that if many oscillation lengths (in matter) fit within the resonance layer the adiabatic approximation (12.32) is valid.

Values of ξ computed from Eq. (12.33) are indicated in Fig. 12.8 for several vacuum mixing angles θ. In general sharp level crossings as in Fig. 12.8(a) are non-adiabatic, while avoided level crossings as in Fig. 12.8(d) are highly adiabatic. In the limit of no mixing ($\theta = 0$) the levels cross with no interaction, as illustrated earlier in Fig. 12.2. From Fig. 12.8, the MSW resonance can occur under approximately adiabatic conditions, even for relatively small values of the vacuum mixing angle [for example, case (b)]. As will be shown in Section 12.9, solar conditions correspond approximately to the highly avoided level crossing in Fig. 12.8(d), for which $\delta r_R \gg L_m^R$. Thus the MSW resonance is expected to be encountered adiabatically in the Sun, which optimizes the chance of resonant flavor conversion according to the discussion in Box 12.2. A simple classical oscillation example illustrating the MSW resonance and its adiabatic constraint is described in Box 12.3.

> **Box 12.3** **Matter-Wave Tuning Forks**
>
> The MSW resonance has been likened to the interaction of two tuning forks, one of which has a variable length [154]. If a vibrating tuning fork with variable length is brought close to another tuning fork and the length of the first varied to match the second, the vibration of the first fork can be almost completely resonantly transferred to the second, but only if the frequency of the variable fork is changed slowly enough (adiabatic condition). Likewise, because the additional potential for the neutrino–matter interaction affects only the electron neutrino and it depends on density, a ν_e matter wave in the Sun has a variable frequency while a ν_μ matter wave has roughly a constant frequency. If as the ν_e frequency varies with density it becomes equal to that of the ν_μ at some depth in the Sun, this will lead to a resonance between the two matter waves and resonant conversion from one flavor to the other at that depth. As for the tuning fork example, flavor conversion can occur with significant probability only if the adiabatic condition is satisfied.

12.8 MSW Neutrino Flavor Conversion

In the general case neutrino flavor conversion in the Sun must be investigated by integrating Eq. (12.31) numerically because of the off-diagonal terms arising from the density gradient, but it has just been argued that the adiabatic approximation (12.32) should be fulfilled very well for the Sun. Hence, we now solve for flavor conversion of solar neutrinos by the MSW mechanism assuming the adiabatic approximation.

12.8.1 Flavor Conversion in Adiabatic Approximation

The adiabatic conversion of neutrino flavor in the Sun is illustrated in Fig. 12.10. An electron neutrino is produced at Point 1 near the center of the Sun and propagates radially outward to Point 2. Detection is assumed to average over many oscillation lengths so that the interference terms are washed out and our concern is with the classical (time-averaged) probability, as described in Section 11.5.3. The probability to be detected at Point 2 in the $|\nu_e\rangle$ flavor eigenstate is then given by the generalization of Eq. (11.16) [138],

$$\bar{P}(\nu_e \to \nu_e) = (1 \ \ 0) \begin{pmatrix} \cos^2\theta_m(2) & \sin^2\theta_m(2) \\ \sin^2\theta_m(2) & \cos^2\theta_m(2) \end{pmatrix}$$
$$\times \begin{pmatrix} 1 & 0 \\ 0 & 1 \end{pmatrix} \begin{pmatrix} \cos^2\theta_m(1) & \sin^2\theta_m(1) \\ \sin^2\theta_m(1) & \cos^2\theta_m(1) \end{pmatrix} \begin{pmatrix} 1 \\ 0 \end{pmatrix}, \tag{12.34}$$

where $\theta_m(i) \equiv \theta_m(t_i)$ and the row vector (1 0) and corresponding column vector denote a pure ν_e flavor state. Evaluating the matrix products and using standard trigonometric identities gives for the probability to remain a ν_e (Problem 12.11),

$$\bar{P}(\nu_e \to \nu_e) = \tfrac{1}{2}\left[1 + \cos 2\theta_m(t_1)\cos 2\theta_m(t_2)\right]. \tag{12.35}$$

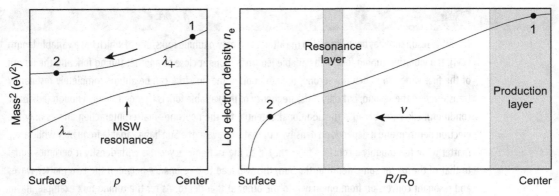

Fig. 12.10 Schematic illustration of adiabatic flavor conversion by the MSW mechanism in the Sun. An electron neutrino is produced at Point 1, where the density lies above that of the MSW resonance, and propagates radially outward to Point 2, where the density lies below that of the resonance. The width of the resonance layer is assumed to be much larger than the matter oscillation length in the resonance layer, justifying the adiabatic approximation of Eq. (12.32). The widths of resonance and production layers are not meant to be to scale in this diagram.

This result is valid (if the adiabatic condition is satisfied) for Point 2 anywhere outside Point 1,[4] but in the specific case that Point 2 lies at the solar surface $\theta_m(t_2) \to \theta$ and the classical probability to detect the neutrino as an electron neutrino when it exits the Sun is

$$\bar{P}(\nu_e \to \nu_e) = \tfrac{1}{2}\left(1 + \cos 2\theta \cos 2\theta_m^0\right) \qquad \text{(at the solar surface)}, \tag{12.36}$$

where θ is the vacuum mixing angle and $\theta_m^0 \equiv \theta_m(t_1)$ is the matter mixing angle at the point of neutrino production.

12.8.2 Adiabatic Conversion and the Mixing Angle

The remarkably concise result (12.36) has a simple physical interpretation. Because of the adiabatic assumption (12.32) the mass matrix for a neutrino propagating down the solar density gradient is diagonal and a neutrino produced in the λ_+ eigenstate remains in that eigenstate until it reaches the solar surface, with the flavor conversion resulting only from the change of mixing angle between the production point and the surface. Thus, in adiabatic approximation the classical probability $\bar{P}(\nu_e \to \nu_e)$ is independent of the details of neutrino propagation and depends only on the mixing angles at the point of production and point of detection.

Example 12.1 Assuming a vacuum mixing angle of $33.5°$, an energy of 10 MeV, a vacuum mass-squared splitting $\Delta m^2 = 7.5 \times 10^{-5}$ eV^2, and a central electron density calculated from the Standard Solar Model, from Eq. (12.20) the matter mixing angle at the center

---

[4] Although the MSW resonance is assumed to lie between Points 1 and 2 in Fig. 12.10, Eq. (12.35) is valid in adiabatic approximation whether that is the case or not.

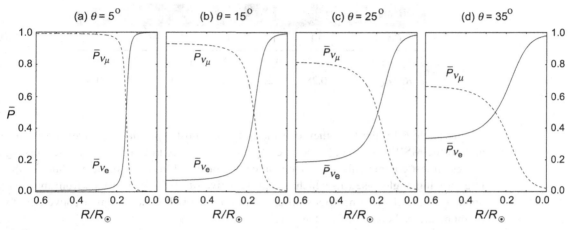

Fig. 12.11 MSW flavor conversion versus fraction of solar radius for four values of the vacuum mixing angle θ. Calculations are classical averages (Section 11.5.3) in adiabatic approximation using Eq. (12.35), assuming $\Delta m^2 = 7.6 \times 10^{-5}$ eV^2 and $E = 10$ MeV. The exponential density approximation of Fig. 12.9(a) was used and neutrinos were assumed to be produced in a ν_e flavor state at the center (right side of diagram at $R/R_\odot = 0$). Solid curves show the classical electron-neutrino probability $\bar{P}_{\nu_e} \equiv \bar{P}(\nu_e \to \nu_e)$ and dashed curves show the corresponding classical muon-neutrino probability $\bar{P}_{\nu_\mu} \equiv \bar{P}(\nu_e \to \nu_\mu)$.

of the Sun is determined to be $\theta_m^0 \sim 76°$ [see Fig. 12.7(a)]. Thus an electron neutrino produced at the center of the Sun has a probability to be a ν_e when it exits the Sun of

$$\bar{P}(\nu_e \to \nu_e) = \tfrac{1}{2}\left(1 + \cos 2\theta \cos 2\theta_m^0\right)$$
$$= \tfrac{1}{2}\left[1 + \cos(2 \times 33.5°)\cos(2 \times 76°)\right] = 0.33,$$

and a probability $1 - \bar{P}(\nu_e \to \nu_e) = 0.67$ to be a ν_μ. We conclude that because of the MSW resonance only $\sim \tfrac{1}{3}$ of 10-MeV electron neutrinos produced in the Sun will still be electron neutrinos when they exit the Sun.

Flavor conversion by the MSW mechanism for a 2-flavor model in adiabatic approximation is illustrated for four different values of the vacuum mixing angle θ in Fig. 12.11. In these figures the MSW resonance occurs at the radius corresponding to the intersection of the solid and dashed curves. Figure 12.11(d) approximates the situation expected for the Sun.

12.8.3 Resonant Conversion for Large or Small θ

As shown in Fig. 12.4(a), for either large or small vacuum mixing angles θ the matter mixing angle θ_m approaches $\tfrac{\pi}{2}$ near the center of the Sun and becomes equal to θ at the surface. Hence neutrinos are produced near the center in a flavor eigenstate that is an almost pure mass eigenstate, but they evolve to a flavor mixture characterized by the vacuum

Table 12.1 Energy dependence of solar ν flavor conversion for $\theta = 35°$

E (MeV)	14	10	6	2	1	0.70
P_{ν_e} (surface)	0.33	0.33	0.34	0.40	0.47	0.50
$R^R/R_\odot$	0.28	0.25	0.20	0.10	0.03	0.0

mixing angle θ by the time they exit the Sun. The most rapid flavor conversion occurs around the MSW resonance where the P_{ν_e} and P_{ν_μ} curves intersect. For smaller mixing angles almost complete flavor conversion can be obtained in the resonance, while for the large mixing angle case (d), which is representative of 10 MeV neutrinos from the Sun, about $\frac{2}{3}$ of the electron neutrinos produced in the core will undergo flavor conversion to muon neutrinos before leaving the Sun.

12.8.4 Energy Dependence of Flavor Conversion

Figure 12.9(b) indicates that there is a substantial energy dependence associated with the flavor conversion. For example, repeating the calculation of Fig. 12.11(d) for a range of neutrino energies E gives the electron neutrino survival probabilities P_{ν_e}(surface) displayed in Table 12.1, along with the fractional solar radius $R^R/R_\odot$ where the MSW resonance occurs for that energy.[5] Consulting the solar neutrino spectrum displayed in Fig. 10.3(b), we see that the MSW effect leads to an overall suppression of expected ν_e probabilities in the 30%–50% range, with larger suppression associated with higher-energy neutrinos. Comparing this with the experimental neutrino anomalies that are summarized in Table 10.3 suggests a resolution of the solar neutrino problem that will be elaborated further in the remainder of this chapter.

12.9 Resolution of the Solar Neutrino Problem

Observations, coupled with the preceding oscillation and MSW matter resonance formalism, have resolved the solar neutrino problem. Comparison with solar neutrino data indicates that electron neutrinos are being converted to other flavors by neutrino oscillations, that if all flavors of neutrinos are detected the solar neutrino deficit disappears, and that the favored scenario is MSW resonance conversion in the Sun, but for a *large* vacuum mixing angle solution. Let us now describe the observations leading to these remarkable conclusions.

---

[5] The lowest energy for which the MSW resonance can occur of 0.7 MeV in Table 12.1 differs somewhat from the 1.6 MeV inferred from Fig. 12.9 and Problem 12.3. The discrepancy is because Table 12.1 (and Fig. 12.11) have for convenience used the exponential approximation illustrated by the dashed line in Fig. 12.9(a) for the central density.

12.9.1 Super-K Observation of Flavor Oscillation

High-energy cosmic rays hitting the atmosphere generate showers of mesons that decay to muons, electrons, positrons, and neutrinos. The Super Kamiokande (Super-K) detector in Japan was used to observe neutrinos produced in these atmospheric cosmic-ray showers. Super-K found that the ratio of muon neutrinos plus antineutrinos to electron neutrinos plus antineutrinos was only 64% of that predicted by the Standard Model [92], and explained the discrepancy as resulting from oscillation of muon neutrinos into another flavor. Analysis suggests that the oscillation partner of the muon neutrino is not the electron neutrino, so ν_μ is oscillating with the tau neutrino [or possibly with an unknown flavor of neutrino that does not undergo normal weak interactions but does participate in neutrino oscillations (*sterile neutrinos*)]. The best fit to the data suggested a mixing angle close to maximal (a *large mixing angle solution*) and a mass squared difference in the range $\Delta m^2 \simeq 5 \times 10^{-4}$ –6×10^{-3} eV^2. The large mixing angle indicates that the mass eigenstates are approximately equal mixtures of the two weak flavor eigenstates.

12.9.2 SNO Observation of Neutral Current Interactions

The Super-K results cited above indicate conclusively the existence of neutrino oscillations and thus of physics beyond the Standard Model, but the detected oscillations do not appear to involve the electron neutrino and so cannot be applied directly to the solar neutrino problem. However, a water Cherenkov detector in Canada yielded information about neutrino oscillations that *is* directly applicable to the solar neutrino problem.

SNO and heavy water: The Sudbury Neutrino Observatory (SNO), sited in an active nickel mine north of Lake Huron, was a water Cherenkov detector like Super Kamiokande that could detect neutrinos in the usual way by the Cherenkov light emitted from

$$\nu + e^- \rightarrow \nu + e^- \quad \text{(elastic scattering)}. \tag{12.37}$$

This reaction can occur for any flavor neutrino but ν_μ and ν_τ scattering are strongly suppressed relative to that for ν_e. However, SNO differed from Super-K in that it contained heavy water (water enriched in ^{2}H) at its core. The heavy water was important because of the deuterium that it contains. In regular water, the relatively low-energy solar neutrinos signal their presence only by elastic scattering from electrons as in Eq. (12.37), and to get sufficient Cherenkov light for reliable detection the neutrino energy has to be greater than about 5–7 MeV. However, because deuterium (d) contains a weakly bound neutron, it can break up when struck by a neutrino in two ways. Through the *weak neutral current*, any flavor neutrino can initiate

$$\nu + d \rightarrow \nu + p + n \quad \text{(neutral current)}, \tag{12.38}$$

but only electron neutrinos can initiate the *weak charged current reaction*

$$\nu_e + d \rightarrow e^- + p + p \quad \text{(charged current)}. \tag{12.39}$$

Both of these reactions have much larger cross sections than those for elastic neutrino–electron scattering, so SNO could gather events at high rates using a small heavy-water volume. In addition, the energy threshold could be lowered to 2.2 MeV, the binding energy of the deuteron. Finally – and most importantly – because neutral currents are flavor-blind reaction (12.38) allowed SNO to see the *total neutrino flux of all flavors*, not just that of the electron neutrinos.[6]

The total solar neutrino flux: Because of its energy threshold, SNO saw primarily ^{8}B solar neutrinos. The initial SNO data confirmed results from the pioneering solar neutrino experiments of Davis and others: a strong suppression of the electron neutrino flux was observed relative to that expected in the Standard Solar Model, assuming no new physics in the Standard Model of elementary particle physics [21]. Specifically, SNO found that only about $\frac{1}{3}$ of the expected ν_e were being detected. However, SNO went further. By analyzing the flavor-blind weak neutral current events, the total flux of all neutrinos in the detector was shown to be almost exactly that expected from the Standard Solar Model. Figure 12.12(a) illustrates the flux of neutrinos from the ^{8}B reaction, based on SNO results from the three reactions in Eqs. (12.37)–(12.39). The best overall fit indicates that $\frac{2}{3}$ of the Sun's electron neutrinos have changed flavor by the time they reach Earth. Table 12.2 summarizes the comparison of SNO results and the Standard Solar Model: although the electron neutrino flux is only 35% of that expected in a model without oscillations, the neutrino flux summed over all flavors is 100% of that predicted by the Standard Solar Model for electron neutrino emission, within the experimental uncertainty.

SNO mixing solution: For solar neutrinos it is common to assume a 2-flavor mixing model and to report confidence-level contours in a 2-dimensional plane with Δm^2 on one axis and $\tan^2\theta$ on the other. Figure 12.12(b) shows the best-fit confidence-level contours for parameters based on SNO data, which suggest that the solar neutrino problem is solved by ν_e–ν_μ flavor oscillations with

$$\Delta m^2 = 6.5^{+4.4}_{-2.3} \times 10^{-5}\,\text{eV}^2 \qquad \theta = 33.9^{+2.4°}_{-2.2°}. \tag{12.40}$$

This large-mixing-angle solution implies that ν_e is almost an equal superposition of two mass eigenstates, separated by no more than a few hundredths of an eV.

12.9.3 KamLAND Constraints on Mixing Angles

KamLAND (Kamioka Liquid Scintillator Anti-Neutrino Detector) was sited in the same Japanese mine cavern that housed Kamiokande, the predecessor to Super Kamiokande. It used phototubes to monitor a large container of liquid scintillator, looking specifically

---

[6] Thus, the essence of the SNO method was to detect neutrinos in *two parallel ways,* one sensitive to ν_e and one sensitive to the total flux of all flavors. The reaction (12.38) can be detected when the neutron produced is captured in an (n, γ) reaction on a nucleus in the water. This measurement requires extremely low background and the deep-underground siting of SNO led to a cosmic-ray muon background that was about 1000 times smaller than that of Super Kamiokande. The heavy water was contained in a 12-meter diameter acrylic sphere in the center of SNO and was borrowed from the Canadian nuclear reactor program; it was estimated to have a value of 300 million Canadian dollars.

Table 12.2 SNO and Standard Solar Model (SSM) neutrino fluxes [21, 22]				
SSM ν_e flux	SNO ν_e flux	SNO ν_e/SSM	SNO all flavors	SNO all/SSM
5.05 ± 0.91	1.76 ± 0.11	0.348	5.09 ± 0.62	1.01

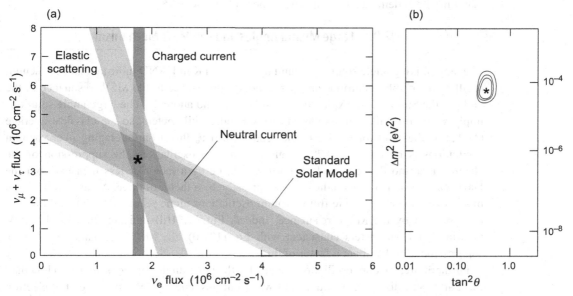

Fig. 12.12 (a) Flux of ^{8}B solar neutrinos detected for various flavors by SNO [21]. The three bands correspond to the results from the three reactions in Eqs. (12.37)–(12.39), with the band widths indicating one standard deviation. The bands intersect at the point indicated by the star, which implies that about $\frac{2}{3}$ of the Sun's ^{8}B neutrinos have changed flavor between being produced in the core of the Sun and being detected on Earth. The Standard Solar Model band is the prediction for the ^{8}B flux, irrespective of flavor changes. It tracks the neutral current band, which represents detection of all flavors of neutrino coming from the Sun. Adapted with permission for Q. R. Ahmad et al. (SNO Collaboration), *Phys. Rev. Lett.*, **89**, 011301. Copyrighted by the American Physical Society, 2002. (b) 2-flavor neutrino oscillation parameters determined by SNO [22]. The 99%, 95% and 90% confidence-level contours are shown, with the star indicating the most likely values. The best fit corresponds to the large-angle solution given in Eq. (12.40). Adapted with permission for Q. R. Ahmad et al. (SNO Collaboration), *Phys. Rev. Lett.*, **89**, 011302. Copyrighted by the American Physical Society, 2002.

for $\bar{\nu}_e$ produced during nuclear power generation in a set of Japanese and Korean reactors located within a few hundred kilometers of the detector. The antineutrinos were detected from inverse β-decay in the scintillator: $\bar{\nu}_e + p \rightarrow e^+ + n$. From the power levels in the reactors, the expected antineutrino flux at KamLAND could be modeled accurately. A shortfall of antineutrinos relative to the expected number was observed and this could be explained assuming (anti)neutrino oscillations with a large-angle solution having [19, 80]

$$\Delta m^2 = 7.58^{+0.14}_{-0.13} \times 10^{-5} \text{ eV}^2 \qquad \tan^2 \theta = 0.56^{+0.10}_{-0.07} \qquad (12.41)$$

(statistical uncertainties only), which corresponds to $\theta \sim 36.8°$ for the vacuum mixing angle.[7] Combining the solar neutrino and KamLAND results leads to a solution [19]

$$\Delta m^2 = 7.59 \pm 0.21 \times 10^{-5} \, \text{eV}^2 \qquad \tan^2 \theta = 0.47^{+0.06}_{-0.05}, \qquad (12.42)$$

implying a vacuum mixing angle $\theta \sim 34.4°$, which is a *large mixing angle solution* (recall that θ has been defined so that its largest possible value is $45°$).

12.9.4 Large Mixing Angles and the MSW Mechanism

The large mixing angle solutions found by SNO and KamLAND indicate that the vacuum oscillations of solar neutrinos are of secondary importance to the MSW resonance in the body of the Sun itself in explaining the solar neutrino anomaly. The large-angle solutions imply vacuum oscillation lengths of a few hundred kilometers, so the classical average (11.12) applies and for $\theta \sim 34°$ the reduction in ν_e flux from averaging over vacuum oscillations is by about $\sim 57\%$. Thus for vacuum oscillations the suppression of the electron neutrino flux detected on Earth would be by a factor of less than two, but the Davis chlorine experiment indicates a suppression by a factor of three. A flavor conversion more severe than is possible from vacuum oscillations alone seems required, and this can be explained by the MSW resonance. Indeed, Fig. 12.11(d) indicates that for 10 MeV neutrinos and parameters consistent with Eq. (12.40), the MSW resonance leads to a suppression of ν_e by about a factor of three.

Furthermore, vacuum oscillation lengths for the large-angle solutions are much less than the Earth–Sun distance, which would wash out any energy dependence of the electron neutrino shortfall. Since the observations indicate that such an energy dependence exists (see Table 10.3), and the MSW effect implies such an energy dependence (see Fig. 12.9 and Table 12.1), the MSW effect is implicated as the primary source of the neutrino flavor conversion responsible for the "solar neutrino problem." That is, the MSW effect converts from $\frac{1}{2}$ to $\frac{2}{3}$ (depending on neutrino energy) of ν_e into other flavors within the Sun, and these populations are then only somewhat modified by vacuum oscillations before the solar neutrinos reach detectors on Earth. (If the neutrinos pass through the Earth on the way to the detectors there will be some flavor modification in the Earth, but that is expected to be a small effect.) The predicted energy dependence for MSW flavor conversion has been confirmed by *Borexino*, which is a liquid scintillator experiment in the Gran Sasso underground laboratory featuring very strong background suppression, thus allowing the neutrino flux from PP-I, PP-II, and PP-III to the measured separately.

12.9.5 A Tale of Large and Small Mixing Angles

The MSW effect was invoked originally to explain how a *small mixing angle* could explain the electron solar neutrino deficit, since initial theoretical prejudice favored a

---

[7] The oscillation properties of neutrinos and antineutrinos of the same generation are expected to be related by CPT symmetry. Thus the large-angle KamLAND solution for electron antineutrinos may be interpreted as corroboration of the large-angle solution found for solar neutrinos.

small vacuum mixing angle and Fig. 12.11 indicates that the MSW resonance can generate almost complete flavor conversion, even for small vacuum mixing angles. However, data now indicate that the MSW effect is the solution of the solar neutrino problem, but for a *large mixing angle solution*. Thus, in this tale a correct physical idea, with some initially incorrect assumptions, led eventually to a surprising resolution of a fundamental problem, with far-reaching implications for both astrophysics and elementary particle physics.

Background and Further Reading

Portions of this chapter have been adapted from an introduction to solar neutrinos by Guidry and Billings [102]. The book by Bahcall [30] is a good introduction to both the Standard Solar Model and to the significance of solar neutrinos. A general introduction to the Standard Model of elementary particle physics may be found in Collins, Martin, and Squires [74] or in Guidry [97], and a summary of important neutrino physics is given in the reviews by Haxton and Holstein [113, 114]. Overviews of the interaction of electron neutrinos with matter and the associated MSW effect that assume some knowledge of quantum field theory may be found in Blennow and Smirnov [49], Kuo and Pantaleone [138], and Smirnov [203]. The solar neutrino problem is reviewed in Haxton, Robertson, and Serenelli [115].

Problems

12.1 Restore factors of c in Eq. (12.27) to obtain a formula for the MSW resonance critical electron number density n_e^R in units of cm^{-3}. Assuming that $\Delta m^2 = 7.5 \times 10^{-5}$ eV^2, $E = 10$ MeV, and $\theta = 35°$, compute n_e^R. Using the Standard Solar Model results given in Table 10.1, at what fraction of the central density does the resonance occur and approximately at what radial distance from the center is this density realized in the Sun? For this estimate assume that the mass density and electron density track each other over the range of interest.

12.2 Using the density information in Fig. 12.9, estimate the value of ℓ_m at the center of the Sun. What is the value of $\chi = L/\ell_m$ there if $E = 10$ MeV and $\Delta m^2 c^4 = 7.5 \times 10^{-5}$ eV^2?***

12.3 The MSW resonance can occur only if an electron neutrino passes through a region with density equal to the critical density n_e^R after its production. Derive a formula for the minimum neutrino energy E_{min} that can encounter the resonance for fixed θ, ΔE^2, and maximum (central) electron density n_e^0 for a star. Using the Standard Solar Model with vacuum mixing angle $\theta = 35°$ and $\Delta m^2 c^4 = 7.5 \times 10^{-5}$ eV^2, what is the minimum energy of an electron neutrino that could be affected substantially by the MSW resonance for the Sun?***

12.4 The additional effective squared mass gained by electron neutrinos interacting with the electrons of a medium is given by $A = 2EV = 2\sqrt{2}G_F E n_e$, where V is given in Eq. (12.1). Derive a formula for A in terms of the mass density ρ, neutrino energy E, and electron fraction Y_e (ratio of electrons to nucleons), and evaluate for a 10 MeV neutrino at the center of the Sun using data from the Standard Solar Model. Compare this calculated increase in effective mass squared A with the difference Δm^2 between vacuum mass eigenstates that are dominantly electron and muon neutrino flavors in Fig. 11.3.***

12.5 Using Eq. (12.33) and the exponential parameterization of the solar electron number density given in Fig. 12.9(a), show that the adiabatic condition $\delta r_R \gg L_m^R$ is well satisfied for realistic values of the parameters in the Sun.

12.6 Some authors give the relationship between the vacuum mixing angle θ and the MSW matter mixing angle θ_m as

$$\sin^2 2\theta_m = \frac{\sin^2 2\theta}{(\cos 2\theta - \chi)^2 + \sin^2 2\theta},$$

where $\chi = L/\ell_m$. Show that this is equivalent to the relationship given in (12.20). Prove that at the MSW resonance the mixing length in matter is given by $L_m = L/\sin 2\theta$.

12.7 (a) Demonstrate that a similarity transform $S^{-1}MS$ on a matrix M by a matrix S does not change the eigenvalues of the matrix M. *Hint*: The determinant of a matrix product is the product of determinants for the individual matrices. (b) Show that a 2×2 matrix can be made traceless by adding a multiple of the unit matrix to it; find the required multiple for a general 2×2 matrix.

12.8 Prove that the eigenvalues of the mass matrix (12.14) are given by Eq. (12.16). *Hint*: See Problem 11.5.***

12.9 Show by inserting explicit forms for the matrices in Eq. (12.19) that for neutrinos the matter mixing angle θ_m is related to the vacuum mixing angle θ by Eq. (12.20),

$$\tan 2\theta_m = \frac{\sin 2\theta}{\cos 2\theta - \chi},$$

where $\chi = L/\ell_m$.***

12.10 Show that Eq. (12.30) leads to Eq. (12.31) for neutrino propagation through a region of changing density.***

12.11 Prove that in adiabatic approximation the classical probability (averaged over oscillations) that an electron neutrino produced at Point 1 near the center of the Sun would be detected as an electron neutrino at Point 2 is given by Eq. (12.35). *Hint:* Start with Eq. (12.34).***

12.12 Demonstrate that if all matter is removed so that the matter mixing angle becomes equal to the vacuum mixing angle, $\theta_m(t) \to \theta$, Eq. (12.34) yields the classical average in vacuum given by Eq. (11.15).

12.13 Supply the missing steps in deriving Eq. (12.5) from Eq. (12.4), assuming ultrarelativistic neutrinos.***

12.14 Prove that Eq. (12.12) follows from Eq. (12.11).***

12.15 Show that Eq. (12.12) can be written in the traceless form (12.14) by subtracting a multiple of the unit matrix and using the definitions given in Eq. (12.15). *Hint*: The solution of Problem 12.7(b) is useful for this problem.

13 Evolution of Lower-Mass Stars

Life on the main sequence is characterized by the stable burning of core hydrogen to helium in hydrostatic equilibrium. While on the main sequence the inner composition of a star is changing but there is little outward evidence of this until about 10% of the original hydrogen is exhausted and the star experiences a rapid evolution away from the main sequence. Although the period of stellar evolution after the main sequence is short compared with the time spent on the main sequence, post main sequence evolution is more complex than main sequence evolution and is relevant for understanding a number of important topics in astrophysics. Accordingly, we turn now to a discussion of stellar evolution after the main sequence and the associated timescales introduced by ignition of thermonuclear fuels more massive than hydrogen. In this chapter the post main sequence fate of lower-mass stars [which will be taken to be stars with initial main sequence (ZAMS) mass less than about $8\,M_\odot$] will be discussed, while in Chapter 14 the evolution of higher-mass stars with initial mass $M \gtrsim 8\,M_\odot$ will be considered.

13.1 Endpoints of Stellar Evolution

The primary distinction between lower-mass and higher-mass stars for our purposes will be that the endpoint of stellar evolution is (1) a white dwarf plus expanding planetary nebula for a lower-mass star, or (2) a core-collapse event that leaves behind a neutron star or black hole, plus expanding debris (supernova explosion)—or a black hole with no significant additional remnant (failed supernova)—for a higher-mass star.[1] Most simulations indicate that the minimum ZAMS mass required for the stellar core to collapse at the end of a star's life is in the 8–$10\,M_\odot$ range, with some uncertainty because of gaps in our understanding of stellar evolution, and because a massive star can lose mass to stellar winds during main-sequence evolution. For this discussion the boundary between higher-mass and lower-mass stars will be taken as $M \sim 8\,M_\odot$, with some fuzziness implied because of the above uncertainties. Note that this definition of lower-mass stars includes the stars with very low main sequence mass ($M \lesssim 0.5\,M_\odot$). Although theoretically these stars are expected to evolve to helium white dwarfs, the pace of evolution for them is so slow that few could have

---

[1] At a more technical level it is common to make more than two mass-range distinctions in discussing stellar evolution, but for our purposes dividing stars into those that evolve to some form of white dwarf (which we shall term lower-mass) and those that evolve to some form of core-collapse event (which we shall term higher-mass) seems most natural.

left the main sequence since the birth of the Universe. Hence there is limited observational evidence and the present discussion will mention them only in passing.

13.2 Shell Burning

An important characteristic of post main sequence evolution is the establishment of *shell burning sources*, as illustrated in Figs. 13.1 and 13.2. As the initial core hydrogen is depleted a thermonuclear ash of helium builds up in its place. This ash is inert under hydrogen burning conditions because much higher temperatures and densities are necessary to ignite helium. However, as the core becomes depleted in hydrogen there remains a concentric shell in which the hydrogen concentration and the temperature are

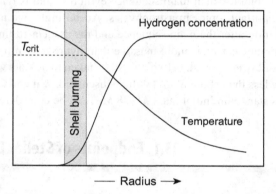

Fig. 13.1 Conditions for hydrogen shell burning. Below the shell there is no hydrogen fuel to burn and above it the temperature is too low to burn hydrogen.

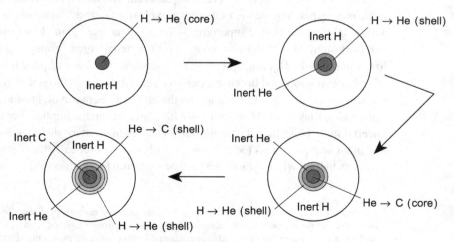

Fig. 13.2 Schematic illustration of successive shell burnings for stars after the main sequence.

both sufficiently high to support hydrogen burning (Fig. 13.1). This is termed a *hydrogen shell source*. As the core contracts after exhausting its hydrogen fuel the temperature and density rise and this may eventually ignite helium in the core. As the helium burns in the core a central ash of carbon is left behind that is inert because much higher temperatures are required to burn it to heavier elements. This is termed *core helium burning*. However, just as for hydrogen, once sufficient carbon ash has accumulated in the core the helium burning will be confined to a concentric shell surrounding the inert core; this is termed a *helium shell source*.

If the star is sufficiently massive the preceding scenario may be repeated for successively heavier core and shell sources, as illustrated in Fig. 13.2. Shell and core sources need not be mutually exclusive. More massive stars may have at a given time only a core source, only a shell source, multiple shell sources, or a core source and one or more shell sources, and these sources can have complex instabilities and interactions. In the schematic example of Fig. 13.2 the third stage involves both a helium core source and a hydrogen shell source, and Fig. 13.6 below illustrates successive burning stages involving core and shell sources in various combinations. The *mirror principle* discussed in Box 13.1 is a useful qualitative guide for understanding shell sources.

Box 13.1 Mirror Response of Mass Shells

An important aspect of stellar evolution in the presence of shell energy sources is termed the *mirror principle*. Experience indicates that shell sources tend to produce "mirror" motion of mass shells above and below them, as illustrated below.

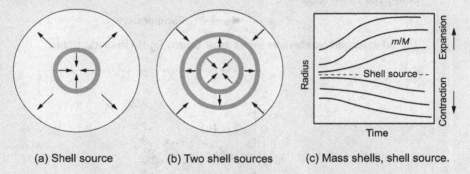

(a) Shell source   (b) Two shell sources   (c) Mass shells, shell source.

For example, if there is a single shell source the mass layers below the shell source tend to contract because it has no central energy source and the mass layers above the shell source tend to expand because contraction of the core releases heat, as illustrated in (a) and (c). For two shell sources, each tends to mirror the mass shells above and below, as illustrated in (b). In the absence of core burning, with two shell sources the core tends to contract, so by the mirror principle the layers above the inner shell source tend to expand (moving the second shell source further outward). Applying the mirror principle to the outer shell source, the layers outside the outer shell source will tend to contract. Motion in the HR diagram that is observed in late stellar evolution simulations often can be understood simply in terms of this principle of mirrored motion when there are shell sources.

13.3 Stages of Red Giant Evolution

As discussed in Section 2.3, globular clusters have HR diagrams differing substantially from those for stars near the Sun or for open clusters. In Section 2.6 this was interpreted as evidence that globular clusters are old and that these differences are connected with the time evolution of star populations. Those qualitative remarks may now be placed on a much firmer footing. The most distinctive features of the HR diagrams for old clusters are (1) absence of main sequence stars above a certain luminosity, and (2) loci of

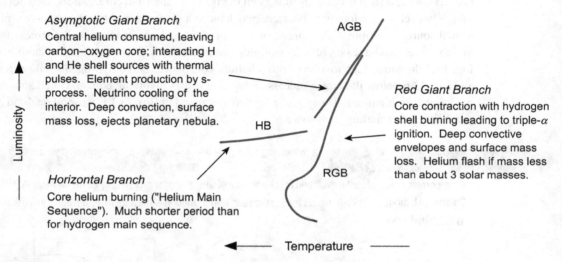

Fig. 13.3 Schematic giant branches in an evolved cluster. Compare Fig. 13.4 for an actual cluster.

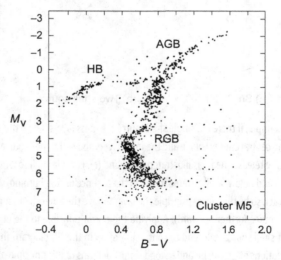

Fig. 13.4 Giant branches for the globular cluster M5 in Serpens. Compare with the schematic branches of Fig. 13.3. Redrawn from www.messier.seds.org/xtra/leos/msf4.html by Leos Ondra.

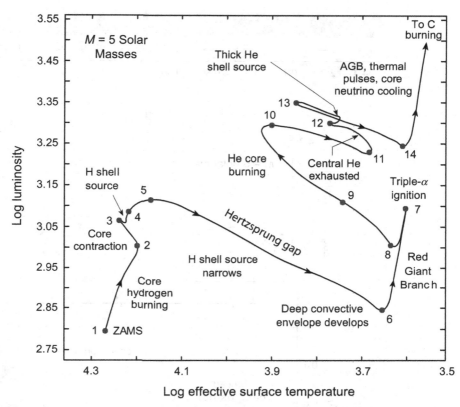

Fig. 13.5 Evolution from the main sequence to the asymptotic giant branch for a metal-rich 5 solar mass star (adapted from Ref. [125]). Luminosities are in solar units and the effective temperature is in kelvin. The times to evolve through each numbered interval are given in Table 13.1. Adapted from Icko Iben, Jr., *Annual Review of Astronomy and Astrophysics*, **5**, 57–62b (1967), https://doi.org/10.114b/annurev.aa.05.090167.003035.

enhanced populations in the giant region termed the *red giant branch* (RGB), the *horizontal branch* (HB), and the *asymptotic giant branch* (AGB), that are illustrated schematically in Fig. 13.3 and for an actual cluster in Fig. 13.4. Regions of enhanced population in the HR diagram indicate that the individual stars of a population spend proportionally larger fractions of their lives there. As we will now discuss, the red giant, horizontal, and asymptotic branches can be identified with distinct periods of post main sequence evolution that are short compared with the time spent on the main sequence, but long compared with the stages in between.

As a representative case, consider the calculated evolution of a 5 solar mass star, as illustrated in Figs. 13.5 and 13.6, and Table 13.1. Beginning at the Zero Age Main Sequence (ZAMS), the star converts hydrogen to helium and this ongoing change in the core composition causes a small upward drift on the HR diagram for the main sequence period (characterized by core hydrogen burning). As its hydrogen is depleted the core contracts and eventually a *hydrogen shell source* is established. These events signal a rapid departure from the main sequence that will now be followed in some detail.

Table 13.1 Duration of intervals for Fig. 13.5 in units of 10^6 yr [125]										
1→2	2→3	3→4	4→5	5→6	6→7	7→8	8→9	9→10	10→11	11→13
64.4	2.2	0.14	1.2	0.80	0.50	6.0	1.0	9.0	2.0	∼2.5

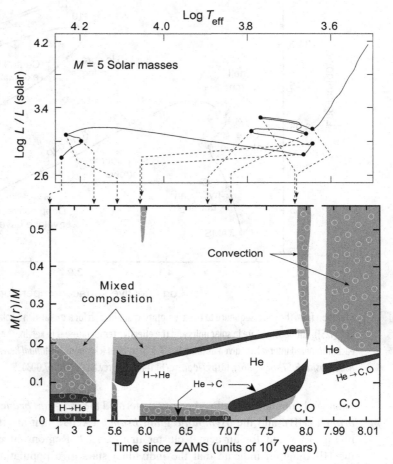

Fig. 13.6 Evolution of a 5 solar mass star after the main sequence. Darkest shading indicates regions of energy production and regions with circles are convective. Note the breaks in scale for the time axis. Adapted from Ref. [134]. Adapted by permission from Springer Nature, *Evolution Through Helium Burning: Massive Stars*, Kippenhahn, R. and Weigert A. COPYRIGHT (1990).

13.4 The Red Giant Branch

Over time the hydrogen shell source burns its way outward, leaving in its wake a helium-rich ash. Because the sole thermonuclear energy source at this point is in a concentric shell, the core cannot maintain a thermal gradient and it equilibrates in temperature. Such *isothermal cores* are characteristic of stars that have only shell energy sources.

13.4.1 The Schönberg–Chandrasekhar Limit

As the core grows because of the shell burning it is supported primarily by the pressure of the helium gas, which is typically nondegenerate and nonrelativistic. However, there is a limit to the mass of an isothermal core of helium gas that can be supported by the gas pressure. This *Schönberg–Chandrasekhar limit*, which should not be confused with the *Chandrasekhar limit* for white dwarf masses to be discussed in Section 16.2.2, is given by (see Problem 13.1)

$$M_c \simeq 0.37 \left(\frac{\mu_{\text{env}}}{\mu_c}\right)^2 M, \tag{13.1}$$

for an isothermal core of ideal helium gas, where M is the total mass of the star, M_c is the mass of the isothermal core, μ_c is the mean molecular weight in the core, and μ_{env} is the mean molecular weight in the envelope. The growth of an isothermal helium core to this size typically requires that about 10% of the original hydrogen be burned, which is one basis for the earlier qualitative statement that significant evolution from the main sequence commences when this amount of hydrogen has been consumed.

When the Schönberg–Chandrasekhar limit is reached in the core it can no longer support itself or the layers pressing down around it against gravity; the core begins to contract on a Kelvin–Helmholtz timescale, which is slow compared with the dynamical timescale but rapid compared to the nuclear burning timescale that has governed the time spent on the main sequence. The contraction will proceed until ignition of helium fusion in the core provides a stabilizing pressure gradient, or until interior densities are reached where the electron gas becomes degenerate. Provided that the core mass does not exceed about 1.4 $M_\odot$ (see the discussion in Section 16.2.2), the electron degeneracy pressure stops the contraction, but only after the core has become much hotter and more dense, and substantial gravitational energy has been released.

13.4.2 Crossing the Hertzsprung Gap

Much of the energy released in the contraction of the isothermal core is deposited in the envelope, which expands and cools, enlarging and reddening the photosphere. Thus the star evolves quickly to the right in the HR diagram. The region between the main sequence and the red giant branch (roughly between points 5 and 6 in Fig. 13.5) contains few stars and is called the *Hertzsprung gap*. A given star evolves so quickly through this region in comparison with its overall evolutionary timescale that there is little chance of observing it in the Hertzsprung gap. From Table 13.1, the 5 $M_\odot$ star spends only 8×10^5 yr in the Hertzsprung gap, compared with $\sim 9 \times 10^7$ yr for all evolutionary stages in the table. As the temperature of the envelope decreases the opacity increases and the temperature gradient exceeds the adiabatic gradient. Thus the star becomes convective in much of its envelope, with the convection typically extending from just outside the hydrogen-burning shell to the surface. As we will discuss further in Section 13.10, this deep convection can bring some nuclear-processed material to the surface, where it can be emitted from the star in the winds described below.

The continued evolution to the red giant region may be viewed as something like the inverse of the contraction on the Hayashi track of fully convective protostars to the main sequence that was discussed in Section 9.8: the almost fully convective star ascends the Hayashi track in reverse to the red giant region. The corresponding evolution in Fig. 13.5 is on the red giant branch between the points labeled 6 and 7. While on the red giant branch the greatly expanded star can exhibit significant envelope mass loss through strong winds, with mass-loss rates as large as 10^{-6} $M_\odot$ per year observed for some RGB stars.

13.5 Helium Ignition

The triple-α reaction will be triggered when the core temperature approaches $\sim 10^8$ K (see Fig. 6.6).[2] The onset of helium burning corresponds to the cusp shown in Fig. 13.5 at point number 7, and signals the end of red giant branch evolution. The ignition of the core helium is qualitatively different for stars above and below about 2 $M_\odot$, as will now be elaborated.

13.5.1 Core Equation of State and Helium Ignition

More massive stars typically have larger core temperatures than less massive stars at all stages of their evolution. Simulations indicate that more massive stars have high enough central temperatures to evolve all the way to helium burning without their cores becoming electron-degenerate. Under these conditions the onset of core helium burning is probably a rather smooth and orderly process. On the other hand, for stars of about 2 $M_\odot$ or less the core electrons will have become extremely degenerate before the triple-α sequence ignites. The Schönberg–Chandrasekhar instability is a property of an ideal gas equation of state. If the core becomes degenerate it will be isothermal (because there is no central energy source, and because degenerate matter is highly heat-conductive), but electron degeneracy pressure will support the core and the Schönberg–Chandrasekhar instability will not develop.

As discussed in Chapter 3, the equations of state for ideal gases and degenerate gases differ fundamentally in the relationship between temperature and pressure: for an ideal gas the pressure is proportional to temperature but for a degenerate gas the pressure is essentially independent of the temperature, because it derives primarily from the Pauli exclusion principle, not thermal motion. As will now be discussed, this degenerate

---

[2] The lightest stars will presumably not be able to initiate the triple-α reaction in their abbreviated red giant phases because of low initial mass and because of mass loss after expanding off the main sequence. These stars will lose their envelopes and their unburned helium cores will be left behind as helium white dwarfs.

core is stable in the absence of thermonuclear reactions but is spectacularly *unstable* if thermonuclear burning is initiated in it.

13.5.2 Thermonuclear Runaways in Degenerate Matter

Thermonuclear reactions that are triggered under degenerate-electron conditions lead to violent energy release because the reaction assumes the character of a positive-feedback runaway:

1. Ignition of the reactions releases large amounts of energy, which quickly raises the local temperature. In a normal explosion governed by an equation of state nearer to that of an ideal gas the rise in temperature causes a corresponding rise in pressure that tends to separate and cool the reactants, limiting the explosion. This does not happen under degenerate conditions because the pressure initially is not increased by the sharp rise in temperature (see the discussion in Box 19.1 and Box 3.8).
2. Since charged-particle reactions have very strong temperature dependence, the rise in temperature causes a rapid increase in the reaction rates; this in turn raises the temperature further and thus the reaction rates increase, and so on in an exponentiating runaway.
3. Because of the large (metal-like) thermal conductivity of degenerate matter, a thermonuclear runaway triggered locally in a star under degenerate conditions spreads rapidly through the rest of the degenerate matter.

This runaway continues until enough electrons are excited to states above the Fermi surface by the high temperatures to lift the degeneracy of the electron gas. The equation of state then tends to that of an ideal gas and the resulting increase of pressure with temperature moderates the reaction.

13.5.3 The Helium Flash

When such a thermonuclear runaway occurs under degenerate conditions for the triple-α reaction it is termed a *helium flash*. Simulations show that stars of less than about 2 $M_\odot$ that ignite helium burning will probably do so under degenerate conditions. Simulations indicate further that the resulting helium flash ignites the entire core of the star within seconds, that the temperature can rise to more than 2×10^8 K before the runaway begins to moderate, and that the energy release during the short flash can approach $10^{11}\ L_\odot$ (comparable to the luminosity of a galaxy)! However, these simulations also indicate that this extremely violent event probably has little directly visible external effect because the enormous energy release is almost entirely absorbed in expanding the envelope. Once the degeneracy of the core is lifted following the helium flash, or following the more orderly initiation of the triple-α reaction in heavier nondegenerate cores, the core helium burns steadily to carbon at a temperature of about 1.5×10^8 K. This signals the beginning of the horizontal branch phase of red giant evolution.

13.6 Horizontal Branch Evolution

The horizontal branch (HB) of Fig. 13.3 corresponds to a period of stable core helium burning that is in many ways analogous to the core hydrogen-burning main sequence (this corresponds to points 8–10 in Fig. 13.5). Indeed, this period is sometimes termed the *helium main sequence*.

13.6.1 Life on the Helium Main Sequence

Life on the helium-burning main sequence is a time of near hydrostatic equilibrium for the same reasons that the hydrogen-burning main sequence corresponds to approximate hydrostatic equilibrium. However, for a given star its helium-burning main sequence is much shorter than its hydrogen-burning main sequence, for two reasons:

1. The nuclear burning timescale for helium is intrinsically shorter than for hydrogen because burning of helium to carbon and oxygen releases only about 10% of the energy per unit mass released by burning hydrogen to helium.
2. The red giant star is more luminous than when it was on the main sequence (see Fig. 13.5). Sustaining this luminosity in equilibrium demands rapid fuel consumption.

For a particular star, these factors can make the time spent on the helium main sequence an order of magnitude less than the time spent on the hydrogen-burning main sequence. The following example illustrates.

Example 13.1 For evolution of the 5 $M_\odot$ star shown in Fig. 13.5 the simulation indicates that approximately 10^7 years are spent on the horizontal branch (see Table 13.1), while the same star spends about 7×10^7 years on the main sequence. So in this case the time on the main sequence is about seven times greater than the time spent on the helium main sequence (horizontal branch).

As discussed in Section 6.3.4, helium burning is extremely sensitive to temperature. Thus, for the same reasons that CNO burning on the main sequence tends to produce convective cores (see Section 7.10.1), on the helium main sequence core helium burning tends to occur under convective conditions. The star remains on the horizontal branch while there is helium fuel in its core to burn, steadily accumulating a carbon–oxygen ash in the center.

13.6.2 Leaving the Horizontal Branch

When the core helium is exhausted, the core contracts, a thick helium-burning shell is established, and the convection in the core is quenched. The star now has two shell sources: the broader helium-burning shell at the boundary of the C–O core, and the thin hydrogen-burning shell lying above it at the base of the hydrogen envelope. In accordance with the mirror property for the motion of mass shells around shell sources described in Box 13.1,

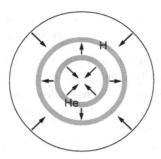

Fig. 13.7 Mirror principle applied to helium and hydrogen shell sources during horizontal-branch evolution.

the core inside the helium source contracts, the inert helium layer outside the helium shell source expands, pushing the hydrogen shell source to larger radius, and the outer layers of the star contract (Fig. 13.7). This causes the star to move to the left on the HR diagram and represents the evolution between points 11 and 13 in Fig. 13.5.

The helium shell source narrows and strengthens as the core compresses. Layers above the helium shell source expand and cool, turning off the hydrogen shell source that was burning above the helium shell source (temporarily; it will reignite later), leaving only a single active shell source. The degeneracy of the core increases as the core contracts and the core becomes isothermal since degenerate matter is highly conductive and there is no central energy source. In accordance with Box 13.1 the star with its single shell source contracts inside the helium source and expands outside it, drifting quickly to the right in the HR diagram until it approaches the Hayashi track (point 14 in Fig. 13.5). This signals the transition to the asymptotic giant branch (AGB) of Fig. 13.3.

13.7 Asymptotic Giant Branch Evolution

In many respects the evolution on the asymptotic giant branch now mimics that following the establishment of the first hydrogen shell source after core hydrogen was depleted on the main sequence. However, the star now has an electron-degenerate C–O core and two potential shell sources rather than one. (The hydrogen source turned off because of the helium shell source but it will re-ignite in later evolution on the asymptotic giant branch.) The temporary quenching of the hydrogen shell source leaves only a single (He) shell source and the outer layers continue to expand rapidly. The star again increases in luminosity and radius and moves into the red giant region as earlier, but at even higher luminosities on the asymptotic giant branch of Fig. 13.3; it now becomes a bright red giant or supergiant star. The corresponding evolution in Fig. 13.5 is from point 14 and beyond. This may be viewed approximately as a continuation of the ascent on the RGB along the Hayashi track that was interrupted by ignition of the core helium source and stabilization for a time on the horizontal branch. If the star is sufficiently massive the growing carbon core may ignite eventually, but if $M \lesssim 4\text{–}5\,M_\odot$ this is not likely and all subsequent thermonuclear energy production will be from the shell sources. A number of important features characterize asymptotic giant branch evolution:

1. The brightness of the star, and thus the height of the ascent on the AGB, is found to depend primarily on the mass of the C–O core and not the envelope mass [176].
2. The shell sources exhibit instabilities called *thermal pulses*.
3. Shell sources in AGB stars are thought to be the primary site for the slow neutron capture or *s-process*.
4. Stars in the red giant region often exhibit *large mass loss*. This is particularly true for AGB stars.
5. *Deep convective envelopes* form in the AGB phase that can dredge elements synthesized in the interior up to the surface, where they can be distributed to the interstellar medium by winds from the surface.

Each of these important aspects of asymptotic giant branch evolution will now be discussed in more depth.

13.7.1 Thermal Pulses

The AGB period is characterized by the presence of both hydrogen and helium shell sources. However, these shell sources exhibit instabilities and a complex interrelationship such that they are unable to coexist in equilibrium and at any one time often only one of the two shell sources is burning. These instabilities in AGB stars lead to what are called *thermal pulses* or *helium shell flashes*.

Assume an inert C–O core surrounded by an inert He layer, with a hydrogen shell source at the base of the hydrogen layer above adding to the He layer, as illustrated in Fig. 13.8(a). There is no energy source inside the hydrogen shell source so the He layer grows and compresses. This eventually ignites the base of the helium layer, giving an inner He shell source and an outer H shell source. But the expansion of the layers above the hot He shell source lowers the temperature at the base of the hydrogen envelope and turns off the H shell source, leaving the star with a single He shell source, as in Fig. 13.8(b). The hot helium shell source produces a steep temperature gradient and deep convective motion develops that reaches down to the vicinity of the He shell source, as illustrated in Fig. 13.8(c). This convection mixes burning products from earlier evolution into the surface layers. The He shell source burns outward, leaving a growing C–O core behind. It eventually extinguishes because of insufficient temperature at larger radius, but not before the proximity of the hot He source re-ignites the shell source at the base of the hydrogen layer, leading to the situation in Fig. 13.8(d). The hydrogen shell source burns outward, leaving behind a new layer of helium and the cycle is repeated, but now with a larger C–O core.

In the complicated tango between the shell sources that defines the cycle in Fig. 13.8, the hydrogen shell source burns in rather stable fashion but the He shell source can be very unstable because of the strong temperature dependence for He burning, and because thin-shell sources are inherently unstable, as described in Box 13.2. Thus when the helium shell

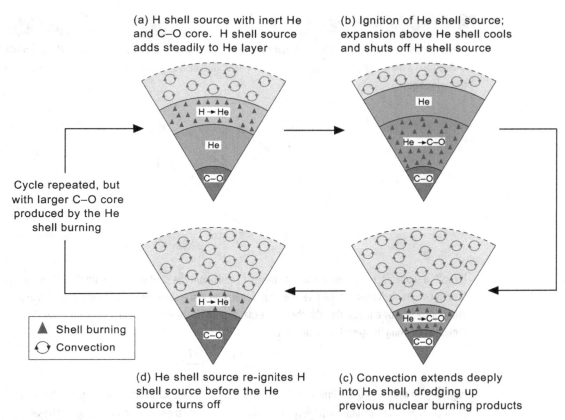

Fig. 13.8 Schematic illustration of the relationship between H and He shell burning that leads to thermal pulses in an AGB star (not to scale). Adapted from figures in Ref. [176].

ignites it can burn explosively in a short-lived shell flash (thermal pulse). The resulting shell luminosity may reach $\sim 10^8 \, L_\odot$, but the energy is almost all absorbed in the outer layers of the star. Simulations indicate that an AGB star can undergo as many as hundreds of thermal pulses before the envelope is eroded away by mass loss (see Section 13.7.4). The duration of a thermal pulse is very short, so it is difficult to catch an AGB star undergoing one.

About a quarter of AGB stars are predicted to undergo one final helium shell flash after hydrogen burning has ceased. This *late thermal pulse* occurs after the star has ejected most of its envelope as a planetary nebula and is settling into the white dwarf phase (see Section 13.9). Computer simulations of this event suggest that in such a star the helium shell can re-ignite and the small remaining hydrogen envelope can be convectively mixed into the helium shell, leading to additional rapid hydrogen-driven flash burning and renewed mass ejection. The evolution in the Hertzsprung–Russell diagram of a star that may have been captured undergoing a late thermal pulse while cooling from the AGM to white dwarf phase is described in Box 13.3.

> **Box 13.2**   **Stability of Thin Shell Sources**
>
> Consider a thin shell source in a star of radius R and mass M. The source is assumed to be in thermal equilibrium and to have a mass Δm, density ρ, temperature T, and thickness $L = r - r_0 \ll R$, as illustrated in the following figure.
>
>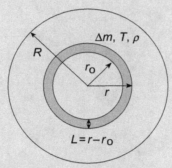
>
> If this is a single energy-producing shell in thermal equilibrium the shell is stable, with the rate of energy flow out of the shell equal to the rate of energy generation in the shell. However, if the energy-generation rate is increased by a fluctuation the shell will expand, pushing the layers above it outward. As explored in Problem 13.4 using the generic equation of state
>
> $$\frac{dP}{P} = \alpha \frac{d\rho}{\rho} + \beta \frac{dT}{T}$$
>
> with $\alpha \geq 0$ and $\beta \geq 0$ (see Problem 4.16), stability of the shell for such fluctuations requires that $4L/r > \alpha$. But α is positive and finite, so for very thin shells $L/r \to 0$ and the stability condition cannot be satisfied. This is called the *thin-shell instability*; it may be significant for AGB stars since they often develop thin shell sources. Physically the stability requirement $4L/r > \alpha$ implies that if a shell source is narrow enough the temperature *increases* upon expansion, which is strongly destabilizing and sets the stage for a thermonuclear runaway. Therefore, in many respects a thin shell source behaves like a degenerate gas with regard to thermal stability, even if the gas in the shell source is not degenerate. Thin He shell sources are particularly unstable because helium is a very explosive thermonuclear fuel (see Section 6.3.4).

13.7.2 Slow Neutron Capture

Figure 13.9 summarizes the abundances observed for the elements as a function of atomic mass number. It was shown in Chapter 6 that the elements up to iron can be produced by fusion reactions and by nuclear statistical equilibrium in stars. But what of the elements beyond iron? They cannot be produced in the same way because the Coulomb barriers become so large that extremely high temperatures would be required to force heavier nuclei to react. These high temperatures would produce a bath of photons having such high energy that they would photodisintegrate any heavier nuclei that were formed. Thus, other mechanisms must be responsible for producing heavier elements. One possibility is the

Box 13.3   Sakurai's Object and Thermal Pulses

Late thermal pulse events in asymptotic giant branch stars are expected to be rare, with a predicted rate of only about one per decade in our galaxy.

Thermal Pulses of V4334 Sgr

V4334 Sgr (*Sakurai's Object*) may be a star caught undergoing a late thermal pulse. Since discovery in 1996, it has exhibited rapid evolution on the HR diagram, accompanied by substantial mass ejection. Data (dots) and model simulation (curves) of the evolution in the HR diagram are illustrated in the following figure [104],

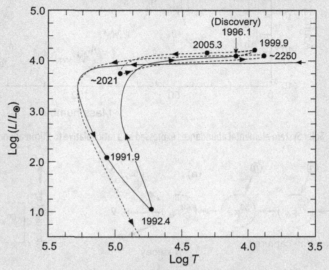

Reprinted from "The Real-Time Stellar Evolution of Sakurai's", Marcin Hajduk et al., *Science*, **308**(5719), 231–233, 2005, DOI:10.1126/science.1108953

with a solid curve indicating the prediction of the model for evolution before discovery and a dashed curve indicating the prediction for evolution afterwards.

Rapid Looping in the HR Diagram

In this evolution two loops in the HR diagram are expected. The first is fast and associated with rapid burning of hydrogen ingested into the helium layer. The second, slower loop corresponds to the helium flash. These loops imply variations in the surface temperature by factors of ten on timescales of tens to hundreds of years, and the observed effective surface temperature increased by about a factor of two in just the ten years following the discovery of the star in 1996. These changes are remarkably fast compared with standard timescales for stellar evolution.

capture of neutrons on nuclei to build heavier nuclei (see Box 5.2). Because neutrons are electrically neutral they do not have a Coulomb barrier to overcome, permitting reactions to take place at low enough temperatures that the newly formed heavy nuclei will not be dissociated immediately by high-energy photons.

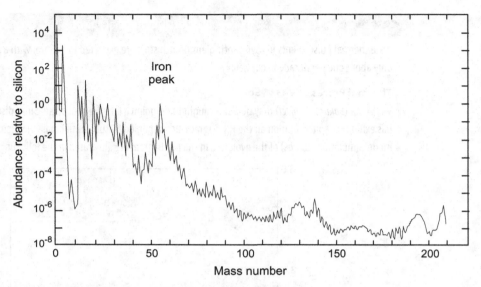

Fig. 13.9 Solar System elemental abundances expressed as a ratio relative to silicon abundance [183].

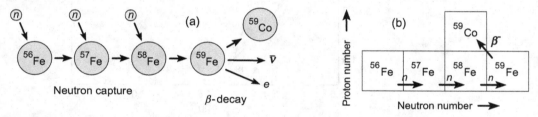

Fig. 13.10 Example of slow neutron capture and β-decay in the s-process. (a) Schematic. (b) Flow in the neutron–proton plane.

Two basic neutron capture processes are thought to produce heavy elements, the slow neutron capture or *s-process* and the rapid neutron capture or *r-process*. Although the astrophysical sites for these neutron capture reactions have not been confirmed, it is widely believed that the s-process takes place in AGB stars (for example, see Ref. [64]) and the r-process in neutron star mergers, core collapse supernova explosions, and possibly in jets produced by rapidly rotating collapsed objects. The s-process will be discussed here and the r-process will be addressed in Section 20.5.

The s-process: The s-process is a sequence of neutron-capture reactions interspersed with β-decays where the rate of neutron capture is *slow on a timescale set by the β-decays;* Figure 13.10 illustrates. In this example, ^{56}Fe subject to a low neutron flux captures three neutrons sequentially to become ^{59}Fe. But as the iron isotopes become neutron rich they become unstable against β^- decay. In this example it is assumed that the neutron flux is such that ^{59}Fe is likely to β-decay to ^{59}Co before it can capture another neutron. Now the ^{59}Co nucleus can absorb neutrons and finally β-decay to produce an isotope of the next atomic number (nickel), and so on. By this process, heavier elements can be built up slowly if a source of neutrons and the seed nuclei (iron in this example) are available. Because of

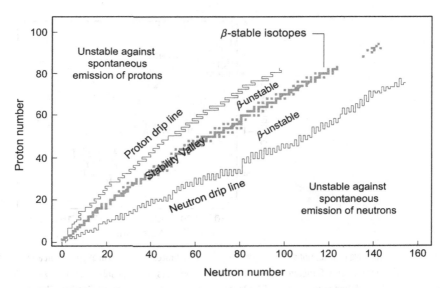

Fig. 13.11 Valley of β-stability (shaded boxes). Isotopes lying in this valley are stable against β-decay. The "drip lines" mark the boundaries for spontaneous emission of protons or neutrons. Isotopes outside the stability valley are increasingly unstable against decay by β^+ emission as the proton number is increased, and unstable against β^- emission as the neutron number is increased.

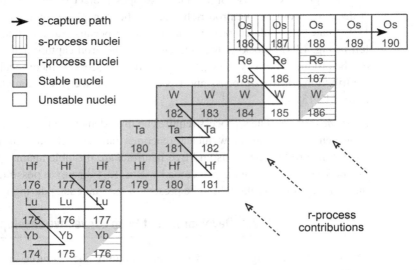

Fig. 13.12 The s-process path (zigzag line) in the Yb–Os region. Gray boxes indicate β-stable isotopes. The s-process path stays close to the β-stability valley. The r-process contributions come by β^- decay from the r-process path in Fig. 20.21.

the competition from β-decay, it is clear that the s-process can build new isotopes only near the valley of β-stability illustrated in Fig. 13.11.

In Fig. 13.12 the s-process path is shown for the Yb–Os region. This figure also illustrates the competition between the s-process and r-process. As we will see in

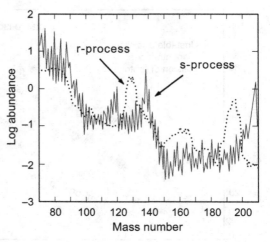

Fig. 13.13 Relative contributions of the s-process (——) and r-process (- - - -) to heavy-element abundances [25]. Reprinted from "Nuclear Structure Aspects in Nuclear Astrophysics," A. Aprahanian, K. Langanke, M. Wiescher, *Progress in Particle and Nuclear Physics*, **54**, 533–613, copyright (2005), with permission from Elsevier.

Section 20.5, the r-process populates very neutron-rich isotopes that then β^- decay toward the stability valley. Some isotopes (for example, ^{186}Os or ^{187}Os) can be populated only by the s-process because other stable isotopes protect them from r-process populations β-decaying from the neutron-rich side of the n–p plane. Other isotopes (for example, ^{186}W) can be populated only by the r-process because an unstable isotope lies to their left in Fig. 13.12, blocking the s-process slow neutron capture path. Many other isotopes can be produced both by the s-process and the r-process. A theoretical estimate for relative contributions of the s-process and r-process to heavy element abundances is summarized in Fig. 13.13.

Neutron sources: The s-process requires neutron densities greater than $\sim 10^6$ neutrons per cubic centimeter. Only a few nuclear reactions that are likely to occur in stars under normal conditions produce neutrons and free neutrons are unstable against β-decay, so neutrons for the s-process are not easy to come by. Box 13.4 discusses possible neutron sources for the s-process that are thought to be present in red giant stars during the AGB phase.

13.7.3 Development of Deep Convective Envelopes

As illustrated in Fig. 13.6, once a thin helium shell source develops the resulting temperature gradients drive very deep convection extending down to the region of the shell sources. The mixing associated with this deep convection is central to understanding nuclear-processed material associated with surfaces and winds for red giant stars (see Section 13.10).

13.7.4 Mass Loss

Once stars leave the main sequence they can experience large mass losses, particularly in the AGB and RGB phases. This is indicated most directly by the observation of gas clouds

> **Box 13.4** **Neutron Sources for the s-Process**
>
> A free neutron β-decays with a mean life of about 15 minutes. Thus, for neutron capture reactions to operate in stars on timescales larger than this there must be a source that makes neutrons available continuously for extended periods. For the slow capture process it is thought that two reactions that can occur in AGB stars are primarily responsible for supplying the neutrons:
>
> $$^4\text{He} + {}^{13}\text{C} \longrightarrow {}^{16}\text{O} + n \qquad {}^4\text{He} + {}^{22}\text{Ne} \longrightarrow {}^{25}\text{Mg} + n$$
>
> The $^{13}\text{C}(\alpha, n)^{16}\text{O}$ reaction is expected to provide the bulk of the neutron captures at low neutron densities ($\lesssim 10^7$ cm^{-3}). It occurs primarily between thermal pulses under radiative transport conditions, in the region immediately below the layers homogenized by deep convection in the preceding pulse (third dredge-up; see Section 13.10). The $^{22}\text{Ne}(\alpha, n)^{25}\text{Mg}$ reaction is of secondary importance and occurs primarily at higher temperatures during thermal pulses, where it exposes material that was produced by neutron irradiation from the $^{13}\text{C}(\alpha, n)^{16}\text{O}$ reaction and subsequently entrained in the thermal pulse to further neutron processing [64].

with outwardly directed radial velocities of 5–30 km s^{-1} near such stars. The mass-loss rate $\dot{m}$ may be described approximately in terms of the semiempirical expression [108]

$$\dot{m} \simeq -A \frac{LR}{M} \, M_\odot \, \text{yr}^{-1}, \tag{13.2}$$

where $A \sim 4 \times 10^{-13}$ is a constant, L is the luminosity, R is the radius, and M is the mass of the star. Thus, the mass-ejection rate increases for larger luminosity, larger radius, and smaller mass, as would be expected for mass loss from the surface of a luminous object[3] with a surface gravitational field determined by its mass and radius. Therefore, for RGB and AGB stars the rapid increase of radius and luminosity leads to increased mass loss, and as the star sheds its matter the decreased residual mass reduces the gravitational potential and can further accelerate the loss. Although the detailed mechanism is not well understood, it is clear empirically that the mass loss can increase by orders of magnitude relative to that associated with normal stellar winds in the RGB and AGB phases.

Example 13.2 For RGB stars mass losses of $10^{-6} \, M_\odot \, \text{yr}^{-1}$ have been recorded, while for AGB stars the losses can approach $10^{-4} \, M_\odot \, \text{yr}^{-1}$. If these rates were sustained, a red giant star would eject its mass on a timescale that is tiny compared to its overall lifetime.

13.8 Ejection of the Envelope

In the AGB phase the envelope of the star is consumed from within and from without: the surface is ejecting mass at a rate governed by Eq. (13.2), while the C–O core is

---

[3] The luminosity of AGB stars is often near the Eddington limit of Eq. (9.15), as shown in Problem 13.5.

growing internally as the shell sources burn outward, with the surface mass loss being more important than growth of the core by orders of magnitude. This rapid loss of the envelope from surface mass ejection while the core grows at very small comparative rates has two important implications for explaining observational characteristics of late-time stellar evolution:

1. The envelope is lost rapidly, leaving behind a core of C and O (or Ne and Mg for more massive AGB stars). The rapid loss of the envelope implies that a range of initial masses will leave behind cores of almost the same mass. This is significant because white dwarf masses are observed to be concentrated in a narrow range near $M \simeq 0.6 \, M_\odot$.
2. The ejected envelope is a natural candidate for producing *planetary nebulae*, which are commonly observed phenomena in late stellar evolution.

Hence the normal outcome of AGB evolution is ejection of the star's envelope as a planetary nebula, leaving behind a bare C–O or Ne–Mg core that will cool to form a white dwarf. The point at which the star leaves the AGB branch of the HR diagram is determined by the mass of the envelope at the end of helium core burning and the rate of mass loss from winds. Late in the AGB phase mass loss can increase dramatically for a short period called the *superwind phase* (which, as for other mass-loss phases, is not well understood). From this point onward the evolution of the core and the envelope may be considered separately.

13.9 White Dwarfs and Planetary Nebulae

As the core of the dying AGB star compresses it follows the evolutionary track sketched in Fig. 13.14. This takes it to much higher temperatures than are encountered for normal stars in the HR diagram. It finally cools to the white dwarf region with attendant decrease in luminosity. This high-temperature excursion is a result of retained thermal energy and

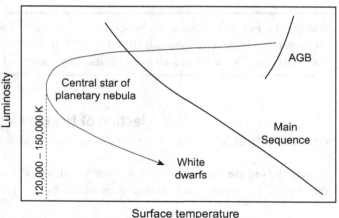

Fig. 13.14 Schematic evolution of the stellar core after the asymptotic giant branch.

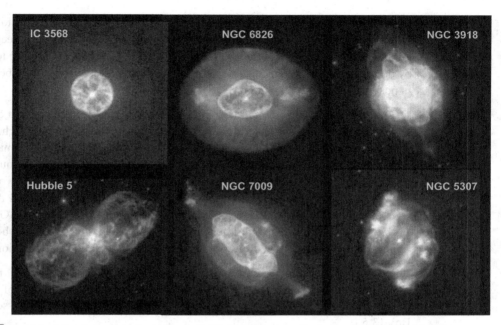

Fig. 13.15 A variety of planetary nebulae imaged by the Hubble Space Telescope. Clearly the ways in which dying AGB stars eject their envelopes can be quite complex.

gravitational compression, since the core is no longer capable of producing energy by thermonuclear processes.

The remnants of the ejected envelope recede from the star. When the temperature of the bare core reaches about 35,000 K a fast wind, probably associated with radiation pressure from the hot core, accelerates the last portion of the envelope to leave, forming a shockwave that proceeds outward and defines the inner boundary of the emitted cloud. As the temperature of the central star climbs, the spectrum is shifted far into the UV and this bath of high-energy photons from the central star ionizes the hydrogen in the receding envelope; the resulting recombination reactions between ions and electrons emit visible light and account for the luminosity and the often beautiful colors associated with the planetary nebula. Fig. 13.15 illustrates some of the complex morphologies that can result from these processes. The core and the planetary nebula now proceed on their separate ways, the core cooling slowly to a white dwarf, the planetary nebula expanding and growing fainter, eventually to merge into the interstellar medium and enrich it with gas and dust containing new elements synthesized in late red giant evolution.

13.10 Stellar Dredging Operations

Red giant stars exhibit abundances of isotopes in their surfaces and winds that only could have been produced by nuclear burning in the core or in shell sources deep within the star (see Ref. [64] for a summary and literature references). Since post main sequence evolution in the red giant phases involves episodes of deep convection, it is logical to assume that the observed nuclear-processed material is brought to the surface by such deep convective mixing. This mechanism of transporting the products of nuclear processing to the surface by deep convection is termed a *dredge-up*. Three dredge-up episodes have been identified in post main sequence evolution:

1. *First dredge-up* is thought to occur as the star develops deep convection driven by the hot hydrogen shell source prior to triple-α ignition on the red giant branch.
2. *Second dredge-up* can occur early in AGB evolution for intermediate-mass stars as a result of convective gradients generated by the narrowing helium shell source.
3. *Third dredge-up* is more difficult to produce in simulations than the first two, but appears necessary to understand surface isotopic abundances for many evolved AGB stars. It is thought to be associated in a complex way with thermal pulses in AGB evolution, through deep convection that extends at least periodically into the region between the H and He shell sources (see Ref. [64] for an overview of current investigations).

These various dredge-up episodes are not well understood but are essential to explaining observations like carbon stars (stars with a greater surface abundance of carbon than oxygen), and the abundance of interstellar carbon dust grains, as discussed further in Box 13.5.

Box 13.5 **Stars, Soot, and Interstellar Dust**

On Earth chemical burning of carbon produces the black powdery material called *soot*. Soot is composed of complex and varied compounds whose basic building blocks are *polycyclic aromatic hydrocarbons (PAHs)*, which are hydrocarbons having multiple-ring structures.[a] Relatively broad mid-infrared emission features observed in various nebulae indicate that interstellar dust is composed largely of clumps of PAHs. It is thought that these form in the outflows from carbon-rich AGB stars by an analog of soot formation in terrestrial environments [214, 217], and that the mid-IR emission features result from IR fluorescence of PAHs containing $\sim$ 50 molecules that have been pumped (excited) by far-ultraviolet radiation [214].

There is little direct observational support for this hypothesized origin of PAHs in outflows from carbon-rich AGB stars. This is probably because PAHs are abundant in these outflows but AGB stars have surfaces too cool to emit the UV photons required to pump IR fluorescence of the PAHs and they are effectively invisible. (IR features characteristic of PAHs have been observed in a binary system where the carbon-rich outflow from an AGB star is illuminated by UV photons from a hot blue companion star [214].) There *is* strong evidence for PAH spectral features in the planetary nebulae emitted by dying AGB stars. This is thought to be caused by the increasing effective temperature of the central object on its way to becoming a white dwarf, which produces a flux of UV photons that excite the PAHs in the receding nebula into IR fluorescence. Thus the clear presence of PAHs in the descendants of carbon-rich AGB stars provides strong circumstantial evidence that such stars are a major source of PAHs, and thus of interstellar dust grains.

[a] "Aromatic" refers to organic compounds that achieve enhanced stability by delocalizing electrons over multiple chemical bonds, which lowers their kinetic energy by uncertainty-principle arguments. The terminology arose because the first such compounds studied often had a distinctly pleasant aroma.

13.11 The Sun's Red Giant Evolution

What does the preceding discussion have to say about our own star? The Sun will evolve into a red giant on the way to shedding its envelope to become a C–O white dwarf. This evolution will have large consequences for Earth. (The Sun has 5 billion years left on the main sequence, so there is plenty of time to prepare!) Figure 13.16 illustrates a simulation describing the predicted expansion of the Sun in the beginning of its red giant phase. The calculation up to 12×10^9 yr is the same as that in Fig. 10.9 but Fig. 13.16 extends the red giant evolution to the point where the expanding Sun engulfs the orbit of Earth. The top panel shows the Sun and Earth's orbit drawn to scale at various stages of the evolution. At present the Earth's orbital radius is 214 times the radius of the Sun, so the Sun

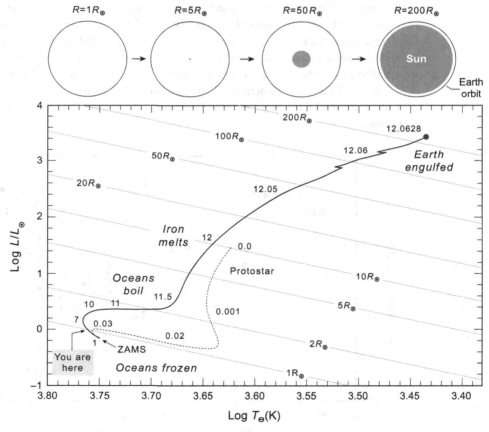

Fig. 13.16 Evolution of the Sun showing projected implications for Earth as the Sun evolves into a red giant, adapted from Iben [130]. Times in units of 10^9 years are shown beside the curve. The parallel diagonal lines join points of constant stellar radius. Protostar evolution is indicated by the dotted curve, beginning from when the protostar has collapsed to a radius 10 times that of the present Sun. The top panel shows Earth's orbit and the Sun drawn to scale at various stages of the evolution.

is largely invisible on the scale of the Earth's orbit. In this simulation the Sun expands to the size of Earth's present orbit 12.0628×10^9 yr after the time marked zero in the protostar collapse.

As indicated in Fig. 13.16, the Sun's later evolution will have dramatic implications for Earth [130]. The oceans presently are mostly liquid, except for ice near the poles. If the Sun's luminosity were decreased somewhat, the oceans would freeze. Conversely, when the Sun's luminosity increases as it begins to leave the main sequence Earth's equilibrium temperature will rise and before the Sun reaches twice its present radius the oceans of Earth will boil away. The temperature will continue to rise and by the time the Earth celebrates its 12×10^9-yr birthday the Sun's radius will have increased by a factor of 10 over the present value so that the Sun will appear as a large, increasingly red (the spectral class will have shifted from the present G2 to $\sim$ K2) ball subtending a 5° angle on the sky, the solar luminosity will have increased to about 40 times the current value, and temperatures on Earth will have reached the melting point of iron. From this point the evolution will become rapid (by astrophysical standards). It will take the Sun about 7 billion years to increase its radius from the present value to 10 times that, but in the ensuing $\sim$ 60 million years the solar radius will increase by an additional factor of more than 20 as the Sun begins to ascend the red giant branch, engulfing the Earth a little over 12 billion years after formation of the Solar System and causing it to spiral inward and be incinerated because of the friction with the Sun's outer envelope (thereby slightly increasing the metallicity of the Sun).

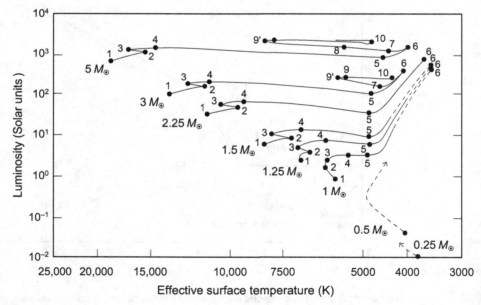

Fig. 13.17 Evolution off the main sequence for stars of 5 $M_\odot$ or less, adapted from simulations in Ref. [125]. Evolutionary times between the numbered points are given in Table 10.4. Dashed curves are theoretical estimates. Adapted From "Stellar Evolution within and off the Main Sequence," Icko Iben, Jr., *Annual Review of Astronomy and Astrophysics*, **5**(1), 571–626, 1967.

13.12 Overview for Low-Mass Stars

An overview of evolution after leaving the main sequence for various stars in the 0.25–5 $M_\odot$ range is given in Fig. 13.17. All but the lightest evolve into the red giant region in these simulations but they exhibit mass-dependent differences: (1) evolution is faster, with possible looping and switchbacks for the heavier stars, and (2) as mass increases the post main sequence motion on the HR diagram is increasingly horizontal and to the right, rather than the highly vertical ascent seen for lower-mass stars. These differences with increasing mass will receive sharp focus in Chapter 14, where we shall take up evolution of the most massive stars.

Background and Further Reading

See Böhm-Vitense [52]; Hansen, Kawaler, and Trimble [107]; Harpaz [108]; Kippenhahn, Weigert, and Weiss [134]; Phillips [169]; Prialnik [176]; and the review in Busso, Gallino, and Wasserburg [64]. The two-volume book by Iben [130] is a comprehensive summary of decades of pioneering stellar evolution simulations.

Problems

13.1 The Schönberg–Chandrasekhar instability for isothermal helium cores is important for understanding the evolution of red giant stars. To investigate the physics of isothermal stellar cores it is useful to apply the virial theorem to the isothermal core of a star rather than to the entire star.

(a) Repeat the derivation of the virial theorem given in Section 4.5 but now extend the integration over mass only from 0 to $M(R_c)$, where R_c is the assumed radius of an isothermal core. Show that the the resulting virial theorem for the isothermal core is

$$P_c V_c = \int_0^{M(R_c)} (P/\rho) dm + \tfrac{1}{3}\Omega_c,$$

where P_c is the pressure at the boundary of the isothermal core and Ω_c is the gravitational energy generated by the sphere of radius R_c.

(b) Assuming ideal-gas behavior, show that the preceding result can also be expressed as

$$P_c V_c - \int_0^{V_c} P dV = \tfrac{1}{3}\Omega_c,$$

where V_c is the volume of the core, and also as

$$3(\gamma - 1)U_c - 3P_c V_c + \Omega_c = 0,$$

where U_c is the total internal energy and γ is the adiabatic index of the core.

(c) Use the result from part (a), parameterize the gravitational potential by

$$\Omega_c = -f \frac{GM_c^2}{R_c},$$

where f is a parameter of order one accounting for the distribution of matter in the star, and assume the gas to be ideal and isothermal to show that

$$P_c = \frac{3}{4\pi} \frac{kT_c M_c}{\mu_c R_c^3} - \frac{fGM_c^2}{4\pi R_c^4},$$

where T_c is the temperature of the isothermal core and μ_c is the mean molecular weight within the isothermal core (assumed to be constant).

(d) Show from the preceding results that there is a maximum pressure that the isothermal core can generate. By setting the derivative of P_c with respect to M_c equal to zero, show that the corresponding radius is

$$R_c = \frac{2\alpha G M_c \mu_c}{3kT_c}$$

giving a maximum pressure

$$P_c^{max} = \left(\frac{81}{64\pi f^3 G^3}\right) \frac{(kT_c)^4}{\mu_c^4 M_c^2}.$$

Part (d) shows that the maximum pressure that the isothermal core can generate scales as the inverse square of the mass of the core. This pressure must balance the pressure of the layers overlying the isothermal core for stability. Hence, there may be a maximum core mass beyond which the pressure of the isothermal core can no longer balance the pressure from the overlying layers. Exceeding this mass for an isothermal helium core precipitates the Schönberg–Chandrasekhar instability.***

13.2 Assume that a star is formed with the mass fractions $X = 0.70$, $Y = 0.28$, and $Z = 0.02$, and that the star is completely ionized. If at the end of main sequence evolution all hydrogen in the core has been converted to helium and the core is isothermal, estimate the Schönberg–Chandrasekhar limit assuming the composition outside the core to be unchanged from that at formation of the star. *Hint*: The approximate formulas of Section 3.4.4 are useful for this problem.

13.3 Assuming a typical neutron capture cross section on a seed nucleus to be 100 mb, estimate the number density of neutrons required under red giant conditions to give a mean life for neutron capture on a given seed nucleus of one year.

13.4 Consider the thin shell source illustrated in Box 13.2.

(a) Use the equation of hydrostatic equilibrium expressed in Lagrangian form to get an expression for the pressure in the shell in terms of an integral over the mass coordinate.

(b) Perturb the shell by shifting its radial coordinate outward by a small amount $\delta r/r_0$, leading to a small change in pressure δP,

$$P \to P + \delta P \qquad r \to \left(1 + \frac{\delta r}{r_0}\right) r.$$

Find an expression for the new pressure in the shell after the small perturbation, and show that this solution implies a variation of pressure with density given by

$$\frac{dP}{P} = -4\frac{dr}{r_0} = \frac{4L}{r_0}\frac{dp}{p}.$$

Hint: The perturbation is assumed small, so expansions are justified.

(c) Use that δr and δP are small, and the general equation of state discussed in Problem 4.16,

$$\frac{dP}{P} = \alpha\frac{d\rho}{\rho} + \beta\frac{dT}{T},$$

to show that the thermal stability condition (T must decrease if the shell expands) can be satisfied only if $4L/r > \alpha$, where α is a finite positive number.

Thus, if the shell source is too thin it will be unstable, even if the gas in it is not degenerate.***

13.5 A semiempirical expression originally proposed by Paczyński for the luminosity of an AGB star may be written as

$$L_{\text{AGB}} = 6 \times 10^4 \left(\frac{M_c}{M_\odot} - 0.5\right) L_\odot \qquad (M_c > 0.5\, M_\odot),$$

where M_c is the mass of the carbon–oxygen core [176]. Compare this luminosity for an AGB star with the corresponding Eddington luminosity assuming a C–O core mass of $0.6\, M_\odot$ and a total mass of $1\, M_\odot$.***

13.6 Use the calculated evolution of a $5\, M_\odot$ star displayed in Fig. 13.5 to argue quantitatively why it is very difficult to observe stars evolving through the Hertzsprung gap. *Hint:* See Table 13.1.

13.7 Use interpolation from Fig. 13.16 to estimate the time since formation, effective surface temperature, spectral class, and luminosity for the Sun when its radius has increased to 5, 10, 50, and 200 $R_\odot$ during its ascent to the red giant branch. Assume for present purposes that the correlation between surface temperature and spectral class for main sequence stars remains approximately valid in this evolution.

13.8 Let M_{to} be the stellar mass corresponding to the luminosity of the turnoff point in a star cluster (Fig. 2.9 and the relationship between mass and luminosity discussed in Section 1.6). Derive an expression for the ratio of the number of white dwarfs to the total number of main sequence stars as a function of M_{to} (and thus of the age of the cluster) assuming that:

1. The lowest-mass star that can form has mass M_1.
2. All stars in the cluster with initial mass $M \geq M_{\text{SN}} \sim 8\, M_\odot$ have already undergone a core collapse supernova explosion.
3. Because the time on the main sequence is long compared with all later stages, when a star leaves the main sequence it instantaneously becomes a white dwarf or explodes as a core collapse supernova.
4. The simple Salpeter power law (9.16) is a reasonable approximation for the initial mass function of the cluster.

Make a plot of the fraction of white dwarfs as a function of the turnoff point mass assuming that $M_1 = 0.1\, M_\odot$ and $M_{\text{SN}} = 8\, M_\odot$.

14 Evolution of Higher-Mass Stars

In this chapter we shall address the evolution of higher-mass stars, which will be defined to be those stars with a zero age main sequence mass of 8 $M_\odot$ or more. As indicated by the discussion of the initial mass function in Section 9.12, such stars are much less common than their lower-mass siblings, but we shall see that they can sometimes have a more dramatic impact. In many respects these stars go through a similar evolution as the lower-mass stars described in Chapter 13, but there are some critical issues that are unique to high-mass stars. We shall first summarize these features in the following section and then proceed to discuss them in more detail in subsequent sections.

14.1 Unique Features of More Massive Stars

Higher-mass stars exhibit a set of characteristics that are not commonly found in lower-mass stars. These include:

1. The same burning stages as for lower-mass stars are encountered, but additional advanced burning stages of heavier fuels are initiated that are not accessible to lower-mass stars. If the star is massive enough it will proceed all the way to formation of an iron core.
2. The evolution through all stages occurs more rapidly and at greater luminosity.
3. Nucleosynthesis occurring in evolution after the main sequence produces heavier and more varied elements than those synthesized for less-massive stars.
4. Neutrino emission becomes increasingly important in more advanced burning stages, with core-cooling dominated by neutrinos for carbon burning and beyond.
5. The luminosity on the main sequence and after is often close to the Eddington limit and remains relatively constant after the main sequence. Thus the evolution after the main sequence for massive stars is very horizontal on the HR diagram.
6. Mass loss by strong stellar winds can be significant, even on the main sequence.
7. The central temperatures are high and the core electrons typically remain nondegenerate despite the high density until the latest burning stages.
8. The iron core formed in the last stages of main sequence evolution for massive stars is supported by electron degeneracy pressure and is inherently unstable if it grows beyond a critical mass of about $1.2 - 1.3\ M_\odot$. This implies that the endpoint of stellar evolution will be fundamentally different for a massive star relative to that for a lower-mass star.

Each of these issues will be addressed in this chapter or in the discussion of core collapse supernova explosions in Chapter 20. It is useful to begin with the consequences of advanced burning stages that are accessible only to massive stars.

14.2 Advanced Burning Stages in Massive Stars

The post main sequence evolution of $9\,M_\odot$ and $15\,M_\odot$ stars is shown in Fig. 14.1. This evolution is extremely rapid, as illustrated in Table 10.4. Because of the sequential advanced burning stages described in Chapter 6, massive stars near the ends of their lives build up the layered structure depicted schematically in Fig. 14.2. If the star has a mass $M \gtrsim 8M_\odot$, simulations indicate that successively heavier fuels can be burned as the star compresses and heats up, until an iron core is formed in the center. The iron core cannot produce energy by fusion (the curve of binding energy peaks in the iron region; see Fig. 5.1), but simulations indicate that by the time the iron core forms its electrons have become highly degenerate, so the core is supported initially against collapse by electron degeneracy pressure. However, as the silicon layer surrounding the iron core undergoes reactions it produces more iron and the central iron core grows more massive. Beyond a critical mass of about $1.2 - 1.3\,M_\odot$ (depending on composition) the core will become gravitationally unstable and collapse. This will be described in some detail in Chapter 20. In the present chapter we will concentrate on understanding the evolution of high-mass stars prior to encountering the gravitational instability of the iron core.

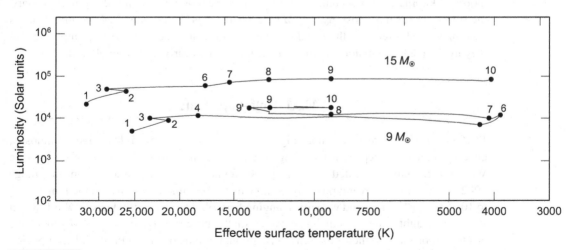

Fig. 14.1 Evolution off of the main sequence (labeled by 1) for high-mass stars, adapted from simulations in Ref. [125]. The evolutionary times between the numbered points are given in Table 10.4. Adapted From "Stellar Evolution within and off the Main Sequence," Icko Iben, Jr., *Annual Review of Astronomy and Astrophysics*, **5**(1), 571–626, 1867.

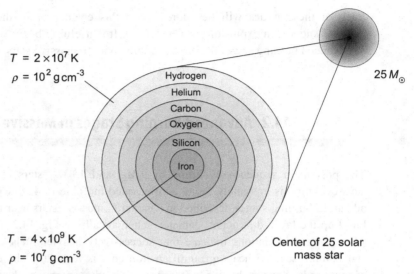

Fig. 14.2 Central region of a 25 solar mass star late in its life. See also Table 6.3. This central region is only a few thousand kilometers in radius and lies at the center of a supergiant star.

14.3 Envelope Loss from Massive Stars

As was discussed in Section 13.7.4, rapid mass loss is characteristic of evolution after the main sequence for many stars. However, massive stars may go through stages where they expel large portions of their envelopes into space at velocities as large as several thousand kilometers per second, even when they are relatively young [204]. In such stars, the timescale for mass loss $\tau_{\rm loss} \sim M/\dot{M}$, where M is the mass and $\dot{M}$ the rate of mass loss, may be shorter than their main sequence timescales [see Eq. (10.5)], implying that they may have lost a major fraction of their initial mass early in their evolution.

14.3.1 Wolf–Rayet Stars

Wolf–Rayet (WR) stars are hot, high-mass stars characterized by large luminosity, envelopes strongly depleted in hydrogen, and rapid mass loss. A survey of 64 galactic Wolf–Rayet stars presented in Ref. [158] found most masses to lie in the range 10–20 $M_\odot$, but with several possibly as large as 50 $M_\odot$, mass-loss rates in the range 10^{-6} to 10^{-4} $M_\odot$ yr^{-1}, and wind velocities ranging from about 700 to 3100 km s^{-1}. Wolf–Rayet stars are thought to be the remains of stars initially more massive than $\sim 30\,M_\odot$ that have ejected all or most of their outer envelope through winds or interactions with companion stars, exposing the hot helium core (which makes them strong UV emitters). The envelopes of Wolf–Rayet stars typically contain 10% or less hydrogen by mass, with individual stars exhibiting different levels of envelope stripping.

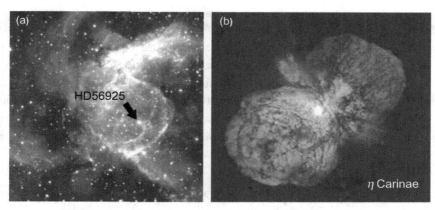

Fig. 14.3 (a) Wolf–Rayet star HD56925 surrounded by remnants of its former envelope. (b) η Carinae, surrounded by ejected material.

Figure 14.3(a) shows a wind-blown shell of gas that has been expelled from the Wolf–Rayet star HD56925 (marked by the arrow.) The nebula contains shockwaves associated with interaction of the wind and the interstellar medium, and is glowing from excitation of previously expelled material. Another example of Wolf–Rayet mass loss is discussed in Box 14.1. This class of stars will also assume a place of prominence in the discussion of gamma-ray bursts in Chapter 21.

14.3.2 The Strange Case of η Carinae

Figure 14.3(b) shows an extreme example of mass loss from a young star: the supermassive, highly unstable star, η Carinae, which is presently ejecting mass at a rate of about 10^{-3} solar masses per year. Elemental abundances in the nebula surrounding η Carinae are consistent with this being the supergiant phase of a 120 $M_\odot$ star that has evolved with very large mass loss while on the main sequence and afterwards.[1]

14.4 Neutrino Cooling of Massive Stars

Emission of neutrinos as a mechanism to transport energy from the stellar interior was discussed in Section 7.11. In the advanced burning stages outlined in Chapter 6, neutrino

---

[1] η Carinae may be a member of a binary or triple-star system. *Luminous blue variables or LBV* are luminous blue stars exhibiting irregular, sometimes violent, variability. The most famous examples are η Carinae, which underwent an enormous eruption in the 1840s [see Fig. 14.3(b)], and P Cygni (see Fig. 9.1), which had a similar episode in the 1600s. LBVs may be an important stage in massive-star evolution. One idea is that they may represent the mechanism by which a massive star ejects most of its hydrogen envelope on the way to becoming a Wolf–Rayet star. The absence of red supergiants with luminosities comparable to the most luminous O stars has been cited in support of this idea, since simulations indicate that extensive LBV mass loss can prevent the star from undergoing normal evolution into a red supergiant [158].

Box 14.1 The Dusty Pinwheel of WR 104

Formation of dust grains is common around cool red giant stars (Box 13.5). It is less common for hot stars but some Wolf–Rayet stars exhibit circumstellar dust clouds. This is puzzling because the intense radiation from these hot stars is generally inimical to formation of dust [225]. (Understanding dust formation around hot Wolf–Rayet stars may be relevant also for dust formation in even hotter environments, such as novae or supernovae.) It has been argued that the formation of dust requires an increase in gas density [225] that could result from the collision of the WR wind with the wind of a high-mass binary companion [220] (see also Section 18.5).

Dust Emission from WR104

Dust can absorb UV photons and re-radiate in the IR. One star observed extensively [218, 219] in the IR is WR 104, which is a Wolf–Rayet star 2300 pc away in Sagittarius, with a high-mass (class OB) companion that is resolved visually [221]. It is unusual in that it is emitting dust in an Archimedian spiral, as illustrated below.

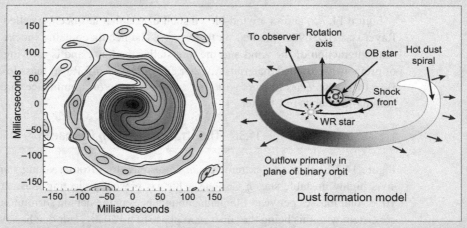

Adapted from "The Prototype Colliding-Wind Pinwheel WR104," Tuthill et al. (Keck Observatory and University of Sydney), *Astrophysical Journal*, **675**(1) © 2008, The American Astronomical Society. All rights reserved.

Adapted by permission from *Nature*, "A Dusty Pinwheel Nebula around the Massive Star WR104," Tuthill et al. (Keck Observatory and University of Sydney). Copyright (1999).

The left figure is a composite of dust observations centered on WR 104 [219]. The right figure illustrates a model for the spiral dust pattern [218]: Increased gas density in the shock produced by the collision of the WR and OB companion winds condenses dust grains that are flung out in a spiral pattern by the revolving system.

Drawing a Bead on Earth?

WR 104 is also (in)famous for the threat it might pose to Earth. A Wolf–Rayet star can create a gamma-ray burst (GRB) focused narrowly along its rotation axis (see Chapter 21), which for WR 104 is pointed very nearly at Earth. If such a burst from WR 104 struck the Earth it would decimate the ozone layer, with catastrophic environmental consequences. However, it seems unlikely that WR 104 would create a GRB that could survive its dusty environment and also be aimed precisely at us, so Earth is (probably) safe.

cooling assumes increased importance because the conditions in stars leading to these burning stages often involve extreme energy-production rates in regions deep within stars having high photon opacity. Then neither radiative nor convective transport can remove the energy fast enough to maintain hydrostatic equilibrium, but the high-temperature, high-density environment is at the same time conducive to neutrino production and the material is still largely transparent to neutrinos that are produced. Hence the very stability of stars undergoing advanced burning depends fundamentally on neutrino cooling. This in turn implies that the properties of late stellar evolution and the types of remnants that result are bound up inextricably with neutrino cooling of the star.

These considerations also are valid for advanced burning in lower-mass stars. For example, neutrino cooling on the AGB branch for a $5\,M_\odot$ star was mentioned in Chapter 13, cooling of hot young white dwarfs and pre-white dwarfs will be found in Box 16.1 to be dominated by plasma neutrino emission, and it was remarked in earlier chapters that from carbon burning and beyond the dominant mode of cooling in stellar evolution becomes neutrino emission (see also Example 7.7, Table 7.2, and Problem 14.4). But neutrino cooling assumes particular importance for high-mass stars, which can access all the advanced burning stages that have been discussed in Chapter 6 so that their evolution depends increasingly on neutrino emission as these stages are encountered. Indeed, it could be argued that the culmination of massive-star evolution in a core collapse supernova (Chapter 20) is the ultimate example of massive-star neutrino cooling, since almost all of the prodigious supernova explosion energy appears in the form of neutrino emission.

14.4.1 Local and Nonlocal Cooling

Below temperatures of about 5×10^8 K stars are cooled dominantly by radiation and convection, for which the net rate of heat removal depends on temperature gradients. Thus the cooling at a given point in the star by radiative diffusion or convection is *nonlocal*, in that it depends on conditions at the point but also on conditions in the surrounding region [26]. In contrast, neutrino cooling is highly *local*, since the energy carried by a neutrino produced at a point is removed from the star at nearly the speed of light with little probability to interact with any of the rest of the star. Thus neutrino cooling depends only on the conditions at the point of production and not on spatial derivatives evaluated at that point.

14.4.2 Neutrino Cooling and the Pace of Stellar Evolution

Neutrino emission begins to dominate the energy-removal budget in stars when temperatures exceed about 10^9 K and densities are sufficiently low that the electrons are not too degenerate (see Figs. 7.13–7.14). However, it should be noted that the terminology "neutrino cooling" is apt when applied to white dwarfs or neutron stars, but is something of a misnomer for stars undergoing thermonuclear burning in hydrostatic equilibrium. Instead of cooling the star, the rapid energy loss from neutrino emission stimulates increased thermonuclear rates that are required to keep the star in equilibrium, so neutrino "cooling" actually *accelerates* burning and the pace of stellar evolution for massive stars.

14.5 Massive Population III Stars

An interesting and exotic aspect of massive star evolution concerns the first generation of stars that formed in the Universe (Population III; see Section 1.9.2). Observational evidence suggests that the first stars began forming several hundred million years after the big bang. These stars would have been hydrogen and helium stars with negligible metals, since they formed from material produced by the big bang. Simulations indicate that because of the absence of metals these stars likely were very massive, with 100–1000 $M_\odot$ being common. Because of their large mass, these stars would have evolved quickly and most would have exploded as pair-instability supernovae (described in Box 20.2) within several million years of their birth, thus seeding the Universe with heavier elements up to iron.

At the *recombination transition* in the early Universe, which occurred at a redshift $z \sim 1100$ (some 380,000 years after the big bang), electrons combined with protons to make neutral hydrogen and the Universe became transparent to visible photons. There were no stars yet, so the ensuing period until stars formed is sometimes called the *dark ages*. Observations indicate that beginning at redshift $z \sim 20$ and continuing to $z \sim 6$ (roughly from 500 million years to almost a billion years after the big bang) the neutral hydrogen was reionized in the *reionization transition*. It is widely believed that Pop III stars were instrumental in this reionization of the Universe. The spectra of high-redshift quasars indicate that there were heavy elements present in the Universe during reionization, which could have come only from stars, and because of their large masses stars in this first generation would have been hot and would have bathed their neighborhoods with ionizing UV radiation. No conclusive observational evidence exists for Pop III stars. Some candidates have been proposed for star clusters found in faint galaxies at large redshift ($z \geq 6$) [205], but the observations are difficult and thus conclusions are necessarily qualified.

14.6 Evolutionary Endpoints for Massive Stars

Stars having $M \lesssim 8\, M_\odot$ are all thought to end their lives with their cores evolving to some form of white dwarf (helium, carbon–oxygen, or neon–magnesium), and their envelopes ejected as planetary nebulae. In contrast, the most massive stars appear ordained to one of three qualitatively different fates more dramatic than becoming white dwarfs, with all three initiated by gravitational collapse of the star's core:

1. The majority of stars having $M \gtrsim 8\, M_\odot$ will eject the outer layers of the star violently in a core collapse supernova (see Chapter 20), with the central regions crushed gravitationally into a *neutron star* that is stabilized by neutron degeneracy pressure.
2. For some fraction of core collapse events the mass of the gravitationally collapsed central region will be too large for neutron degeneracy pressure to halt the infall and the star will collapse instead to a *black hole*.

3. For the special case of very massive stars ($M \sim 130 - 250\ M_\odot$) and low metallicities, the star can destroy itself in a *pair-instability supernova,* which leaves behind no compact remnant. This was probably the fate of many Pop III stars and the mechanism will be considered in Box 20.2

Neutron stars will be discussed in Section 16.7 and black holes in Chapter 17. There is abundant direct observational evidence for the former, and a wealth of indirect observational evidence for the latter. Thus, there is strong reason to believe that both of these endpoints can issue from the evolution of massive stars.

14.6.1 Observational and Theoretical Characteristics

The neutron star and black hole scenarios for the outcome of core collapse might have different observational characteristics. For the former it is rather certain that gravitational waves and a burst of neutrinos will be emitted from the supernova explosion, the ejected outer layers will produce an expanding supernova remnant, and the residual neutron star will cool primarily by neutrino emission, perhaps manifesting itself as a pulsar. For the latter, the outcome is less-well understood. A direct collapse to a black hole is expected to produce a burst of neutrinos, gravitational waves, and possibly a γ-ray burst, but any ejected supernova remnant might range from similar to that for collapse to a neutron star to the case of no ejected remnant at all, which will be discussed below.[2]

It is also unclear theoretically what the mechanisms are that distinguish whether the product of stellar core collapse is a neutron star or a black hole. As will be discussed in Chapter 20, although detection of the neutrino burst from Supernova 1987A gives strong confidence in the basic core collapse mechanism, only recently have quantitative simulations of that mechanism incorporating the most realistic physics begun to yield robust explosions, and even these simulations leave many unanswered questions. Thus, although it is expected conceptually that for higher-mass stars the heaviest are likely to produce black holes and the lightest to produce neutron stars, the hard evidence backing this up, and the quantitative features that distinguish these scenarios from each other, are not yet firmly established by realistic simulations and observations.

14.6.2 Black Holes from Failed Supernovae?

There is considerable interest in whether massive stars can collapse directly to black holes, without ejection of a significant remnant and without a large increase in optical luminosity. Such events are called *failed supernovae,* since they produce a black hole, gravitational waves, and presumably neutrinos, but few of the other characteristics associated with core collapse supernovae that will be discussed in Chapter 20. It has long been thought that

---

[2] The observational outcome depends on whether part of the star's envelope attains escape velocity as an expanding nebula, or whether the envelope is dragged into the black hole through accretion. The issue also can be complicated by angular momentum of the core, which may delay the formation of a black hole through centrifugal effects. This will be discussed further in Chapter 21.

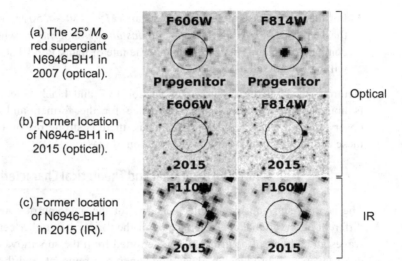

Fig. 14.4 Observational evidence for a failed supernova [20]. Hubble Space Telescope optical and IR images of the region surrounding the 25 $M_\odot$ red supergiant star N6946-BH1 in NGC 6946 through different telescopic filters labeled FabcW. The galaxy NGC 6946 is about 7 Mpc away and exhibits solar metallicity. It is known as the Fireworks Galaxy because of the abnormally high rate of supernova explosions observed there. (a) In these optical images obtained in July, 2007, N6946-BH1 is the dark spot at the center of the circles, which have radii of 1 arcsec. (b) in optical images of the same region obtained in October, 2015, N6946-BH1 has disappeared. (c) In 2015 very faint IR emission was observed consistent with the former position of N6946-BH1. Adapted from S. M. Adams et al., "The Search for Failed Supernovae with the Large Binocular Telescope: Confirmation of a Disappearing Star," *Monthly Notices of the Royal Astronomical Society*, **468**(4), 4968–4981, 2017, with permission by Dr Adams and Oxford University Press.

such could be the fate of very high-mass stars of low metallicity but more recent evidence suggests that it might occur also for massive red supergiants of solar metallicity [20].

A survey by the Large Binocular Telescope, supplemented by followup observations using the Hubble Space Telescope and Spitzer Space Telescope, identified a strong candidate for a failed supernova that is illustrated in Fig. 14.4. In 2007 the star designated N6946-BH1, identified from systematics as a 25 $M_\odot$ red supergiant, appears as a dark spot in Hubble Space Telescope optical images [center of the circles in Fig. 14.4(a)]. In 2009 this star underwent a weak optical outburst, brightening to $L \geq 10^6 L_\odot$ but then fading to a luminosity less than its pre-outburst luminosity over a matter of months. Images obtained in 2015 [shown in Fig. 14.4(b)] indicate that N6946-BH1 has disappeared from view in the optical (with an upper limit on optical luminosity five magnitudes less than that of the progenitor), but Fig. 14.4(c) indicates faint infrared emission at the former location of N6946-BH1. The total bolometric luminosity of N6946-BH1 as of 2017 is much less than that of the progenitor, suggesting that the star did not survive the 2009 event. After systematic analysis to rule out competing explanations such as obscuration by dust, a stellar merger, or N6946-BH1 being an exotic kind of variable star, it was concluded that these observations are most consistent with the 25 $M_\odot$ red supergiant undergoing a failed supernova and collapsing directly to a black hole, with faint residual IR activity from weak accretion on the black hole. If this interpretation is correct, it represents the first observation of a failed supernova and the first direct observation of black-hole birth.

14.6.3 Gravitational Waves and Stellar Evolution

As will be discussed in Chapter 22, the detection of gravitational waves thought to be emitted from the merger of two $\sim 30\,M_\odot$ black holes in a galaxy more than 400 Mpc away is perhaps the strongest observational evidence that black holes with masses comparable to those of massive stars exist. This new window on the Universe will presumably have large implications for our understanding of late stellar evolution, since supernovae, neutron stars, and black holes figure prominently in the kinds of events that can produce detectable gravitational waves. This is particularly true if *multimessenger astronomy* with gravitational waves becomes commonplace, as will be described in Section 22.5.

14.7 Summary: Evolution after the Main Sequence

An overview of post main sequence evolution for stars of higher and lower masses has been given in this chapter and in Chapter 13. A summary of evolution off the main sequence for $1\,M_\odot$, $5\,M_\odot$, and $25\,M_\odot$ stars is shown in Fig. 14.5. This is a quantitative version of the schematic diagram in Fig. 10.8, illustrating that lower-mass stars evolve to formation of white dwarfs with emission of the envelope as a planetary nebula, but higher-mass stars evolve quickly to catastrophic collapse of the star's core, leaving behind a neutron star or black hole, and possibly ejecting the envelope as an expanding supernova remnant. This will be recognized as yet another installment in the ongoing saga that *for stars, mass is destiny*.

14.8 Stellar Lifecycles

Having summarized the evolution of lower-mass and higher-mass stars in Fig. 14.5, we conclude this chapter by noting that the evolution of stars leads to extensive recycling of stellar material. Each star ties up a certain amount of mass at its birth. As the star evolves, some of that mass is returned to the interstellar medium to participate in future star formation by stellar winds, planetary nebulae, supernova explosions, and so on, while some becomes locked away in compact objects: white dwarfs, neutron stars, and black holes. The birth, evolution, and death of successive generations of stars[3] has three general consequences for a galaxy:

---

[3] This is a complex process because each generation of stars has a range of lifetimes spanning many orders of magnitude. Thus successive generations overlap strongly and aren't neatly separated in time, not unlike human generations, though the range of lifetimes within a generation is much larger for stars than for humans.

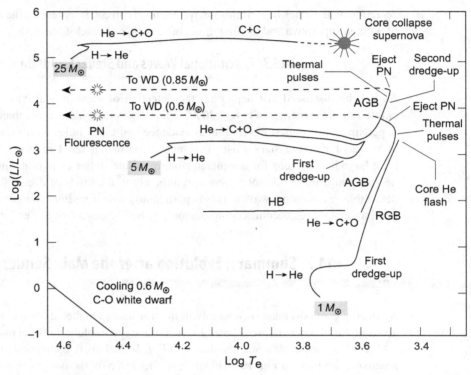

Fig. 14.5 A summary of late stellar evolution for stars of several main sequence masses, adapted from Ref. [126]. This is a quantitative version of the cartoon shown in Fig. 10.8. Abbreviations: HB (horizontal branch), RGB (red giant branch), AGB (asymptotic giant branch), PN (planetary nebula), and WD (white dwarf). Nuclear reactions in various burning stages are indicated. A discussion of evolution for stars with $M < 8\,M_\odot$ may be found in Chapter 13 and that of more massive stars is given in the present chapter. Adapted from Iben, I., Jr., *Quarterly Journal of the Royal Astronomical Society* (ISSN 0035-8738), **26**, 1–34, March 1985. Copyright Royal Astronomical Society. Provided by the NASA Astrophysics Data System.

1. The amount of gas available to make stars decreases as more of it becomes locked in white dwarfs, neutron stars, black holes, brown dwarfs, and very low-mass stars that have main sequence lifetimes much longer than the present age of the Universe.[4]
2. Over time the luminosity declines as massive, bright stars die more quickly and the population is increasingly dominated by less-massive, long-lived, fainter stars (while at the same time the color of light from the galaxy shifts to longer wavelengths).
3. The composition of gas in the galaxy becomes enriched in metals as nuclear-processed material is returned from stars to the interstellar medium by winds and explosions.

Thus, successive generations of stars typically have higher metallicities. However, the metal content does not increase uniformly with time. From metallicities of stars with different ages in the Milky Way it may be estimated that the mass fraction of heavy

---

[4] This discussion assumes an isolated galaxy. Galaxy collisions complicate things (for example, triggering intense episodes of star formation called *starbursts*), but the conclusions are valid when averaged over galaxies.

elements Z increased by much more early in the history of the galaxy than it has more recently (see Problem 14.7) [176]. The contribution to metallicity also is not uniform with star mass. Massive stars are rare but they are the primary source of metallicity increase because they eject large amounts of processed mass as winds and explosions on a relatively short timescale.

Since the fraction and composition of stellar material returned to the interstellar medium depends strongly on the mass of a star, the initial mass function discussed in Section 9.12 is important in understanding the recycling of stellar material. This is illustrated in Problem 14.6, where you are asked to estimate the fraction of material returned to the interstellar medium during the evolution of a single generation of stars.

Background and Further Reading

See Arnett [26]; Hansen, Kawaler, and Trimble [107]; Kippenhahn, Weigert, and Weiss [134]; and Prialnik [176] for general introductions to massive stars. The importance of mass loss for the evolution of massive stars has been reviewed by Smith [204].

Problems

14.1 Hotter stars on the upper main sequence can exhibit significant mass loss because of strong winds. Typical rates are $\dot{m} \simeq 10^{-7}\ M_\odot\ \mathrm{yr}^{-1}$, but in hot, massive Wolf–Rayet stars (see Section 14.3.1) the rate of mass loss can be much greater. Reference [158] analyzes 64 Wolf–Rayet stars. They give a semi-empirical formula to describe the rate of mass loss in these stars:

$$\dot{m} = 1 \times 10^{-11} \left(\frac{L}{L_\odot}\right)^{1.29} Y^{1.7} Z^{0.5}\ M_\odot\ \mathrm{yr}^{-1},$$

where L is the luminosity, and in the emitted material Y is the helium mass fraction and Z is the mass fraction of metals (which can be determined observationally).

(a) Using the data in Ref. [158], calculate the rate of mass loss predicted by the preceding formula for the star labeled 136 in Table 2 of Ref. [158] and compare with the observed rate.

(b) For the star in Table 5 of Ref. [158] with the largest rate of mass loss, how many years would it take for the star to emit all its mass through its wind, assuming (unrealistically) that the present rate were to be sustained?

(c) Assume for purposes of this exercise that the star in Table 5 of Ref. [158] with the highest rate of mass loss is a member of a binary system, with the companion star having the same mass as it (but to not be a Wolf–Rayet star), with circular orbits and an orbital period of 10 days. Use the results of Problem 18.9 to estimate the shift in binary orbital period over a year's time, assuming it to be caused entirely by the mass loss from the Wolf–Rayet star resulting from a spherically symmetric wind that is assumed to interact negligibly with the companion star.

14.2 The iron core of a massive main sequence star that is destined to become a core collapse supernova is produced by silicon burning (see Section 6.5.2). The rate-determining step in silicon burning is the photodisintegration of ^{28}Si, with the remarkably strong temperature dependence displayed in Fig. 6.9. This rate may be parameterized in the form given in Eqs. (D.2)–(D.4) of Appendix D. There generally are two significant components contributing to the sum over R_k for Si photodisintegration. However, around $T_9 \sim 3$ where silicon burning first becomes important the rate may be approximated by a single component R_k with the parameters given in the following table [182]

Reaction	p_1	p_2	p_3	p_4	p_5	p_6	p_7
^{28}Si$(\gamma,\alpha)^{24}$Mg	15.758	−129.56	−51.6428	68.4625	−3.86512	0.208028	−32.0727

Derive a formula for the temperature exponent of the corresponding Si photodisintegration reaction rate and evaluate it for $T_9 \sim 3$.

14.3 It was argued in Chapter 6 that the heaviest stars can produce isotopes up to the iron group ($Z \sim 25$) in their cores. Show that for a plasma of iron-group isotopes the mean molecular weight is about 2 amu. *Hint*: For isotopes of interest in this context, typically $Z \sim N$.

14.4 For the late stages of Si burning in a massive star, assume an ^{56}Fe core of radius 5000 km, density 10^{10} g cm^{-3}, and temperature 10^{10} K, and that the average electron neutrino energy is $E_\nu \sim 15$ MeV. Assuming the scattering to be dominated by coherent neutral-current interactions, estimate the neutrino cross section and the neutrino mean free path. Estimate the neutrino luminosity associated with the core. Compare this with the expected surface photon luminosity, assuming the star to be a red supergiant at this point. *Hint*: Use Fig. 7.14(b) and rough approximations to estimate the neutrino luminosity from the core.***

14.5 In a core collapse supernova (see Chapter 20) the density may reach 10^{14} g cm^{-3} with temperatures of order 10^{10} K or greater. Under these conditions, electron neutrinos are produced copiously and may have energies in the 10 MeV range or greater. Use the simple formula (7.48) to estimate the cross section and the mean free path for electron neutrinos of energy 10 MeV for this case.***

14.6 Consider a single generation of stars with a maximum initial mass M_2 and minimum initial mass M_1 for stars in the population. Assume that all stars with initial mass $M > M_{SN}$ have exploded as core collapse supernovae, that the initial mass function can be approximated by the Salpeter form (9.16), and that stars with mass less than $M_{MS} \sim 0.7 \, M_\odot$ have not had time to evolve off the main sequence since the stars were formed. Using reasonable assumptions and approximations, estimate the fraction of the initial mass that evolution of this population has returned to the interstellar medium.***

14.7 Assume the Milky Way to be about 13 billion years old and that the metallicity of the Sun is representative of stars its age in the galaxy. Given that the metal fraction by mass of the oldest observed stars in the galaxy is $Z \sim 3 \times 10^{-4}$ and that of the youngest is $Z \sim 4 \times 10^{-2}$, what can be said quantitatively about how the metal content of the galaxy has changed with time?***

15 Stellar Pulsations and Variability

A commonplace of modern astronomy that would have been quite perplexing for ancient astronomers is that many stars exhibit variations in brightness. In some cases these variations are asynchronous and in others they are highly periodic. They may be so small as to require precise instruments to detect them, or sufficiently large that they are easily visible to the naked eye. These *variable stars* may be classified into three broad categories. (1) *Eclipsing binaries*, for which the luminosity of the system is altered by eclipses (see Section 1.5.4). (2) *Eruptive and exploding variables*, which brighten suddenly and eject mass because of a disruption or partial disruption of the star. Novae (Chapter 19) and supernovae (Chapter 20) are dramatic examples. (3) *Pulsating variables*, which undergo pulsations that alter the brightness periodically, without disrupting the overall structure of the star; Cepheid variables are a well-known example. This chapter examines in more depth this latter category and the reasons that some stars become unstable against pulsations. We shall find that pulsational instabilities are common in particular stages of stellar evolution, and that they are caused in most cases by a relatively subtle interplay among heat flow, ionization, and photon opacities in stars.

15.1 The Instability Strip

Some classes of pulsating variable stars and their characteristics are given in Table 15.1. A strong clue to their nature is that these pulsating variable stars are found in specific regions of the Hertzsprung–Russell diagram, as illustrated in Fig. 15.1. For example, many types of pulsating variables are confined to a narrow, rather vertical band in the HR diagram called the *instability strip*. This suggests that there is a fundamental mechanism operating in a variety of stars in different luminosity classes, but over a relatively narrow range of surface temperatures, that leads to pulsational instability. Conversely, most stars observed at a given time are not pulsating, which suggests that pulsational instability requires a special situation not fulfilled by most stars over most of their lives.

15.2 Adiabatic Radial Pulsations

At the simplest level, stellar pulsation may be investigated in terms of oscillations that are adiabatic and linear in the displacement, and that maintain spherical symmetry for the star.

Table 15.1 Pulsating variable stars (adapted from Ref. [77])

Variable type	Period	Population	Mode[†]
Long-period variables	100–700 d	I, II	R
Classical Cepheids	1–50 d	I	R
Type-II Cepheids	2–45 d	II	R
RR Lyrae stars	1.5–24 hr	II	R
δ Scuti stars	1–3 hr	I	R, NR
β Cephei stars	3–7 hr	I	R, NR
ZZ Ceti stars	100–1000 s	I	NR

[†] R = Radial; NR = Non-radial

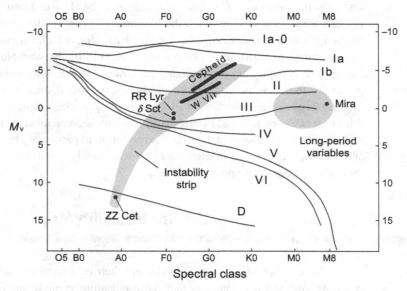

Fig. 15.1 Schematic illustration of the instability strip and the region of long-period red variables in the HR diagram [76]. With the exception of the long-period variables, most variable stars are found within the instability strip. The ZZ Ceti stars are pulsating white dwarfs that also are classified as DAV stars, according to the nomenclature outlined in Box 2.1.

Such an analysis has much in common with the study of small-amplitude vibrations in other physical systems, treating the pulsation as free radial vibrations with gas compressibility playing the part of a spring constant. These investigations find that stars disturbed slightly from spherical hydrostatic equilibrium exhibit discrete vibrational frequencies that are called *radial acoustic modes*. It is convenient to work in Lagrangian coordinates, where $m(r)$ is the independent variable and corresponds to the mass contained within a radius r; see Section 4.3. Then if the pressure, radial coordinate, and density are expanded to linear order as time-dependent oscillations around the equilibrium values (which are denoted by a subscript zero),

15.2 Adiabatic Radial Pulsations

$$P(m,t) = P_0(m)\left(1 + \delta P(m)e^{i\omega t}\right),$$

$$r(m,t) = r_0(m)\left(1 + \delta r(m)e^{i\omega t}\right),$$

$$\rho(m,t) = \rho_0(m)\left(1 + \delta\rho(m)e^{i\omega t}\right),$$

the radial displacement $\delta r(m)$ may be described by the differential equation

$$\frac{d^2(\delta r)}{dr_0^2} + \left(\frac{4}{r_0} - \frac{\rho_0 g_0}{P_0}\right)\frac{d(\delta r)}{dr_0} + \frac{\rho_0}{\Gamma_1 P_0}\left[\omega^2 + (4 - 3\Gamma_1)\frac{g_0}{r_0}\right]\delta r = 0, \qquad (15.1)$$

where the adiabatic exponent Γ_1 is defined in Eq. (3.39), ω is the adiabatic oscillation frequency, and $g_0 \equiv Gm/r_0^2$. We must solve this equation with two boundary conditions, one at the center of the star and one at the surface. At the center we require $d(\delta r)/dr_0 = 0$. The simplest physically reasonable surface boundary condition is to require $\delta P P_0 = 0$, though more complicated ones can be used. Reference [134] may be consulted for solutions and a derivation of Eq. (15.1).

Most intrinsically variable stars appear to be pulsing in radial acoustic modes, which correspond to standing waves within the star. The *fundamental mode* has no nodes (points of zero motion) between the center and surface, implying that the stellar matter involved in the vibration all moves in the same direction at a given time. The *first overtone* has one node between the center and the surface, meaning that the matter moves in one direction outside this node and in the opposite direction inside this node at a given phase of the pulsation. Likewise, higher overtones with additional nodes and more complex motion may be defined. Just as for musical instruments and other acoustically vibrating systems, a star may exhibit several modes of oscillation at once. The physical motion of the gas in radial stellar pulsations is largest in the fundamental mode and is considerably smaller in the first overtone. In higher overtones the motion of the gas in an oscillation cycle is even smaller. Observed pulsating variable stars appear to be oscillating primarily in the fundamental mode and/or the first overtone, with higher overtones not contributing substantially to the observed pulsation.

Example 15.1 It is thought that most Classical and Type II Cepheids oscillate in the fundamental mode, while RR Lyrae stars oscillate in either the fundamental mode or first overtone (or both). For long-period red variables the evidence is less conclusive and they may pulsate in either the fundamental mode or first overtone.

We may expect that realistic pulsations of variable stars will be more complicated than the simple linear, adiabatic analysis of the preceding paragraphs would indicate. For example, both the rate of energy production and the rate of internal energy transport could be modified by pulsations, so it may be expected that there will be deviations from adiabatic behavior in real stars. In particular, we must ask: what energy input sustains the long-term pulsations of a variable star?

15.3 Pulsating Variables as Heat Engines

Arthur Eddington examined whether stellar pulsations could be explained by free radial oscillations but realized that dissipation would damp the oscillations too quickly. For example, he estimated that Cepheid pulsations would be damped out in $\sim 10^4$ years, absent some mechanism to amplify and sustain the oscillations. Thus the steady, long-term pulsing of a Cepheid variable cannot be due to a one-time excitation of eigenmodes and cannot be adiabatic; these oscillations must be *driven* in some way. Conversely the observation that the Sun vibrates, but only with very small amplitude oscillations, indicates that in a star like the Sun there is no mechanism to drive sustained oscillations and vibrational excitations damp out quickly because of gas viscosity. These ideas led Eddington to propose that pulsating variable stars must correspond to a form of *heat engine,* continuously transforming thermal energy into the mechanical energy of the pulsation.

Although real stars are not adiabatic, it will turn out that the pulsation may often be approximated as *nearly adiabatic,* because instabilities grow on a timescale that is long relative to the time for one pulsation. Over one acoustic oscillation cycle (which occurs essentially on a *dynamical timescale* since it is related to the time for a sound wave to travel through the star), the heat exchanged is small because energy transfer occurs on a much longer *Kelvin–Helmholtz timescale.* Therefore, after a single acoustic cycle the star returns almost – but not quite – to the initial state. The "not quite" measures the lack of reversibility and therefore the non-adiabaticity of the process. With this as introduction, let us now investigate the significance of non-adiabatic effects in sustaining stellar pulsation.

15.4 Non-adiabatic Radial Pulsations

For each layer of the star a net amount of work is done during a pulsation cycle that must be equal to the difference of the heat flowing into that layer and that flowing out. Qualitatively it may be expected that if the oscillation is to be self-sustaining for a single layer there must be a mechanism whereby heat enters the layer at high temperature and leaves it at low temperature.[1] A sustained oscillation for a significant part of the star then requires that a set of layers have some level of coherence in the phase of these driven oscillations. We shall now show that these assertions can be justified more quantitatively using basic ideas from thermodynamics.

15.4.1 Thermodynamics of Sustained Pulsation

Many of the features required to sustain stellar pulsation follow from simple considerations based on the first and second laws of thermodynamics. Following Clayton [71], we work

---

[1] If layers driving the pulsation absorb energy near maximum compression, oscillations will be amplified. This is similar to the reason that it is optimal to fire the spark plug near the end of the compression stroke in an internal combustion engine.

in Lagrangian coordinates, assume the system to be nearly adiabatic, and consider initially a single radial mass zone in the star. By the first law, for a pulsation cycle the net change in the heat Q for a mass zone is a sum of contributions from changes in the internal energy U and the work W done on its surroundings during the pulsation,

$$dQ = dU + dW. \tag{15.2}$$

After a complete oscillation cycle it is assumed that the internal energy U returns to its original value so that the work done over the cycle is entirely contributed by the change in Q,

$$W = \oint dQ. \tag{15.3}$$

To drive oscillations, the gas must do positive work on its surroundings, meaning that it must absorb some net heat. However, the system is assumed to be nearly adiabatic, so that the gas returns essentially to its original state after one cycle. Therefore, in zeroth order there is no net change in entropy and

$$\oint dS = \oint \frac{dQ}{T} = 0. \tag{15.4}$$

Now suppose that during the cycle the system is perturbed by a small periodic variation in the temperature T of the form

$$T = T_0 + \Delta T(t),$$

where ΔT is zero at the beginning and end of the cycle. Then from Eq. (15.4)

$$\oint \frac{dQ(t)}{T_0 + \Delta T(t)} = 0.$$

Assuming the variation in T to be small, the denominator of the integrand may be expanded to first order, giving

$$\oint dQ(t)(T_0 - \Delta T(t)) = 0,$$

or upon rearrangement,

$$\oint dQ(t) = \oint \frac{\Delta T(t)}{T_0} dQ(t). \tag{15.5}$$

Then from Eqs. (15.3) and (15.5), the work done in one pulsation cycle is

$$W = \oint \frac{\Delta T}{T_0} dQ. \tag{15.6}$$

For the cyclic integral on the right side of (15.6) to give a net positive value (so that the mass zone does work on its surroundings over one cycle and can therefore drive an oscillation), it is clear that ΔT and dQ must have the *same sign* over a major part of the cycle. That is, heat must be absorbed ($dQ > 0$) when the temperature is increasing in the cycle ($\Delta T > 0$), and released ($dQ < 0$) when temperature is decreasing in the cycle ($\Delta T < 0$).

The preceding discussion has concentrated on the behavior of a single mass zone. Oscillation of the entire star means that some zones may do positive work and other zones may do negative work within a pulsation cycle. Thus, the condition for amplifying and sustaining oscillation of the entire star is that

$$W = \sum_i W_i = \sum_i \oint \left(\frac{\Delta T}{T_0}\right)_i dQ_i > 0, \tag{15.7}$$

where i labels the mass zones of the star. (Strictly this sum is an integral over the continuous mass coordinate, but in practical numerical simulations the zones are normally discretized.) Now let's ask whether the requirement Eq. (15.7) can be realized in stars.

15.4.2 Opacity and the κ-Mechanism

One way to favor sustained oscillations is to arrange a situation where opacity *increases* as the gas in a layer is compressed. Then the radiative energy outflow can be trapped more efficiently by the layer (it begins to "dam up" the outward energy flow) and this can push it and layers above it upward, until the layer becomes less opaque upon expansion and the trapped energy is released, permitting the layer to fall back and initiate another cycle. If a sequence of layers one above the other behaves in this general way, a sustained oscillation could be set up. Conversely, if compressing the layer *decreases* opacity the heat flow works against the oscillation and tends to damp it rather than sustain it. Normally stellar radiative opacities do not increase with compression of the gas. From the Kramers form (7.21), the opacity κ is proportional to $\rho T^{-3.5}$. Compression of a layer increases both ρ and T but the temperature dependence of κ is much stronger than the density dependence and a gas described by a Kramers opacity tends to experience a *decrease* in opacity under compression. Thus a star exhibiting the usual opacity behavior has a built-in damping mechanism that stabilizes it against pulsations. This explains why most stars are *not* pulsating variables.

However, there is a special situation for which the opacity could be expected to increase with compression. If a layer contains *partially ionized gas,* a portion of the energy flowing into it can go into additional ionization. Since this energy is absorbed into internal electronic excitations, it is not available to increase the temperature in the layer. Thus, if there is sufficient ionization during the compression portion of the pulsation cycle the effect on the opacity of the small rise in temperature can be more than offset by the effect of the increase in density and compression can increase the opacity. Conversely, electron–ion recombination in the decompression portion of the cycle can release energy and lead to a decreased opacity. Then, in partial ionization zones a layer can absorb heat during compression when the temperature is high and release it during expansion when the temperature is low, thereby setting the stage for a sustained oscillation as described in Section 15.4.1. This heat-engine mechanism for driving oscillations through ionization-dependent opacity effects is called the κ-*mechanism.*

15.4.3 Partial Ionization Zones and the Instability Strip

The κ-mechanism provides a possible way to drive stellar oscillations, but where is the κ-mechanism expected to operate? For most stars there are two significant zones of partial ionization, corresponding to the possible stages of ionization for hydrogen and helium:

> **Box 15.1** **Temperature Boundaries for the Instability Strip**
>
> The physical radius for the hydrogen and helium partial ionization zones within a given star will depend strongly on the effective surface temperature of that star. For stars with higher temperatures the ionization zones will be near the surface and there will be insufficient mass in the partially ionized layers to drive sustained oscillations. On the other hand, if the surface temperature is too low, convection will operate in the outer layers and undermine the κ-mechanism.[a] This suggests that there will be an optimal range of stellar surface temperatures for which the ionization zones are deep enough to drive sustained oscillations by coupling to the fundamental and overtones of the characteristic vibrational frequencies (which determines the higher-temperature end of the optimal range), but for which the convection is not strong enough to invalidate the mechanism (which determines the lower-temperature end of the optimal range). That is, these qualitative arguments indicate that pulsating variables should be found in localized regions of the HR diagram, as already suggested by the instability strip in Fig. 15.1.
>
> [a] A radiative opacity was assumed in the preceding argument, which leads crudely to a "dam" or "trapping" mechanism: if opacity increases under compression in a gas layer, that layer will tend to obstruct the outward flow of energy through it and heat up. The relationship between pulsation and convection is complex but detailed simulations generally show that when convective energy transport supplants radiative transport, it tends to interfere with this trapping effect and thus to damp stellar pulsations.

(i) The *hydrogen ionization zone*, where hydrogen is ionizing (H I $\to$ H II) and helium is undergoing first ionization (He I $\to$ He II). This region is broad and typically has a temperature in the range 10,000–15,000 K.

(ii) The *helium ionization zone*, where second ionization of helium (He II $\to$ He III) occurs, typically at a temperature around 40,000 K.

From the preceding discussion, one or both of these ionization zones may be expected to drive the pulsations of many variable stars. As discussed in Box 15.1, the radial location of the hydrogen and helium ionization zones in stars of particular surface temperatures, and the onset of convection near the surface for stars with surface temperatures that are too low, are the determining factors in producing the instability strip of Fig. 15.1.

Example 15.2 For classical Cepheids (and most variables found in the instability strip), the pulsation is caused by the κ-mechanism, primarily by forcing of the fundamental mode in the helium ionization zone. On the other hand, the long-period red variables (large AGB stars like Mira) and ZZ Ceti stars (which are a class of variable white dwarfs that have hydrogen envelopes) are thought to be driven by hydrogen ionization zones.

In Fig. 15.2 opacities that are expected for stars are plotted as a function of temperature and pressure. The shaded regions correspond to conditions that are expected to damp oscillations and the lighter regions correspond to conditions in which the κ-mechanism can be realized. The dashed line indicates the relationship between temperature and pressure expected for a 7 $M_\odot$ Cepheid variable. The helium ionization region crossed by the dashed line near $\log T = 4.6$ is thought to be the primary driver of classical Cepheid oscillations.

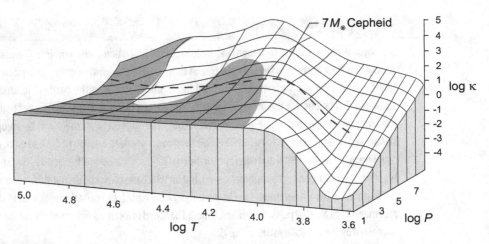

Fig. 15.2 Rosseland mean opacity versus pressure and temperature [134]. Shaded surface areas correspond to conditions that damp the pulsation and unshaded areas to conditions that excite the pulsation. The interior temperature–pressure relationship expected for a 7 $M_\odot$ Cepheid variable is indicated by the dashed line. The helium ionization zone primarily responsible for sustaining Cepheid pulsations is located in the vicinity of $\log T \simeq 4.6$. Adapted by permission from Springer Nature, *Non-adiabatic Spherical Pulsations*, Kippenham, R., Weigert, A., and Weiss, A. copyright (2012).

15.4.4 The ε-Mechanism and Massive Stars

Before the κ-mechanism was proposed it was suggested that pulsations could be driven by variations in energy production caused by radial oscillations. This was called the ε-*mechanism*. The ε-mechanism can enhance oscillations if energy production increases upon contraction – a condition that usually is satisfied. Oscillations can alter the thermonuclear energy production through density and temperature variations, but this is important only in the more central regions of the star where energy is produced. The problem then is that in those regions the amplitudes of fundamental modes and overtones are small, making it difficult for changes there to drive oscillations strongly enough to sustain pulsation. Thus the ε-mechanism is not important for most variable stars, where the pulsation is not because of temporal variations in energy production but rather is due to temporal variations in the efficiency of transporting that energy. However, the ε-mechanism may be important for stability in very massive stars, where oscillations coupled to variations in energy production deep inside may generate pulsations causing the star to shed mass (see Section 9.11 and footnote 1 about luminous blue variables in Section 14.3.2).

15.5 Non-radial Pulsation

For the variable stars in Table 15.1 that are labeled NR, the mode of pulsation is not spherically symmetric. The corresponding oscillations are called *non-radial modes*. Stars exhibiting non-radial pulsation include the δ Scuti stars, β Cephei stars, and ZZ Ceti stars. In addition, although our own Sun is not presently classified as a variable star (it presumably will become variable if it passes through the instability strip in the HR diagram after leaving the main sequence), it undergoes weak non-radial pulsations that are the target of the helioseismology observations described in Section 10.2. Non-radial pulsations will not be considered further here but an introduction may be found in Kippenhahn, Weigert, and Weiss [134] or Hansen, Kawaler, and Trimble [107].

Background and Further Reading

Stellar pulsations represent a rather complex and technical subject. In the present discussion we have deliberately avoided as much formalism as possible, concentrating instead on a qualitative introduction emphasizing the most important physical ideas. Other accessible and pedagogical discussions may be found in Carroll and Ostlie [68]; Clayton [71]; and Percy [167] (who places particular emphasis on the contribution of amateur astronomers to the database for variable stars). Somewhat more involved treatments may be found in Hansen, Kawaler, and Trimble [107]; Kippenhahn, Weigert, and Weiss [134]; and Padmanabhan [163]. A classic reference giving a comprehensive discussion is Cox [77].

Problems

15.1 Estimate the period for radial oscillations of a pulsating variable star by determining the time for sound waves to cross the diameter of the star, assuming the density to be constant.

15.2 Use the result obtained in Problem 15.1 to estimate the period of a classical Cepheid variable having $M = 7\,M_\odot$ and $R = 80\,R_\odot$.

15.3 Starting from Eq. (4.9) expressed in Lagrangian coordinates, examine small fluctuations about hydrostatic equilibrium by adding to the pressure and radial coordinates (which are functions of mass m in the Lagrangian description) a small time-dependent deviation,

$$r(m,t) = r_0(m)(1 + \delta r(m,t)) \qquad P(m,t) = P_0(m)(1 + \delta P(m,t)),$$

inserting this into the previous equation, and linearizing the resulting modified equations by expanding small quantities and retaining only linear terms. Show that the deviation in pressure δP obeys the differential equation

$$\frac{\partial(\delta P)}{\partial m} = \frac{1}{4\pi r_0 P_0}\left(\frac{Gm}{r_0^3}(\delta P + 4\delta r) - \frac{\partial^2(\delta r)}{\partial t^2}\right),$$

where subscript zeros denote the equilibrium values of r and P for a given enclosed mass m.***

16 White Dwarfs and Neutron Stars

Red giants eventually will consume all their accessible nuclear fuel. After ejection of the envelope as a planetary nebula the cores of these stars shrink to the very dense objects called *white dwarfs*. An even more dense object termed a *neutron star* can be left behind after the evolution of a more massive star terminates in a core collapse supernova explosion. We know already from the discussion in Chapter 3 that white dwarfs and neutron stars will have properties very different from normal stars because their extreme densities imply the importance of quantum mechanics and special relativity in defining their equations of state. As a consequence their behaviors will sometimes be quite different from objects such as normal stars that contain approximately ideal gases. In this chapter we address the structure and properties of these highly compact endpoints of stellar evolution in more depth. In addition we shall discuss *pulsars,* which are rapidly spinning neutron stars, and *magnetars,* which are spinning neutron stars with anomalously large magnetic fields.

16.1 Properties of White Dwarfs

Let's begin by getting acquainted with a white dwarf from the local neighborhood. The bright star Sirius in Canis Major is in reality a binary, with a brighter component labeled Sirius A and a fainter companion star labeled Sirius B. Sirius A is a spectral-class A main sequence star but Sirius B is a white dwarf.[1] Sirius B is not particularly dim (visual magnitude 8.5), but it is not easy to observe because it is so close to Sirius A. Its spectrum and luminosity indicate that it is hot (about 25,000 K surface temperature) but very small [124]. The spectrum contains pressure-broadened hydrogen lines, implying a surface environment with much higher density than that of a normal star. Assuming the spectrum to be blackbody and using the well-established distance to Sirius (it is relatively nearby, so its parallax of 0.38″ is measured precisely), the luminosity of Sirius B implies that it has a radius of only about 5800 km. But the Sirius system is a visual binary with a very well-studied orbit (the period is slightly less than 50 years, so about three full orbits have been observed since its discovery). Therefore, Kepler's laws may be used to infer that the mass of Sirius B is 1.02 $M_\odot$, implying that a white dwarf like Sirius B packs the mass of a

---

[1] Sirius B is the nearest and brightest white dwarf, and will be used at times as illustration. But it is in some respects not so representative because its mass of about 1.02 $M_\odot$ is much larger than the average mass of about 0.58 $M_\odot$ observed for white dwarfs.

star into an object the size of the Earth. These inferences concerning Sirius B allow some immediate estimates that will shed light on the nature of white dwarfs even before carrying out any detailed analysis.

16.1.1 Density and Gravity

White dwarfs contain roughly the mass of the Sun in a sphere the size of the Earth, so the average density is the density of the Sun multiplied by the cube of the ratio of the Solar to Earth radii. The Sun is of order 100 times larger than the Earth and the Solar average density is of order 1 g cm^{-3}, so white dwarfs have densities in the vicinity of 10^6 g cm^{-3}. For Sirius B the average density calculated from the observed mass and radius is about 2.5×10^6 g cm^{-3}. The gravitational acceleration and the escape velocity at the surface for Sirius B are [124]

$$g = \frac{Gm}{R^2} \simeq 3.7 \times 10^8 \text{ cm s}^{-2} \qquad \frac{v_{\text{esc}}}{c} = \sqrt{\frac{2Gm}{Rc^2}} \simeq 0.02, \qquad (16.1)$$

respectively, indicating that the gravitational acceleration g is almost 400,000 times larger than at the Earth's surface but that $v_{\text{esc}} \ll c$, so general relativity effects are small and Newtonian gravity can be used in initial approximation.

16.1.2 Equation of State

The preceding discussion suggests that hydrostatic equilibrium under Newtonian gravitation is adequate as a first approximation for the structure of white dwarfs. What about the microphysics of the gas? Is a Maxwell–Boltzmann description valid, or will the quantum statistical properties of the gas play a crucial role, and will electron velocities be describable classically or will velocities become relativistic?

Average electron velocities and special relativity: Assume initially that velocities are nonrelativistic and that electrons are the primary source of internal pressure for the white dwarf, and for simplicity take the white dwarf to be composed of a single kind of nucleus having atomic number Z, neutron number N, and atomic mass number $A = Z + N$. Then the average electron velocity is $\bar{v}_e = \bar{p}/m_e$ where $\bar{p}$ is the average momentum and m_e is the electron mass. By the uncertainty principle, the average momentum may be estimated as

$$\bar{p} \simeq \Delta p \simeq \hbar/\Delta x \simeq \hbar n_e^{1/3}, \qquad (16.2)$$

where n_e is the electron number density. The gas may be expected to be completely ionized at the temperature and pressure characteristic of a white dwarf and the corresponding electron number density is

$$n_e = \left(\frac{\text{number } e^-}{\text{number nucleons}}\right) \left(\frac{\text{number nucleons}}{\text{unit volume}}\right) = \left(\frac{Z}{A}\right) \left(\frac{\rho}{M_u}\right). \qquad (16.3)$$

Therefore, the average electron velocity may be approximated by

$$\frac{\bar{v}_e}{c} = \frac{\hbar n_e^{1/3}}{m_e c} = \frac{\hbar}{m_e c} \left(\frac{Z\rho}{AM_u}\right)^{1/3} \simeq 0.25,$$

where a density of 10^6 g cm^{-3} was used and where it was assumed that $A = 2Z$, as would be true for ^{12}C, ^{16}O, or ^{4}He, which are the primary constituents of most white dwarfs. (Matter composed of isotopes for which $Z = N$ is termed *symmetric matter.*) This is only an order of magnitude estimate but it suggests that on general grounds, electron velocities will become relativistic (a significant fraction of c) for higher-density white dwarfs.

Average separation of electrons and degeneracy: The average spacing between electrons may be estimated as $d \simeq n_e^{-1/3} \simeq 1.5 \times 10^{-10}$ cm, where $Z/A = 0.5$ and $\rho = 10^6$ g cm^{-3} have been used. The de Broglie wavelength of electrons in the gas is on average

$$\bar{\lambda}_e = \frac{h}{p} = \frac{h}{\bar{v}_e m_e} \simeq 9.6 \times 10^{-10} \text{ cm}.$$

Because the separation of particles in the gas is less than their de Broglie wavelength, by arguments similar to those given in Chapter 3 it may be concluded that the electron gas will be degenerate, provided that the temperature is not too high. For a degenerate fermion gas the Fermi energy in $\hbar = c = 1$ units is $E_f = (p_f^2 + m^2)^{1/2}$, and the gas will remain degenerate as long as the Fermi energy is much larger than the characteristic energy kT of particles in the gas. From the preceding equation $E_f \geq m_e c^2 = 0.511$ MeV, implying that a temperature $T = E/k = 0.511$ MeV$/k \simeq 6 \times 10^9$ K is required to break the degeneracy. The properties of white dwarfs indicate that their interior temperatures are typically in the range 10^6–10^7 K, so white dwarfs contain cold, degenerate gases of electrons, which may be described approximately using polytropic equations of state having the form $P = K\rho^\gamma$, where $\gamma = \frac{5}{3}$ in the limit of nonrelativistic degenerate electrons and $\gamma = \frac{4}{3}$ in the limit of ultrarelativistic degenerate electrons.

The ions and photons: While the electrons may be expected to be degenerate and to become relativistic at higher densities, the ions are much more massive than the electrons. They are neither relativistic nor degenerate, and are well described by an ideal gas equation of state. Because the ions move slowly, they contribute little to the pressure. However, they carry most of the mass and most of the heat energy stored in the white dwarf is associated with motion of the ions. Finally, photons in the white dwarf constitute an ultrarelativistic gas approximated by a Stefan–Boltzmann equation of state, $P = \frac{1}{3} a T^4$. Thus, the picture that emerges for a white dwarf is of a dense object that may be hot in the normal sense but that is cold in a quantum-mechanical sense, for which the mechanical properties (exemplified by the pressure, which is generated primarily by the degenerate electrons) are decoupled from the thermal properties (which are associated primarily with the ions at normal temperatures).

16.1.3 Ingredients of a White Dwarf Description

Summarizing the results described above, it is suggested that an approximate description of a white dwarf may be afforded by a theory for which

1. The stable configurations correspond to hydrostatic equilibrium under Newtonian gravitation.

2. The ions carry most of the mass and store most of the thermal energy of the white dwarf, but the electrons are responsible for the bulk of the pressure.
3. The electron equation of state will be that of a cold degenerate gas, conveniently approximated in the polytropic form $P = K\rho^\gamma$ with $\gamma = \frac{5}{3}$ for nonrelativistic and $\gamma = \frac{4}{3}$ for ultrarelativistic electrons, respectively.
4. Ions are nonrelativistic and may be described by an ideal gas equation of state.
5. Photons are described by a Stefan–Boltzmann equation of state.
6. Because the degenerate electron gas is primarily responsible for the pressure but its equation of state does not depend on temperature, the thermal and mechanical properties of the white dwarf are decoupled.
7. As density increases the average velocity of the electrons increases and special relativity becomes increasingly important, corresponding to a transition $P \propto \rho^{5/3} \to P \propto \rho^{4/3}$ in the electron equation of state.

We turn now to a theoretical description embodying these ideas in a simple formalism. In all cases an ideal gas equation of state will be assumed for the ions and a Stefan–Boltzmann equation of state for the photons. For the electron equation of state, first an analytical approximation will be considered in terms of solutions to the Lane–Emden equations described in Section 8.4, and then a simple model that incorporates a more realistic equation of state allowing arbitrary levels of relativity and degeneracy will be investigated numerically.

16.2 Polytropic Models of White Dwarfs

Since white dwarfs are in hydrostatic equilibrium with degenerate electrons supplying the pressure, we may expect that solutions of the Lane–Emden equation (8.8) with polytropic index $n = \frac{3}{2}$, corresponding to $\gamma = (n+1)/n = \frac{5}{3}$, are relevant for the structure of low-mass white dwarfs where electron velocities are nonrelativistic. Likewise, it may be expected that in more massive white dwarfs the electrons become relativistic and the corresponding structure is related to a Lane–Emden solution with polytropic index $n = 3$, corresponding to $\gamma = \frac{4}{3}$.

16.2.1 Low-Mass White Dwarfs

First consider a low-mass white dwarf. Assuming a $\gamma = \frac{5}{3}$ (that is, $n = \frac{3}{2}$) polytropic equation of state, Eq. (8.13) then gives

$$MR^3 = \text{constant.} \tag{16.4}$$

This surprising result implies that, contrary to the behavior of normal stars, increasing the mass of a low-mass white dwarf causes its radius to *shrink*. This behavior is a direct consequence of the degenerate electron equation of state. Inserting constants in Eq. (8.10),

16.4 Cooling of White Dwarfs

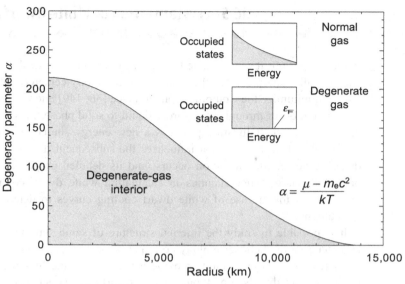

Fig. 16.5 The calculated degeneracy parameter $\alpha \equiv (\mu - m_e c^2)/kT$ of Eq. (16.14) versus radius for a white dwarf simulation, where μ is the electron chemical potential and m_e the electron mass. A similar equation of state as for Fig. 16.1 was used in the calculation. We see that the electron gas is highly degenerate except very near the surface of the star. Shown inset are state occupation profiles for a normal gas and a degenerate fermionic gas, with ε_F the Fermi energy.

Box 16.1   **Neutrino Cooling of White Dwarfs**

White dwarfs can cool by emission of neutrinos from the interior as well as through photons emitted from the surface. From Fig. 7.14 and the typical densities and interior temperatures encountered for white dwarfs, the dominant source of neutrino cooling is expected to be plasma neutrinos emitted from the central region, for white dwarfs with surface temperatures greater than about 25,000 K [228]. Direct neutrino emission from specific stars has been observed thus far only for two cases: neutrinos from the Sun and the neutrino burst detected from Supernova 1987A. It has been proposed that neutrino emission might be observed *indirectly* by studying the effect of neutrino cooling on pulsations of young, hot, variable white dwarfs [228]. The DBV white dwarfs (white dwarfs with a helium atmosphere that are pulsating variables; see the classification discussed in Box 2.1) have effective surface temperatures around 25,000 K, so they are thought to cool largely through emission of plasma neutrinos. Simulations indicate that the rate of change in the observed pulsation period versus time is affected significantly by neutrino emission, suggesting that changes observed in the pulsation period of a suitable DBV white dwarf could be used to infer the rate of neutrino cooling.

As discussed further in Box 16.1, white dwarfs may cool by neutrino emission from hot, dense central regions in addition to cooling by photon emission from the surface. This is in fact thought to be the dominant source of cooling for young, hot white dwarfs and occurs primarily through emission of plasma neutrinos from the deep interior.

16.5 Crystallization of White Dwarfs

In the early 1960s it was predicted that as the plasma in a white dwarf cools it may become energetically favorable for the ions to form a body-centered cubic (BCC) crystalline lattice to minimize the Coulomb repulsion (see [36, 149] and references therein).[3] This is expected to occur through a first-order liquid to solid phase transition. The corresponding latent heat of crystallization provides a new energy source supplementing the thermal energy stored in the ions that influences the subsequent thermal evolution of the white dwarf. Whether this transition occurs, and its detailed properties if it does, constitutes one of the largest uncertainties in calculating white dwarf cooling. This in turn has implications for the use of white dwarf cooling curves to determine the age of stellar populations.

It is possible to study the internal structure of some stars through *asteroseismology*, by extending the helioseismology concepts of Section 10.2 to other stars (see Ref. [106] for a review). These methods provide a way to test the hypothetical crystallization of cooling white dwarfs. For typical white dwarfs theory suggests that crystallization in the core begins when the surface temperature decreases to 6000–8000 K, but in more massive white dwarfs crystallization is expected to set in at a higher surface temperature. Thus asteroseismology on massive white dwarfs is a promising source of evidence for crystallization.

Asteroseismology of the pulsating DAV white dwarf BPM 37093 has been used to infer its internal structure [149].[4] This star represents a particularly favorable case because its mass of $1.1\,M_\odot$ is the largest known for a DAV white dwarf. The oscillations of this and other pulsating white dwarfs correspond to non-radial gravity waves (g-modes; see Section 10.2.1), which represent oscillations with a restoring force provided by gravity. If the core of a white dwarf becomes solid because of the crystalline phase transition, the difference in density at the solid–liquid core boundary is very small so the mechanical properties of the white dwarf are not altered significantly and the effect on evolution of the white dwarf is expected to be minimal. However, formation of a crystalline core may have a *significant effect on the star's pulsations* because the additional shear in the solid relative to the liquid causes a mismatch between interior and exterior waves at the core boundary, and the exterior waves are almost completely reflected by the boundary. Hence, the non-radial g-modes *cannot penetrate the solid–liquid interface*, the white dwarf's observable pulsations become linked to g-modes confined to the non-crystalline liquid region outside the core, and the size of the crystalline core exerts a potentially observable effect on the pulsations of the star [149, 152]. From analysis of the observed pulsation frequencies it was

---

[3] Strictly this presumes infinite ionic masses but the quantum zero-point energy associated with uncertainty-principle fluctuations is estimated to be much less than the Coulomb repulsion. Hence the classical approximation of a fixed lattice is expected to be a very good one.

[4] As discussed in Box 2.1, DAV is the classification for a pulsating white dwarf that has a hydrogen atmosphere. White dwarfs of class DAV are also known as ZZ Ceti variables (see Table 15.1 and Fig. 15.1). The star BPM 37093 (V886 Centauri in variable-star nomenclature) is in Centaurus at a distance of about 53 ly. It has a surface temperature of a little less than 12,000 K and a mass of $1.1\,M_\odot$.

concluded that BPM 37093 has a core of crystallized carbon and oxygen containing about 90% of the white dwarf's mass [149]. A different analysis of BPM 38093 observational data concluded that the crystalline mass most likely lies between 32% and 82% [58]. In either case there is credible evidence that a substantial fraction of the white dwarf has entered the crystalline phase predicted by theory.[5]

16.6 Beyond White Dwarf Masses

The preceding discussion of limiting masses for white dwarfs assumes that all pressure derives from electrons. However, if the Chandrasekhar mass is exceeded and the system collapses gravitationally, eventually a density will be reached where the nucleons, which also behave as fermions under these conditions, will begin to produce a strong degeneracy pressure. Whether this nucleon degeneracy pressure can halt the collapse depends on the mass of the collapsing object. Calculations indicate that for a mass less than 2–3 solar masses (depending on the equation of state), the collapse converts essentially all protons into neutrons through the weak interactions, producing a neutron star. The degeneracy pressure of the neutrons halts the collapse at neutron-star densities and radii approximately 500 times smaller than characteristic white dwarf radii. Calculations, and general considerations concerning strong gravity that are independent of details, indicate that for masses greater than this even the neutron degeneracy pressure cannot overcome gravity and the system collapses to a black hole. These considerations indicate also that white dwarfs, stabilized by electron degeneracy pressure, and neutron stars, stabilized by neutron degeneracy pressure, are the *only* possible stable configurations lying between normal stars and black holes. Therefore, let us now consider the properties of neutron stars.

16.7 Basic Properties of Neutron Stars

Neutron stars were predicted in 1933 by Walter Baade (1893–1960) and Fritz Zwicky (1898–1974) as a possible end result of what would now be called a core collapse supernova, and Robert Oppenheimer (1904–1967) and George Volkov (1914–2000), building on work by Richard Chase Tolman (1881–1948), worked out and solved equations describing their general structure and properties in 1939. However, they were not taken very seriously by observers until the discovery of radio pulsars in the 1960s pointed to rapidly spinning neutron stars as their most likely explanation.

---

[5] Most white dwarfs are rich in carbon so crystallized white dwarfs have been referred to whimsically as "diamonds in the sky." Accordingly, BPM 37093 has been nicknamed "Lucy" by some, in reference to the famous Beatles song "Lucy in the Sky with Diamonds."

> **Box 16.2**   **Electron Capture and Neutronization**
>
> The formation of a neutron star results from a process called *electron capture* (a form of β-decay), which can follow the core collapse of a massive star late in its life to produce a supernova. The process is also called *neutronization,* because it destroys protons and electrons to create neutrons. The basic reaction is $e^- + p^+ \rightarrow n + \nu_e$, where the protons can be either free or bound in nuclei. It is slow under normal conditions (because it is mediated by the weak interaction) but faster in the high density and temperature environment produced by core collapse in a massive star.[a] In the resulting supernova explosion the enormous amount of energy released gravitationally leads to expulsion of the outer layers of the star, leaving behind a dense, hot remnant. As neutronization proceeds the neutrinos escape, carrying off energy and leaving behind the neutrons. Because neutrons carry no charge, there is no electrical repulsion as in normal matter and the core remnant can collapse to very high density once it has become mostly neutrons. The formation of actual neutron stars is more complex than this and they are not composed entirely of neutrons, but this simple picture captures the basic idea.
>
> [a] At a fundamental level the weak interaction strength scales linearly with the density and quadratically with the temperature. At normal temperatures and densities the strength is small, but at extremely high temperatures and densities the weak interactions are much stronger than under normal conditions.

16.7.1 Sizes and Masses

Most neutron stars have masses of 1–2 solar masses and diameters of 15–20 km. Very loosely (a more precise discussion follows below), a neutron star packs the mass of a normal star like the Sun into a volume of about 10 km in radius. From this we may estimate immediately an average density of order 10^{14} g cm^{-3} for neutron stars (it can actually be about an order of magnitude larger than that). Thus, they have enormous densities that are similar to those encountered in the nucleus of the atom. In fact, in certain ways (but not all), a neutron star may be likened to a gigantic atomic nucleus. These very large densities imply strong gravitational fields and the possibility of significant general relativistic deviations from Newtonian gravity.[6] The mechanism leading to formation of a neutron star is described qualitatively in Box 16.2, and in more detail in Chapter 20.

Because neutron stars are tiny it might be expected that they would be very difficult to detect. In fact, you are asked to demonstrate in Problem 16.4 that neutron stars have luminosities that are comparable to that of stars like the Sun because they have very high surface temperatures (of order 10^6 K). Because of the high temperature, the light emitted peaks in the extreme-UV and soft X-ray portion of the spectrum, so neutron stars are readily visible to X-ray observatories. An X-ray image of a neutron star at the center of an expanding supernova remnant is shown in Fig. 16.6.

---

[6] Our initial discussion will depend primarily on Newtonian gravity, which is sufficient for a qualitative description. However, a quantitative treatment requires general relativity, since the escape velocity at the surface of a neutron star is a significant fraction of the speed of light. An introduction to the general relativistic description of neutron stars may be found in Chapter 10 of Ref. [100], and in Section 16.8.1 we will summarize briefly this description.

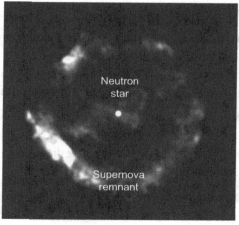

Fig. 16.6 Chandra X-ray observatory image of a neutron star in the center of an expanding supernova remnant. This neutron star is also a pulsar (see Section 16.9). It is believed that the neutron star and the expanding remnant surrounding it were produced by a supernova seen on Earth in 386 AD by Chinese observers.

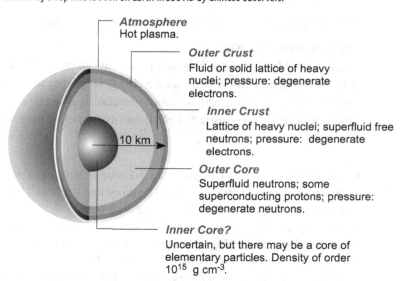

Fig. 16.7 The schematic internal structure of a typical neutron star.

16.7.2 Internal Structure

The internal structure of a neutron star can be divided into the following general regions, which are illustrated in Fig. 16.7.

1. The atmosphere is thin and consists of very hot, ionized gas.
2. The outer crust is only a few hundred meters thick and consists of a solid lattice or a dense liquid of heavy nuclei. The dominant pressure is from electron degeneracy and the density is not high enough to favor neutronization.

> **Box 16.3** **Neutron Stars are Bound by Gravity**
>
> A neutron star has some resemblance to a 20-km diameter atomic nucleus. However, there is at least one big difference. Gravity is negligible on a nuclear scale and a nucleus is held together by effective nuclear forces that derive from the *strong interactions*, but a neutron star is bound by *gravity*, and the strength of that binding is such that the density of neutron stars is even greater than that of nuclear matter.
>
> **How Does the Weakest Force Produce the Most Dense Object?**
> How, then, is it possible that the weakest force (gravity) can produce an object more dense than atomic nuclei, which are held together by a diluted form of the strongest force known? The answer has to do with the *range and sign* of the forces involved:
>
> 1. Gravity is weak but it is long-ranged and always attractive.
> 2. The nuclear force is short-ranged, acting only between near neighbors.
> 3. The normally attractive nuclear force becomes repulsive at short distances. In fact, a neutron star would explode if gravity were suddenly removed, because the neutrons have been forced so close together by gravity that the average nuclear force between them is repulsive.
>
> Thus gravity is weak but long-ranged, and always acts with the same sign.
>
> **The Tortoise and the Hare**
> This is a kind of *Tortoise and Hare* fable, with a correspondingly predictable outcome: gravity is weak, but relentless and always attractive (in the absence of dark energy acting on cosmological scales). Over large enough distances and long enough time, gravity – that plodding tortoise of forces – always wins. That is why a neutron star can be compressed to the highest material density known by the most feeble of interactions.

3. The inner crust is ~ 1 km thick. The pressure is higher and the lattice of heavy nuclei is now permeated by free superfluid neutrons that begin to "drip" out of the nuclei. The pressure is still mostly from degenerate electrons.
4. The outer core is primarily superfluid neutrons that supply most of the pressure through neutron degeneracy. This region gives the neutron star its name.
5. The structure of the inner core is less certain because we know less about how matter behaves under the intense pressure at the center. It might even consist of a solid core of particles more elementary than nucleons (pions, hyperons, quarks, ...).

Although much of a neutron star consists of closely packed neutrons and thus has some resemblance to a giant atomic nucleus, it is important to remember that it is gravity, not the nuclear force, that holds a neutron star together. This is discussed further in Box 16.3.

16.7.3 Cooling of Neutron Stars

As will be discussed further in Chapter 20, neutron stars form from the innermost material left behind in a core collapse supernova (provided that the center does not collapse to a black hole). The *protoneutron star* formed in the supernova is initially very hot and bloated (typically with a temperature $\sim 10^{11}$ K and a radius some 30% larger than the final neutron star that it will become), and is still being powered by accretion from the part of the envelope that did not escape the star in the explosion. As the accretion tapers off the nascent neutron star cools rapidly by neutrino emission (see Fig. 20.12).

Example 16.2 In high-energy astrophysics temperatures are often quoted in energy units, with the corresponding temperature in kelvin given by $T = E/k$, where k is Boltzmann's constant. A characteristic temperature for a protoneutron star formed in the core collapse of a 20 $M_\odot$ star is ~ 50 MeV [79], from which

$$T \simeq \frac{50\,\text{MeV}}{8.617 \times 10^{-11}\,\text{MeV}\,\text{K}^{-1}} = 5.8 \times 10^{11}\,\text{K}$$

for the corresponding temperature in kelvin.

Once the protoneutron star has cooled and shrunk enough in radius to resemble a neutron-star proper, it continues to cool both by X-ray emission from the surface and neutrino emission from the interior. As will be discussed in Section 16.7.4, the protons in the interior likely become superconducting and the neutrons superfluid very quickly as the temperature drops. Hence the interior becomes a very good conductor of heat and its temperature is uniform. But in a thin layer near the surface the material is composed of atomic nuclei rather than a nucleonic superfluid, which is a much poorer conductor of heat and the temperature drops off rapidly at the surface. In a situation reminiscent of the blanket over a metal ball picture invoked earlier in this chapter to describe the temperature profile of a white dwarf, the interior is much hotter than the (X-ray emitting) photosphere: typical estimates are that the interior is initially at a temperature of $\sim 10^9$ K while the photosphere has a temperature of only a few times 10^6 K.

Under these conditions the surface X-ray luminosity will be quite large (see Problem 16.4), but since even dense neutron star matter is largely transparent to neutrinos,[7] neutrino emission from the interior will be *much more efficient* at transporting energy than X-ray emission from the photosphere and the young neutron star cools primarily by neutrino emission. This situation will continue until the neutron star is quite cold; only then will X-ray emission from the photosphere begin to rival neutrino emission in cooling the star. In more massive young neutron stars the primary neutrino-cooling mechanism is the direct Urca process operating at the center (see Box 7.5). In less massive neutron stars the dominant cooling is from slower neutrino processes such as neutrino bremsstrahlung and the modified Urca process, also described in Box 7.5. In either case the neutron star cools sufficiently on a 100,000-year timescale that the surface temperature is reduced to less than 10^6 K, suppressing the emission of X-rays and rendering it largely invisible if it is an isolated neutron star.

[7] Conversely, it will be seen in Chapter 20 that in the protoneutron star forming at the center of a core collapse supernova the neutrino mean free path can become less than a meter and initially neutrinos can be trapped for of order 10 seconds before they are able to escape. A primary reason is temperature. It is much higher in the protoneutron star, and the strength of the neutrino interaction with matter scales as the square of the temperature.

16.7.4 Evidence for Superfluidity in Neutron Stars

Just as for certain systems in condensed matter – though for different microscopic reasons – in neutron stars the neutrons and protons can exhibit properties of essentially zero resistance to mass flow (superfluidity) or to charge flow (superconductivity). (For convenience we will sometimes term both effects superfluidity.) This can have strong influence on the rotational and magnetic properties of the neutron star, as well as its rate of cooling.

The Cas A neutron star: Chandra X-ray Observatory first-light observations discovered a compact object at the center of the Cassiopeia A supernova remnant. It was subsequently identified as the neutron star left over from the supernova explosion, which is estimated to have occurred in the year 1681 ± 19. The corresponding age of about 330 years makes the Cas A neutron star the youngest known. Evidence for superfluidity from cooling of the Cas A neutron star is presented in Fig. 16.8, where comparison with theory indicates that

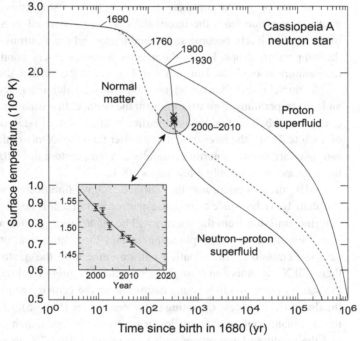

Fig. 16.8 Cas A neutron star cooling compared with calculations [122, 164, 165]. Curves indicate theory: "Normal matter" (dashed curve) assumes no superfluidity, the solid curve labeled "Proton superfluid" assumes only the protons to be superfluid, and the solid curve labeled "Neutron–proton superfluid" assumes both protons and neutrons to be superfluid. Temperatures predicted by the neutron–proton superfluid model are marked for several years beginning 10 years after the birth of the neutron star in ∼1680. Data points (×) were obtained by the Chandra X-ray Observatory from 1999–2010 and suggest rapid cooling. Adapted from *Proceedings of Science–PoS (Confinement X)*, **260**, "The Hottest Superfluid and Superconductor in the Universe: Discovery and Nuclear Physics Implications," Ho, W. C. G. et al. Copyright (2012) by the author(s) under the terms of the creative Commons Attribution (Non-Commercial) Share Alike Licence.

> **Box 16.4**   **Superfluidity in Helium-3 and Neutron Stars**
>
> At zero pressure and zero magnetic field a gas of ^{3}He atoms (commonly termed helium-3) becomes a liquid if the temperature is lowered to about 1 K. If the temperature is lowered further to 2–3 millikelvin, two distinct superfluid phases that are termed the A phase and B phase appear (see Chapter 7 of Ref. [24] for an introduction to the superfluid phases of helium-3).
>
> **Dense Fermi Liquids**
> Because of the odd number of spin-$\frac{1}{2}$ nucleons, the total spin of helium-3 is $\frac{1}{2}$. Thus, to the degree that the influence of any internal structure can be ignored in interactions, which is a good approximation at such low temperatures, the helium-3 is found to behave very much like a strongly interacting gas of fermions (see Problem 16.11). Because the correlations are strong, helium-3 is commonly termed a *dense Fermi liquid*, and is a specific example of a *quantum liquid*, which is a liquid in which the effects of both quantum mechanics and quantum statistics are important. In this example, the Fermi–Dirac statistical properties of the helium-3 are central to its behavior.
>
> **Dense Quantum Liquids in Neutron Stars**
> The relevance of this interesting physics from condensed matter to the present astrophysical discussion is that in neutron stars it is thought that the spin-$\frac{1}{2}$ neutrons can become superfluid and the spin-$\frac{1}{2}$ protons can become superconducting for interior temperatures below around 10^8–10^{10} K because of the very high density. This is observable in effects on the moment of inertia for rotating neutron stars (for example, the pulsar glitches discussed in Section 16.9.4), and in the cooling rates for young neutron stars discussed in Section 16.7.4. Thus, the interiors of neutron stars are likely dense quantum liquids exhibiting physics that may have some relationship to that of superfluid helium-3.

observed cooling curves for neutron stars are strongly influenced by the presence of both neutron and proton superfluidity.

A possible superfluid phase transition: The theoretical curves suggest substantial differences between neutron stars with "normal" matter and those containing superfluids beginning about 40 years after the birth of the neutron star. In these models, proton superconductivity sets in soon after the neutron star is formed, which suppresses neutrino emission and causes the cooling rate to be slower than that for normal matter. Then, when the core neutrons also become superfluid around the year 1930 the crust is predicted to cool very quickly for several hundred years. The rapid drop of surface temperature observed by Chandra between 1999 and 2010 has been interpreted as a phase transition to superfluid neutrons in the core (though alternative explanations have been proposed) [122, 164, 165]. Superfluidity in neutron stars is discussed further in Box 16.4.

16.8 Hydrostatic Equilibrium in General Relativity

The discussion of neutron stars has been based primarily on Newtonian gravity. This is adequate at a qualitative level but gravity for neutron stars is of sufficient strength that a quantitative description of them requires general relativity (GR), with their structure determined by solving the Einstein equations for their dense-matter interior. This task is beyond our present scope. It is taken up in Ref. [100], to which the reader is directed for more details. However, before leaving this introduction to neutron stars, let us sketch briefly without proof how hydrostatic equilibrium is modified by general relativity in neutron stars.

16.8.1 The Oppenheimer–Volkov Equations

Stable neutron stars are in hydrostatic equilibrium, with gravity balanced against pressure-gradient forces, just as was found in Chapter 4 for normal stars. However, when gravity is derived from general relativity rather than from the Newtonian theory the corresponding equations for hydrostatic equilibrium are modified in a non-trivial way. As discussed in Chapter 10 of Ref. [100], by assuming the interior of neutron stars to consist of a perfect fluid (no shear or viscosity effects) the general relativistic equations for hydrostatic equilibrium can be written in the form

$$4\pi r^2 dP(r) = \frac{-m(r)dm(r)}{r^2}\left(1+\frac{P(r)}{\varepsilon(r)}\right)\left(1+\frac{4\pi r^3 P(r)}{m(r)}\right)\left(1-\frac{2m(r)}{r}\right)^{-1} \quad (16.15)$$

$$dm(r) = 4\pi r^2 \varepsilon(r) dr. \quad (16.16)$$

where P is pressure, $\varepsilon(r)$ is energy density, and units have been chosen so that the gravitational constant G is equal to one. Equation (16.15) expresses hydrostatic pressure balance for a fluid in general relativity and Eq. (16.16) expresses the conservation of mass–energy; together they are termed the *Oppenheimer–Volkov* equations (or sometimes the *Tolman–Oppenheimer–Volkov equations*). It is instructive to compare the Oppenheimer–Volkov equations with their Newtonian counterparts.

16.8.2 Comparison with Newtonian Gravity

For Newtonian gravity the equations of hydrostatic equilibrium are given in Lagrangian form by Eqs. (4.12) and (4.15), which in $G=1$ units become

$$4\pi r^2 dP = -\frac{mdm}{r^2} \qquad dm = 4\pi r^2 \rho dr. \quad (16.17)$$

Comparing these equations with Eqs. (16.15)–(16.16) indicates that the formulation of hydrostatic equilibrium in the general relativistic solution is equivalent to that in the Newtonian gravity solution if the energy density is substituted for the mass density, $\rho c^2 \to \varepsilon$, except for three correction factors (in parentheses) in the GR version (16.15) that depend on the pressure and the mass, and represent general relativity effects causing deviations from Newtonian gravitation. They have the following consequences.

1. In stars described by Newtonian gravity, ε is dominated by the rest mass of the baryons and the baryons don't contribute significantly to the pressure (which is dominated by

electrons). Thus $P(r)/\varepsilon(r) \sim 0$, and $P(r)/M(r) \sim 0$, and the first two correction factors in Eq. (16.15) are approximately unity.
2. Conversely, the first two correction factors generally exceed unity as the star becomes more massive because *pressure couples to gravity in general relativity* but not in Newtonian gravity.
3. The final correction factor in Eq. (16.15) is approximately unity for stars described by Newtonian gravity, but becomes greater than one as the mass increases and gravity becomes stronger.

Hence all three correction factors will generally be greater than one in more massive objects with greater pressure. One of the most important consequences following from these differences between Newtonian and general relativistic gravity is that gravity is stronger and is enhanced by coupling to pressure in the general relativistic description. This will imply ultimately that there are fundamental limiting masses for stable strongly gravitating objects because if the mass is large enough no amount of pressure will be able to prevent their gravitational collapse to a black hole.

16.9 Pulsars

The year 1967 saw the discovery of something remarkable in the sky: an object that appeared to be pulsing on and off with a period of about one second. Shortly, other such "pulsars" of even faster variation were discovered and the fastest now known (the *millisecond pulsars*) pulse on and off with frequencies that exceed 700 times per second. There is some variety in pulsars, but they exhibit several common characteristics:

- They have well-defined periods that challenge timing from the best atomic clocks.
- Measured periods range from tens of seconds down to 1.4 ms.
- The period of a given pulsar increases very slowly with time.

What could cause this rather remarkable behavior? We will now make the case that only a rapidly spinning neutron star can do the job.

16.9.1 The Pulsar Mechanism

The observational details for pulsars are inconsistent with an actual pulsation for known astronomical objects, but a rotating object could *appear to pulse* if it had a way to emit light in a beam that rotated with the source (just as a lighthouse appears to flash as the beam sweeps over an observer). What kind of object would be consistent with observed pulsar periods?

A dense object is required: Simple calculations show that only a very dense object could rotate fast enough and not fly apart because of the rapid rotation. Even a white dwarf is not dense enough. The minimum rotational period for a typical white dwarf would be several seconds; for shorter periods it would disintegrate. But a neutron star is so dense that it could

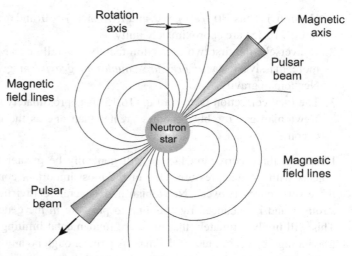

Fig. 16.9 The pulsar lighthouse mechanism. The pulsing effect arises because the rotation and magnetic axes are not aligned, causing one or both beams to sweep over the observer as the neutron star rotates.

rotate more than a thousand times a second and still hold together. This qualitative inference, augmented by much more detailed considerations, leads to the conclusion that the only plausible explanation for pulsars is that they are rapidly spinning neutron stars, with a mechanism to beam radiation in a kind of lighthouse effect that is illustrated in Fig. 16.9.

The lighthouse mechanism: The rapidly spinning electromagnetic field of the pulsar accelerates electrons away from the surface near the magnetic poles and these accelerated electrons can produce radiation by the synchrotron effect. Because of the synchrotron mechanism and the high velocity of the particles, the radiation produced is beamed strongly in the direction of electron motion. These beams of radiation rotate with the star but the magnetic axis need not coincide with the rotation axis (recall that Earth's rotational and magnetic axes have different orientations also) and the beams can gyrate around the rotation axis, as illustrated in Fig. 16.9.

If these gyrating beams sweep over the Earth they act similarly to a lighthouse and an observer on Earth sees flashes of light. Thus, the neutron star appears to be pulsing, even though it is neither pulsing nor is it really a star. Notice that this lighthouse mechanism means that not all rotating neutron stars will be seen as pulsars. Only if they are favorably oriented will the beams sweep over the Earth and give a pulsing effect.

16.9.2 Pulsar Magnetic Fields

The spin rate of a pulsar decreases slowly as it radiates away its energy, because of coupling to its magnetic field. This change is small but can be measured precisely and can be used to estimate the strength of the magnetic field associated with the neutron star. It is found that some pulsars contain the strongest magnetic fields known in our galaxy, and many of their basic properties are thought to derive from these fields. Some typical magnetic field strengths for various objects are listed in Table 16.1. The table indicates that the largest

Table 16.1 Characteristic magnetic field strengths

Object	Strength (gauss)†
Earth's magnetic field	0.6
Simple bar magnet	100
Strongest sustained laboratory fields	4.5×10^5
Strongest pulsed laboratory fields	10^7
Strong magnetic stars	10^4–10^5
Maximum field for ordinary stars	10^7
Radio pulsar	10^{10}–10^{12}
Magnetars	10^{12}–10^{15}

†Magnetic fields in astrophysics are often given in gauss; 1 tesla = 10^4 gauss.

known magnetic fields are associated with radio pulsars and magnetars (see Section 16.10), which both involve rotating neutron stars. We conclude that strong magnetic fields are likely to be common for neutron stars, although deducing that is more difficult if the neutron star is not observed as a pulsar or magnetar.

16.9.3 The Crab Pulsar

The first pulsar was found by Jocelyn Bell Burnell and Anthony Hewish in 1967. The most famous pulsar was discovered shortly after that. The *Crab Pulsar* lies in the Crab Nebula (M1), about 7000 lightyears away in Taurus. It rotates about 30 times a second, emitting a double pulse during each rotation in the radio through gamma-ray spectrum. In visible light the Crab Pulsar appears as a magnitude 16 star near the center of the nebula but stroboscopic techniques reveal it to be pulsing, as shown in Fig. 16.10(a). Both the image sequence at the top and the lightcurve display the "double pulsing" of the Crab: in each cycle there is a strong primary pulse followed by a much weaker secondary pulse. This double pulsing effect can be explained by the lighthouse model if the beam from one magnetic pole sweeps more directly over the Earth but the beam from the other pole does so only partially. Most pulsars are detectable only by their radio frequency radiation but a few (like the Crab) pulse strongly in other wavelength bands.

16.9.4 Pulsar Spindown and Glitches

In some pulsars *glitches* are observed where the spin rate jumps suddenly to a slightly higher value and then continues its slow decline. Three glitches for the Vela Pulsar are illustrated in Fig. 16.10(b). These glitches are evidence that some internal rearrangement of the neutron star has altered its rotation rate by a small amount. One proposal is that "starquakes" occur in the dense crust, causing the neutron star to contract slightly and thus to spin faster by angular momentum conservation. A second theory proposes that angular momentum stored in circulation of an internal superfluid liquid (one that exhibits no frictional effects) is suddenly transferred to the crust, altering the rotation rate. The possibility of superfluidity in neutron stars was discussed in Section 16.7.4 and in Box 16.4.

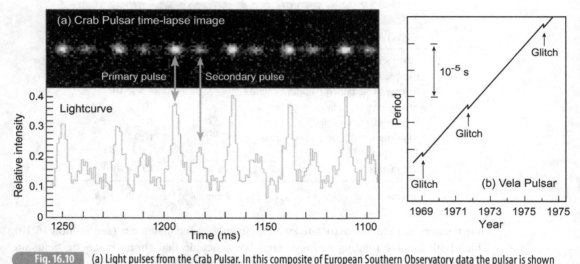

Fig. 16.10 (a) Light pulses from the Crab Pulsar. In this composite of European Southern Observatory data the pulsar is shown in a time lapse image at the top and the lightcurve is displayed at the bottom on the same timescale. (b) Glitches in the period of the Vela Pulsar.

16.9.5 Millisecond Pulsars

Since the spin of a pulsar slows with time, it may be expected that the fastest pulsars are the youngest. For example, the Crab Pulsar is young (less than 1000 years), and pulses 30 times a second. However, this reasoning breaks down for the fastest pulsars known, which have millisecond periods. About 300 such *millisecond pulsars (MSP)* are known in the galaxy and in globular clusters [181]. The fastest one known pulses 716 times per second [116], implying a 20-km wide neutron star that is spinning as fast as a kitchen blender. For many of these very fast pulsars there is evidence that they are *old, not young*, as would be expected for the fastest spin rates. This evidence consists primarily of (1) the rate at which the pulsar spin is slowing and (2) the astrophysical environment where the millisecond pulsars are found.

For example, the first millisecond pulsar discovered, PSR 1937+21, has a very high spin rate but the rate is decreasing very slowly.[8] This slow spindown rate implies that it has a weak magnetic field and is old. (Older pulsars should have weaker fields and these should be less effective than younger, stronger fields in braking the pulsar spin by electromagnetic coupling.) Also, many millisecond pulsars are found in globular clusters, which contain an old population of stars. Therefore, they are not likely to be sites of recent supernova

---

[8] In the name PSR 1937+21 the PSR indicates a pulsar, the first part of the number gives the right ascension in hours and minutes, and the second part of the number gives the declination (preceded by sign) in degrees. The Crab Pulsar is PSR 0531-21 in this naming system. The letter B (1950.0 epoch coordinates) or J (2000.0 coordinates) may be prepended to the number to indicate the coordinate system used; for example PSR J0337+1715.

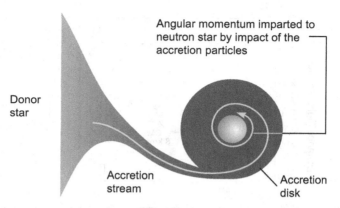

Fig. 16.11 The spinup mechanism for producing millisecond pulsars. Accretion in binary systems is discussed more extensively in Chapter 18.

explosions that could have produced young pulsars since core collapse supernovae occur in very short-lived, massive stars.

Binary spinup: The most plausible explanation for the seeming contradiction that the fastest pulsars appear to be very old is that millisecond pulsars have been "spun up" at some point in their history to higher rates. The mechanism proposed to do this involves mass transfer in binary systems that adds angular momentum to the neutron star, as illustrated in Fig. 16.11. In effect, this *binary spinup* accretion mechanism (also termed *MSP recycling*) transfers angular momentum from the orbital motion of the binary to rotation of the neutron star. Later in the evolution of the system, after the neutron star has been spun up to high rotational velocity the primary star may become a supernova and disrupt the binary system, leaving the rapidly spinning but old neutron star as a lone millisecond pulsar that defies the systematics expected for the evolution of isolated neutron stars.

The pulsar–WD–WD triplet PSR J0337+1715: A rather exotic example of a millisecond pulsar is the triple-star system PSR J0337+1715, which contains a millisecond radio pulsar of period 2.3 ms and two white dwarfs, with orbits shown in Fig. 16.12. This is a *hierarchical triple-star system,* meaning that two of the stars are relatively close to each other and the other is much further away: in PSR J0337+1715 the ratio of the periods associated with the outer and inner binaries is ~ 200. Such hierarchical systems can have long periods of dynamical stability. PSR J0337+1715 has an interesting evolutionary history that is sketched in Fig. 16.13 [181, 210]. According to the scenario outlined in the caption, which is based on a self-consistent, semi-analytic analysis constrained by the current observations [210], this system underwent a common envelope (CE) phase (Section 18.3), three periods of Roche lobe overflow (Section 18.2), a supernova (Section 20.3), and two low-mass X-ray binary (LMXB) episodes (Section 18.6.2) to arrive at the present configuration of a completely degenerate (neutron star + WD + WD) triple-star system in which the neutron star is a millisecond pulsar. As the authors of Ref. [210] stress, this explanation for how PSR J0337+1715 came to be stretches current understanding of stellar evolution and stellar interactions to the limit. Hence PSR

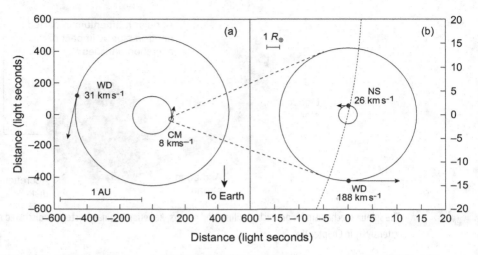

Fig. 16.12 Orbits of the triplet hierarchical system PSR J0337+1715 [181]. (a) Orbits of the outer white dwarf (WD) and the center of mass (CM) for the inner white dwarf and neutron star pair. (b) Left side scaled up by a factor of 30 to show the orbits for the inner white dwarf and neutron star (NS). Arrows indicate orbital velocities for the center of mass of the inner binary and the individual white dwarfs and neutron star. All orbits lie almost in the same plane, are nearly circular, and have a tilt angle $i \sim 39°$ relative to the line of sight. Adapted by permission from Springer Nature: *Nature*, "A Millisecond Pulsar in a Stellar Triple System," Ransau et al., Copyright (2014).

J0337+1715 should prove to be an excellent laboratory to study many aspects of stellar evolution that are not currently well understood, such as common envelope phases and binary spinup. In addition, this system is extremely promising as a test of the strong equivalence principle of general relativity because of the large gravitational acceleration of the inner pulsar–WD binary by the outer white dwarf, and the precise timing afforded by the pulsar.[9]

16.9.6 Binary Pulsars

Several binary star systems are known in which both components are neutron stars and one component is observed as a pulsar (*binary pulsars*), or both components are observed as pulsars (*double pulsars*) [63, 212].

Formation of neutron-star binaries: Binary neutron stars are of considerable interest for stellar physics because of the question of how such systems could form. Either a binary star system survives two successive supernova explosions to form the neutron stars (see Chapter 20), without disrupting the binary gravitationally, or two free neutron stars in a dense cluster capture gravitationally into a binary. Neither scenario is easy to pull off

---

[9] In this context the strong equivalence principle asserts that the neutron star and inner white dwarf should fall in the same way in the gravitational field of the outer white dwarf, despite their having very different gravitational binding energies. Any deviation from this behavior would signal a breakdown of general relativity. This is discussed more extensively in Section 10.4.6 of Ref. [100].

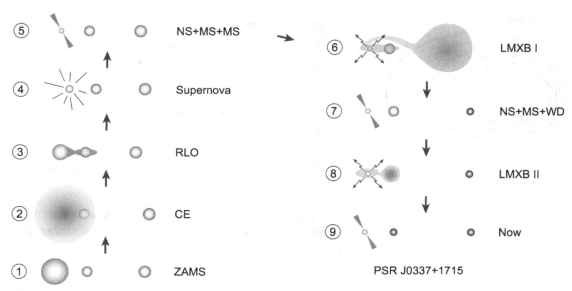

Fig. 16.13 Evolutionary history of PSR J0337+1715 [210]. Referring to the sequence numbers: (1) The triple-star system formed (ZAMS) with a 10 $M_\odot$ spectral class B star and two $\sim$ 1 $M_\odot$ stars, with one lighter star near the B star and one more distant. (2) The massive star evolved quickly and expanded, initiating a common-envelope (CE) phase in which the closest star was engulfed and the more distant one only partially so. (3) After shedding its hydrogen envelope the helium core of the massive star expanded to fill its Roche lobe and accreted onto the nearest other star by Roche lobe overflow (RLO). (4) The massive star underwent a core collapse supernova, which failed to unbind the two companions. (5) The neutron star left behind (NS) became a pulsar, while the other two stars continued their main sequence (MS) evolution. (6) The outermost star was the most massive of the two lighter stars at formation. It evolved to fill its Roche lobe and began spilling mass onto the two inner stars, with the accretion on the neutron star producing a low-mass X-ray binary (LMXB I). The spinup of the pulsar in this accretion phase was likely minimal. (7) After the outer star shed its envelope it became a white dwarf (WD), leaving the system with a pulsar, a white dwarf, and a main sequence star. (8) The remaining main sequence star expanded to fill its Roche lobe and began a second phase of accretion onto the neutron star, again producing a low-mass X-ray binary (LMXB II). During this accretion phase the pulsar was spun up to near its current 2.3 ms period. (9) The donor star for LMXB II shed all of its envelope and became a white dwarf, leaving the present configuration of PSR J0337+1715: a millisecond pulsar and two white dwarfs in hierarchical orbits. Adapted by permission from Springer Nature, "The First Gravitational-Wave Source from the Isolated Evolution of Two Stars in the 40–100 Solar Mass Range," Belczynski et al., Copyright (2016).

and each appears to be possible only under very special conditions – nevertheless, binary neutron stars may be rare but they exist.

Laboratories for testing general relativity: Binary pulsars and double pulsars are of great value in their own right as exotic endpoints of stellar evolution, but the most important consequence of binary pulsars and double pulsars is that they are tools providing extremely precise tests of the general theory of relativity [212]. This follows because pulsar periods have atomic-clock precision, so the discovery of one (better yet, two) pulsars in a binary system permits exquisite timing and precise tests of gravitational theory. For example, as

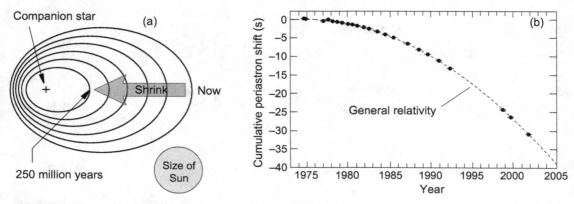

Fig. 16.14 (a) Orbit of the Binary Pulsar and its decay by gravitational wave emission, drawn to scale with the Sun shown for comparison. (b) Shift of periastron time because of gravitational wave emission. Dots with error bars indicate data; the curve is the prediction of general relativity.

illustrated in Fig. 16.14(a), the orbital semimajor axis for the *Binary Pulsar* (or *Hulse–Taylor binary*) [212] is observed to decay by about three millimeters per revolution, in precise accord with the decay of the orbit because of the emission of gravitational wave radiation (waves corresponding to ripples in spacetime; see Chapter 22) that is predicted by general relativity. Likewise, as illustrated in Fig. 16.14(b), the time of closest approach (periastron) between the two neutron stars has been observed to shift, in precise agreement with the predictions of general relativity. A more extensive discussion of the Binary Pulsar and tests of general relativity using pulsar timing may be found in Ref. [100].

The precise tracking of the Binary Pulsar orbit was the first compelling (although indirect) proof that gravitational waves exist. With the detection of a gravitational wave produced in the merger of two black holes by LIGO (Laser Interferometer Gravitational-Wave Observatory) in 2015 [12], the evidence became direct that gravitational waves – the last major untested prediction of Einstein's general relativity – exist and can be observed. This confirmation came almost exactly 100 years after gravitational waves were predicted by Einstein to be a necessary consequence of general relativity [81, 82] (though Einstein doubted that it would ever be possible to detect them, and even at times had doubts about whether they were physical). A concise review of gravitational waves in the context of stellar evolution is given in Chapter 22 and a much more extensive discussion of gravitational waves and their production in astrophysical phenomena may be found in Ref. [100].

16.10 Magnetars

Neutron stars can exhibit very strong magnetic fields. However, a few neutron stars have been found to emit bursts of high-energy photons suggesting the presence of magnetic fields that are anomalously large, even by neutron star standards. Observationally, these are called *soft gamma-ray repeaters* (SGR), or more colloquially *magnetars*. In this designation, "soft" means that the gamma-rays are of relatively low energy (in fact, they lie more in the X-ray portion of the spectrum), and "repeater" means that the bursts of gamma-rays can recur, unlike ordinary gamma-ray bursts, which have higher energy and have not been observed to recur. In these rotating neutron stars it has been proposed that the enormous magnetic fields act as a kind of brake, slowing the rotation of the star. This slowing of the rotation disturbs the interior structure of the neutron star and "starquakes" release energy periodically into the surrounding gas, rearranging the magnetic field and causing bursts of high-energy photons to be emitted.

Background and Further Reading

Standard references for white dwarfs and neutron stars include Shapiro and Teukolsky [200], and Glendenning [95]. A pedagogical introduction to neutron stars may be found in Silbar and Reddy [201] and a review in Lattimer and Prakash [140]. Neutron star cooling is reviewed in Yakovlev and Pethick [232]. General relativistic descriptions of neutron stars and the tests of general relativity using pulsars in multiple-star systems are discussed in Guidry [100].

Problems

16.1 Following the discussion in Section 16.2.3, show that the Chandrasekhar limiting mass for gravitational stability of a white dwarf is of order one solar mass by the following argument. Assume that the energy balance of the star is dominated by a competition between Fermi energy of the (assumed ultrarelativistic) electrons and the gravitational energy of the nucleons. Show that both forms of energy have the same dependence on the radius of the white dwarf and that for a sufficiently large mass the total energy is negative, implying instability against collapse. Estimate this critical mass.***

16.2 Use general arguments as in Section 16.2.3 (not Lane–Emden formulas) to show that for a nonrelativistic white dwarf supported by electron degeneracy pressure and having mass M and radius R, the product MR^3 is constant. Estimate R if $M = 0.6\,M_\odot$.***

16.3 A neutron star may be expected to have an "atmosphere" of gas at its surface, just as for a normal star. Using Newtonian gravity and typical neutron star parameters, estimate the scale height for the atmosphere (vertical distance for the pressure to decrease by e^{-1}) for a neutron star. Compare this with the radius of the neutron star and contrast with the corresponding quantities for the Earth.

16.4 Because neutron stars are so small, they might be expected to be dim and difficult to observe. Use their basic properties to estimate the luminosity of a typical neutron star and show that in fact they are expected to be comparable in luminosity to a star like the Sun, but for photons of much shorter wavelength than for the Sun. Thus they are readily detected by X-ray observatories.***

16.5 Unlike for the neutron stars considered in Problem 16.4, white dwarfs are generally of low luminosity. Use the properties of white dwarfs to estimate the luminosity of one having the same mass and surface temperature as the Sun. Repeat for Sirius B.

16.6 Use the known properties of white dwarfs to show that for the fastest pulsars a white dwarf in hydrostatic equilibrium cannot pulsate fast enough to account for the observed periods, and a white dwarf spinning fast enough to account for the period would require velocities greater than the speed of light.

16.7 The number density n for free electrons in the momentum interval p to $p+dp$ obeys

$$dn = \frac{8\pi p^2 dp}{(2\pi\hbar)^3}.$$

By integrating this up to the Fermi momentum, show that for a white dwarf consisting of a single isotope of atomic number Z and mass number A the Fermi momentum is

$$p_f = \hbar \left(\frac{3\pi^2 \rho}{M_u} \frac{Z}{A}\right)^{1/3},$$

where ρ is the mass density and M_u is the atomic mass unit, and that the dimensionless parameter $x \equiv p_f/m_e c$ can be expressed using convenient units as

$$x = 9.89 \times 10^{-3} \left(\frac{Z}{A}\right)^{1/3} \left(\frac{\rho}{\text{g cm}^{-3}}\right)^{1/3}.$$

Estimate x for typical conditions in more massive white dwarfs. Do the values found for x indicate that the corresponding electrons are relativistic? *Hint*: See Silbar and Reddy [201].

16.8 In our discussion it was assumed that the energy density carried by the ions in a white dwarf is much larger than the energy density carried by the electrons, and that the ions can be treated nonrelativistically but the electrons may be relativistic. Justify this quantitatively by the following considerations. Show that with these assumptions (and the further one that the temperature is low enough to neglect the contribution of photons to the energy density), the total energy density for a white dwarf is

$$\varepsilon = \varepsilon_{\text{ion}} + \varepsilon_{\text{electron}} = \left(\frac{A}{Z}\right) n M_u c^2 + \frac{m_e^4 c^5}{8\pi^2 \hbar^3}\left((2x^3+x)(1+x^2)^{1/2} - \sinh^{-1} x\right),$$

where $x \equiv p_f/m_e c$ and the energy density contributed by an electron of momentum p is

$$\varepsilon_{\text{electron}}(p) = \sqrt{p^2 c^2 + m_e^2 c^4}.$$

Show that for typical white dwarf conditions the first term in the equation above for ε is much larger than the second term, confirming that the rest mass of the ions dominates the total energy density. *Hint*: See Silbar and Reddy [201].

16.9 The neutronization (electron capture) reaction $p+e^- \to n+\nu$ has $Q = -0.78$ MeV. Thus it can occur only when an additional 0.78 MeV of energy is supplied from kinetic energy, primarily of the electrons. Consider high-density, cold, neutral matter containing only neutrons, protons, and electrons (no composite nuclei). Estimate a threshold density above which the system can be expected to convert protons into neutrons spontaneously by $p + e^- \to n + \nu$. *Hint:* The maximum electron energy is the Fermi energy, which increases with density.

16.10 For a carbon–oxygen white dwarf with the pressure supplied by a degenerate electron gas, estimate the mass density above which the electrons should be considered to be relativistic.

16.11 Consider the discussion in Box 16.4. Assume helium-3 at the superconducting transition temperature of 2–3 mK to be a quantum gas of fermions (the ^{3}He atoms), with a mass density of 81 kg m^{-3}. Calculate the Fermi wavenumber $k_{\rm f} = p_{\rm f}/\hbar$, Fermi velocity $v_{\rm f}$, and Fermi energy $\varepsilon_{\rm f}$, and show that the helium-3 fermion "gas" at these temperatures is nonrelativistic and highly degenerate in the sense discussed in Section 3.8. In reality, the interactions in low-temperature helium-3 are sufficiently large that the system is commonly termed a *dense Fermi liquid,* rather than a Fermi gas. As discussed in Box 16.4, this dense Fermi liquid may have some similarities to the interior of neutron stars.***

16.12 In the interior of white dwarfs the electrons are highly degenerate but near the surface the density drops rapidly and an equation of state similar to that appropriate for the surface of normal stars ($\sim$ ideal gas) is a reasonable approximation. Estimate the scale height for a white dwarf atmosphere compared with that of a normal star like the Sun assuming ideal gas equations of state. What implication to you think this comparison has for observed spectral lines in white dwarfs relative to more normal stars?

16.13 Assume that the electron gas inside a white dwarf is degenerate and nonrelativistic. Show that even though the density is very high the electrons can be assumed to be essentially free. *Hint*: Compare the kinetic energy with the Coulomb interaction for the electrons.

17 Black Holes

In preceding chapters we saw that the endpoints for stellar evolution grow increasingly bizarre as stars increase in mass. For lighter stars the final chapter of stellar evolution was found to involve white dwarfs having incredible density by Earth standards, stabilized against further collapse by electron degeneracy pressure. For more massive stars the endpoint was found to be neutron stars with densities exceeding that even of atomic nuclei, stabilized against the crush of gravity by neutron degeneracy pressure. In this chapter we consider the strangest endpoint of all: modern gravitational theory, and a wealth of observational evidence, supports the notion that some massive stars treat even neutron degeneracy pressure with disdain, collapsing right through the neutron star stage until the mass of the star is concentrated at a point singularity, which surrounds itself with a one-way spacetime membrane called the *event horizon* that lets light and matter in, but forbids their escape. This most extreme consequence of gravity is called a *black hole*. For a full understanding of black holes, general relativity is essential. A systematic introduction to general relativity is outside the present agenda, but in this chapter some essential concepts and a few formulas will be imported to allow a meaningful qualitative discussion of black holes as a possible endpoint of stellar evolution.

17.1 The Failure of Newtonian Gravity

The standard theory of gravity proposed originally by Newton is a remarkably good description of the Universe (as should be abundantly clear from preceding chapters!). It gives predictions for most gravitational phenomena that for all practical purposes are in exact agreement with observations (and with the corresponding predictions of general relativity). However, there is a small set of phenomena for which general relativity gives the correct prediction but Newtonian gravity fails. These failures of Newtonian gravity typically share some combination of three characteristics:

1. Gravity becomes extremely strong, by measures that we shall quantify shortly.
2. Characteristic velocities approach the speed of light. This can occur in weak gravity, but is almost a given in the presence of large accelerations produced by strong gravity.

3. Even if neither (1) nor (2) is especially well satisfied, a particular measurement may require sufficient precision that even for weak gravity and low speeds the deviations of general relativity from Newtonian gravity become manifest.[1]

If any of these conditions is fulfilled, the predictions of Newtonian gravity begin to fail at some level and in the extreme case where all are true general relativity becomes the only viable theory of gravity. Black holes tend to fall into this latter category. Although Newtonian concepts are of some utility, they often are unreliable or even in downright error where the physics of black holes is concerned. Accordingly, we turn now to a description of black holes in terms of general relativity.

17.2 The General Theory of Relativity

The general theory of relativity (GR) may be thought of as resulting from the implementation of two general principles: (1) the *principle of general covariance* and (2) the *principle of equivalence*, in a 4-dimensional spacetime having a geometry that differs fundamentally from ordinary (euclidean) geometry.

17.2.1 General Covariance

The essential idea of both special and general relativity is an extremely powerful principle: the laws of physics should not depend on the coordinate system in which they are formulated and so should be unchanged by transformation to a new coordinate system. The basic difference between special and general relativity then is just that in general relativity the laws are formulated to be invariant under the most general possible transformations between coordinate systems, while special relativity requires invariance only under a more restricted set of transformations that are *between inertial frames*. Since general relativity is invariant under transformations even in the absence of global inertial frames, it can describe gravity. The formulation of physical laws such that they retain the same form under transformation between arbitrary coordinate systems is called *general covariance*.

17.2.2 The Principle of Equivalence

The fundamental insight that allowed Einstein to generalize special relativity to a theory of gravity embodied in general relativity began with the idea known since the time of Galileo that objects of different mass fall at the same rate in a gravitational field. This is one formulation of the *(weak) equivalence principle*. An alternative formulation is that the

---

[1] Perhaps the best-known example is the Global Positioning System (GPS), where the timing precision required to determine position with meter-level accuracy implies that even the relativistic corrections for low velocity in Earth's weak gravitational field are substantial. The GPS system would not even work without accounting for the special relativistic dilation of time caused by relative motion of satellite and receiver, and the general-relativistic dilation of time caused by the receiver being in a stronger gravitational field than the satellite.

inertial mass of an object, corresponding to the mass m in Newton's second law, $F = ma$, is measured to be equivalent to the *gravitational mass* of that same object, corresponding to the mass m in the gravitational law $F = GmM/r^2$, to extremely high precision. Starting from this insight, Einstein was led to propose that it is impossible locally to distinguish the effect of gravity from the effect of an arbitrary acceleration. This is called the *(strong) equivalence principle*, which henceforth will be termed simply the *equivalence principle*.

Furthermore, Einstein reasoned that since the acceleration of an object by gravity was independent of the mass or any other characteristic of the object, the effect of gravity *cannot be a property of objects in spacetime* and therefore must be a property of spacetime itself. This led Einstein eventually to the central thesis of general relativity: that spacetime is *curved*, and that gravity is not a force but rather corresponds to the motion of *free particles* in a curved spacetime. In this view the Earth is in orbit around the Sun, not because of a force acting between them, but because the gravitational field of the Sun curves the spacetime around it and the Earth follows freely a curved path in that curved spacetime. This means that general relativity is a theory about the *geometry of spacetime*.

17.2.3 Curved Spacetime and Tensors

In a 4-dimensional possibly curved spacetime manifold the coordinates of a spacetime point P are given by the *4-vector* x^μ,

$$x^\mu \equiv (x^0, x^2, x^3, x^4) = (ct, x, y, z) \qquad (\mu = 0, 1, 2, 3), \tag{17.1}$$

where x, y, and z are spatial coordinates, t is the time coordinate, and c is the speed of light. The most powerful and useful mathematical implementation of general relativity is in terms of objects called *rank-n tensors*, which carry a total of n upper and lower indices (when evaluated in a basis), obey particular transformation laws, and may be viewed mathematically as functions of n vectors into the real numbers. Loosely, tensors are the extension of vectors to objects that generalize the vector transformation law and may have more than one index; indeed, a vector may be viewed as an example of a rank-1 (carrying only one index) tensor. Tensors are the natural mathematical framework for general relativity because they implement automatically the principle of general covariance: If an equation written in terms of tensors is valid in one coordinate system, it is guaranteed to be valid in any other possible coordinate system.

The geometry of spacetime is described by a rank-2 tensor called the *metric tensor*, $g_{\mu\nu}$, which can be viewed as the *source of the gravitational field*. Thus the problem in general relativity is "simple": just determine the metric tensor for the space, which then determines the complete effect of gravity. But not so fast! Not only does the gravitational "force" acting on mass and energy in spacetime result from the curvature of spacetime, but that same mass and energy acts on spacetime to curve it. This implies that general relativity is a *highly nonlinear theory* (to determine the metric you must already know the metric). Thus the equations of general relativity can be written in a concise fashion using the mathematical elegance of tensors, but they are extremely difficult to solve.

Table 17.1 Gravitational strengths R/r_c at the surface of some objects

Object	R(km)	M(kg)	ρ(g cm^{-3})	g(m s^{-2})	r_c(km)	R/r_c
Earth	6378	6×10^{24}	5.6	9.8	9.2×10^{12}	6.9×10^{-10}
White dwarf	5500	2.1×10^{30}	$\sim 10^6$	4.6×10^6	1.9×10^7	2.8×10^{-4}
Neutron star	10	2×10^{30}	$\sim 10^{14}$	1.3×10^{12}	67.5	0.15

17.2.4 Curvature and the Strength of Gravity

Our primary concern in this chapter will be with *strong gravity*, but what does that mean in this context? In general relativity light follows a curved path in a gravitational field and the stronger the gravitational field the more curved the path. A *radius of gravitational curvature* r_c may be obtained by fitting a circle to a local part of the curved path. This gives $r_c = c^2/g$, where g is the gravitational acceleration and c the speed of light. A natural measure of gravitational strength at the surface of a gravitating object may then be formed from the ratio of a characteristic distance scale for the object such as the radius to the gravitational radius of curvature,

$$\frac{R}{r_c} = \frac{GM}{Rc^2} = \frac{\text{Actual radius}}{\text{Light curvature radius}}, \qquad (17.2)$$

where $g = GM/R^2$ was used. Then weak gravity is characterized by $GM/Rc^2 \ll 1$ but if $GM/Rc^2 \gtrsim 1$ a gravitational field may be characterized as strong. (For example, if $r_c \sim R$ gravity can put light into orbit around the object, which is pretty strong!) Most gravitational fields are weak by the natural measure of Eq. (17.2). Table 17.1 gives some examples. You may tend to think of Earth's gravity as relatively strong, especially when climbing stairs! But from Table 17.1, it corresponds to a paltry $R/r_c \sim 10^{-9}$. Even a white dwarf has only $R/r_c \sim 10^{-4}$, which is weak on the natural scale set by light curvature (though enormous by Earth standards), so that Newtonian gravity is still a rather good approximation. But for gravity at the surface of a neutron star or at the event horizon of a black hole, the gravitational curvature radius and actual radius will be comparable and a correct description of gravity requires general relativity.

17.3 Some Important General Relativistic Solutions

In general relativity the rank-2 metric tensor $g_{\mu\nu}$ is both the source of the gravitational field and the description of the geometry of spacetime. Thus the task is to determine $g_{\mu\nu}$, which is generally dependent on the spacetime coordinates, for a given situation. But this is a quite non-trivial task. In Newtonian physics the metric is fixed and specified implicitly at the beginning of a problem. It corresponds to the flat (euclidean) spatial coordinates and the time, which is assumed in Newtonian physics to be a globally defined quantity that is the same for all observers. In contrast, in general relativity the metric is not known beforehand: it is the solution of the problem, so the framework of spacetime in which the problem is formulated is itself unknown at the beginning.

17.3.1 The Einstein Equation

The way that this highly nonlinear problem is solved in general relativity is that it can be shown that the solutions must obey the *Einstein equation(s)*,

$$R_{\mu\nu} - \tfrac{1}{2} g_{\mu\nu} R = \frac{8\pi G}{c^4} T_{\mu\nu}. \tag{17.3}$$

In this expression the indices μ and ν each range over the labels for the spacetime dimensions (0, 1, 2, 3), and $R_{\mu\nu}$ and R are rank-2 and rank-0 tensors called the *Ricci tensor* and the *Ricci scalar*, respectively, that can be constructed from the metric tensor $g_{\mu\nu}$ and its first derivatives, and that describe the *curvature of spacetime*. On the right side, G is the gravitational constant, c is the speed of light, and $T_{\mu\nu}$ is a rank-2 tensor called the *stress–energy tensor* (or the *energy–momentum tensor*) that describes the coupling of gravity to matter, energy, and momentum. Because of the indices, each term in Eq. (17.3) can be viewed as a matrix with 16 components, but only 10 of them are independent because all terms in Eq. (17.3) are symmetric under exchange of indices. Hence this deceptively simple expression actually represents 10 coupled, nonlinear, partial differential equations that must be solved to determine the effect of gravity. In the general case analytical solutions are out of the question but in some cases of physical interest the problem has a high degree of symmetry and this reduces the problem to solving a much smaller set of equations that is still often formidable, but tractable.

Often only the gravitational solution outside some mass responsible for producing the gravitational field is of physical interest; for example, the gravitational field outside the mass distribution of a star. Then, if the exterior region is assumed to be a vacuum it can be shown that the Einstein equation reduces to the *vacuum Einstein equation*,

$$R_{\mu\nu} = 0. \tag{17.4}$$

Don't be fooled by the seeming triviality of this equation either! Because of the nonlinearity and the tensor indices, the vacuum Einstein equation is also extremely difficult to solve in the general case.

17.3.2 Line Elements and Metrics

In the following some solutions of the Einstein equations (17.3) or (17.4) will be introduced that will be of utility in the subsequent discussion. Such solutions are often called "spacetimes," which makes sense because the solution literally specifies the geometry of the corresponding space and time. Instead of giving the metric tensor that corresponds to the solution it is common to express solutions in terms of the *line element* ds^2 [with a standard notation $ds^2 \equiv (ds)^2$], which is related to the metric tensor $g_{\mu\nu}$ by

$$ds^2 = \sum_{\mu=0}^{3}\sum_{\nu=0}^{3} g_{\mu\nu}dx^\mu dx^\nu \equiv g_{\mu\nu}dx^\mu dx^\nu, \qquad (17.5)$$

where in the last step the *Einstein summation convention* (ubiquitous in discussions of relativity) has been introduced: any index repeated twice on one side of an equation, once in a lower position and once in an upper position, implies a summation over that index.[2] In this expression dx^α indicates a differential of the spacetime coordinate given in Eq. (17.1) and ds is the length of an infinitesimal line segment. Given the definition (17.5), specifying $g_{\mu\nu}$ determines ds^2 and vice versa, so often ds^2 is just called "the metric," when in reality one means the square of the line element corresponding to a specific metric tensor.

17.3.3 Minkowski Spacetime

Let's warm up with the "trivial" case. The simplest possibility is that there are no gravitational fields so that spacetime has no curvature at all (flat spacetime). Then general relativity reduces to special relativity, the resulting 4-dimensional manifold is called *Minkowski spacetime*, or just *Minkowski space* for short, and the metric corresponds to the line element

$$ds^2 = -c^2 dt^2 + dx^2 + dy^2 + dz^2. \qquad (17.6)$$

Notice the crucial point that the time-like component $c^2 dt^2$ has a sign *opposite* that of the three space-like components dx^2, dy^2, and dz^2. A metric for which the terms in the line element do not all have the same sign is called an *indefinite metric*; such metrics are characteristic of our physical spacetime in both flat and curved space. Thus 4-dimensional spacetime has a very different geometry than 4-dimensional euclidean space (where all metric coefficients would have the same sign), even though both are flat and correspond to spaces with no intrinsic curvature. In fact, the relative negative sign between space and time coordinates in the Minkowski metric is the source of all the "strange" behavior associated with special relativity: space contraction, time dilation, relativity of simultaneity, the "twin paradox," and so on, all derive from the indefinite Minkowski metric.

The metric must be used to compute physical observables, which illustrates another fundamental difference between relativity and Newtonian physics. In a Newtonian description coordinates may be themselves physical quantities. For example, the value of r in spherical coordinates is a distance that could be measured. In general (and special) relativity, space and time coordinates are just *labels*, without direct physical significance. *Physical quantities must be computed using the metric* and are generally not given directly by values of coordinates. This is illustrated by the metric itself: $(ds^2)^{1/2}$ measures the physical length of an infinitesimal line segment; by inspection this distance is not given directly by any of the coordinates, but rather is in general a non-trivial mixture of contributions from space and time coordinates.

---

[2] Whether a tensor index is in an upper or lower position is mathematically and physically meaningful, but it will be sufficient for present purposes just to remember that fact, without going into details of why.

Example 17.1 Consider the time coordinate t in Eq. (17.6). In Newtonian theories t is a direct measure for all observers of the passage of time. In Minkowski space the metric (17.6) may be used to show (see Problem 17.1) that the *proper time* τ, which is defined to be the time measured by a clock carried by an observer in his inertial frame and is related to the distance interval by $d\tau^2 = -ds^2/c^2$, is related to the coordinate time t appearing in Eq. (17.6) by

$$dt = \gamma d\tau \qquad \gamma \equiv \left(1 - \frac{v^2}{c^2}\right)^{-1/2}, \qquad (17.7)$$

where v is the magnitude of the velocity and γ is termed the *Lorentz γ-factor*. The proper time that elapses between coordinate times t_1 and t_2 is then

$$\tau_{12} = \int_{t_1}^{t_2} \left(1 - \frac{v^2}{c^2}\right)^{1/2} dt. \qquad (17.8)$$

The proper time interval τ_{12} is shorter than the coordinate time interval $t_2 - t_1$ because the square root in the γ-factor of Eq. (17.8) is always less than one. For constant velocity, (17.8) yields

$$\Delta\tau = \left(1 - \frac{v^2}{c^2}\right)^{1/2} \Delta t, \qquad (17.9)$$

which is just the time dilation equation of special relativity.

Thus time dilation in special relativity is a direct consequence of the Minkowski metric (17.6), deriving specifically from the difference in signs between the timelike and spacelike components. In a similar manner the Minkowski metric may be used to derive the space contraction effect and other standard features of special relativity.

17.3.4 Schwarzschild Spacetime

If the spacetime is curved, gravitational fields are present and the Minkowski metric no longer applies. The simplest solution in that case is obtained by assuming the spacetime where the solution is valid to be devoid of matter, pressure, and fields [so that Eq. (17.3) reduces to Eq. (17.4)], independent of time, and spherically symmetric in the spatial coordinates.[3] This solution of the vacuum Einstein equation is called the *Schwarzschild spacetime*, and has the metric

$$ds^2 = -\left(1 - \frac{2M}{r}\right)dt^2 + \left(1 - \frac{2M}{r}\right)^{-1} dr^2 + r^2 d\theta^2 + r^2 \sin^2\theta d\varphi^2, \qquad (17.10)$$

---

[3] That is, some time-independent, spherical distribution of mass is assumed to produce a gravitational field, but the Schwarzschild solution is valid only *outside* the mass distribution responsible for the field. For a spherical star, this solution would be valid beyond the radius of the star. For the spherical black holes to be discussed below, all the mass that is the source of the gravitational field has been crushed into a singularity at the center of the black hole. Then the Schwarzschild solution is assumed valid anywhere not too near the singularity.

where t is a time coordinate, r is a radial coordinate, θ and φ are the usual spherical angular coordinates, and M is the single parameter of the theory, which may be interpreted in the weak-field limit as the mass responsible for the gravitational field. In this equation another standard convention of the relativity formalism has been introduced: the equation is expressed in a special set of units where the speed of light c and the gravitational constant G are numerically equal to one, so G and c do not appear explicitly in the equations (this is explained further in Appendix B).

As for Minkowski space, the coordinates (t, r, θ, φ) are just labels and physical quantities must be computed from the metric. Our specific interest in the Schwarzschild solution in this chapter is that it predicts the existence of a very unusual situation if the mass M is compressed into a region smaller than the Schwarzschild radius r_s defined by

$$r_s = \frac{2GM}{c^2}, \tag{17.11}$$

where for this expression the c and G factors have been reinserted. By computing observables using the metric, it is found that in this case the radius r_s defines an *event horizon*, and that as a consequence of the extreme curvature of spacetime at the event horizon, matter or light can fall through the horizon but once inside nothing can escape, not even light. This solution is the simplest example of a *black hole*.

17.3.5 Kerr Spacetime

The Schwarzschild black hole described above is spherically symmetric and has no angular momentum. It is very useful to illustrate the general properties of black holes, but if black holes form from the gravitational collapse of stars it is expected that they will have angular momentum (because the original star is likely to have at least some spin). Assuming axial symmetry, the general relativistic solution giving more realistic black holes that are deformed and spinning is called the *Kerr spacetime*. It is specified in terms of *Boyer–Lindquist coordinates* (t, r, θ, φ) by the metric

$$ds^2 = -\left(1 - \frac{2Mr}{\rho^2}\right) dt^2 - \frac{4Mra\sin^2\theta}{\rho^2} d\varphi dt$$
$$+ \frac{\rho^2}{\Delta} dr^2 + \rho^2 d\theta^2 + \left(r^2 + a^2 + \frac{2Mra^2\sin^2\theta}{\rho^2}\right)\sin^2\theta d\varphi^2 \tag{17.12}$$

with the definitions

$$a \equiv J/M \qquad \rho^2 \equiv r^2 + a^2\cos^2\theta \qquad \Delta \equiv r^2 - 2Mr + a^2. \tag{17.13}$$

This gives a 2-parameter family of solutions in terms of the parameters a (or equivalently J) and M, where in the weak-field limit J may be interpreted as angular momentum and M as the mass. As for preceding examples, the coordinates are labels without direct physical significance and the metric must be used to calculate observables in the Kerr spacetime. The most important features of Kerr black hole properties for the present discussion are:

1. The Kerr spacetime has an event horizon and a region outside its event horizon called the *ergosphere* where a particle could enter and still escape, carrying off part of the rotational angular momentum and rotational energy of the black hole.
2. If the rotational energy and angular momentum are removed completely from a Kerr black hole, what remains is a Schwarzschild black hole, from which no additional mass or energy can be removed. In a very loose sense, the Kerr black hole may be thought of as a "rotationally-excited state" of the Schwarzschild black hole, which is the "ground state" (state of lowest possible energy) for a black hole of given mass [179].
3. The spinning black hole drags the surrounding spacetime with it as it rotates. This is called *dragging of inertial frames*, or more tersely *frame dragging*, and anything in that spacetime will be dragged with it. Thus objects near the black hole will be dragged with the rotation of the black hole *even if no angular force acts between the object and the black hole*.[4]
4. It may be shown that there is a maximum possible angular momentum for a Kerr black hole of mass M that is given by

$$J_{\max} = M^2. \qquad (17.14)$$

Kerr black holes having $J = J_{\max}$ are called *extremal Kerr black holes*. Because black holes are likely born with some angular momentum, and accretion from a companion through an accretion disk can spin them up over time, it is expected that near-extremal Kerr black holes could be relatively common.

Schwarzschild black holes are a special case of Kerr black holes corresponding to $J = 0$ and it may be assumed that any real black holes are Kerr black holes, usually with $J \neq 0$. In principle black holes could be electrically charged, which corresponds to yet other solutions of the Einstein equations. However, it is generally thought that any black holes formed in realistic astrophysical processes would be quickly charge-neutralized, so our interest here will be solely in uncharged black holes.

17.4 Evidence for Black Holes

With an understanding that black holes are intrinsically objects that must be described by general relativity and armed with a qualitative understanding of concepts from general relativity, let us now summarize some of the observational evidence supporting the thesis that black holes exist. Their very name suggests that they are difficult to observe directly, but if black holes are not isolated they should often be accreting matter and interacting gravitationally with nearby masses, which could have observable consequences. There are

---

[4] For example, in Section 21.8 it will be shown that frame dragging of the spacetime around a Kerr black hole can wind up a magnetic field and spiral it off the polar axes of the rotating black hole as relativistic jets. This is believed to be a leading candidate for producing the jets observed for gamma-ray bursts (where the black hole is of stellar mass), and active galactic nuclei and quasars (where the black hole has masses of millions to billions of $M_\odot$).

in fact strong reasons to believe in the reality of black holes, based on observations in three categories involving accretion of matter or gravitational interactions with other objects.

1. Binary star systems that are strong X-ray sources and where there is gravitational evidence for a massive unseen companion.
2. Detection of gravitational waves with properties suggesting that they originated in the merger of two black holes.
3. Observational anomalies in the centers of many galaxies, where very large masses (millions to billions of solar masses) inferred from star velocities exist, often accompanied by evidence for enormous energy generation in the core of the galaxy.

Our primary interest here is in black holes with masses comparable to those of stars that are potential endpoints for stellar evolution (which will be termed *stellar black holes*), so let us concentrate on evidence for stellar black holes in categories 1 and 2. At the end of the discussion some evidence for the supermassive black holes in category 3 will be presented for completeness, noting that such supermassive black holes may possibly have formed from the merger of stellar black holes.

17.4.1 Compact Objects in X-ray Binaries

There is appreciable indirect evidence for stellar black holes with masses ~ 5–$50\,M_\odot$, much of it coming from observation of X-ray sources powered by accretion in binary star systems (see Section 18.6). Most such systems are *spectroscopic binaries*, where an unseen compact object (usually a neutron star or black hole) is inferred from periodic Doppler shifts of spectral lines for the visible star. Typically X-ray emission in a spectroscopic binary is caused by significant accretion onto the compact companion, which implies a relatively small separation between components of the binary. Tidal interactions in close binaries tend to circularize elliptical orbits, so our discussion will be considerably simplified but not seriously compromised by assuming circular orbits. Then from Kepler's laws, the *mass function* $f(M)$ may be related to the observed radial velocity curve through

$$f(M) \equiv \frac{(M\sin i)^3}{(M+M_c)^2} = \frac{M\sin^3 i}{(1+q)^2} = \frac{PK^3}{2\pi G} \qquad \text{(circular orbits)}, \qquad (17.15)$$

where K is the semiamplitude and P the period of the radial velocity curve, i is the tilt angle relative to the observer of the orbit, M_c is the mass of the visible companion star, M is the mass of the unseen component, and where the *mass ratio* $q \equiv M_c/M$ has been introduced. The mass function is useful because in Eq. (17.15) the right side is determined by direct observation of the radial velocity curve and the left side is a function of the masses, so the measured velocity curve can be related to the masses in the binary.

The tilt angle i is illustrated in Fig. 17.1(a) and a typical observed velocity curve for a binary system is shown in Fig. 17.1(b). The angle i is generally not known for a spectroscopic binary (except that the presence or absence of eclipses can place some limits on it). Hence, given a measured $PK^3/2\pi G$, solutions of Eq. (17.15) for the mass M of the unseen compact object depend in the most general case on two unknowns, the tilt angle i and the mass of the visible component M_c. As will be seen below, the mass function

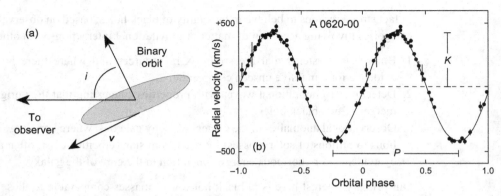

Fig. 17.1 (a) Tilt angle *i* for a binary orbit. (b) Observed radial velocity curve for the spectroscopic binary A 0620–00 [146]. The period is $P = 0.323$ days and the semiamplitude is $K = 433 \pm 3$ km s^{-1}. The orbital phase corresponds to the fraction of one complete orbit. Reproduced from Roger Blandford and Neil Gehrels, *Physics Today*, **52**(6), 40 (1999); https://doi.org/10.1063/1.882697. With the permission of the American Institute of Physics.

in conjunction with some additional information on M_c and i can often place significant constraints on the mass of the unseen compact object.

Example 17.2 Let's compute the mass function for a binary having a period of 5.6 days and a semiamplitude for the radial velocity curve $K = 75$ km s^{-1}. From Eq. (17.15) expressed in convenient units

$$f(M) = \frac{PK^3}{2\pi G} = 1.036 \times 10^{-7} \left(\frac{P}{\text{day}}\right)\left(\frac{K}{\text{km s}^{-1}}\right)^3 M_\odot. \quad (17.16)$$

Inserting $P = 5.6$ days and $K = 75$ km s^{-1} gives $f(M) = 0.245\, M_\odot$.

Suppose that the period P and velocity semiamplitude K have been determined from the observed velocity curve for a spectroscopic binary such as the one in Fig. 17.1(b), and that the quantity

$$F = F(P, K) \equiv \frac{PK^3}{2\pi G} \quad (17.17)$$

has been computed from that information. Then from Eq. (17.15) the unknown compact-object mass M is determined by the cubic equation $M^3 \sin^3 i / (M + M_c)^2 = F$, for which the solution of physical interest is given by the real root

$$M(F, M_c, i) = \left(R + \sqrt{Q^3 + R^2}\right)^{1/3} + \left(R - \sqrt{Q^3 + R^2}\right)^{1/3} - \frac{a}{3}$$

$$R \equiv \tfrac{1}{54}(9ab - 27c - 2a^3) \qquad Q \equiv \tfrac{1}{9}(3b - a^2) \quad (17.18)$$

$$a = -\frac{F}{\sin^3 i} \qquad b = -\frac{2FM_c}{\sin^3 i} \qquad c = -\frac{FM_c^2}{\sin^3 i}$$

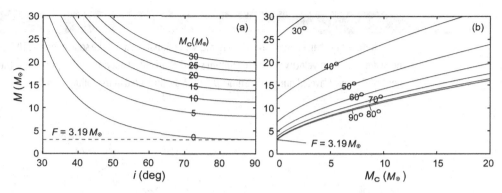

Fig. 17.2 Mass plots using Eq. (17.18) assuming a measured mass function $F = 3.19\,M_\odot$. (a) Mass M of the unseen component versus the tilt angle i for different values of the companion mass M_c. (b) Mass M of the unseen component versus M_c for different values of i. The measured mass function F is seen to set a lower limit on the unseen mass M.

Since F is known from measurement, the mass M of the compact unseen component is a function of two unknowns: the mass of the visible companion M_c and the tilt angle i. If these can be estimated in some way, Eq. (17.18) provides a meaningful constraint on the mass M. In Fig. 17.2 the solution $M(F, M_c, i)$ from Eq. (17.18) is plotted as a function of i and M_c assuming that $F = 3.19\,M_\odot$. These figures illustrate clearly

1. the degeneracy of the unknown mass M with respect to the parameters i and M_c and
2. that the measured value $F = 3.19\,M_\odot$ is the minimum possible mass for the unseen component [corresponding to the limit of $f(M)$ when $M_c = 0$ and $i = \frac{\pi}{2}$].

Point 2 already is a significant constraint but a more precise statement about M is possible if further information can be obtained about M_c and i, as we shall demonstrate below.

17.4.2 Causality Constraints

Another important piece of observational information that can be marshaled to determine whether a spectroscopic X-ray binary harbors a black hole is the causality argument described in Box 17.1. If the X-ray source is observed to vary periodically, the maximum size of the source is limited by the finite speed of light, since some signal must correlate the periodic variation and it cannot travel faster than light. If such considerations point to a very small energy source, typically a black hole or a neutron star is implicated.

17.4.3 The Black Hole Candidate Cygnus X-1

Let's analyze the black hole candidate Cygnus X-1. Optical, X-ray, and RF observations in the 1960s and 1970s [53, 56, 57, 120, 159, 223] determined that Cygnus X-1 is an X-ray source in a binary system consisting of the visible blue supergiant HDE 226868 and an unseen compact companion. The X-ray source flickers with a period of milliseconds. From

Box 17.1 Causality and the Size of Energy Sources

If the luminosity of an energy source is periodic, some signal must tell the source to vary. The signal can travel no faster than light velocity, so the maximum size D of an object varying with a period P is the distance that light could have traveled during that time, $D \sim cP$, as illustrated in the following figure.

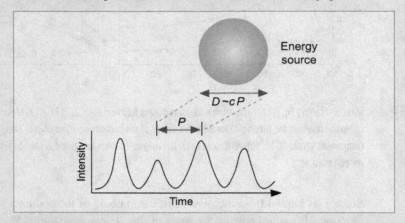

The distances covered by light for various fixed times are summarized in the following table.

Time	km	AU	Parsecs
Year	9.46×10^{12}	63,240	3.07×10^{-1}
Month	7.88×10^{11}	5270	2.58×10^{-2}
Week	1.82×10^{11}	1216	5.90×10^{-3}
Day	2.59×10^{10}	173	8.41×10^{-4}
Hour	1.08×10^{9}	7.21	3.50×10^{-5}
Minute	1.80×10^{7}	0.120	5.84×10^{-7}
Second	3.00×10^{5}	0.002	9.73×10^{-9}
Millisecond	3.00×10^{2}	0.000002	9.73×10^{-12}

This argument places only an *upper limit* on source size and it may be *smaller* than the limit imposed by c. But the argument is powerful because it depends only on causality.

the table in Box 17.1, this suggests that the source size is no more than a few hundred kilometers, ruling out a white dwarf and indicating that the X-rays result from accretion onto either a neutron star or a black hole. An artist's conception is displayed in Fig. 17.3.

Analysis of the observed velocity curve for the blue supergiant indicates a period of $P = 5.6$ days and a semiamplitude $K = 75 \, \text{km s}^{-1}$ (see fig. 2 of Ref. [161]), which gives from Eq. (17.16) $F = PK^3/2\pi G = 0.245$. Inserting this in Eq. (17.18) and plotting M versus i and M versus M_c gives the graphs shown in Fig. 17.4. The spectrum–luminosity class of the blue supergiant is O9.7Iab, which permits its mass to be estimated from stellar systematics as 20–30 $M_\odot$. The tilt angle cannot be measured directly since the orbit is not resolved, but detailed comparison with observed systematics for the system such as

17.4 Evidence for Black Holes

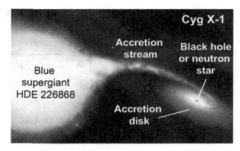

Fig. 17.3 Artist's conception of the high-mass X-ray binary, Cyg X-1. Detailed analysis suggests that the unseen companion is a 10–20 $M_\odot$ black hole.

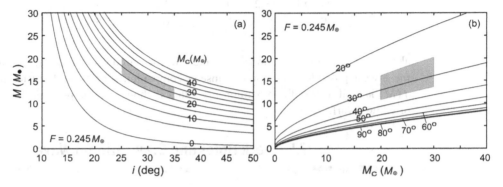

Fig. 17.4 Analysis of masses in the high-mass X-ray binary Cygnus X-1, based on the observed mass function $F = 0.245\ M_\odot$ and Eq. (17.18). (a) Mass of unseen companion M versus tilt angle i for various assumed masses M_c of the supergiant companion. (b) M versus M_c for various tilt angles i. The gray boxes indicate further observational constraints discussed in the text. Notice that the minimum possible mass for the unseen companion is given by the measured mass function $F = 0.245\ M_\odot$, which corresponds to the limit $M_c \to 0$ and $i \to 90°$.

whether eclipses are seen permits it to be estimated as $i = 25\text{–}35°$. These allow the acceptable ranges for a solution to be displayed as the gray boxes in Fig. 17.4. From this it may be concluded from our simple analysis that the mass of the unseen companion lies in the range 10–20 $M_\odot$. A more comprehensive analysis concludes that $i = 27.1 \pm 0.8°$ and $M_c = 19.2 \pm 1\ M_\odot$, implying that $M = 14.8 \pm 1.0\ M_\odot$ [161], which is consistent with our simple estimate. Since no plausible equation of state supports a neutron star with $M > 2\text{–}3\ M_\odot$, it may be concluded with high certainty that the $M = 14.8\ M_\odot$ unseen companion in Cygnus X-1 can only be a black hole. It also may be noted that an extensive analysis [96] has concluded that the black hole in Cyg X-1 has a spin greater than 95% of the maximal value given in Eq. (17.14). Thus, Cyg X-1 may be a *near-extremal Kerr black hole*.

An analysis similar to the one outlined above has been carried out for many X-ray binaries in the galaxy. A summary of the cases that place the mass of the unseen companion well above the maximum mass for a neutron star or white dwarf is shown in Table 17.2. By the preceding types of arguments, these binary systems are assumed to contain a black hole of mass M as the unseen companion. As was noted in Figs. 17.2 and 17.4, even in the

Table 17.2 Black hole candidates in galactic X-ray binaries [48]

X-ray source	P (days)	$f(M)$	$M_c(M_\odot)$	$M(M_\odot)$
Cygnus X-1	5.6	0.24	24–42	11–21
V404 Cygni	6.5	6.26	∼0.6	10–15
GS 2000+25	0.35	4.97	∼0.7	6–14
H 1705−250	0.52	4.86	0.3–0.6	6.4–6.9
GRO J1655−40	2.4	3.24	2.34	7.02
A 0620−00	0.32	3.18	0.2–0.7	5–10
GS 1124−T68	0.43	3.10	0.5–0.8	4.2–6.5
GRO J0422+32	0.21	1.21	∼0.3	6–14
4U 1543−47	1.12	0.22	∼2.5	2.7–7.5

absence of further information on M_c and i the measured value of the mass function defines the lowest possible mass for the unseen companion. For several entries in Table 17.2, $f(M)$ is well above the maximum mass that is thought to be possible for a neutron star or white dwarf.

17.5 Black Holes and Gravitational Waves

As was mentioned in Section 14.6.3, the first direct observation of gravitational waves in 2015 opened a new window on the Universe capable of probing *dark events* that might not be observable using the tools of traditional astronomy. Black holes are the quintessential dark objects of our Universe, so gravitational wave astronomy is particularly well suited to their study. Indeed the first several gravitational waves reported by the LIGO collaboration were each interpreted as resulting from the merger of binary black holes. These gravitational-wave observations not only provide the strongest evidence to date for the existence of black holes with masses comparable to stars, but their detailed interpretation has begun to yield quantitative information about the black holes that were involved in the merger. This in turn establishes a new methodology to study late stellar evolution for massive stars and the black hole endpoint that is one possible outcome of that evolution. Going forward, gravitational wave astronomy may well prove to be the most powerful method at our disposal for the study of black holes, as will be discussed more extensively in Chapter 22.

17.6 Supermassive Black Holes

The center of the Milky Way coincides approximately with the radio source Sgr A*. A 15 $M_\odot$ star denoted S0-2 has been tracked extensively in a Keplerian orbit with Sgr A* near a focus. The positions observed for S0-2 through 2002 are shown in Fig. 17.5, with dates in fractions of a year beginning in 1992. The orbit drawn in Fig. 17.5 corresponds

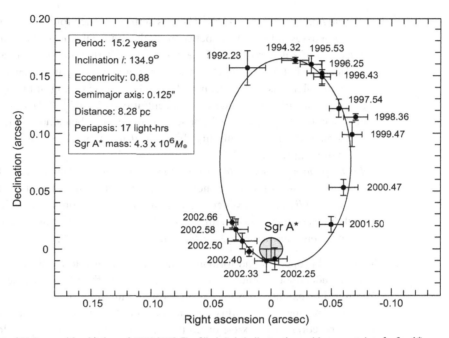

Fig. 17.5 Orbit of S0-2 around Sgr A* through 2002 [197]. The filled circle indicates the position uncertainty for Sgr A* assuming a point mass located at the focus to be responsible for the orbital motion. The star completed this orbit in 2008 and the parameters displayed in the box are those obtained from the completed orbit [94]. Periapsis is the general term for closest approach of an orbiting body to the center of mass about which it is orbiting. Adapted by permission from Springer Nature, *Nature*, "A Star in a 15.2-year Orbit around the Supermassive Black Hole at the Centre of the Milky Way, Schödel, R. et al., Copyright (2002).

to the projection of the best-fit ellipse with Sgr A* at a focus [197]. At closest approach the separation of S0-2 from Sgr A* is only 17 lighthours. From fits to the orbit of S0-2 assuming Keplerian motion, the mass contained inside the orbit is $4.3 \times 10^6 \, M_\odot$ [94], in a region that cannot be much larger than the Solar System (and may be smaller) where there is little luminous mass. The obvious explanation is that the radio source Sgr A* coincides with an approximately 4.3 million solar mass black hole at the center of the Milky Way. Extensive evidence for average star motion indicates that many galaxies contain black holes of $10^6 \, M_\odot$ to $10^8 \, M_\odot$. Whether these form by the merger of many stellar black holes created by stellar core collapse and thus have an indirect relationship to stellar evolution, or whether they were formed by some process independent of stellar evolution like direct collapse from gas clouds, is unknown at present.

17.7 Intermediate-Mass and Mini Black Holes

For completeness, let us remark briefly about two other classes of black holes that either exist or are conjectured to exist, but that might not be connected very directly to issues in stellar evolution:

1. *Intermediate-mass black holes*: There is some evidence for a population of black holes having masses intermediate between stellar and galactic ones (hundreds to tens of thousands of solar masses). The evidence for intermediate-mass black holes has been inconclusive but in 2017 a pulsar was discovered orbiting an unseen mass concentration in the globular cluster 47 Tucanae [135]. The precise timing of the pulsar indicates that the magnitude of the unseen mass concentration is $2200^{+1500}_{-800}\ M_\odot$, which may represent the first conclusive evidence for an intermediate-mass black hole. There is no significant electromagnetic signal from this object, so if it is a black hole it must not be accreting any matter at present.
2. *Hawking (mini) black holes*: From approximations of quantum field theory in strong gravitational fields it may be shown that black holes can radiate their mass over time as *Hawking radiation*. However, the rate of emission is completely negligible except for black holes of incredibly small mass (say the mass of a proton). Such *mini black holes* are commonly termed *Hawking black holes*.

The detailed properties and formation mechanisms for intermediate-mass black holes are not well understood. Hence it is not clear whether they have any connection to stellar evolution. For example, do they form through clumping of stellar-mass black holes into more massive ones, and is this an intermediate step in forming the supermassive black holes at the cores of galaxies, or do they collapse directly from gas clouds? In contrast, there is no observational evidence for Hawking black holes thus far, and if they exist they must have been formed in the incredibly high temperatures and densities of the big bang and not in stellar processes. They are of enormous potential interest for theories of quantum gravity, but can be ignored for the present discussion.

17.8 Proof of the Pudding: Event Horizons

A compelling circumstantial case may be made for the existence of black holes. However, the characteristic of a black hole that distinguishes it from anything else is its *event horizon*, and none of the evidence supporting black holes yet demands the existence of an event horizon. Therefore, irrefutable proof requires imaging the event horizon of a black hole, which is obviously a considerable challenge. The best prospects are for the supermassive black hole at Sgr A* (see Section 17.6 above), which has a Schwarzschild radius about 18 times larger than the radius of the Sun. Thus, seeing the event horizon of Sgr A* requires resolving an object of this size at a distance of about 8 kpc (see Problem 17.7). This may be possible, as very long baseline interferometry with arrays of broadly dispersed radio telescopes is beginning to achieve resolutions of this magnitude.

A hint of what a resolved event horizon might look like comes from detailed analysis of data from the gravitational wave event GW150914 (corresponding to the merger of 29 and 36 $M_\odot$ black holes), which will be described in Chapter 22. A frame extracted from a computer simulation of how the merger might have appeared to an observer a few hundred kilometers away is shown in Fig. 17.6. The jet-black shapes are the event horizons

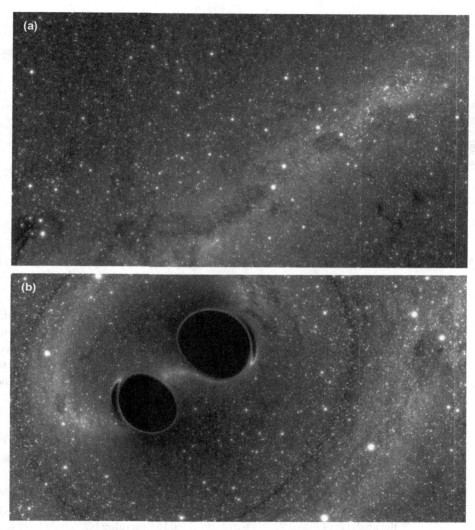

Fig. 17.6 A computer simulation showing what the two black holes might have looked like to a nearby observer just prior to merger in the gravitational wave event GW150914. (a) Background stars in the absence of the black holes. (b) Image including black holes. The ring around the black holes is an *Einstein ring,* which results from strong focusing by gravitational lensing of the light from stars behind the black holes. Image extracted from video in [73]. A more complete sequence of images for the merger is given later in Fig. 22.6.

shadowing all light from behind. All stars are in the background but gravitational lensing in the strongly curved space near the black holes severely distorts their apparent positions in Fig. 17.6(b), and produces the flattened dark features around the event horizons. In Fig. 17.6 the black holes are assumed isolated with no surrounding matter. Hence the image is dominated by the shadowing of the black hole event horizons and strong gravitational lensing effects. In contrast, the black hole at Sgr A* is in a dense cluster of stars and is

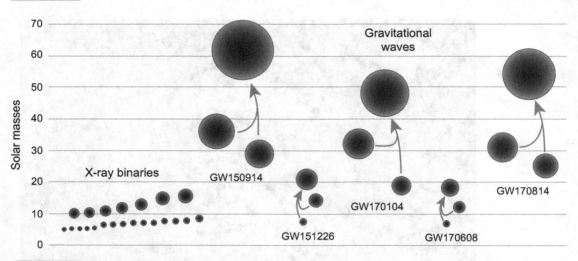

Fig. 17.7　A summary of some black hole masses determined from X-ray binary and gravitational wave data (adapted from Ref. [7]). Arrows indicate black hole mergers.

likely accreting some surrounding matter and producing radiation from this accretion. The dominant observational feature may still be the complete and sharply defined shadowing of background light by the event horizon of the black hole and strong gravitational lensing near the horizon, as suggested by Fig. 17.6, but the open question is how the environment of Sgr A* will distort this picture and whether the event horizon will still be identifiable in sufficiently resolved observations.

17.9　Some Measured Black Hole Masses

The most reliable methods for discovering stellar-size black holes and determining their masses are the mass-function analysis of X-ray binaries described in Section 17.4.1, and the analysis of gravitational waves produced by merging black hole binaries to be described in Chapter 22. Figure 17.7 summarizes masses for more than 35 black holes in the range ∼5–65 $M_\odot$ determined from these two types of analysis. These data from X-ray binaries and gravitational waves from black hole coalescence constitute the strongest evidence currently available for the existence of stellar-size black holes. It would be difficult to account for the properties of these objects through any hypothesis other than that of black holes.

Background and Further Reading

Anything more than a cursory understanding of black holes requires a basic grounding in the theory of general relativity, which in turn requires at least a working knowledge of differential geometry and tensor calculus. Some recommended introductions include Hartle [109]; Foster and Nightingale [87]; and Hobson, Efstathiou and Lasenby [123].

An extensive discussion of black holes presented within the framework of general relativity but at a pedagogical level commensurate with the material in this book is given in Ref. [100]. Parts of the discussion in this chapter have been adapted from that reference.

Problems

17.1 Prove the result of Eq. (17.7) for time dilation starting from the metric (17.6) and that the proper distance s and proper time τ are related by $d\tau^2 = -dt^2/c^2$. ***

17.2 Compute the mass function for a binary having a period of 6 hours and a semiamplitude for the radial velocity curve $K = 400 \text{ km s}^{-1}$.

17.3 The binary system Cygnus X-1 has a total mass $M \sim 30 - 50\, M_\odot$, an orbital period of 5.6 days, and is a strong X-ray source with fluctuations in intensity on a timescale of milliseconds. Assuming the validity of Kepler's laws and that the total mass is $40\, M_\odot$, what is the average separation of the components of the binary and what is the maximum size of the X-ray emitting region based on these observations?

17.4 Show that the binary star mass function $f(M) = PK^3/2\pi G$ may be expressed in convenient units as

$$f(M) = 1.036 \times 10^{-7} \left(\frac{P}{\text{day}}\right) \left(\frac{K}{\text{km s}^{-1}}\right)^3 M_\odot,$$

where P is the binary orbital period in days, K is the semiamplitude of the radial velocity in units of km s^{-1}, and $f(M)$ is in solar mass units [see Eq. (17.15) and Fig. 17.1]. From the data in Fig. 17.1 and Table 17.2, estimate the mass function and thus the lower limit for the mass of the black hole in the X-ray binary A 0620–00.

17.5 Use Kepler's laws to derive the mass function relation in Eq. (17.15), assuming circular orbits for the binary. *Hint:* Generalize the discussion in Section 1.5.2 to circular orbits tilted with respect to the observer as in Fig. 17.1(a).

17.6 The soft X-ray transient A0620–00 is an interacting binary system [a velocity curve was shown in Fig. 17.1(b)] in which a main sequence star is transferring mass to an accretion disk around an unseen compact object [111, 146, 147, 148, 160]. Take the orbital period to be $P = 7.75$ hr and the semiamplitude of the radial velocity curve to be $K = 457 \text{ km s}^{-1}$ [147]. Repeat the analysis of Section 17.4.3 to place limits on the mass of the unseen compact object in A0620–00. Useful observational constraints are (1) the known distance to the system and the absence of observed eclipses of the X-ray source precludes i larger than about 80°, and (2) the companion is determined observationally to be a main sequence star with spectral class lying in the range K7–K2, which from stellar systematics constrains the mass to $0.5\, M_\odot < M_c < 0.8\, M_\odot$.

17.7 What is the Schwarzschild radius of the $4.3 \times 10^6\, M_\odot$ black hole at the center of the Milky Way? What angular resolution is required to resolve it from Earth, taking the distance to the center of the galaxy to be 8 kpc? *Hint:* In reality there would be strong gravitational lensing effects that would affect the answer, but ignore those for a simple estimate of the angular resolution that is required.***

PART III

ACCRETION, MERGERS, AND EXPLOSIONS

18 Accreting Binary Systems

Observations suggest that more than half of the more massive stars in the sky are in multiple-star systems, most in binary stars. When binary components are well separated they have long orbital periods and behave approximately as isolated stars unless a strong wind is blowing from one of them. However, if the semimajor axis of the orbit is small enough, mass may spill directly from one star onto the other during all or part of the orbital period. This is an example of *accretion*. Although accretion does not sound at first blush like a very exciting topic, in fact it is a critical ingredient in many of the most interesting phenomena in astrophysics. It plays this role either through providing a mechanism to initiate such phenomena (for example, in novae or Type Ia supernovae), or as the primary power source for the phenomenon (for example, in supermassive black hole engines that power quasars), or both (for example, in high-mass X-ray binaries). This chapter investigates a number of phenomena in binary star systems that are accretion-driven, and lays the groundwork for more general discussions of accretion in a variety of astrophysical settings.

18.1 Classes of Accretion

For purposes of discussion it is useful to think of binary star accretion as falling into two extreme categories.

1. If the stars are sufficiently close together a gas particle "belonging" to one star may wander far enough from that star to be captured by the gravitational field of the other star. This is termed *Roche-lobe overflow*.
2. Even if the two stars are not close enough for Roche-lobe overflow, mass may be transferred between them if one star has a very strong wind blowing from its surface and the second star captures particles from this wind. This is termed *wind-driven accretion*.

As will be seen, these two methods of accretion tend to involve binary systems having very different total masses, with Roche-lobe overflow favored in low-mass binaries and wind-driven accretion favored in high-mass binaries.

18.2 Roche-lobe Overflow

Let's first address accretion in binary systems through the mechanism of Roche-lobe overflow, with wind-driven accretion to be discussed afterward.

18.2.1 The Roche Potential

Consider the *restricted 3-body problem,* which refers to a 3-body gravitational problem for the special case where two of the masses may be considered to be much larger than a third test mass. Figure 18.1 illustrates. In particular, for the present discussion interest lies in the case where M_1 and M_2 are masses for the two components of a binary star system in revolution around its center of mass and m is the mass of a gas particle in the system. If a coordinate system rotating with the binary axis is employed, the potential acting on the gas particle is termed the *Roche potential* $\Phi_R(r)$, and is given by [89]

$$\Phi_R(r) = -\frac{GM_1}{|r-r_1|} - \frac{GM_2}{|r-r_2|} - \tfrac{1}{2}(\omega \times r)^2, \tag{18.1}$$

where G is the gravitational constant, ω is the frequency for revolution about the center of mass, and the other quantities are defined in Fig. 18.1. An energy surface corresponding to this potential is illustrated in Fig. 18.2. The final term in Eq. (18.1) appears because a *non-inertial coordinate system* rotating with the binary has been chosen. It will give rise to centrifugal and Coriolis (pseudo-) forces. The falloff of the potential surface in Fig. 18.2(a) at increasing distances from the binary pair is a consequence of this term. The rotational frequency entering Eq. (18.1) is given by $\omega = (GM/a^3)^{1/2}\,e$, where a is the semimajor axis, $M = M_1 + M_2$ is the total mass, and e is a unit vector normal to the orbital plane.

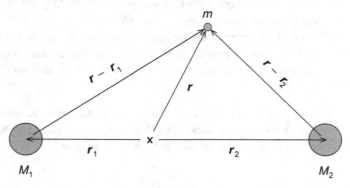

Fig. 18.1 Three-body gravitational interaction. The restricted 3-body problem corresponds to the assumption that m is much smaller than M_1 and M_2.

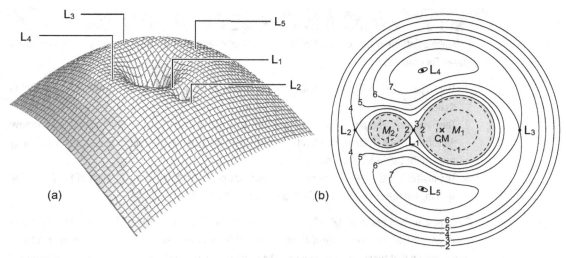

Fig. 18.2 (a) Potential energy surface and Lagrange points L_n for a generic binary system (surface courtesy of John Blondin). (b) Potential contours and the Lagrange points L_n for binary masses M_1 and M_2 with $M_2/M_1 = 0.25$. Dashed contours lie inside the Roche lobes (indicated in gray) and the × marked CM denotes the location of the center of mass. Adapted from Ref. [89].

18.2.2 Lagrange Points

In Fig. 18.2 the five *Lagrange points* L_n associated with the restricted 3-body problem are indicated. These points correspond to the five special points in the vicinity of two large orbiting masses where a third body of negligible mass can orbit at a fixed distance from the larger masses (because at these points the gravity of the two large bodies provides exactly the centripetal forces required for the small mass to revolve with them). As indicated in Figs. 18.2(a) and 18.2(b), the Lagrange points L_1, L_2, and L_3 lie on the line of centers for the two large masses and are points of unstable equilibrium associated with saddle points of the potential. The points L_4 and L_5 are "hilltops" in the potential surface of Fig. 18.2(a) and appear at first glance to also be points of unstable equilibrium. However, for particular ranges of masses for the two large bodies, L_4 and L_5 are actually *stable* equilibrium points because of the Coriolis force associated with the rotating frame. A particle rolling away from the hilltop at L_4 or L_5 experiences a Coriolis force that alters its direction of motion. For favorable values of the parameters, the Coriolis deflection is sufficiently strong to put the particle into an orbit around the Lagrange point.

The L_1 Lagrange point is of particular importance for binary star systems because mass flow between the stars can occur through the L_1 point. The L_2 point is also of more limited interest for binary accretion because mass overflow from star 1 to star 2 can in some cases overshoot and escape the system through the L_2 point. Such Lagrange points are also of considerable interest in the dynamics of natural and artificial objects in the Solar System, as discussed in Box 18.1.

| Box 18.1 | Lagrange Points in the Solar System |

Lagrange points often are significant in the Solar System when one of the large masses is the Sun and one a planet. For example,

1. The Trojan Asteroids lie at the Jupiter–Sun L_4 and L_5 points, 60 degrees ahead and behind Jupiter in its orbit.
2. The Solar and Heliospheric Observatory (SOHO) is parked at the L_1 point and the Wilkensen Microwave Anisotropy Probe (WMAP), Planck satellite, and James Webb Space Telescope have used or will use the L_2 point of the Earth–Sun system.
3. The mythical "Planet X" of science fiction fame was purported to be at the L_3 point of the Earth–Sun system, and therefore always on the opposite side of the Sun from Earth.

Dynamical analysis of the unstable Lagrange points indicates that for the Earth–Sun system the L_1 and L_2 points are unstable on a timescale of about 25 days; thus the observatories parked there require small orbit corrections on that timescale to remain near the Lagrange points. The L_3 point for the Earth–Sun system is dynamically unstable on a 150-day timescale (which bodes ill for Planet X!). The parameters of the Earth–Sun system, as for the Jupiter–Sun system, indicate that the L_4 and L_5 points are stable because of Coriolis forces. No analogs of Jupiter's Trojan asteroids have been found for Earth but there is evidence for enhanced dust concentrations at the L_4 and L_5 points for the Earth–Sun system.

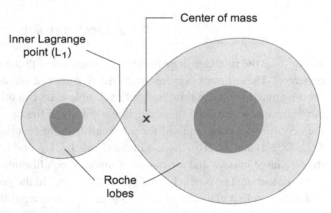

Fig. 18.3 Schematic illustration of Roche lobes and the inner Lagrange point.

18.2.3 Roche Lobes

One contour of the Roche potential intersects itself at the L_1 Lagrange point lying on the line connecting the center of mass for each star. The interior of this figure-8 contour defines a tear-drop shaped region for each star called a *Roche lobe*. The Roche lobes for the potential displayed in Fig. 18.2(b) are shaded in gray and Fig. 18.3 illustrates the Roche lobes for a binary in more schematic form. Roche lobes may be viewed as defining the gravitational domain of each star. A gas particle within the Roche lobe of one star feels a stronger attraction from that star than from the other. It "belongs" gravitationally to the star unless there are instabilities (such as those responsible for winds) that upset the hydrostatic equilibrium. However, the L_1 Lagrange point is a saddle between the potential wells corresponding to the two stars. A gas particle located at L_1 belongs equally to both stars, suggesting that mass transfer can be initiated if a star expands to fill its Roche lobe, thereby placing gas at the inner Lagrange point L_1.

18.3 Classification of Binary Star Systems

The Roche lobes defined in Fig. 18.3 provide a convenient classification scheme for binary systems that is illustrated in Fig. 18.4.

1. *Detached binaries* are binaries where each star is well within its Roche lobe. In the absence of strong winds, there is little chance for mass transfer in this case. Orbits for detached binaries range from the large orbits with periods of hundreds of years found for some visual binaries down to stars that are separated by little more than a solar radius but still lie within their respective Roche lobes.
2. *Semidetached binaries* are binaries in which one of the stars has filled its Roche lobe. Thus, semidetached binaries generally are either accreting through Roche-lobe overflow or are unstable against initiation of such accretion. Semidetached binaries have orbital separations ranging from about a solar radius up to several astronomical units.
3. *Contact binaries* (also called W UMa stars if they are eclipsing) correspond to binaries in which both members of the binary have filled or even overfilled their respective Roche lobes. This may lead to a "neck" between the stars, or to *common envelope evolution* where both stars orbit within the same envelope. Contact binaries are typically separated by two stellar radii or less.

As a star evolves away from the main sequence into the RGB and AGB regions it may increase its size by a factor of 100–1000 (see Fig. 13.16). Thus, as its stellar components evolve detached binaries can morph into semidetached or even contact binaries. In addition, separation of the members of a binary may change because of mass transfer or emission of gravitational radiation, which can alter the classification of the binary with respect to Fig. 18.4. Finally, as will be discussed in Section 18.9, mass transfer can alter the

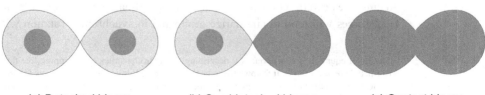

(a) Detached binary   (b) Semidetached binary   (c) Contact binary

Fig. 18.4 Classification of binary star systems according to whether neither, one, or both of the stars fill their Roche lobes.

evolution of the individual members of the binary, confusing the issue of stellar evolution relative to that expected for isolated stars.

18.4 Accretion Streams and Accretion Disks

The preceding discussion has introduced the idea of Roche-lobe overflow in qualitative terms. We now consider mass transfer through Roche-lobe overflow in a more precise manner. To do so requires a quantitative description of the gas flow between stars in a binary system.

18.4.1 Gas Motion

We shall assume the gas motion to be governed by a continuum version of Newton's second law called the *Euler equation*[1]

$$\rho \frac{\partial v}{\partial t} + \rho(v \cdot \nabla)v = -\nabla P + f, \tag{18.2}$$

where v is the velocity, ρ is the density, P is the pressure, and f is the force density acting on the gas (for example, from gravity or an external magnetic field). In a frame rotating with the binary at a frequency ω the Euler equation takes the form [89]

$$\frac{\partial v}{\partial t} + (v \cdot \nabla)v = -\nabla \Phi_{\rm R} - 2\omega \times v - \frac{1}{\rho}\nabla P. \tag{18.3}$$

In general, numerical solutions of this equation are required for a quantitative description of gas flow in accretion. However, we shall now argue that many of the basic features of accretion through Roche-lobe overflow may be understood with only minimal calculation. These features will follow largely from two observations:

1. Mass transfer is extremely likely and highly efficient if a star fills its Roche lobe.
2. In most cases, conservation of angular momentum for the transferred matter implies that an *accretion disk* will form around the primary star.

This discussion will be guided by the presentation in Ref. [89].

18.4.2 Initial Accretion Velocity

Imagine the accretion process as seen from the primary onto which accretion takes place.[2] Tidal forces will tend to circularize orbits and to synchronize rotation with revolution in

---

[1] The Euler equation is appropriate for describing the hydrodynamics of a compressible fluid without viscosity (termed an *inviscid fluid*). It follows from the more general *Navier–Stokes equations* for compressible fluids by neglecting the effect of viscosity and thermal conductivity.
[2] In what follows the star onto which matter accretes will be termed the *primary* and the other star will be termed the *secondary* of the binary. Notice that the primary is not necessarily the brighter star. In fact, for the case of most interest where accretion is onto a compact object the primary will usually be the less bright star.

18.4 Accretion Streams and Accretion Disks

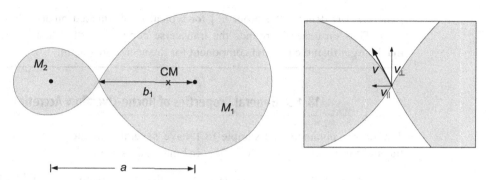

Fig. 18.5 Velocity components for the gas stream in Roche-lobe overflow for a binary system.

close binaries, so we will assume that the primary and the secondary star keep the same faces turned toward each other during the orbital period on a circular orbit. Viewed from the primary the companion appears to be moving across the sky since it makes a complete circuit of the celestial sphere once each binary period. If the Roche lobe of the companion is filled so that matter comes across the L_1 point, it appears to an observer on the primary star to have a large transverse component of motion because of the revolution of the binary system. The relevant geometry is shown on the left side of Fig. 18.5 and the components of velocity perpendicular and parallel to the line of centers for the two stars are illustrated on the right side of Fig. 18.5. In the non-rotating frame, the perpendicular and parallel components of velocity for the stream of gas coming across the L_1 point satisfy $v_\perp \sim b_1 \omega$ and $v_\parallel \leq c_s$, where c_s is the local speed of sound in the vicinity of the L_1 Lagrange point.

Example 18.1 We may make some estimates of the velocity components for the accretion stream by using Kepler's third law (1.20) as follows [89]. The length of the semimajor axis a is given by

$$a = 2.9 \times 10^{11} m_1^{1/3} \left(1 + \frac{m_2}{m_1}\right)^{1/3} \left(\frac{P}{\text{day}}\right)^{2/3} \text{ cm}, \tag{18.4}$$

where $m_1 = M_1/M_\odot$ and $m_2 = M_2/M_\odot$. Taking $b_1 \sim \tfrac{1}{2} a$ and utilizing $\omega = 2\pi/P$ yields

$$v_\perp \simeq b_1 \omega \simeq 105 m_1^{1/3} \left(1 + \frac{m_2}{m_1}\right)^{1/3} \left(\frac{\text{day}}{P}\right)^{1/3} \text{ km s}^{-1}, \tag{18.5}$$

for the perpendicular component of velocity and the local sound speed may be approximated by

$$c_s \simeq 10 \left(\frac{T}{10^4 \text{ K}}\right)^{1/2} \text{ km s}^{-1}. \tag{18.6}$$

For normal stellar envelopes $T \leq 10^5$ K and therefore $v_\parallel \leq c_s \leq 10 \text{ km s}^{-1}$. Thus,

$$v_\perp \sim \mathcal{O}\left(100 \text{ km s}^{-1}\right) \qquad v_\parallel \sim c_s \sim \mathcal{O}\left(10 \text{ km s}^{-1}\right) \tag{18.7}$$

[where $\mathscr{O}(n)$ denotes "of order n"] for typical semidetached binaries having periods of days. Thus on general grounds the transverse component of velocity is expected to be much larger than the parallel component for the accretion stream.

18.4.3 General Properties of Roche-Overflow Accretion

The results obtained in Example 18.1 have several immediate consequences for mass transfer through Roche-lobe overflow in binary star systems:

1. Since $v_\perp \gg v_\parallel$ the situation is as illustrated in the right side of Fig. 18.5, with the velocity at the L_1 point dominated by the perpendicular component. Therefore, gas particles coming across the L_1 point will have *large angular momentum*.
2. The *accretion flow is supersonic* because $|v| = (v_\parallel^2 + v_\perp^2)^{1/2} \gg c_s$. Hence pressure effects will be small because supersonic flow has no time to react to pressure waves limited to traveling at sound speed, and the motion of the gas packets flowing across the L_1 point will be *approximately ballistic* (motion influenced only by gravity).
3. Since $v_\parallel \sim c_s \ll v_{\text{ff}}$, where v_{ff} is the free-fall velocity acquired by the gas particle accelerated in the gravitational field of the primary, initial conditions at L_1 will have little influence on trajectories, leading to a *narrow accretion stream*.

After passing through the L_1 point a gas particle falls essentially freely in the gravitational potential of M_2 with the angular momentum that it had at L_1. Thus, the particle enters an approximately elliptical orbit in the plane defined by revolution of the binary. The set of particles executing elliptical motion in the gravitational field of the primary forms an accretion disk if the transverse velocity of the particles entering the Roche lobe of the primary is sufficiently high that the deflection of the accretion stream causes it to miss the body of the primary. This is not always the case (see Section 18.9.1 and Problem 18.3), but it typically will be in the most interesting situation where the primary is compact.

18.4.4 Disk Dynamics

The preceding discussion introduces the basic features of accretion by Roche-lobe overflow. However, the effect of the accretion disk is more complex than suggested by those considerations. This is primarily because the orbit of a gas particle within the Roche lobe of the primary is not actually a closed ellipse because of gravitational perturbations, most notably those caused by the presence of the secondary mass M_1. This causes deviation from an r^{-1} potential and implies precession of the elliptical particle orbits.

Disk heating and angular momentum transfer: Precession of the particle orbits caused by gravitational perturbations leads to collisions of particles as orbits cross each other. These collisions will heat the gas in the disk, which can then emit energy as electromagnetic radiation. Shockwaves presumably play a leading role in this heating because the velocities in the accretion disk are generally supersonic. As the disk is emitting electromagnetic radiation it has limited opportunity to exchange angular momentum with external objects. Therefore, the timescale for angular momentum transfer out of the disk is expected to be much longer than the timescale for radiating energy from the disk. As a consequence of the mismatch between these timescales, the particles in the disk will tend quickly to nearly circular orbits having the original specific angular momentum of the particle at L_1 (since circular orbits have the lowest energy for a given angular momentum). This orbit may be approximated by [89]

$$\frac{R_{\text{circ}}}{a} = \frac{4\pi^2}{GM_1 P^2} a^3 \left(\frac{b_1}{a}\right)^4 = \left(1 + \frac{M_2}{M_1}\right)\left(\frac{b_1}{a}\right)^4, \tag{18.8}$$

where R_{circ} is the *circularization radius*.

Internal angular momentum transfer: Since the disk radiates energy, some particles must descend lower into the gravitational potential of the primary to conserve energy, which requires losing angular momentum. But the timescale for transferring angular momentum from the disk is long compared with the timescale for radiating energy, so the disk must transfer angular momentum internally: some particles in the disk must spiral inward while other particles spiral outward. This net outward transfer of angular momentum implies that the disk is broadened both inward and outward around the circularization radius. A primary unresolved issue in the physics of accretion disks is the detailed mechanism by which an accretion disk accomplishes this internal redistribution of angular momentum.

An accretion elevator: The picture that emerges then is of particles in the inner portion of the disk that spiral inward slowly on a series of nearly circular orbits of gradually decreasing radius in the binary plane. An accretion disk then may be viewed as a natural elevator that lowers particles gradually in the gravitational field of the primary until they accrete onto the surface of the star, while radiating potentially large amounts of electromagnetic radiation as the gravitational energy is released. Let us make an estimate of the luminosity for the disk associated with this emitted radiation.

Luminosity of the disk: The density of the accretion disk is typically low enough that the self-gravity of the disk material may be ignored in calculations. The particle orbits then tend to circular Kepler orbits with angular velocity $\Omega(R) = (GM_1/R^3)^{1/2}$, where M_1 is the mass of the primary and R is the radius of the orbit. For a Kepler orbit that just grazes the surface of the primary at radius R_* the binding energy of a gas packet of mass ΔM is given by $E_{\text{bind}} = GM_1 \Delta M / 2R_*$, so in equilibrium the total luminosity of the disk is

$$L_{\text{disk}} = \frac{GM_1 \dot{M}}{2R_*} = \tfrac{1}{2} L_{\text{acc}}, \tag{18.9}$$

where $\dot{M}$ is the accretion rate on the primary and L_{acc} is the accretion luminosity defined in Eq. (18.12) below. Thus, we may expect that approximately half of the energy derived from accretion is radiated from the disk as the matter spirals inward toward the primary.

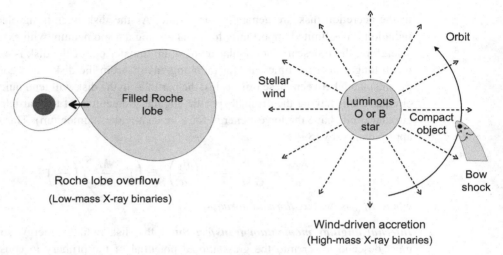

Fig. 18.6 Comparison of Roche-lobe overflow and wind-driven accretion. A bow shock is expected to form in the latter case because the wind flow is highly supersonic. As will be discussed below, low-mass X-ray binaries are typically associated with accretion by Roche-lobe overflow and high-mass X-ray binaries with wind-driven accretion.

18.5 Wind-Driven Accretion

The schematic mechanisms for accretion driven by Roche-lobe overflow and by winds are illustrated in Fig. 18.6. Wind-driven accretion is not as well understood as accretion by Roche-lobe overflow. It is likely that it is particularly important for those binary systems that contain an early spectral class (O or B) star with a neutron star or black hole companion. These systems tend to be luminous sources of X-rays (see Section 18.6.1). The stellar wind from the O or B star is generally both supersonic and intense. The velocity of the wind may be approximated by the escape velocity from the star,

$$v_{\text{wind}} \simeq v_{\text{esc}} = \left(\frac{2GM_*}{R_*}\right)^{1/2}, \qquad (18.10)$$

where R_* is the radius and M_* the mass of the O or B star. This implies wind speeds of order 10^3 km s^{-1} in typical cases – far higher than the sound speed, which is of order 10 km s^{-1}. The rate of mass emission from such hot, luminous stars can be 10^{-6}–10^{-5} $M_\odot$ yr^{-1}.

Because the flow is supersonic the wind particles may be assumed to follow ballistic trajectories, which allows for a simple estimate of the accretion rate on a compact companion (see Problem 18.2). Such estimates indicate that in wind-driven scenarios the accretion is very inefficient, being typically 1000–10,000 times less than the mass-loss rate from the companion. In contrast, Roche-lobe overflow is highly efficient, with close to 100% of the mass loss from one star accreting onto the other star in normal cases. It is only the remarkably high mass-loss rate from the young O or B star, and that energy is emitted largely in the form of X-rays for neutron star or black hole companions, that makes wind-driven accretion easily observable in binary systems.

18.6 Classification of X-Ray Binaries

For binaries with very compact remnants (neutron stars or black holes), persistent binary accretion seems to occur in only two general cases: *High-Mass X-Ray Binaries* (HMXB) and *Low-Mass X-Ray Binaries* (LMXB). These two extremes were displayed earlier in Fig. 18.6 and are discussed more extensively in the following sections.

18.6.1 High-Mass X-Ray Binaries

High-mass X-ray binaries have the following characteristics [163]:

1. Optical counterparts are typically luminous O or B stars and the optical luminosity from the system (dominated by visible and UV from the O or B star) is often larger than the X-ray luminosity.
2. They are commonly found in a young stellar population near the galactic plane.
3. They exhibit regular X-ray emission and transients with variation on timescales of minutes, but no large X-ray bursts.
4. The X-ray spectrum is relatively "hard" (has significant high-energy components), with an effective $kT \geq 15$ keV.
5. HMXBs are thought to consist of a neutron star or black hole, and a high-mass ($\geq 15\ M_\odot$) companion with a strong stellar wind, leading to wind-driven accretion on the compact object. Characteristic accretion rates are $\dot{M} \sim 10^{-10} - 10^{-6}\ M_\odot\ \text{yr}^{-1}$, with wind velocities $v \sim 2000\ \text{km s}^{-1}$.

Such systems are often very luminous X-ray sources and were among the first X-ray binaries discovered in the galaxy. The black hole candidate Cygnus X-1 discussed in Section 17.4.3 is an example of a high-mass X-ray binary. In Cyg XI the separation between the two stars is so small that the wind-driven accretion is probably focused by gravity of the primary into a stream that resembles Roche lobe overflow, even though the blue supergiant HD 226868 is likely not overflowing its Roche lobe.

18.6.2 Low-Mass X-Ray Binaries

In contrast to high-mass X-ray binaries, low-mass X-ray binaries exhibit the following characteristics [163]:

1. LMXB have faint blue optical counterparts and the emission from the accretion disk may dominate over emission from the stars. The optical luminosity is typically less than the X-ray luminosity by a factor of 10 or more, and the non-compact component is normally of spectral class A or later (cooler) in the spectral sequence.
2. They are commonly parts of old stellar populations, spread out of the galactic plane and concentrated toward the galactic center.
3. They produce strong X-ray outbursts.
4. The X-ray spectrum is softer than for HMXB, with an effective $kT \leq 10$ keV.
5. LMXB are thought to be binary systems having a compact object and a low-mass companion ($\leq 2\ M_\odot$), with accretion onto the compact object by Roche-lobe overflow.

The *X-ray bursters* that will be discussed in Section 19.2 are produced by accretion on neutron stars and are examples of low-mass X-ray binaries.

18.6.3 Suppression of Accretion for Intermediate Masses

Thus, HMXB appear to correspond to wind-driven accretion from high-mass companions and LMXB to Roche-lobe overflow accretion from low-mass companions, with few X-ray binaries lying in between. This separation of mass scales can be understood as being caused by strong suppression of accretion onto compact objects from companions in the 2–15 $M_\odot$ range that arises for two reasons: (1) For companion masses lying in this intermediate range the stellar winds from the companion are too weak to drive significant X-ray luminosity from accretion. (2) For companion masses in this range, Roche-lobe overflow accretion is quenched because the mass transfer rates tends to become super-Eddington (see Section 18.7.2), effectively halting the accretion by virtue of the radiation pressure resulting from the accretion.

18.7 Accretion Power

Astrophysical accretion is important for various reasons but the most important one is that it serves as a highly-efficient mechanism for gravitational energy conversion, and thus is potentially a very large source of power.

18.7.1 Maximum Energy Release in Accretion

The energy released by accretion onto an object is given approximately by

$$\Delta E_{\text{acc}} = G\frac{Mm}{R}, \tag{18.11}$$

where M is the mass of the object, R is its radius, and m is the mass accreted. In Table 18.1 the amount of energy released per gram of hydrogen accreted onto the surface of various objects is summarized (see Problem 18.6). From Table 18.1 we see that accretion onto very compact objects is a much more efficient source of energy than the thermonuclear burning of hydrogen to helium. On the other hand, accretion onto normal stars or even white dwarfs is much less efficient than converting the equivalent amount of mass to energy by

Table 18.1 Energy released by hydrogen accretion

Accretion onto	Max energy (erg g^{-1})	Ratio to H burning
Black hole	1.5×10^{20}	25
Neutron star	1.3×10^{20}	20
White dwarf	1.3×10^{17}	0.02
Normal star	1.9×10^{15}	10^{-4}

hydrogen burning. Let us assume for the moment, unrealistically, that all energy generated by conversion of gravitational energy in accretion is radiated from the system (efficiency for realistic accretion will be addressed shortly). Then the accretion luminosity is

$$L_{\rm acc} = \frac{GM\dot{M}}{R} \simeq 1.33 \times 10^{21} \left(\frac{M/M_\odot}{R/{\rm km}}\right)\left(\frac{\dot{M}}{{\rm g\,s}^{-1}}\right) \;{\rm erg\,s}^{-1}, \qquad (18.12)$$

if a steady accretion rate $\dot{M}$ is assumed.

18.7.2 Limits on Accretion Rates

Assuming fully ionized hydrogen and Thomson scattering of photons, the Eddington luminosity is given in Section 9.11.2 as

$$L_{\rm Edd} \simeq 1.3 \times 10^{38}\left(\frac{M}{M_\odot}\right)\;{\rm erg\,s}^{-1}. \qquad (18.13)$$

If the Eddington luminosity is exceeded (in which case the luminosity is said to be *super-Eddington*), accretion will be blocked by the radiation pressure, implying that there is an upper limit to the accretion rate on compact objects. Equating $L_{\rm acc}$ and $L_{\rm edd}$ gives as an estimate for this limiting rate

$$\dot{M}_{\rm max} \simeq 10^{17}\left(\frac{R}{{\rm km}}\right)\;{\rm g\,s}^{-1}. \qquad (18.14)$$

Eddington-limited accretion rates for white dwarfs, neutron stars, and black holes calculated from this formula are given in Table 18.2.

18.7.3 Accretion Temperatures

As shown in Problem 18.10, a crude estimate can be made of the accretion temperature for compact objects by assuming steady accretion at a rate $\dot{M}$ corresponding to the Eddington limit, and assuming that the accreted material equilibrates in a surface layer with a blackbody temperature $T_{\rm acc}$ given by

$$T_{\rm acc} = \left(\frac{GM\dot{M}}{4\pi\sigma R^3}\right)^{1/4}, \qquad (18.15)$$

Table 18.2 Some Eddington-limited accretion rates and temperatures

Compact object	Radius (km)	Max rate (g s^{-1})	$T_{\rm acc}$ (K)	$kT_{\rm acc}$ (eV)	Spectrum
White dwarf	~ 800	8×10^{20}	10^6	~ 100	UV
Neutron star	~ 10	1×10^{18}	$\sim 10^7$	~ 1000	X-ray
10 $M_\odot$ black hole	~ 30	3×10^{18}	$\sim 10^7$	~ 1000	X-ray

where R is the radius and M the mass of the compact object, and σ is the Stefan–Boltzmann constant.[3] The temperatures and corresponding spectral regions for white dwarfs, neutron stars, and a $10\,M_\odot$ black hole are also displayed in Table 18.2. From this table we expect that white dwarf accretion should lead to $T_{\rm acc} \sim 10^6$ K, and neutron star and black hole accretion should lead to $T_{\rm acc}$ in excess of 10^7 K. This corresponds to spectra in the UV to X-ray region for accretion on these objects.

18.7.4 Maximum Efficiency for Energy Extraction

The preceding discussion has considered the energy that is potentially available from accretion. However, the issue of efficiency in extracting this energy has not yet been addressed. For the gravitational energy released by accretion to be extracted, it must be radiated or matter must be ejected at high kinetic energy (for example, in relativistic jets). It may be expected that such processes are inefficient and that only a fraction of the potential energy available from accretion can be extracted to do external work. This issue is particularly critical when black holes are the central accreting object, since they have no "surface" onto which accretion may take place and the existence of an event horizon makes energy extraction acutely problematic.

Accretion efficiencies: From the previous equation for accretion power an efficiency factor η may be introduced through

$$L_{\rm acc} = \frac{GM\dot{M}}{R} = \frac{GM}{Rc^2}\dot{M}c^2 = \eta\dot{M}c^2 \qquad \eta \equiv \frac{GM}{Rc^2}, \qquad (18.16)$$

where R is the effective accretion radius. Thus η measures of the efficiency of converting mass to energy by accretion. For accretion onto a white dwarf or neutron star we may take the radius of the object for R. For accretion on a spherical black hole we may assume that R is some multiple of the Schwarzschild (event horizon) radius, which is given by Eq. (17.11),

$$r_{\rm s} = \frac{2GM}{c^2} = 2.95\left(\frac{M}{M_\odot}\right)\ {\rm km}, \qquad (18.17)$$

since any energy to be extracted from accretion must be emitted from outside that radius. Then for a spherical black hole

$$L_{\rm acc}^{\rm bh} \equiv \frac{r_{\rm s}}{2R}\dot{M}c^2 = \eta\dot{M}c^2 \qquad \eta = \frac{r_{\rm s}}{2R}. \qquad (18.18)$$

For a black hole a typical choice for R is the radius of the innermost stable circular orbit in the Schwarzschild spacetime, which is located at $3r_{\rm s}$.

---

[3] Accretion onto realistic compact objects is more complicated, involving accretion disks with possibly complex dynamics (Problem 18.8 explores a somewhat more realistic scenario). Furthermore, general relativistic effects may not be negligible for accretion onto neutron stars and black holes. Nevertheless, this simple estimate gives the right order of magnitude for accretion temperatures because they are determined primarily by the release of gravitational energy, which causes the system to re-equilibrate at a higher temperature than would be expected in the absence of accretion.

Efficiencies for various processes: For burning of hydrogen to helium the mass to energy conversion efficiency is $\eta \sim 0.007$ (see Section 5.1.4). For compact spherical objects like Schwarzschild black holes or neutron stars, reasonable estimates suggest $\eta \sim 0.1$. For rotating, deformed (Kerr) black holes, it is possible to be more efficient in energy extraction and efficiencies of $\eta \sim 0.3$ might be possible.

Example 18.2 The high efficiency available from accretion onto a supermassive black hole as a power source provides the most convincing general argument that active galactic nuclei (AGN) and quasars must be powered by accretion onto rotating supermassive ($M \sim 10^9 \, M_\odot$) black holes. For example, in Problem 18.4 you are asked to use observed luminosities and temporal luminosity variation to show that a quasar could be powered by accretion of as little as several solar masses per year onto an object of mass $\sim 10^9 \, M_\odot$, and that this mass must occupy a volume the size of the Solar System or smaller.

18.7.5 Storing Energy in Accretion Disks

In addition to being a primary source of power for varied astrophysical phenomena, an accretion disk can function as a *storage reservoir* for energy released on a short timescale that then can meter the original energy release out over a much longer period than the dynamical timescale for direct collapse. For example, the long-period gamma-ray bursts to be discussed in Chapter 21 last as long as many tens of seconds. They are thought to be powered by the collapse of the core of a massive star, with the gravitational energy from the collapse released on a timescale of order one second. It is proposed that the core collapse leads to a rotating black hole surrounded by an accretion disk and emitting ultrarelativistic jets on its rotation axis. The gamma-ray burst is then produced by the jets, partially energized by the accretion of matter from the disk, which spreads part of the collapse energy out over tens to hundreds of seconds to power the long-period gamma-ray burst.

18.8 Some Accretion-Induced Phenomena

Accretion is a primary factor in a number of astrophysical phenomena, either as the initiator or as the primary power source, or as both. Let us summarize briefly some of these phenomena.

1. *Cataclysmic variables* are accreting binary systems in which accretion is onto a white dwarf, with the accretion leading to a variety of outbursts depending on the circumstances. The most spectacular are novae, which will be described in Chapter 19.
2. *X-ray binaries* may be divided into high-mass and low-mass categories. The distinction between them was discussed in Section 18.6.
3. *X-ray bursts* occur in low-mass X-ray binaries. *Type II bursts* are less common and may be associated with fluctuations in the accretion rate. *Type I bursts* are more common

and are characterized by X-ray luminosities that increase by factors of 10 or more over a period of a few seconds. X-ray bursts will be discussed further in Chapter 19.
4. *Type Ia supernovae* have been ascribed to two primary models, both likely involving accretion onto a white dwarf as the initiator of the explosion. In the first, accretion from a nondegenerate star onto a white dwarf triggers a runaway thermonuclear flash in the degenerate white dwarf matter that consumes the entire star. This is called the *single-degenerate scenario*. In the second model a binary white dwarf system spirals together and merges. Near merger one of the white dwarfs likely is tidally disrupted, forming a disk that accretes mass onto the other, triggering a thermonuclear runaway in degenerate white-dwarf matter. This is called the *double-degenerate scenario*. Type Ia supernovae will be discussed further in Chapter 20.
5. *Rotating black hole engines* are thought to power active galaxies and quasars (see Example 17.2 above and Problem 18.4). These central engines produce far more power within a small region than can be accounted for easily by any source of energy other than gravitational. The standard paradigm is that these engines derive significant energy from accretion onto supermassive, rotating black holes.[4] On a stellar scale, gamma-ray bursts are believed to be powered in a similar way by accretion onto a rotating black hole produced either by the core collapse of a massive star, or by merger of two neutron stars (or a black hole and neutron star). This will be discussed further in Chapter 21.

As noted explicitly above, most of these topics will be taken up in more depth in other chapters.

18.9 Accretion and Stellar Evolution

Mass is destiny for stars: generally, the more massive a star is the faster it evolves through all stages of its life. Although the evidence is overwhelming for this hypothesis, there are particular data sets that appear to contradict it. It is instructive that these apparent contradictions often involve the possibility of a star interacting with another star, either through accretion in a binary system or through collisions and mergers in clusters.

18.9.1 The Algol Paradox

One particularly interesting case is the Algol system[5] (see Box 1.6), where the more massive star of spectral class B8 appears to be much less evolved than the less massive K0 companion. However, spectroscopic evidence suggests that weak accretion is occurring in the Algol system (and that the accretion is directly into the body of the primary rather than through an accretion disk; see Problem 18.3 and Fig. 18.7). Therefore, one is led to ask whether it is possible that accretion in the Algol system is distorting the picture of which

---

[4] See Chapter 15 of Ref. [100] for a more comprehensive discussion.
[5] Algol is actually a triple-star system, but two of the three stars are very close together and form the eclipsing binary that is commonly meant by the Algol system.

18.9 Accretion and Stellar Evolution

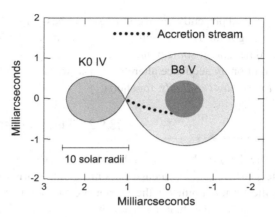

Fig. 18.7 Mass transfer in the Algol system from the K0 star to the B8 star.

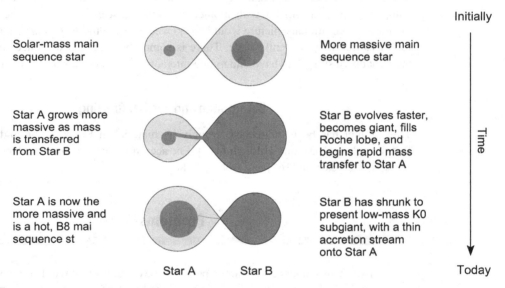

Fig. 18.8 Resolution of the "Algol paradox." The less massive star is more evolved because of an earlier episode of mass transfer in the system.

star is the older star. This "Algol paradox" is thought to be resolved by the evolutionary sequence depicted in Fig. 18.8, which indicates that previous mass transfer has altered the system from what would have been expected for the evolution of isolated stars.

Specifically, it is believed that initially the present spectral class K0 star (denoted by Star B in the figure) was a more massive main sequence star that evolved faster than its less massive companion (the present spectral class B8 star, denoted by Star A in the figure). This initially more massive star evolved off the main sequence, expanded to fill its Roche lobe, and began transferring mass to its companion. Over time sufficient mass transfer occurred to make the companion more massive and the present spectral class K0 star less massive, and presently the accretion stream has diminished to a trickle. It may be expected that in Algol the B8 star will evolve to fill its Roche lobe and begin mass transfer back to the other star at some point in the future. In general, eclipsing variable stars exhibiting an evolutionary sequence altered by mass transfer similar to that sketched here are called *Algol variables* (or *algols*, for short).

18.9.2 Blue Stragglers

An issue that may be related indirectly to the Algol paradox is that of "blue stragglers" in clusters, where main sequence stars more blue (earlier spectral class) and more luminous than the turnoff point for the cluster are observed. According to standard evolutionary models such stars should already have evolved off the main sequence. However, the presence of blue stragglers could be explained if they have not evolved in isolation but instead are stars that have had their masses altered by merger or accretion with a binary companion, or by a collision with another star in the dense environment of the cluster. Then the normal evolutionary picture would be skewed because these stars have not always had the mass that they presently have. There is some observational evidence supporting the view that blue stragglers have undergone interactions with other stars in a cluster.

Background and Further Reading

This chapter has been influenced strongly by material in Frank, King, and Raine [89]; Padmanabhan [163]; and Hilditch [117]. For accretion disks in active galactic nuclei and quasars, see Peterson [168] and Robson [186].

Problems

18.1 Hot main sequence stars are expected to have substantial winds blowing from their surfaces. Estimate the wind speed expected for an O8 main sequence star.

18.2 Consider wind-driven accretion on a neutron star from a companion with a strong, steady wind. Estimate the accretion rate by assuming all wind particles to be captured if the particles cross an area πr_{acc}^2 around the neutron star, with the radius r_{acc} determined by the volume around the neutron star where the gravitational potential is equal to or greater than the kinetic energy of the particles in the wind. Assume the wind to blow radially outward from the companion, to not be deflected by the gravity of the neutron star before capture, and to be supersonic so that pressure effects may be neglected. Show that significant accretion rates – say 10^{-12} $M_\odot$ yr^{-1} or greater – occur only for companion mass-loss rates in excess of 10^{-8} $M_\odot$ yr^{-1} (which is relatively uncommon).***

18.3 Show that for Algol-like systems (binaries consisting of main sequence or subgiant stars with periods of a few days or less) the accretion stream may intersect the body of the primary directly (see Fig. 18.7). On the other hand, show that for compact primaries (white dwarfs, neutron stars, or black holes), this is unlikely and an accretion disk is almost certain to form if Roche-lobe overflow occurs.***

18.4 Active galactic nuclei and quasars can produce luminosities of 10^{47} erg s^{-1} with variations on timescales of days. Use this to show that

(a) If this energy derives from fusion reactions it is necessary that several hundred solar masses be fused per year, but if it derives from accretion (with accretion efficiencies that are theoretically attainable with rotating supermassive black holes), the energy output could be generated by accretion of as little as several solar masses per year.

(b) If the source of the energy powering the AGN is accretion and the source radiates at the Eddington limit, the mass of the central object onto which accretion occurs must be of order $10^9 \, M_\odot$.

(c) The variability timescale and part (b) imply that 10^{47} erg s^{-1} must be produced in a region containing at least $10^9 \, M_\odot$, and this region cannot be substantially larger than the Solar System. *Hint*: What is the fastest speed for transferring information in such a system?

This reasoning is the basis for the usual assumption that quasars and active galaxies must be powered by accreting, supermassive black holes at their centers.***

18.5 Derive the result (18.8) for the circularization radius. *Hint*: Assume the accreted particles to go into a circular orbit determined primarily by the gravitational field of the primary and assume the angular momentum per unit mass at the L_1 point to be preserved in this circular orbit.

18.6 Verify the entries for energy released per gram by accretion and the ratio of the energy released to that obtained from thermonuclear burning of hydrogen for Table 18.1. *Hint*: For the black hole, assume it to be spherical and for accretion to occur from the innermost stable circular orbit with radius three times the Schwarzschild radium r_s.***

18.7 Verify the equations and conclusions of Section 18.4.2 concerning velocities in accreted steams.

18.8 Accretion disks around compact objects are observed to radiate approximately as blackbodies, often at a significant fraction of the Eddington luminosity (9.14). Take as a simple model of an accretion disk around a compact object a thin disk radiating as a blackbody at a fraction η of the Eddington luminosity. Derive an expression for the temperature T of such a disk having radius R. Use this expression to estimate the temperature and therefore peak emission wavelength for an accretion disk around a neutron star, and for Schwarzschild black holes having masses of order $10 \, M_\odot$ and $10^8 \, M_\odot$, respectively. For the neutron star, you may approximate R as comparable to the radius of the neutron star, and for the black hole you may approximate R as comparable to the innermost stable circular radius in a Schwarzschild spacetime, which is given by $R = 6(G/c^2)M$, where M is the mass of the black hole. Use these results to explain the observed wavelength ranges associated with the accretion disks for such objects.

18.9 Mass transfer in a binary system can change the orbital period P of the binary. There are a variety of ways in which this can happen depending, for example, on how the mass is transferred (wind or Roche-lobe overflow), and whether the mass lost by one star is completely captured by the other star (*conservative* mass transfer) or whether some of it escapes the system (*non-conservative* mass transfer). Let us investigate one of the simplest possibilities. Consider a binary with circular orbits in which star 1 is the source of a spherically symmetric wind that causes a mass change rate $\dot{m}_1 < 0$, any interaction of the wind with the other star is neglected (so the mechanism is non-conservative), and the emitting star is assumed to maintain a constant velocity magnitude on its orbit so that all changes in the period come from mass loss.

(a) Show that for this case
$$\frac{\dot{P}}{P} \equiv \frac{dP/dt}{P} = \frac{-2\dot{m}_1}{m_1 + m_2},$$
implying that since $\dot{m}_1$ is negative (the wind causes star 1 to lose mass), the binary period must increase as time goes on.

(b) Observationally, binary orbital periods can be determined to about one part in 10^7. Assume a binary system having a 10-day period, with $m_1 = 20\,M_\odot$ and $m_2 = 10\,M_\odot$. For a wind-driven mass loss rate of $\dot{m}_1 \sim 10^{-7}\,M_\odot\,\mathrm{yr}^{-1}$ that is typical for normal upper main sequence stars, estimate the total binary period change over a time of 10 years and comment on the feasibility of measuring it.

(c) Repeat the preceding analysis but assume the star that is losing mass to be a Wolf–Rayet star (see Section 14.3.1), which can lose mass at rates as high as 10^{-5} to $10^{-4}\,M_\odot\,\mathrm{yr}^{-1}$.

18.10 Assume steady accretion onto a spherical compact object at a rate $\dot{M}$. Real accretion is more complicated, typically involving accretion disks (see Problem 18.8 for a more realistic example), but make a simplified model where the accreting material accumulates in a surface layer that equilibrates as a blackbody at temperature T_{acc} and radiates the excess energy deriving from accretion (assume 100% efficiency for purposes of this exercise). Taking the accretion to occur at the Eddington limit, derive a formula for the temperature T_{acc} in terms of the accretion luminosity and verify the entries in Table 18.2 for accretion onto a representative white dwarf, neutron star, and a spherical $10\,M_\odot$ black hole. You may take as an accretion radius for the black hole radius of the innermost stable circular orbit, which is at three times the Schwarzschild radius $r_s = 2.95(M/M_\odot)$ km. (Although not necessary to work this problem, a more extensive discussion of spherical black holes and associated accretion may be found in Chapters 11 and 15 of Ref. [100].)*** *Hint*: Take the time derivative of Kepler's 3rd law and use that with our assumptions that the velocity of star 1 and the mass of star 2 are constant.

19 Nova Explosions and X-Ray Bursts

Some stars are observed to increase their optical brightness by as much as factors of 10^4–10^5 in a matter of days, and then slowly dim back to obscurity over a period of months. This is called a *nova*. Furthermore, for some stellar X-ray sources the flux can increase suddenly in an *X-ray burst* superposed on the background emission, and then fall quickly back to the background level. What is the nature of these nova and X-ray burst events, and what are the energy sources that power them? Strong clues are provided by the observation that both novae and X-ray bursts seem to be associated with *binary star systems*. In this chapter we shall discuss the characteristics of novae and X-ray bursts, and propose that they are caused by a common mechanism: a *runaway thermonuclear explosion* in degenerate matter triggered by accretion onto a compact object in a binary star system. The primary difference between the two is that the compact object is a white dwarf in the case of the nova and a neutron star in the case of the X-ray burst, and this difference influences the observational characteristics.

19.1 The Nova Mechanism

We begin the discussion by considering the better-understood case of novae. The nature of a nova event is suggested by three key observations.

- Novae seem to be associated with binary systems in which one star is a white dwarf.
- Doppler shifts indicate an expanding shell of gas emitting the light of the nova.
- There are recurrent novae that repeat after some period of time.

Taken together, these observational characteristics suggest the nova mechanism illustrated in Fig. 19.1. A nova can occur in a binary star system for which one star is non-compact (typically a main sequence, subgiant, or giant star) and the other star is a white dwarf. Matter from the first star accretes in a thin layer on the surface of the white dwarf because of the other star filling its Roche lobe and spilling matter onto the white dwarf (typically through an accretion disk), or possibly because of a strong wind from the other star that the white dwarf captures onto its surface (see binary accretion in Chapter 18). Eventually this layer ignites in a thermonuclear explosion under degenerate conditions. The resulting thermonuclear runaway (recall the earlier discussion of the helium flash in red giant stars and see Box 19.1) blows a thin, hot surface layer off into space, causing a large rise in light output from the system while the expanding shell remains optically thick. Figure 19.2(a) shows the shell ejected by Nova Cygni 1992, as imaged by the Hubble Space Telescope

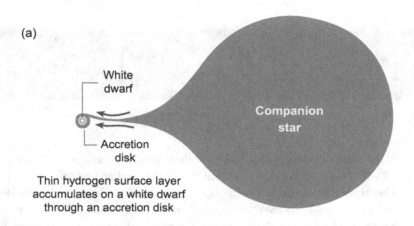
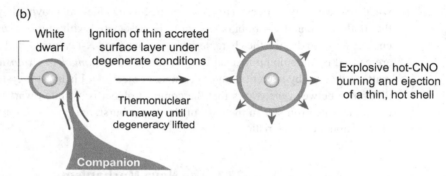

Fig. 19.1 (a) Accretion in a binary system that can lead to a nova outburst. (b) The hydrogen accumulated on the surface of the white dwarf can ignite in a thermonuclear runaway, blowing off a thin, hot, expanding shell that produces the nova outburst.

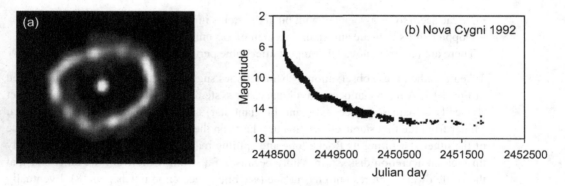

Fig. 19.2 (a) Expanding shell around Nova Cygni 1992, two years after the nova explosion. (b) Visual lightcurve of Nova Cygni 1992 from the AAVSO International Database, plotted for 1991–2002 [8]. Reprinted from www.aavso.org/v1974-eyg-nova-cygni-1992

Box 19.1   Degeneracy and Thermonuclear Runaways in Novae

The dependence of pressure on temperature computed numerically for conditions expected in a nova is illustrated in the following figure.

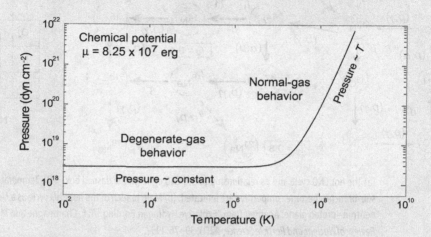

The equation of state allowing arbitrary degeneracy and degree of relativity from the numerical simulations of white dwarf structure in Chapter 16 was used, with the pressure coming dominantly from the electron gas. At low temperatures the electrons are degenerate and the pressure is essentially independent of the temperature. At high temperatures the degeneracy is lifted and the pressure increases with temperature, as expected for an ideal gas. Thus, a thermonuclear reaction ignited in degenerate matter on the surface of the white dwarf becomes a runaway until the temperature rises sufficiently to break the degeneracy and generate a pressure that increases rapidly with the temperature. This pressure blows off the hot burning surface layer in a thin shell that expands rapidly, producing the rising light output of the nova.

---

two years after the explosion was first observed.[1] The corresponding lightcurve is shown in Fig. 19.2(b). Nova Cygni 1992 was visible without a telescope at its peak.

19.1.1 The Hot CNO Cycle

The nova thermonuclear runaway is powered by an extension of the CNO cycle at higher temperatures to a wider set of reactions called the *hot CNO cycle*. Figure 19.3 illustrates the relationship between the CNO and hot CNO cycles. The transition from CNO to hot CNO cycle is initiated by a proton capture reaction on ^{13}N. Whether the hot CNO cycle is

---

[1] It is common to name novae using the word "Nova," followed by the constellation and the year the outburst was first observed on Earth. This nova was observed in the constellation Cygnus in 1992. It is also known by its variable star name V1974 Cygni.

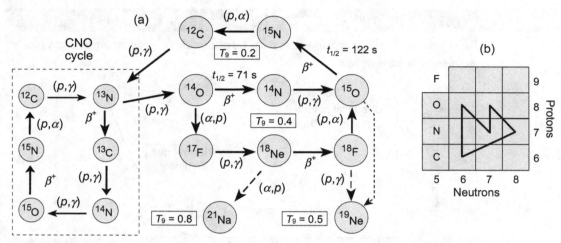

Fig. 19.3 (a) The hot CNO cycle and its relationship to the CNO cycle (in the dashed box) [70]. Temperatures where sub-branches become competitive are indicated. (b) Main branch of the hot CNO cycle as a closed path in the neutron–proton plane. Adapted from "Explosive Hydrogen Burning," A. E. Champagne and M. Wiescher, *Annual Review of Nuclear and Particle Science*, **42**(1), 39–76, 1992.

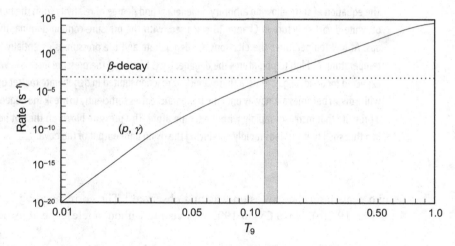

Fig. 19.4 Competition of proton capture and β-decay in breakout from the CNO to hot CNO cycle (see Problem 19.2). The proton capture rate was calculated assuming a density of 100 g cm^{-3} and a hydrogen mass fraction $X = 0.7$. Under these conditions the proton capture leading to hot-CNO breakout begins to complete favorably with β-decay when the temperature reaches the range indicated by the gray box.

populated is a strong function of temperature and its influence on the competition between proton capture and β-decay, as illustrated in Fig. 19.4, Example 5.7, and Problem 19.2.

The β-decay of ^{13}N in the CNO cycle is an internal nuclear process that is independent of temperature. However, ^{13}N also can capture a proton to make ^{14}O, which initiates the breakout into the hot CNO cycle. This reaction has a very strong temperature dependence (see Fig. 6.3), since it is inhibited by a Coulomb barrier. At low temperatures the β-decay wins but for temperatures exceeding $T_9 \sim 0.1$ the proton capture reaction begins to compete strongly and quickly dominates with even small increases in temperature. The rising temperature of the initial nova outburst triggers this breakout into the hot CNO cycle and the nova is largely powered by the corresponding energy that is released. Nuclear burning through the hot CNO cycle is often termed *explosive hydrogen burning* [70].

19.1.2 Recurrence of Novae

The characteristic total energy output of a nova is of order 10^{44-45} erg, which is about 10^{11} times more energy than the Sun produces each second. The duration of the thermonuclear runaway that produces most of this energy is 100–1000 seconds. Despite this large energy release, a nova outburst typically ejects only about 10^{-4} of the mass of the white dwarf, thus leaving the white dwarf largely intact. This is confirmed by the observation of *recurrent novae*, where following the nova outburst the white dwarf begins accumulating accreted material again that eventually will trigger a new nova explosion.

Example 19.1 RS Ophiuchi is a white dwarf and red giant binary, 5000 ly away in Ophiuchus. It has been observed in nova outburst six times since 1898. In its quiet phase RS Ophiuchi has $m_V \sim 12.5$ but in nova outburst this can rise to $m_V \sim 5$.

19.1.3 Nucleosynthesis in Novae

The hot CNO cycle leads to synthesis of new elements. Although the species of elements produced in nova explosions are relatively few in number compared with those produced in other events like supernova explosions, certain isotopes likely owe their existence primarily to nova events. The inferred abundances of elements relative to that for hydrogen in the expanding shell around Nova Cygni 1992 are given in Table 19.1.

19.2 The X-Ray Burst Mechanism

The mechanism for an X-ray burst is thought to be similar to that of a nova, except that the matter accretes onto a neutron star rather than a white dwarf (replace the white dwarf in

Table 19.1 Nova Cygni 1992 abundances relative to hydrogen [155]

He	C	N	O	Ne	Na	Mg	Al	Si	S	Ar	Ca	Fe	Ni
4.5	70.6	50.0	80.0	250.0	37.4	129.4	127.5	146.6	1.0	5.0	46.8	8.0	36.0

> **Box 19.2** **Production of X-Rays**
>
> X-rays are emitted when fast-moving electrons pass close to slow-moving ions and are accelerated. In equilibrium, only if the temperatures are millions of degrees are the electrons moving at high enough velocities to produce X-rays. The higher the temperature, the faster the electrons move. This increases both the energy and the intensity of the X-rays, since collisions become more violent and more frequent at high temperature. An X-ray burst on the surface of a neutron star may last for a few seconds, during which time the temperatures can reach as high as $\sim 10^9$ K. This causes X-rays to be produced in abundance. Most nova events have maximum temperatures in the vicinity of several times 10^8 K or smaller, and this tends to produce light at visible and other longer wavelengths.

Fig. 19.1 with a neutron star).[2] The X-ray burst is triggered by a thermonuclear runaway under degenerate conditions, as for a nova. However, the gravity of a neutron star is much stronger than that of a white dwarf. Thus, matter falling to the surface of the neutron star releases more gravitational energy and the thermonuclear runaway occurs at much higher temperatures and densities than in the nova outburst. This tends to produce X-rays rather than visible light in the explosion, as discussed in Box 19.2.

19.2.1 Rapid Proton Capture

Because the temperatures in an X-ray burst are very high compared with a nova, the hot CNO reaction sequence responsible for powering novae can break out into a much more extensive network of reactions involving competition between proton capture, α-particle capture, and β-decay that is called the *rapid proton capture process* or *rp-process*. The reaction path for the rp-process is illustrated in Fig. 19.5. The energy released in the reactions of the hot CNO cycle and the rp-process are thought to provide the power source for X-ray bursts. The typical duration of the thermonuclear runaway powering the burst is a few seconds, during which time up to 10^{39-40} erg may be released, largely as X-rays. X-ray bursts from a given system are typically highly recurrent, with some repeating on timescales as short as hours.

19.2.2 Nucleosynthesis and the rp-Process

Because of uncertainties in nuclear reaction rates for isotopes difficult to produce under terrestrial conditions and uncertainties in the conditions characterizing the burst, it is not known precisely how high in proton and neutron number the nucleosynthesis in the rp-process can go during an X-ray burst. Because the gravitational field of a neutron star is so strong, even if proton-rich nuclei are synthesized by the rp-process it will be difficult

---

[2] X-ray bursters also may exhibit a more steady X-ray emission upon which the bursts are superposed. The steady emission is likely caused by heating of matter in the accretion disk. Flickering is sometimes observed for the more steady emission, probably because of instabilities in the accretion disk. A more general discussion of X-ray emission from low-mass and high-mass X-ray binaries has been given in Section 18.6.

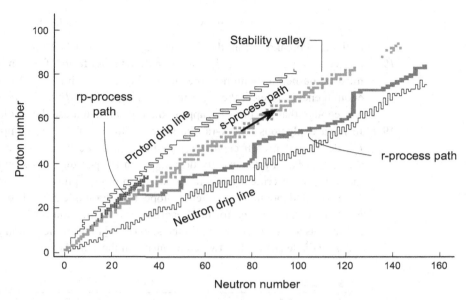

Fig. 19.5 Path for the rp-process. Also shown are the s-process and r-process paths. Light-gray squares along the s-process path indicate the several hundred isotopes that are stable against radioactive decay. Outside the proton and neutron drip lines isotopes are unstable against spontaneous emission of protons or neutrons, respectively. About 7000 isotopes are predicted to exist between the drip lines. As of 2017 about 3300 of these have been observed experimentally.

for them to escape the gravity of the neutron star, though it may be possible for some rp-processed material to be ejected in special circumstances.

Background and Further Reading

Short introductions to accretion, novae, and X-ray bursts may be found in Carrol and Ostlie [68], and Rolfs and Rodney [188]. Many aspects of nova explosions and explosive hydrogen burning are reviewed in Champagne and Wiescher [70].

Problems

19.1 From Fig. 19.2(b), how much brighter was Nova Cygni 1992 near its peak relative to when it later returned to approximately constant brightness?

19.2 Use the Caughlan and Fowler compilations [69, 88] to calculate the rate for the reaction ^{13}N(p, γ)^{14}O, for which $Q = 4.628$ MeV, for temperatures relevant to main sequence CNO burning; including resonant and nonresonant contributions. If the density is 10^3 g cm^{-3}, estimate the temperature at which this reaction begins to compete with the β-decay

$$^{13}\text{N} \to {}^{13}\text{C} + e^+ + \nu_e \qquad (t_{1/2} = 10\,\text{min}),$$

initiating a breakout from the CNO cycle into the hot CNO cycle.***

19.3 Derive a formula as a function of white dwarf mass for the fraction of hydrogen that must be burned to helium in the surface layer of a white dwarf in order to supply enough energy to eject the layer in a nova explosion. *Hint*: Assume the mass to be low enough that Eq. (16.4) is approximately valid, and use the results in Fig. 16.1.

19.4 In nova explosions and X-ray bursts one often finds strong competition between β^+ decay and proton capture (p, γ) reactions. The first has a constant rate and the second has a rate strongly dependent on temperature. Thus, at a given density there will be a critical temperature where proton capture begins to compete with β-decay to depopulate a given isotope (see Example 5.7). The following table gives the parameters p_n for the ReacLib parameterization (D.3) for β-decay of ^{17}F [which has a single component in the sum (D.2)] and for the proton capture reaction $^{17}\text{F}(p, \gamma)^{18}\text{Ne}$ [which has two components in the sum (D.2)].

	p_1	p_2	p_3	p_4	p_5	p_6	p_7
β^+:	−4.538	0.0	0.0	0.0	0.0	0.0	0.0
$(p, \gamma)_1$:	17.29	4.341×10^{-4}	−18.10	0.1471	−0.1550	0.0167	−0.7411
$(p, \gamma)_2$:	3.428	−4.654	−3.999	8.947	−0.6332	0.0402	−5.113

(a) From these data, what is the half-life for β-decay of ^{17}F?

(b) Plot the β-decay rate and the total (p, γ) rate computed from the above table and Eqs. (D.3) and (D.2) as functions of temperature for the range $T_9 = 0.1$ to $T_9 = 1$.

(c) Assuming a constant density of $\rho = 500\,\text{g cm}^{-3}$, estimate the temperature at which the rate for proton capture on ^{17}F to produce ^{18}Ne becomes comparable to that for β-decay of ^{17}F to produce ^{17}O.

20 Supernovae

Supernovae represent the catastrophic demise of certain stars or compact objects. They are among the most violent events in the Universe, releasing as much as $\sim 10^{53}$ erg of energy, much of it in the first second of the explosion. For perspective, the total luminosity of the Sun is only about 10^{33} erg s^{-1} and even a nova outburst releases only of order 10^{45} erg over a characteristic period of a few hundred seconds. There is more than one type of supernova, with two general methodologies for classification: (1) according to the spectral and lightcurve properties, or (2) according to the fundamental mechanism responsible for the energy release. In addition to their intrinsic interest, supernovae are of fundamental importance for a variety of astrophysical phenomena, including element production and galactic chemical evolution, the relationship to some types of gamma-ray bursts, a connection to star formation through energizing and compressing the interstellar medium, a source of gravitational wave emission, the creation of neutron stars and black holes, and applications in cosmology associated with measuring distance through standardizable candle properties. We shall initiate the discussion by considering the taxonomy of these events.

20.1 Classification of Supernovae

The traditional classification of supernovae is based primarily on spectra and lightcurves. Some representative spectra are displayed in Fig. 20.1 and some typical lightcurves are illustrated in Fig. 20.2. In most cases a schematic model can be associated with each class that can account for the observational characteristics of that class. Those models suggest that all supernova events derive their enormous energy from either gravitational collapse of a massive stellar core or a thermonuclear runaway in dense, electron-degenerate matter. The observational characteristics of supernovae derive both from the internal mechanism causing the energy release (for example, collapse of a stellar core) and the interaction of the initial energy release with the surrounding outer layers or extended atmosphere of the star. Thus, some observational characteristics are diagnostic of the explosion mechanism itself, while others are related only indirectly to the explosion mechanism and instead are diagnostics for the state of the star and its surrounding medium at the time of the outburst.

The standard classes of supernovae are illustrated in Fig. 20.3. The primary initial observational distinction is whether hydrogen lines are present in the spectrum, which divides supernovae into Type I (no hydrogen lines) and Type II (significant hydrogen lines). The standard subclassifications then correspond to the characteristics described in

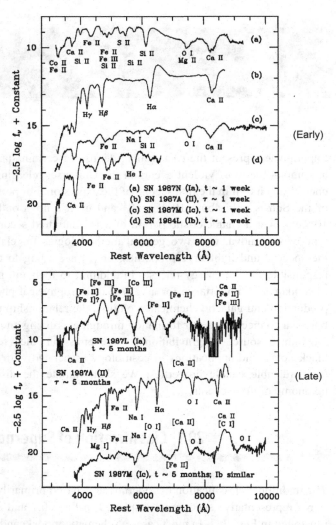

Fig. 20.1 Early-time and late-time spectra for several classes of supernovae [86]. Reproduced from "Optical Spectra of Supernovae," Alexei V. Filippenko, *Annual Review of Astronomy and Astrophysics*, **35**(1), 309–355, 1997.

the following paragraphs. The bottom row of Fig. 20.3 indicates the mechanism thought to be responsible for each class of supernova explosion, and implies that the broad variety of supernova observational characteristics correspond to only two basic explosion mechanisms: gravitational collapse of a massive star core, or a thermonuclear runaway in degenerate matter.

20.1.1 Type Ia

A Type Ia supernova is thought to be associated with a thermonuclear runaway under electron-degenerate conditions in white-dwarf matter. This class of supernovae is sometimes termed a *thermonuclear supernova,* to distinguish it from all other classes

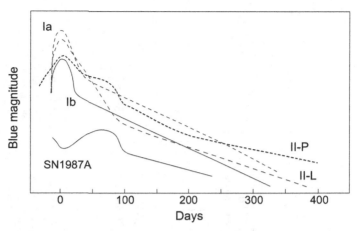

Fig. 20.2 Schematic lightcurves for several representative classes of supernovae [86]. Also shown is the lightcurve for Supernova 1987A, which will be discussed in Section 20.4. Reproduced from "Optical Spectra of Supernovae," Alexei V. Filippenko, *Annual Review of Astronomy and Astrophysics*, **35**(1), 309–355, 1997.

Box 20.1 — **Standard and Standardizable Candles**

A *standard candle* is a source that always has the same intrinsic brightness (a 100-watt light bulb, for example). A *standardizable candle* is a light source that may vary in brightness but that can be standardized (normalized to a common brightness) by some reliable method. Standard candles, or standardizable candles, then permit distance measurement by comparing observed brightness with the standard brightness. Different Type Ia supernovae have similar but not identical lightcurves. Hence they are not standard candles. However, empirical methods have been developed that allow the lightcurves of different Type Ia supernovae to be collapsed approximately to a single curve. Figure 20.4 gives an example. Thus, Type Ia supernovae are standardizable candles.

Type Ia standardizable candles are particularly valuable because their brightness makes them visible at very large distances. The standardizable candle and brightness properties of Type Ia supernovae have made them a central tool in modern cosmology. For example, they are the most direct indicator that the expansion of the Universe is currently accelerating, implying that the Universe is permeated by a mysterious *dark energy* that effectively turns gravity into antigravity.

---

that derive their power from gravitational collapse and not from thermonuclear reactions. No hydrogen is observed but calcium, oxygen, and silicon appear in the spectrum near peak brightness. Type Ia supernovae are found in all types of galaxies and their standardizable candle properties make them a valuable distance-measuring tool (see Box 20.1).

20.1.2 Type Ib and Type Ic

Type Ib and Ic supernovae are thought to represent the core collapse of a massive star that has lost much of its outer envelope because of strong stellar winds or interactions

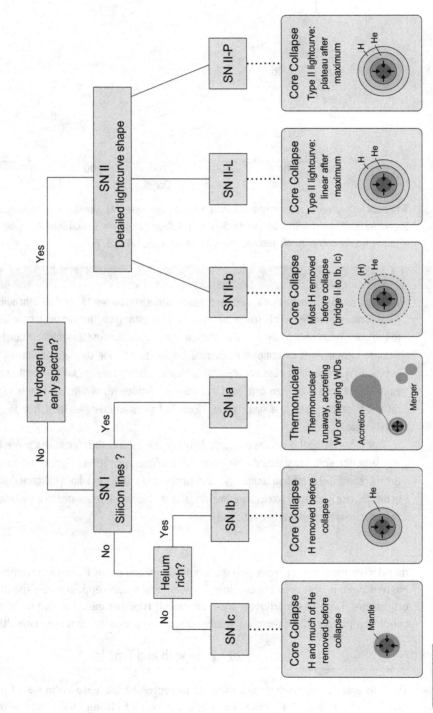

Fig. 20.3 Classification of supernova events. The tree hierarchy classifies according to observational properties. The bottom row describes the mechanism thought to be responsible for the explosion. Note that Type II-n is not shown here but is discussed in the text.

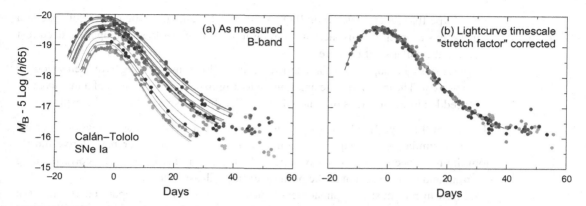

Fig. 20.4 Empirical rescaling of Type Ia supernova lightcurves to make them standardizable candles. (a) B-band lightcurves for low-redshift Type Ia supernovae (Calán–Tololo survey [105]). As measured, the intrinsic scatter is 0.3 mag in peak luminosity. (b) After 1-parameter correction the dispersion is 0.15 mag. Adapted from Ref. [91]. Reproduced from "Dark Energy and the Accelerating Universe," J. A. Frieman, M. S. Turner, and D. Huterer, *Annual Review of Astronomy and Astrophysics*, **46**(1), 385–432, 2008.

with a binary companion (see the *Wolf–Rayet stars* discussed in Section 14.3.1). For both Type Ia and Ib supernovae hydrogen and silicon spectral lines are absent, but helium lines are present for Type Ib supernovae. The distinction between Types Ib and Ic is thought to lie in whether only the hydrogen envelope has been lost before core collapse (Type Ib), or whether most of the helium layer has also been expelled (Type Ic). There is some observational evidence (for example, from the polarization of detected light) that these classes of supernovae involve highly asymmetric explosions. Type Ib and Ic supernovae are found only in spiral galaxies, implying a relationship with regions of strong star formation (since such regions are characteristic of spiral galaxies but not of elliptical galaxies).

20.1.3 Type II

Type II supernovae are characterized by prominent hydrogen lines. They are thought to be associated with the core collapse of a massive star and are found only in regions of active star formation (they seldom occur in elliptical galaxies, for example). Type II supernovae are further subdivided according to detailed spectral and lightcurve properties:

1. *Type II-P*: In the designation Type II-P, the P refers to a plateau in the lightcurve.
2. *Type II-L*: In the designation Type II-L, the L refers to a linear decrease of the lightcurve in the region where a Type II-P lightcurve has a plateau.
3. *Type II-b*: In a Type II-b event the spectrum contains prominent hydrogen lines initially, but the spectrum then transitions into one similar to that of a Type Ia,b supernova. The suspected mechanism is core collapse in a red giant that has lost most but not all of its hydrogen envelope through strong stellar winds, or through interaction with a binary companion. Type II-b supernovae are thus viewed as a link between Type II supernovae

and Type Ib,c supernovae. Type II, Type Ib, and Type Ic all involve core collapse of a massive star, with the distinctions coming in how much of the envelope has been lost before the collapse of the core.

4. *Type II-n*: In this supernova class narrow emission lines and a strong hydrogen spectrum are present. These supernovae are thought to originate in core collapse of a massive star embedded in dense shells of material ejected by the star shortly before the explosion.

Thus, all of the Type-II subcategories, and the Type Ib and Type Ic subcategories, correspond to a similar core collapse mechanism. The observational differences derive mostly from the influence of the outer envelope and surrounding medium on the corresponding spectrum and lightcurve, not in the primary energy-release mechanism.

We now turn from strictly phenomenological classification to a deeper understanding of the fundamental reasons that supernovae explode. As alluded to above, there is substantial evidence that the zoo of observational supernova types can be accounted for with only two basic mechanisms for the central engine driving the explosion: (1) thermonuclear runaways in degenerate matter (*thermonuclear supernovae*), and (2) massive-star core collapse (*core collapse supernovae*). Although these lead to gigantic explosions of similar energy, the source of the energy is fundamentally different. In the first category the energy derives from catastrophic release of nuclear binding energy; in the second category the explosion is powered by catastrophic release of gravitational binding energy.[1]

20.2 Thermonuclear Supernovae

A Type Ia supernova is thought to correspond to a thermonuclear explosion in electron-degenerate, carbon–oxygen white dwarf matter. Thus it differs fundamentally in mechanism from all of the other classes of supernovae in Fig. 20.3. While there is broad agreement that a Type Ia supernova represents the thermonuclear incineration of white dwarf matter in a binary star system, there is considerable uncertainty as to how the explosion is initiated. In the *single-degenerate model* accretion onto an electron-degenerate white dwarf from a nondegenerate star in a binary system triggers the explosion. An alternative mechanism proposes the triggering of a thermonuclear runaway by merger of two white dwarfs in a binary system. This is called the *double-degenerate model,* because it involves two degenerate objects. At present neither model can yet describe all aspects of a Type Ia explosion without making assumptions that are not well-tested by current observations. It could be that both mechanisms (and several possible variations) may be required to account for the observational characteristics of Type Ia supernovae. Substantial current research centers on whether observational constraints on the required progenitor populations (a white dwarf accreting from a nondegenerate companion, or a binary white dwarf system) are consistent with the observed rate of Type Ia explosions.

---

[1] Another proposed mechanism involves some aspects of both core collapse and thermonuclear runaway: massive stars of low metallicity can undergo a *pair-instability supernova,* which is described in Box 20.2.

> **Box 20.2** **Pair-Instability Supernovae**
>
> More massive stars ($M \sim 130$–$250\ M_\odot$) of lower metallicity are predicted theoretically to undergo a *pair-instability supernova*. In very massive stars the radiation pressure is primarily responsible for balancing the enormous gravity, with the gas pressure playing a smaller role. At high temperatures and densities energetic photons can produce electron–positron pairs in abundance, which removes photons and part of the pressure support for the core. If pairs are produced at a high-enough rate the core begins to collapse, which leads to increased pair production and accelerates the collapse. This in turn greatly accelerates thermonuclear burning and leads to a thermonuclear runaway that blows the star apart, without leaving behind a neutron star or black hole. For the most massive progenitors, pair-instability supernovae can be brighter than either Type Ia or core collapse supernovae. It has been proposed that some observed overly luminous supernovae may have been pair-instability supernovae.
>
> For stars with masses less than about 130 $M_\odot$ the pair production rate is not high enough to trigger the above-mentioned runaway. A pair-instability explosion also is unlikely if the metallicity of the star is too high, because this increases the photon opacity and prevents the runaway collapse that initiates the explosion. For stars more massive than about 250 $M_\odot$ photodisintegration of nuclei (see Fig. 20.8) removes pressure support so rapidly that the star collapses to a black hole rather than exploding through thermonuclear reactions. For stars in the mass range ~ 100–$130\ M_\odot$ the pair instability does not lead to a supernova but destabilizes the star sufficiently that it exhibits pulsations leading to large mass ejection. One possible explanation for the eruption of the unstable star η Carinae shown in Fig. 14.3(b) is such an instability.

20.2.1 The Single-Degenerate Mechanism

The Type Ia single-degenerate scenario is illustrated in Fig. 20.5. It is similar to the nova mechanism discussed in Chapter 19, except that in a nova a thermonuclear runaway is initiated in a thin surface layer after a certain amount of accretion and the white dwarf remains largely intact after the layer is ejected in the explosion, but in a Type Ia supernova matter accretes onto the white dwarf over a long period without triggering a runaway. As the matter accumulates nuclear burning is stable and the mass of the white dwarf grows and can eventually approach the Chandrasekhar limit discussed in Section 16.2.2. Central densities become very large and temperature and density fluctuations can trigger a thermonuclear runaway that ignites carbon and then oxygen, and quickly (in a matter of a second or less) burns the entire white dwarf to heavier nuclei, with an enormous release of energy. Thus, unlike a nova or a core collapse supernova that will be discussed further in Section 20.3, a Type Ia supernova explosion does not leave behind a significant remnant. A natural explanation of how the single-degenerate mechanism can lead to the standardizable candle properties of Type Ia supernovae is discussed in Box 20.3.

20.2.2 The Double-Degenerate Mechanism

In the double-degenerate Type Ia mechanism both stars in a binary evolve to the white dwarf stage. This white dwarf binary loses orbital energy from gravitational wave emission

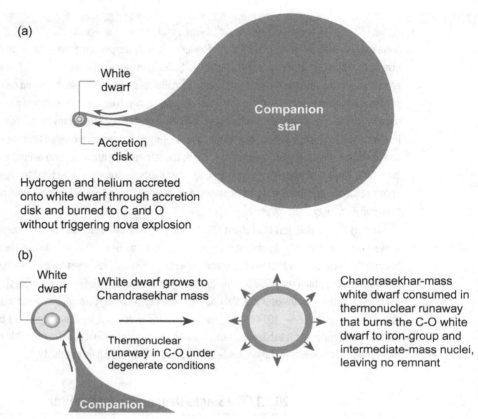

Fig. 20.5 The single-degenerate mechanism for a Type Ia supernova.

Box 20.3   Origin of the Standardizable Candle Property

One feature of the single-degenerate model is that it can give a plausible explanation for the observation that all Type Ia explosions are similar in intrinsic brightness (the standardizable candle property discussed in Box 20.1 that is crucial to modern cosmology). The Chandrasekhar mass is almost the same for all white dwarfs, so if the white dwarf that explodes is always near the Chandrasekhar mass it makes sense that the total energy produced by different Type Ia events is similar. In contrast, for the double-degenerate model there is no obvious reason for the sum of the masses of the two white dwarfs that merge to be similar in different events. However, this may be an oversimplified analysis. The later-time Type Ia lightcurve is largely determined by how much ^{56}Ni is produced in the explosion. Thus the standardizable candle property could result from any mechanism that causes a similar amount of ^{56}Ni to be made in all Type Ia explosions.

and other processes, causing the white dwarfs to gradually spiral together. After a very long inspiral the white dwarfs will merge, either directly, or by one star tidally disrupting the other and rapidly accreting its matter. This initiates a thermonuclear runaway in the merged material and the rest of the Type Ia explosion mechanism is presumably similar to the preceding description for the single-degenerate mechanism.

> **Box 20.4** **Thermonuclear Burn Fronts**
>
> In the Type Ia explosion there is a thermonuclear burn corresponding to conversion of carbon and oxygen fuel into heavier elements by nuclear reactions that release large amounts of energy. This burn involves energy and temperature scales far beyond our everyday experience, but it shares many qualitative properties with ordinary chemical burning.
>
> **Disparate Distance Scales**
>
> A *burn front* proceeds through the white dwarf, with "cooler" (a highly relative term!) unburned fuel in front and hot burned products (ash) behind. This burn front can be remarkably narrow – as little as millimeters thick. Thus there are two extremely different distance scales characterizing the explosion: the size of a white dwarf, which is of order 10^4 km, and the width of the burn front that consumes the white dwarf matter, which can be billions of times smaller. This presents severe difficulties in accurately modeling Type Ia explosions, since standard numerical solutions of the equations governing the explosion cannot handle such disparate scales easily.
>
> **Deflagration and Detonation Waves**
>
> In thermonuclear and ordinary chemical burning there is an important distinction associated with the speed of the burn front. If the burn front advances through the fuel at a speed less than the local speed of sound in the medium (subsonic), it is termed a *deflagration wave*. In a deflagration, fuel in front of the advancing burn is heated to the ignition temperature by conduction of heat across the burn front (recall that matter described by a degenerate equation of state is a very good thermal conductor, much like a metal). On the other hand, if the burn front advances at greater than the speed of sound in the medium (supersonic) it is called a *detonation wave*. In a detonation a shockwave forms and the fuel in advance of the burn front is brought to ignition temperature by shock heating. Generally detonation is much more violent than deflagration.
>
> **Observational Signatures**
>
> Deflagrations and detonations produce different isotopic abundance signatures in the ash that is left behind. The detailed observational characteristics of Type Ia supernovae (in particular, the elemental abundances detected in the expanding debris) could be accounted for most naturally if it is assumed that part of the burn is a deflagration and part of it is a detonation. This is a difficulty for the theory because general considerations suggest that the explosion starts as a deflagration and it is not easy to get the burn in computer simulations to transition to a detonation without making significant untested assumptions. Thus, the proposed Type Ia mechanism is plausible in outline, but there are details that leave some doubt about how much is understood about the mechanism of these gigantic explosions.

20.2.3 Thermonuclear Burning in Extreme Conditions

Because of the gigantic energy release in a small region over a very short period of time, the conditions in a Type Ia explosion are extreme, to say the least. Simulations indicate that temperatures in the hottest parts can approach 10^{10} K, with densities as large as 10^9 g cm^{-3}, and the temperature may change at rates of order 10^{17} K s^{-1} in the burn front. The physics of the Type Ia explosion presents a number of issues that are difficult to deal with in the large numerical simulations that are required to model such events, as discussed in Box 20.4.

20.2.4 Element and Energy Production

The energy released in a Type Ia supernova derives primarily from the burning of carbon and oxygen to heavier nuclei. If the explosion lasts long enough to achieve nuclear statistical equilibrium (NSE), the characteristic final products of this burning will be

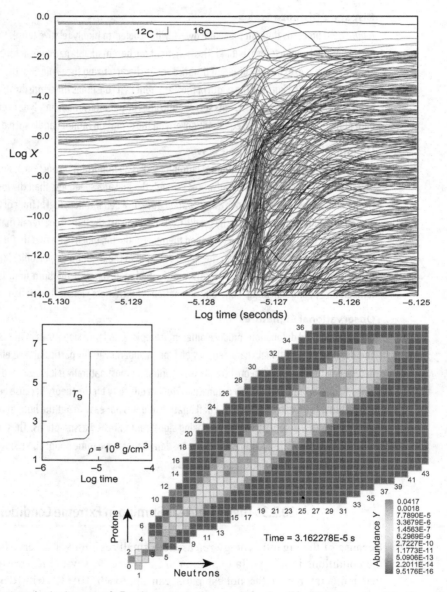

Fig. 20.6 Element production in a zone of a Type Ia supernova. Upper: mass fractions X for 468 isotopes near time of maximum burning. Lower: abundances Y near the end of maximum burning. Inset: variation of temperature with time (density is almost constant over this period).

iron-group nuclei. An example of network evolution under conditions typical of the Type Ia explosion in the deep interior of the white dwarf is illustrated in Fig. 20.6. In this calculation the initial temperature was $T_9 = 2$, the initial density was $\rho = 1 \times 10^8 \, \text{g cm}^{-3}$, and the initial composition was equal mass fractions of ^{12}C and ^{16}O.

The explosion is initiated by carbon burning, which raises the temperature quickly (see figure inset) and leads to burning of oxygen and all the reaction products produced by carbon and oxygen burning. The rapid temperature increase is associated with coupling of the large energy release from the thermonuclear burning to the fluid of the white dwarf, which is described by hydrodynamics. This sudden rise in temperature increases the rate of nuclear reactions in the network rapidly, leading to a thermonuclear runaway in the electron-degenerate matter. Within less than 10^{-5} s from initiation of the explosion the isotopic species in the network have increased from two to almost five hundred, with significant population of the iron group of nuclei already evident. As the thermonuclear flame burns through the white dwarf the carbon and oxygen fuel in each region is burned in a fraction of a second, and the entire white dwarf is consumed on a timescale of about a second.

20.2.5 Late-Time Observables

A Type Ia supernova is expected to leave no remnant behind (except the companion donor star in the single-degenerate scenario). The primary late-time observables are the supernova lightcurve and the motion and spectrum of the expanding supernova remnant. An example of a Type Ia lightcurve was shown in Fig. 20.2 and a Type Ia supernova remnant is shown in Fig. 20.7(a). Spectroscopy of a Type Ia supernova remnant can determine the elements in the expanding debris and their radial velocity. The radial velocity is in turn correlated with how deep in the explosion the element was produced (higher velocity is expected to come from deeper). Such measurements indicate that many

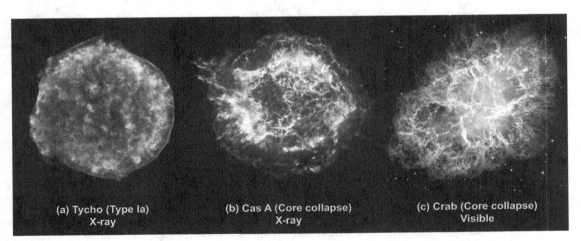

Fig. 20.7 Supernova remnants. (a) Tycho's supernova of 1572 in X-rays; it was a Type Ia explosion. (b) The Cas A supernova remnant in X-rays. It was a core collapse supernova that occurred about 300 years ago. (c) The Crab Nebula, which is the remains of the core collapse supernova of 1054, in visible light.

intermediate-mass isotopes such as those of Si are produced in addition to iron-group nuclei [207], suggesting that the explosion is not a pure detonation (which would produce mostly iron-group nuclei).

20.3 Core Collapse Supernovae

A core collapse supernova is one of the most spectacular events in nature and is likely the source of at least some of the heavy elements that are produced in the rapid neutron capture or r-process (see Section 20.5). Considerable progress has been made over the past several

---

Box 20.5   **Supernova 1987A**

The Tarantula Nebula is a star-forming region in the Large Magellanic Cloud, a satellite galaxy of the Milky Way visible in the Southern Hemisphere. Some 163,000 years ago the core of a mag 12 blue supergiant star in the Tarantula — Sanduleak —69 202 — imploded, producing a burst of neutrinos and a shockwave that reached the surface several hours later, sending most of the star's mass hurtling into space and generating a billion-fold increase in luminosity. Time passed ... On February 23, 1987 on Earth, 163, 000 ly away, detectors searching for something else entirely saw an unexpected burst of $\sim$20 neutrinos. A clue to their origin was not long in coming. Three hours later, light from the explosion arrived and that night observers in Chile and New Zealand were startled to find a "new star," visible to the naked eye, in the Tarantula. The progenitor (right) and supernova (left) are shown below.

Thus did SN 1987A announce the demise of Sk —69 202 (which could no longer be found after the supernova dimmed). The first nearby supernova since the invention of the telescope, SN 1987A has been studied extensively, confirming most and modifying some of our understanding of core collapse supernovae (see Section 20.4).

decades in understanding the mechanisms responsible for such events. This understanding was tested both qualitatively and quantitatively by the observation of Supernova 1987A (see Box 20.5 and Section 20.4) in the nearby Large Magellanic Cloud – the brightest supernova observed from Earth since the time of Kepler.

20.3.1 The "Supernova Problem"

The observations of Supernova 1987A and its aftermath provide compelling evidence that a core collapse supernova represents the death of a massive star in which an electron-degenerate iron core of approximately 1.2 solar masses collapses on timescales of tens of milliseconds. This gravitational collapse is reversed as the inner core exceeds nuclear densities because of the properties associated with the stiff nuclear equation of state and a pressure wave reflects from the center of the star and propagates outward, steepening into a shockwave as it passes into increasingly less dense material. However, the most realistic simulations of this event indicate that the shockwave loses energy rapidly as it propagates through the outer core and stalls into an *accretion shock* (a standing shockwave at a constant radius) within several hundred milliseconds of the bounce at a distance of several hundred kilometers from the center. Thus the "prompt shock" does not blow off the outer layers of the star and fails to produce a supernova. This is the "supernova problem": there is good evidence that the basics are understood, but the details fail to work robustly in the simple form described above. In the remainder of this chapter we introduce important modifications of the simple prompt shock mechanism and summarize the progress that has been made on this problem.

20.3.2 The Death of Massive Stars

As a consequence of the advanced burning stages that are described in Chapter 6, a massive star near the end of its life builds up a layered structure as was depicted in Fig. 14.2. The iron core that forms in the central region of the star grows as the silicon layer surrounding it burns to iron. Because the iron core cannot undergo exothermic fusion reactions it must be supported against gravity by electron degeneracy pressure and a modified form of the discussion in Section 16.2.2 is applicable. Electron degeneracy can support the iron core against gravitational collapse only if the iron-core mass remains below the Chandrasekhar limit, which is approximately 1.2–1.3 solar masses for typical iron cores, depending on the electron fraction of the core (see Example 16.1).

When the iron core exceeds this critical mass it begins to collapse on a *dynamical timescale* (which is only tens of milliseconds for the dense iron core – see Problem 20.2). At the point where the collapse begins the iron core of a representative 25 $M_\odot$ star has a mass equal to the Chandrasekhar mass of about 1.2 solar masses, a diameter of several thousand kilometers,[2] a core density of $\sim 6 \times 10^9 \,\text{g cm}^{-3}$, a core temperature

---

[2] This is a miniscule fraction of the total diameter. The massive star is typically a supergiant by this time, with its tenuous outer layers spread over a volume that would encompass much of the inner Solar System if it were placed at the position of the Sun.

> **Box 20.6** **Entropy of the Iron Core**
>
> The entropy of the iron core in a pre-supernova star is approximately one per baryon per Boltzmann constant. This is remarkably low. For example, the entropy of the original main sequence star that produced this iron core was probably about 15 in these units.
>
> **The Core of the Massive Star Tends to Order**
> It might seem contradictory for the core entropy to decrease as the star burns its fuel. However, the star is *not a closed system*: as nuclear fuel is consumed, energy leaves the star in the form of photons and neutrinos.[a] In the process, the nucleons in the original main sequence star are converted to iron nuclei, which represent a relatively low-entropy form of matter. For example, in ^{56}Fe the 26 protons and 30 neutrons are extremely ordered compared with 56 free nucleons in the original star because they are constrained to move together as constituents of a single iron nucleus.
>
> **The Universe as a Whole Tends to Disorder**
> Thus, the core of the star becomes *more ordered* as the nuclear fuel is consumed. However, the entire Universe tends to *greater disorder*, as required by the second law of thermodynamics, because the star radiates energy in the form of photons and neutrinos as it builds its ordered core. This is not unlike biological evolution, where chemical processes assemble locally highly ordered objects (living things), while at the same time the entire Universe tends to greater disorder.
>
> [a] Central regions are cooled by neutrino emission for massive stars in late burning stages; see Section 7.11.

of $\sim 6 \times 10^9$ K, and an entropy per baryon per Boltzmann constant $S \sim 1$. As discussed in Box 20.6, this is a very small entropy, which has consequences for the subsequent collapse of the core.

20.3.3 Sequence of Events in Core Collapse

The collapse of the iron core as the Chandrasekhar limit is exceeded triggers a sequence of events that will occur in an elapsed time of less than a second. Let us outline this scenario.

Initiating the collapse: When the mass of the iron core exceeds the Chandrasekhar limit it begins to collapse under the influence of gravity. This collapse is accelerated by two factors that are accentuated as the temperature and density of the collapsing core rise, *iron photodisintegration* and *neutronization*.

As the core heats up, high-energy γ-rays are produced in copious amounts. These photodisintegrate iron-peak nuclei, as illustrated in Fig. 20.8. Although hundreds of isotopes are produced from the photodisintegration of the iron, the most abundant species at equilibrium under these conditions is alpha particles, and only alpha particles, neutrons, and protons have abundances larger than 10^{-3}. The photodisintegration of iron is highly endothermic, as indicated in Fig. 20.8(b); for example, ^{56}Fe $\rightarrow 13\alpha + 4n$ has a Q-value of -124.4 MeV (see Problem 20.4). This decreases the kinetic energy of the electrons in the core, which lowers the pressure and hastens the collapse.

As the density and temperature increase in the core, the rate for the electron capture reaction $p^+ + e^- \rightarrow n + \nu_e$, is greatly enhanced. This *neutronization reaction* (recall the

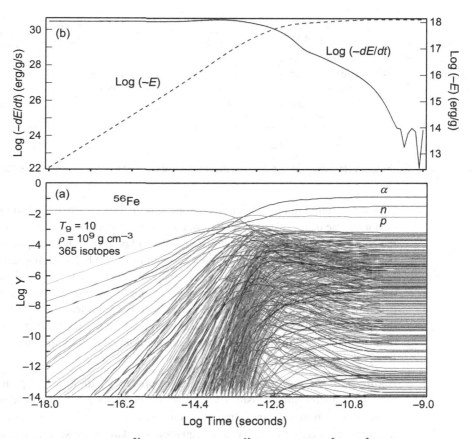

Fig. 20.8 Simulated photodisintegration of ^{56}Fe at a temperature of 10^{10} K and density of 10^9 g cm^{-3}. (a) Initially only pure ^{56}Fe is present; after $\sim 10^{-12}$ s the original ^{56}Fe has been transformed into 365 isotopes with non-zero populations, but the only species with abundances in excess of 10^{-3} are alpha particles, neutrons, and protons. (b) The rate of energy absorption for this very endothermic reaction.

discussion in Box 16.2) decreases the electron fraction Y_e of the core and thus decreases the electronic contribution to the pressure. The neutrinos produced in this reaction easily escape the core during the initial phases of the collapse because their mean free path is much larger than the initial radius of the core. These neutrinos carry energy with them, decreasing the core pressure and accelerating the collapse even further. They also deplete the lepton fraction (defined analogous to the electron fraction, but for all leptons, which in this context means mostly electrons, positrons, neutrinos, and antineutrinos).

Initial infall and neutrino trapping: The accelerated core collapse proceeds on a timescale of tens of milliseconds, with velocities that are significant fractions of the free fall velocities. The core separates into an *inner core* that collapses subsonically and homologously,[3] and an *outer core* that collapses largely in free fall, with a velocity exceeding the local

---

[3] Recall from Section 9.5.1 that a homologous collapse is "self-similar": it can be described by changing a scale factor.

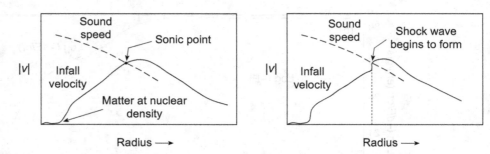

Fig. 20.9 The sonic point during collapse (left) and the beginning of shock wave formation following the bounce (right).

velocity of sound in the medium (it is supersonic). This collapse is rapid on the timescales characteristic of most stellar evolution, but it is slow compared with the reaction rates and the core is approximately in equilibrium during all phases of the collapse. This implies that the *entropy is constant*, and the highly ordered iron core before collapse ($S \simeq 1$) remains ordered during the collapse. As the collapse proceeds and temperature and density rise, neutrino interactions eventually become so strong because of coherent neutrino scattering (Section 7.11.3) that the neutrino mean free path becomes less than the radius of the core, the time for neutrinos to diffuse outward becomes longer than the characteristic timescale of the collapse, and the neutrinos are effectively trapped in the imploding core (with a mean free path that may be a fraction of a meter).

Bounce and shock formation: The collapse proceeds with low entropy and the nucleons remain in nuclei until densities where nuclei begin to touch. The collapsing core now begins to resemble a gigantic "macroscopic nucleus" containing a nearly degenerate Fermi gas of nucleons with a very stiff equation of state because nuclear matter is highly incompressible. At this point, the pressure of the nucleons begins to dominate that from the electrons and neutrinos. Somewhat beyond nuclear density the incompressible core of nearly degenerate nuclear matter rebounds violently as a pressure wave reflects from the center of the star and proceeds outward.

The bounce wave steepens into a shockwave as it travels outward through material of decreasing density (and thus decreasing sound speed), with the shockwave forming near the boundary between the subsonic inner core and supersonic outer core (at the *sonic point* illustrated in Fig. 20.9). In the simplest picture this shockwave would eject the outer layers of the star, resulting in a supernova explosion. This is called the *prompt shock mechanism*. The gravitational energy released in the collapse is about 10^{53} erg, and the typical observed energy of a supernova (the expanding remnant plus photons) is about 10^{51} erg.[4] Thus, only

---

[4] This defines a unit of energy commonly used in supernova discussions that originally was called a *foe:* 1 foe ≡ 10^{51} ergs, with the name deriving from the first letters of fifty-one ergs. This unit is now termed a *bethe*, in honor of the remarkable physicist Hans Bethe (1906–2005), who won the 1967 Nobel prize in Physics for illuminating how nuclear reactions powered the stars and made seminal contributions to many other fields of science, including supernova physics. Bethe (whose name was pronounced the same as "beta") continued to do science until close to his death at age 98, inspiring a joke that went "The nuclear physicists were wrong; there is no Bethe decay!"

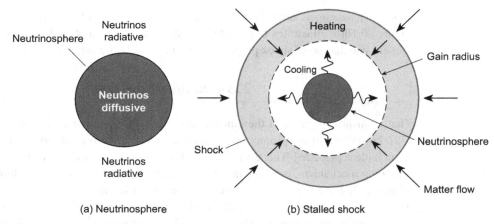

Fig. 20.10 (a) The neutrinosphere. (b) Conditions at time of shock stagnation in a core collapse supernova. The neutrinosphere and the gain radius (defined in the text) are indicated.

about 1% of the gravitational energy need be released in the form of light and kinetic energy to account for the observed properties of supernovae.

Death of the prompt shock: The simple prompt shock mechanism as described above runs afoul of the details (where, recall, the devil is known to reside): realistic calculations suggest that the prompt shock dissipates energy rapidly as it progresses through the outer core for two primary reasons:

1. *Dissociation*: The shockwave dissociates Fe nuclei as it passes through the outer core and this highly endothermic reaction saps it of a large amount of energy.
2. *Neutrino emission*: As the shockwave passes into increasingly less dense material the mean free path for the trapped neutrinos increases until the neutrinos can once again be freely radiated from the core. This further deprives the shock of its vitality by lowering the pressure behind it.

The radius at which the neutrinos change from diffusive to radiative (free-streaming) behavior is termed the *neutrinosphere* [by analogy with the photosphere of a regular star; see Fig. 20.10(a)]. Technically, the neutrinosphere is an imaginary sphere beyond which a neutrino on average will suffer less than one interaction before escaping from the star, which is similar to the definition of the photosphere for photons. When the shockwave penetrates the neutrinosphere a burst of neutrinos is emitted from the core, carrying with it large amounts of energy (most of the energy released in the gravitational collapse resides in neutrinos).

As a result of energy losses caused by dissociation and neutrino emission, the most realistic calculations indicate that the prompt shock stalls into an accretion shock before it can exit the core, unless the original iron core contains less than about 1.1 solar masses. In a typical calculation, the accretion shock forms at about 200–300 km from the center of the star within about 10 ms of core bounce. Since there is considerable agreement that SN 1987A resulted from the collapse of a core having more than 1.1 solar masses, the prompt shock mechanism is unlikely to be a generic explanation of core collapse supernovae. Figure 20.10(b) illustrates the conditions characteristic of a stalled accretion shock in a modern calculation of shock propagation in a core collapse supernova.

20.3.4 Neutrino Reheating

That neutrinos might eject the outer layers of a star in a supernova is an old suggestion [72], but failure of the prompt mechanism to yield supernova explosions revived interest in such ideas [45, 226]. This evolved into what is now termed the *delayed shock* or *neutrino reheating* mechanism, in which the stalled accretion shock is re-energized through heating of matter behind the shock by neutrinos produced in the region interior to the shock. This raises the pressure sufficiently to impart an outward velocity to the stalled shock on a timescale of approximately one second and the reborn shock then disrupts the outer envelope of the star, producing the supernova explosion. The schematic mechanism for the supernova event thus becomes the two-stage process depicted in Fig. 20.11.

All neutrino and antineutrino flavors are produced copiously in the hot, dense region near the center of the collapsed core. Predicted luminosities are shown in Fig. 20.12. In discussing neutrinos interacting with matter behind the shock it is useful to introduce two characteristic radii. The first already has been encountered in Fig. 20.10(a): the *neutrinosphere* marks a boundary between diffusion and free streaming for neutrino propagation. The second is associated with the observation that the neutrino interactions

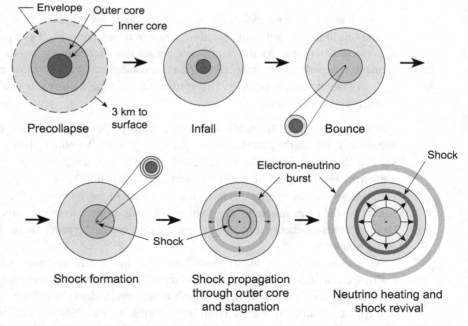

Fig. 20.11 Neutrino reheating mechanism for a supernova explosion (after Bruenn [60]). Figures are approximately to scale; the surface of the star would be 3 km from the center on this scale.

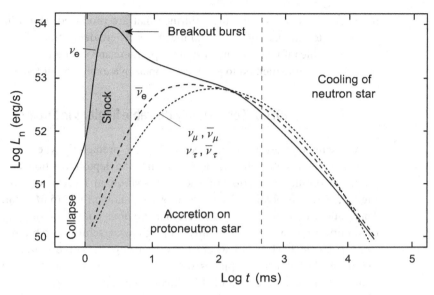

Fig. 20.12   Neutrino luminosities in a core collapse supernova.

with the shocked matter could either cool or heat the matter. General considerations suggest that there is always a radius outside of which the net effect of the neutrino interactions is to heat the matter. This break-even radius, beyond which the neutrino interactions become effective in increasing the pressure behind the shock is termed the *gain radius*. The neutrinosphere and gain radius are indicated schematically in Fig. 20.10(b). Shock revival is favored by *deposition of neutrino energy between the gain radius and the shock.*

The result of a large number of calculations is that the neutrino reheating helps, but generally does not produce successful explosions without artificial boosts of the neutrino luminosities that are not easy to justify. This suggests that there are additional ingredients in the supernova mechanism that must be included to obtain a quantitative description. One possibility is that convection in the region interior to the shock may affect it mechanically and may alter the neutrino spectrum and luminosity in a non-negligible fashion. Let us now turn to a discussion of how convection might affect supernova explosions.

20.3.5 Convection and Neutrino Reheating

In Section 7.6 the general conditions under which stars can become unstable toward convective overturn were discussed. It may be conjectured that substantial convection inside the stalled shock could influence the possibility of neutrinos re-energizing the shock, and affect quantitative characteristics of a re-energized shock. In particular, to boost the stalled shock it is necessary for the neutrinos to deposit energy behind the shock front, outside the gain radius but inside the shock. Convective motion inside the shock front could, by overturning hot and cooler matter, cause more neutrino production and alter the neutrino spectrum. The convection could also move neutrino-producing matter beyond the neutrinosphere, so that the neutrinos that are produced would have a better chance to propagate into the region behind the shock where deposition of energy would have the most favorable influence in increasing the pressure and re-energizing the shock. This would provide a possible method to produce a supernova explosion of the required energy.

20.3.6 Convectively Unstable Regions in Supernovae

Armed with our results from Section 7.6 for predicting when a region may become convectively unstable, let us now examine the lepton fraction and entropy gradients produced during the period of shock stagnation in typical supernova calculations. In Fig. 20.13 results obtained by Bruenn [60] are shown for a 15 $M_\odot$ star, 6 ms after bounce. There are two obvious regions where our previous considerations suggest the possibility of significant convective instability: (1) a Schwarzschild unstable region near the shock front at about 50–100 km where there is a large negative entropy gradient, and (2) a region inside the neutrinospheres where both the entropy and the lepton fraction exhibit strong negative gradients and thus favor convection driven by the Ledoux instability. These arguments identify regions that favor convective motion but whether such regions develop convection, the timescale for that convection, and the quantitative implications for supernova explosions are dynamical questions that can be settled only by detailed calculations incorporating realistic physics.

Many simulations now have demonstrated that convection plays a significant role in the core collapse supernova problem. An example is shown in Fig. 20.14, which exhibits spectacular convection below the shock. Such violent and large-scale convection breaks spherical symmetry and cannot be treated with approximations like mixing-length theory (see Section 7.9); it can be modeled adequately only by using numerical multidimensional hydrodynamics simulations. Current state-of-the-art simulations include

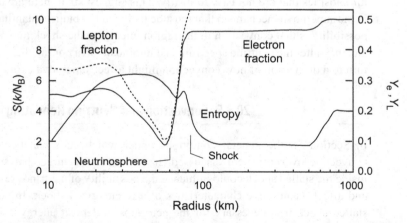

Fig. 20.13 Lepton fractions and entropy 6.3 ms after bounce in a supernova calculation. Regions that may be expected to be particularly favorable for convective motion are described in the text. The progenitor had a mass of 15 $M_\odot$, and the calculation is described in Ref. [60].

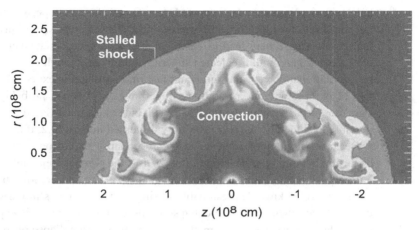

Fig. 20.14 2D core collapse simulation exhibiting violent convection behind a stalled shock [150]. The r coordinate is the distance from the z axis. Entropy in grayscale, with white maximum and dark gray minimum. The shock is beginning to be distorted by the convection beneath it. In modern calculations a *standing accretion shock instability (SASI)* develops associated with deformations of the shock that can be significant in producing successful explosions.

2- or 3-dimensional hydrodynamics and better approximations for neutrino transport than earlier work, and are beginning to produce explosions that look somewhat realistic. However, a completely successful model of core collapse supernovae will likely require both 3-dimensional hydrodynamics and a complete treatment of neutrino transport. It has not been possible to include both in current codes because of inadequate computing power, but it is thought that the next generation of massively parallel supercomputers[5] may provide sufficient computational power to determine whether these ingredients lead to a successful model of core collapse supernovae, or whether they point to new physics not yet incorporated into the models.

20.3.7 Remnants of Core Collapse

Like a Type Ia supernova, a core collapse supernova is expected to eject an expanding supernova remnant, as illustrated in Figs. 20.7(b) and 20.7(c). Unlike a Type Ia explosion, a core collapse supernova is expected to leave behind also a compact remnant – either a neutron star or a black hole. Present understanding suggests that less-massive progenitors lead to neutron stars but for more-massive stars the end result is a black hole, produced either immediately, or with a time delay corresponding to accretion on a remnant protoneutron star causing it to collapse to a black hole. For increasingly-massive black

---

[5] The next generation is expected to be *exascale systems*, with speeds $\sim 10^{18}$ floating point operations per second (1 *exaflop*). The fastest computers available as this is written in 2017 are capable of 10–100 petaflops ($10-100 \times 10^{15}$ floating point operations per second). Optimistically the exascale for scientific computing could be reached by the early 2020s.

hole progenitors it may be expected that less of the envelope is ejected and more of it falls back into the black hole. For masses above about 30 $M_\odot$, current simulations indicate that core collapse may lead to *complete fallback* of the outer layers of the star, leaving only a black hole with no ejected supernova remnant, though even in that case there will be gravitational waves and significant neutrino emission is expected. This *direct collapse* to a black hole without a traditional supernova explosion is probably the general fate of stars more massive than about 30 $M_\odot$ (with black hole masses in the $\sim$ 100–250 $M_\odot$ range possibly excluded by the pair instability discussed in Box 20.2, if metallicities are low).

Because simulations indicate that core collapse explosions are asymmetric (see the distortion beginning to develop in Fig. 20.14), the compact remnant is expected to receive a *natal kick* in the explosion. Neutron stars have been observed with space velocities as large as $\sim$1000 km s^{-1}, presumably arising from natal kicks in the supernova explosion that produced them. For core collapse in more massive stars it is expected that the natal kick is less severe, since less matter is ejected. For the collapse of massive cores directly to black holes it is often assumed to be zero. The natal kick and amount of ejected matter affect strongly whether a binary remains bound if one of the stars undergoes core collapse (see the discussion of massive-binary evolution in Section 22.4.2).

20.4 Supernova 1987A

Supernova 1987A (see Box 20.5) has been the most-studied supernova because of its proximity to Earth. This section summarizes how the core collapse mechanism outlined in preceding sections has fared in the light of SN 1987A data. In this summary, it is important to remember the distinction made in connection with Fig. 20.3 between classification of supernovae with respect to spectral and lightcurve characteristics, and classification with respect to explosion mechanism.

20.4.1 The Neutrino Burst

Arguably the most important result from SN 1987A was detection of the neutrino burst, which was consistent qualitatively and quantitatively (within relatively large errors because of low statistics) with the core collapse mechanism. Neutrinos detected in the Kamiokande II and IMB water Cherenkov detectors are shown in Fig. 20.15. Only 20 neutrinos in total were seen but the general background expected in this plot is very low. This low background, systematic analysis to rule out the burst being created by a cosmic ray shower, and the coincidence of the burst with light from SN1987A (offset by about three hours, as expected) leaves little doubt that these neutrinos originated in the supernova. The observation of the neutrinos in Fig. 20.15 makes it virtually certain that a neutron star or black hole was produced by SN 1987A with the release of $\sim 10^{53}$ erg of gravitational energy, thus confirming the basic core collapse mechanism. Most other observational characteristics of SN 1987A that will now be discussed are related to the properties and

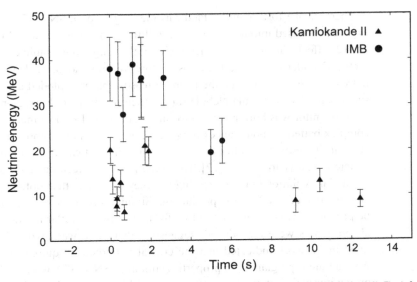

Fig. 20.15 Neutrino burst from SN 1987A detected in two water Cerenkov detectors (data from Refs. [27, 47, 118]). The inferred direction (with large errors) was consistent with origin of the burst in the Large Magellanic Cloud. This means that the neutrinos passed through the Earth en route to the detectors, which were located in Earth's Northern Hemisphere.

evolution of the envelope of the progenitor star and are only indirectly connected to the explosion mechanism.

20.4.2 The Progenitor was Blue!

The progenitor of Supernova 1987A came as a surprise for many because it was widely (though not uniformly) believed at the time that supernova explosions resulted from core collapse in red supergiant stars,[6] not blue supergiants like Sk −69 202, and because the early lightcurve of SN 1987A (shown in Fig. 20.2) deviated substantially from that expected for a Type II supernova. For example, the initial rise was quite slow and the luminosity did not peak until 80 days after the explosion, and SN 1987A was ∼ 100 times less luminous than a typical Type II supernova.

---

[6] A standard argument was that, assuming the usual model that supernova remnants and lightcurves are the result of a shockwave that disrupts the outer layers of the star after core collapse, the properties of observed bright supernovae in distant galaxies required that the shockwave pass through mostly low-density matter on its way out of the star, and red supergiants are of much lower density than blue supergiants (see Table 2.4) [27]. As will be discussed further below, this was partially observational bias because core collapse in blue supergiants is expected to produce supernovae that are much less luminous than those originating in red supergiants. It should be noted that some theoretical work before SN 1987A considered the possibility that supernovae of lower luminosity might originate in blue supergiants (see [27] for references) but the lack of observational data for such events, and that most stellar evolution calculations of the time favored explosion in the red supergiant phase, caused this possibility to be discounted in favor of the red supergiant progenitor paradigm.

Theoretical efforts to understand why the progenitor of SN 1987A was blue when the star exploded focused initially on two possibilities: (1) extensive mass loss in prior evolution, and (2) effects due to the low metallicity of the Large Magellanic Cloud (LMC). While Sk −69 202 underwent some mass loss before the supernova, evidence suggests that it was not extensive enough to be the primary reason that it exploded as a blue rather than red supergiant. The Tarantula Nebula (also called the 30 Doradus region) of the LMC where the progenitor was born has a composition generally lower in metals than the Sun, with a complex pattern of abundances for specific elements. For example, the oxygen abundance is about three times smaller than that of the Sun and carbon and nitrogen are even more deficient relative to the Sun [27]. These composition effects can have complex implications for stellar evolution but two are obvious: they can affect the photon opacities and they can alter the rate of CNO energy production. Simulations indicated that blue progenitors could be produced without large mass loss for metallicities similar to those of the LMC but not all properties were reproduced, suggesting that more was involved than the metallicity. Subsequent work indicated that the crucial ingredients required to produce a supernova from a blue supergiant with properties similar to SN 1987A were (a) low metallicity, (b) a progenitor mass not much greater than $\sim 20\,M_\odot$, (c) mass loss of no more than a few solar masses in prior evolution, and (d) a tuned prescription for convection [27].

Because Sk −69 202 exploded as a blue supergiant the radius of the envelope was much smaller than for a supernova in a red supergiant, as illustrated in Fig. 20.16. While a 20 $M_\odot$ red supergiant has a radius comparable to the size of the Earth's orbit, the radius of Sk −69 202 when its core collapsed was only about 20% of the radius of Earth's orbit. Thus there was much less envelope for the shockwave to plow through, and the delay between emission of the initial neutrino burst and the sudden increase in light output when the shock reached the surface was only about three hours for SN 1987A. For a core collapse supernova in a red supergiant of similar mass the time for the shock to reach the surface is likely several times larger than that, with a correspondingly longer delay between emission of the initial neutrinos and the sudden increase in photon luminosity.

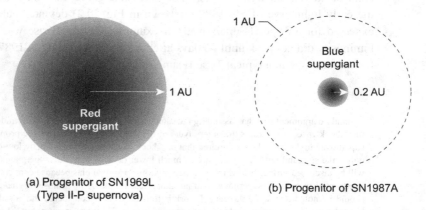

Fig. 20.16 (a) Size of a typical red supergiant supernova progenitor at explosion; (b) size of the blue supergiant progenitor of Supernova 1987A at explosion.

The much more compact nature of the blue supergiant Sk −69 202 relative to a red supergiant of comparable mass illustrated in Fig. 20.16 also provides a basic explanation for the abnormally low luminosity of SN 1987A relative to other Type-II supernovae (see Fig. 20.2). The primary energy budget of a core collapse supernova may be divided into (1) production of neutrinos, (2) ejection of the envelope by the shockwave, and (3) powering the lightcurve. The neutrino emission dominates the energy budget but it is a property of the core and not the envelope, so it is similar in the two cases. For a compact blue supergiant the envelope lies in a deeper gravitational potential than for a red supergiant and more energy must be expended to eject it, leaving less energy for the subsequent light emission. Thus supernovae that explode in the blue supergiant phase are much less luminous than those exploding in the red supergiant phase (partially explaining why at the time of SN 1987A understanding of supernovae was dominated by data from the explosion of red supergiants in other galaxies – they were easier to see).

20.4.3 Radioactive Decay and the Lightcurve

Initially the lightcurve of a Type II supernova is powered by the shockwave but at later times it derives its energy from radioactive decay of isotopes produced in the explosion. Figure 20.17 illustrates for SN 1987A. The lightcurve is for optical photons but the energy causing the optical emission at later times is supplied primarily by radioactive decay. From the shape and height of the lightcurve the isotopes produced and their abundances may be inferred. The initial part of the lightcurve for SN 1987A is accounted for if the explosion

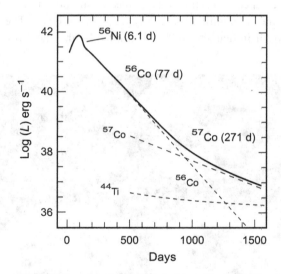

Fig. 20.17 The lightcurve of Supernova 1987A 1500 days after the explosion (solid curve), adapted from Ref. [209]. The dominant radioactive decays powering the lightcurve at different times are indicated above the lightcurve. The rate of decay for the isotopes powering the lightcurve are indicated by the dashed curves. Adapted from "The Energy Sources Powering the Late-Time Bolometric Evolution of SN 1987A," published in *Astrophysical Journal, Part 2, Letters* (ISSN0004-637X), **384**, L33–L36, 1992. © 1992. The American Astronomical Society. All rights reserved.

produced 0.075 $M_\odot$ of ^{56}Ni, which decays by ^{56}Ni $\to$ ^{56}Co $+ e^+ + \nu_e + \gamma$ with a 6.1 days half-life. Initially the optical depth for gamma-rays was high and the energy released in the ^{56}Ni decay produced the early bump observed in the lightcurve. Soon after peak luminosity the lightcurve was increasingly dominated by decay of the ^{56}Co daughter of ^{56}Ni through ^{56}Co $\to$ ^{56}Fe $+ e^+ + \nu_e + \gamma$, which has a half-life of 77 days. Because of the slowly decreasing optical depth for gamma-rays and X-rays, the rate of light production soon becomes dominated by the rate of energy production and the slope of the lightcurve is then determined by the half-life of the radioactive decay that is powering it.

The explosion also produced a much smaller amount of radioactive ^{57}Co, which decays with a 271 days half-life. After about 1000 days enough ^{56}Co had decayed away that ^{57}Co decay became dominant and the lightcurve assumed the shallower slope determined by the half-life of ^{57}Co. In 2017, 30 years after the explosion, the ^{56}Ni, ^{56}Co, and ^{57}Co had all decayed away and the lightcurve was being powered by the decay of the ^{44}Ti produced in the explosion, which has a 47-year half-life.

20.4.4 Evolution of the Supernova Remnant

Evolution of the expanding remnant of SN 1987A has been studied extensively at multiple wavelengths, as illustrated in Figs. 20.18 and 20.19. In Fig. 20.18 a time lapse of Hubble Space Telescope images from 1994 to 2016 depicting the collision of the SN 1987A blast wave with a ring of matter emitted by the progenitor before the supernova is shown. Figure 20.19 shows a multiwavelength composite from 2017. In the center dust forming in the supernova remnant is imaged at submillimeter wavelengths by the Atacama Large Millimeter/submillimeter Array (ALMA) in Chile. The locus of a ring of matter about a lightyear in diameter that was emitted by the star before the explosion is indicated. This

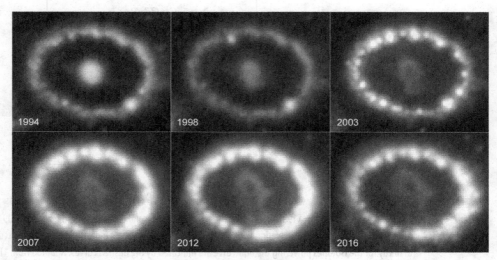

Fig. 20.18 Images from 1994 to 2016 showing the collision of the SN 1987A shockwave with a ring of matter emitted by the progenitor before the supernova explosion. See also Fig. 20.19.

20.4 Supernova 1987A

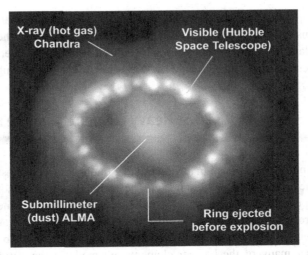

Fig. 20.19 Multiwavelength composite 30 years after SN 1987A. In the center dust forming in the supernova remnant is imaged at submillimeter wavelengths by ALMA. The locus of a ring of gas emitted by the star before the explosion is indicated. The brightest clumps in the ring indicate visible light captured by the Hubble Space Telescope that was emitted from the collision of the shockwave from the supernova with the ring. The more diffuse glow concentrated outside the ring represents X-rays imaged by the Chandra X-ray Observatory.

ring is thought to have been produced by a wind emitted late in the life of the pre-supernova star (at least 20,000 years before the explosion) that collided with a slower wind emitted in an even earlier red giant phase. (Two other larger and fainter rings that were also formed before the supernova are not shown.) It was illuminated initially by a flash of UV light produced by the supernova, which ionized the ring and has caused it to glow for decades since then because of electron recombination.

Beginning in the early 2000s the ring began to brighten further as the shockwave from the explosion reached it. The brightest regions indicate visible light emitted from this collision and captured by the Hubble Space Telescope. The more diffuse glow corresponds to X-rays emitted from hot gas produced in the collision and imaged by the Chandra X-ray Observatory. Concentration of X-rays outside the ring suggests that the shockwave has now passed through the ring and into the less dense matter beyond.

20.4.5 Where is the Neutron Star?

A mystery concerning SN 1987 is the compact remnant. The observed burst of neutrinos is a sure sign that a neutron star or black hole was formed, since gravitational collapse to a compact remnant is the only plausible way to release the energy to make the neutrinos. From stellar systematics it is estimated that Sk −69 202 had a mass of about 18 $M_\odot$ when its core collapsed, and core collapse simulations indicate that for a progenitor of that mass the compact remnant should be a neutron star. However, no clear evidence for a neutron star has been found, despite extensive searches. Various explanations have been proposed, none supported conclusively by data. The most plausible are that the neutron star is obscured by dust and not accreting, making it difficult to see, or that the compact object formed was a black hole and not a neutron star (either directly, or by later fallback of matter on an initial neutron star), which would not be visible if it isn't accreting matter.

20.5 Heavy Elements and the r-Process

An important question having broader implications than just that for astrophysics concerns the origin of the heaviest nuclei. They cannot be made by normal charged-particle reactions in equilibrium in stars because the peak of the binding energy curve occurs for the iron-group nuclei, and because of Coulomb barrier effects. It was noted earlier that (uncharged) neutron capture reactions could circumvent the Coulomb barrier problem. It is thought that many of the heavier elements are made in the rapid neutron capture or r-process that is illustrated in Fig. 20.20. The r-process is similar to the s-process discussed in Section 13.7.2, except now it is assumed that there is a high flux of neutrons and that they can be captured rapidly compared with the rate for β-decay. As illustrated by the theoretical r-process path shown in Fig. 20.21, this tends to take the population up and very far to the neutron-rich side of the chart of the nuclides before it begins to β-decay back toward the stability valley. Thus the r-process can populate many neutron-rich isotopes out of the stability valley that cannot be reached by the s-process. Furthermore (unlike for the s-process), the path illustrated in Fig. 20.21 can populate isotopes beyond the *actinide gap* in the stability valley found near lead and bismuth, thus accounting for the production of actinide species like uranium that are observed in nature.

The astrophysical site of the r-process has been an enduring mystery. A large neutron flux is required, which is consistent with only a few known possibilities. In addition,

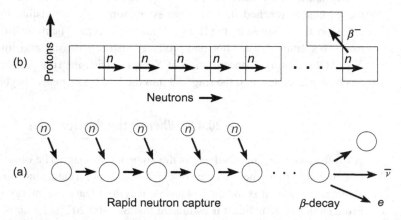

Fig. 20.20 (a) A schematic representation of the rapid neutron capture or r-process. (b) Characteristic r-process path in the neutron–proton plane. The s-process occurs at free-neutron densities $\sim 10^6$ neutrons cm^{-3}, but in the r-process the density is $\sim 10^{20}$ neutrons cm^{-3} and neutron capture is very rapid compared with the rate of β decay.

20.5 Heavy Elements and the r-Process

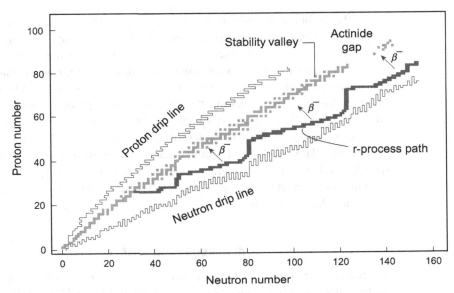

Fig. 20.21 Path for the r-process expected from theory. Nuclei produced along the r-process path will undergo rapid β^- decay back toward the stability valley (see Fig. 13.12).

Box 20.7 **Timescales for the r-Process**

Supernovae and neutron star mergers imply different timescales for r-process nucleosynthesis. A supernova requires a massive star to evolve to gravitational instability of its core, which occurs essentially instantaneously on cosmic timescales. A merger requires a neutron star binary to form by two successive supernova explosions in a massive binary, or by capture of one neutron star by another, and the binary must then spiral together by emission of gravitational waves on a much longer timescale. Thus, the r-process associated with mergers has an inherent *time delay*. The delay timescale depends strongly on initial conditions for formation of the binary and in general can be billions of years, but it has been argued that there is a population of fast-merger binaries that can merge on timescales of 10^8 yr or less. An open question then is whether binary mergers can account for r-process nuclides observed in low-metallicity stars, which likely formed early in galactic history.

---

there are timescales associated with production of r-process nuclei with observable implications that are discussed in Box 20.7. The leading candidates are (1) core collapse supernovae, (2) merger of neutron stars, and (3) jets that may be produced in mergers or in core collapse of massive, rapidly rotating stars, with the first two options being the strongest candidates. Until recently observational data and simulations were inadequate to distinguish conclusively among these possibilities and it is possible that the abundances of r-process nuclei receive contributions from more than one of these possible sites (or sites as yet unknown). A large step may have been taken in unraveling the r-process mystery by the observation of electromagnetic radiation in coincidence with gravitational waves from a neutron star merger, as described below and in Section 22.6.

One theme for understanding the origin of r-process nuclei is to ask whether observations suggest that they were produced in a few rare events (neutron star mergers are relatively rare, occurring maybe only once every million years in a large galaxy), or instead were produced in many more common events (core collapse supernovae are much more common than neutron star mergers, occurring about once every 100 years in a large galaxy). Some evidence had been accumulating that at least some r-process nuclei were produced in rare events [90]. The neutron star merger leading to gravitational wave GW170817 and associated gamma-ray burst to be described in Section 22.6 gives direct evidence for the production of large amounts of r-process nuclei in a single rare event. This has led to much speculation that neutron star mergers are the primary site of the r-process, though there are open questions about whether mergers can account for all r-process observations because of the time-delay issues discussed in Box 20.7.

Background and Further Reading

Mezzacappa [150] and Janka [129] give reviews of core collapse supernovae, and the earlier history of the problem was reviewed in Bethe [44] and Brown [59]. An introduction to simple estimates for core collapse supernova physics may be found in Cardall [66]. Portions of the core collapse supernova discussion were adapted from Guidry [98] and the r-process is reviewed in Thielemann, Eichler, Panov, and Wehmeyer [213]; and in Frebel and Beers [90]. The evolution and explosion of massive stars is reviewed in Woosley, Heger, and Weaver [231], and a discussion of what was learned from Supernova 1987A may be found in Arnett et al. [27]. An overview of the Type Ia supernova problem is given in Guidry and Messer [9, 103].

Problems

20.1 How much thermonuclear energy is produced if a C–O white dwarf of $1.4\,M_\odot$ is burned to ^{56}Ni? Compare this energy with the gravitational binding energy of the white dwarf, assuming uniform density and a mass of $10.4M$.

20.2 Estimate the dynamical timescale for collapse of the iron core of a $25\,M_\odot$ star. Assume the density of the core to be $5 \times 10^9\,\text{g cm}^{-3}$ at the beginning of collapse.***

20.3 How much gravitational energy is released if the iron core of a massive star collapses to neutron-star size? Assume the core to be of uniform density $5 \times 10^9\,\text{g cm}^{-3}$ with a radius of 500 km, and that it collapses to a uniform sphere of radius 10 km. Compare the energy released in this collapse with the total gravitational binding energy of the star before collapse, assuming it to be a supergiant of $15\,M_\odot$ and radius 2 AU before the collapse begins (you may use the uniform-density formula for the estimate).

20.4 When the ^{56}Fe core of a massive star becomes gravitationally unstable, two things greatly accelerate its collapse: (1) photodisintegration of ^{56}Fe is extremely endothermic and absorbs energy that can no longer be used for pressure support, and

(2) rapid neutronization, $p + e \to n + \nu_e$ undermines stability further because the major pressure support comes from the degenerate electrons that are disappearing, and the core is initially transparent to neutrinos so each emitted neutrino carries its energy out of the core at nearly the speed of light. Estimate the amount of energy absorbed by these two processes if the ^{56}Fe core is entirely photodisintegrated to α particles and neutrons, and in addition all of the available electrons undergo neutronization reactions with the protons. Assume that the core is electrically neutral, that $Q = -124.4$ MeV for ^{56}Fe $\to 13\alpha + 4n$, and that each emitted neutrino carries off average energy of 10 MeV. In a core collapse supernova the primary part of this energy release from photodisintegration and neutronization occurs in a fraction of a second. How long would the Sun need to radiate at its present luminosity to account for the energy loss estimated above for photodisintegration and neutronization of an iron core?***

20.5 Basic properties of the r-process suggest that successive neutron captures must occur on a timescale of microseconds. Estimate the neutron number density required to produce this capture rate, assuming a typical neutron capture cross section to be 100 mb. *Hint*: The solution to Problem 13.3 will be useful in this estimate.

20.6 A Type Ia supernova is powered by a thermonuclear runaway in electron-degenerate white dwarf matter. To see how such an instability could arise, use the approximate relationship between the central pressure and central density for a Lane–Emden polytropic solution given by Eq. (8.14), and the generic equation of state

$$\frac{dP}{P} = \alpha \frac{d\rho}{\rho} + \beta \frac{dT}{T}$$

(where α and β are non-negative) discussed in Problem 4.16, to show that thermonuclear burning is unstable at the center of a white dwarf described by a polytropic equation of state. *Hint*: Use Eq. (8.14), and that stability requires that changes in density and changes in temperature have the same sign.

20.7 The absolute bolometric magnitude for the progenitor of SN 1987A was estimated in Ref. [27] to be -7.8. Spectroscopically it was a luminous B3 supergiant with an estimated surface temperature of 16,000 K. What was its luminosity and radius?

20.8 The bright red supergiant Betelgeuse is a relatively nearby star that could undergo a core collapse at any time. Use the SIMBAD database [2] to retrieve the parallax and the ultraviolet magnitude U, the blue magnitude B, and the visual magnitude V for this star. Determine its $U - B$ and $B - V$ color indices and its distance from Earth. Given that the (large) bolometric correction for Betelgeuse is -2.15, what is its luminosity? If the effective surface temperature is 3500 K, what is its radius? Other things being equal, how much larger would you expect the neutrino flux on Earth to be relative to that for Supernova 1987A when Betelgeuse goes supernova?

20.9 What was the likely host galaxy, supernova spectal type (see Fig. 20.3), and spectroscopic redshift for SN 1969L? *Hint*: Use the SIMBAD astronomical database [2].

21 Gamma-Ray Bursts

Earth's atmosphere absorbs high-frequency photons strongly, so systematic observation of the heavens at X-ray and gamma-ray wavelengths had to await the space age and orbiting observatories high above the absorptive atmosphere. In earlier sections of this book X-rays emitted by objects of astronomical interest have been considered in various contexts, but we have said little about gamma-rays, the highest-energy photons of all. Because gamma-rays are energetic, they can be produced only in rather unusual and often violent events. Therefore, the realization beginning in the 1960s that gamma-rays (and X-rays) could be seen coming from many sources in the sky was a revelation, suggesting that our Universe was much less sedate and orderly than had often been assumed. The most mysterious of the gamma-ray sources were *gamma-ray bursts,* which were first observed in the 1960s but began to be understood only in the 1990s. As will be discussed in this chapter, it is now believed that they represent the violent death of a certain class of massive stars, or the nearly as violent demise of merging neutron stars. As such, they are an important part of the story of late stellar evolution, in addition to being of high intrinsic interest because they are among the most energetic events that occur in the Universe.

21.1 The Sky in Gamma-Rays

When seen from space the sky glows in gamma-rays, in addition to the other more familiar wavelengths. Figure 21.1(b) shows the continuous glow of the gamma-ray sky, as measured from orbit by the Compton Gamma-Ray Observatory (CGRO). In addition to the steady gamma-ray flux illustrated in Fig. 21.1, sudden bursts, which can be as short as tens of milliseconds and as long as several minutes, are observed. Figure 21.2 displays the time profile for a typical burst event. Gamma-ray bursts were discovered unexpectedly in the 1960s by gamma-ray detectors aboard satellites testing the feasibility of monitoring for nuclear explosions that violated nuclear test bans treaties. Quite surprisingly, these satellites began to see strong bursts of gamma-rays that did not look like nuclear weapons tests. These gamma-ray bursts (GRBs) were for several decades a great puzzle but, as will now be discussed, newer observations have led to a much deeper understanding of these remarkable events.

About one burst a day is observable somewhere in the sky by orbiting observatories. Figure 21.3 shows the position of 2704 gamma-ray bursts recorded by the Compton Gamma-Ray Observatory. The highly isotropic distribution of GRB events over a broad range of fluences (energy received per unit area) in this figure argues strongly that they

21.1 The Sky in Gamma-Rays

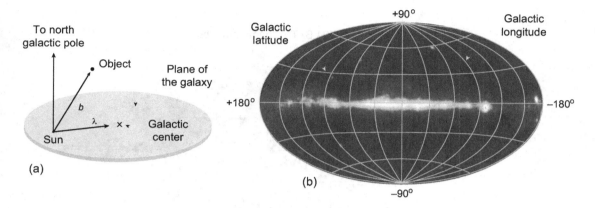

Fig. 21.1 (a) The galactic coordinate system. The angle *b* is the galactic latitude and the angle λ is the galactic longitude, which are related to right ascension and declination by standard spherical trigonometry. (b) The sky at gamma-ray wavelengths in galactic coordinates, with white the most intense and black the least intense. The diffuse horizontal feature at the galactic equator is from gamma-ray sources in the plane of the galaxy. Bright spots to the right of center in the galactic plane are galactic pulsars and brighter spots above and below the plane of the galaxy are quasars far outside the galaxy.

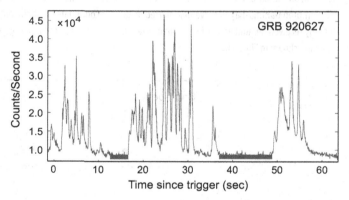

Fig. 21.2 Time profile of a gamma-ray burst [172]. Reprinted figure with permission from Tsvi Piran, *Rev. Mod. Phys.*, **76**, 1143. Published January 28, 2005. Copyright (2005) by the American Physical Society. DOI: https://doi.org/10.1103/RevModPhys.76.1143

occur at cosmological distances – hundreds of megaparsecs or greater. The origin of gamma-ray bursts far outside our galaxy will be confirmed more directly below from redshifts of spectral lines observed in the aftermath of gamma-ray bursts. Figure 21.4 illustrates that there appear to be two classes of gamma-ray bursts:

1. *Short-period bursts*, which last less than two seconds and exhibit harder (higher-energy) spectra.
2. *Long-period bursts*, which have softer (lower-energy) spectra and typically last from several seconds up to several hundred seconds.

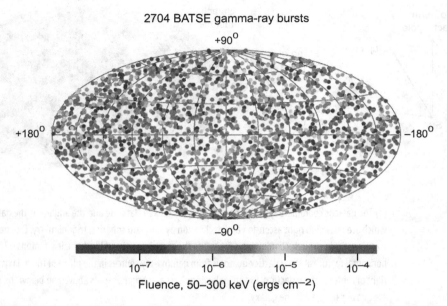

Fig. 21.3 Location on the sky of 2704 gamma-ray bursts recorded by the Burst and Transient Source Experiment (BATSE) of the Compton Gamma-Ray Observatory, plotted in galactic coordinates with the grayscale indicating the *fluence* (energy received per unit area) of each burst. These bursts are observed superposed on the continuous gamma-ray emission shown in Fig. 21.1(b).

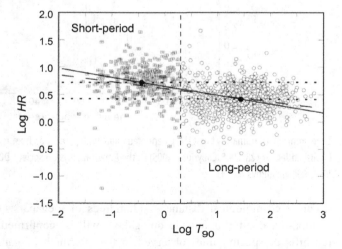

Fig. 21.4 Hardness *HR* (a parameter measuring the propensity to contain higher-energy photons) of the spectrum versus duration of the burst, illustrating the separation of the GRB population into long, soft bursts and short, hard bursts [178]. The parameter T_{90} is the time from burst trigger for 90% of the burst energy to be collected. Reprinted figure with permission from Tsvi Piran, *Rev. Mod. Phys.*, **76**, 1143. Published January 28, 2005. Copyright (2005) by the American Physical Society. DOI: https://doi.org/10.1103/RevModPhys.76.1143

21.2 Localization of Gamma-Ray Bursts

These two classes of gamma-ray bursts share many common features but their differences suggest that they arise from two different mechanisms. Let us now turn to a discussion of what these mechanisms could be.

21.2 Localization of Gamma-Ray Bursts

The first step in understanding what causes gamma-ray bursts was to pin down the astrophysical environment in which they originate. Could they be associated with known galaxies or with specific events like supernova explosions, for example? BATSE observations in the 1990s had angular resolutions of several degrees, so it was difficult to know exactly where to point telescopes to find evidence associated with the gamma-ray burst at other wavelengths. Help in this regard came from a satellite looking not at gamma-rays, but at X-rays.

In the late 1990s it became possible to correlate some gamma-ray bursts with other sources in the visible, RF, IR, UV, and X-ray portions of the spectrum because of a Dutch–Italian satellite called BeppoSAX that was capable of localizing X-ray transients following a gamma-ray burst with arc-minute resolution. This permitted other instruments to look quickly at the burst site at multiple wavelengths, and for the first time transient sources ("afterglows") at other wavelengths could be correlated with a burst. Figure 21.5(a) shows

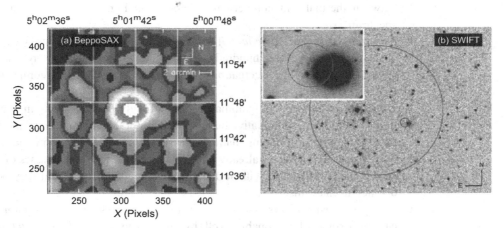

Fig. 21.5 (a) First localization of an X-ray afterglow for a gamma-ray burst by the satellite BeppoSAX. (b) Optical association of short-period GRB 050509B with a large elliptical galaxy at a redshift of $z = 0.225$ by SWIFT [93]. The larger circle is the error circle for the Burst Alert Telescope (BAT) on SWIFT. The smaller circle is the error circle for the X-Ray Telescope (XRT), which was slewed to point at the event when alerted by the BAT. The XRT error circle is shown enlarged in the inset at the upper left, suggesting that the GRB occurred on the outskirts of the large elliptical galaxy (dark oval) partially overlapped by the XRT error circle. Reproduced by permission from Springer Nature: *Nature*, "A Short X-Ray Burst Apparently Associated with an Elliptical Galaxy at Redshift $z = 0.225$," **437**(7060), N. Gehrels et al., Copyright (2005).

an X-ray transient observed by BeppoSAX following a long-period gamma-ray burst and a localization for a short-period burst by the SWIFT satellite is illustrated in Fig. 21.5(b).

Redshifted spectral lines were observed in the transients after the burst, which for the first time began to allow distances to be estimated to gamma-ray bursts. Assuming the redshifts to be Hubble law redshifts associated with the expansion of the Universe, these observations show conclusively that gamma-ray bursts are occurring at cosmological distances. Thus they must emit enormous power at gamma-ray wavelengths, which raises challenging questions concerning the source of that power.

21.3 Generic Characteristics of Gamma-Ray Burst

Based on the original BATSE data and a series of observations that became possible once the afterglow transients of gamma-ray bursts could be localized on the sky and studied at various wavelengths, it is now agreed that gamma-ray bursts have the following characteristics:

1. *Cosmological origin*: The isotropic distribution of gamma-ray bursts in Fig. 21.3 suggested a cosmological origin, which was confirmed by redshift measurements on emission lines in GRB afterglows. As of 2018, the largest known spectroscopic redshift for a gamma-ray burst is $z = 8.2$ for GRB 090423.[1]
2. *Nonthermal spectrum*: The spectrum is not thermal. Box 21.1 describes the difference between thermal and nonthermal radiation and Fig. 21.6 illustrates a typical GRB spectrum.
3. *Duration and time structure*: The lengths of individual bursts vary from about 0.01 seconds to several hundred seconds, and their time structure can range from smooth to millisecond fluctuations (with the latter implying a compact source; see Box 17.1).
4. *Ultrarelativistic jets*: The gamma-rays are strongly beamed, implying emission from tightly collimated, ultrarelativistic jets. Furthermore, the gamma-rays must suffer little interaction with surrounding matter before escaping, as discussed in Section 21.4.
5. *Two classes of bursts*: As already noted, there appear to be two classes of bursts: long-period and short-period, with sufficient differences to suggest that they occur through distinct mechanisms.
6. *Afterglows and fireballs*: The transients (afterglows) observed after gamma-ray bursts can be explained reasonably well by the *relativistic fireball model* illustrated in

---

[1] The naming convention for gamma-ray bursts is of the form GRB followed by three 2-digit numbers indicating the year, month, and day of observation, respectively. For example, GRB 090423 was observed by the SWIFT satellite on April 23, 2009. The redshift $z = 8.2$ for GRB 090423 implies that the gamma-ray burst occurred only about 600 million years after the big bang! It will be shown below that the likely cause of the gamma-ray burst was core collapse of a massive star. This implies that the Universe was making stars within a few hundred million years after the big bang (see Section 14.5), which has significant implications for ideas of how structure formed in the Universe.

Box 21.1 — Nonthermal Emission

The Planck law describes *thermal emission* from a hot gas in equilibrium. The resulting *blackbody spectrum* peaks at some wavelength and falls off rapidly at longer and shorter wavelengths, with the position of the peak moving to shorter wavelength as the temperature is increased (Wien law). Light from normal stars and galaxies is dominantly thermal in character.

Synchrotron Radiation

Nonthermal emission has a spectrum exhibiting increased intensity at long wavelengths. The most common example is *synchrotron radiation*, where high-velocity electrons in a strong magnetic field follow a spiral path around the field lines, radiating strongly beamed light. Synchrotron radiation is *polarized* because it is emitted in a narrow beam in the local plane of the electron's spiral path; see the figure below left. The figure below right contrasts nonthermal emission with a thermal (blackbody) spectrum characteristic of 6000 K.

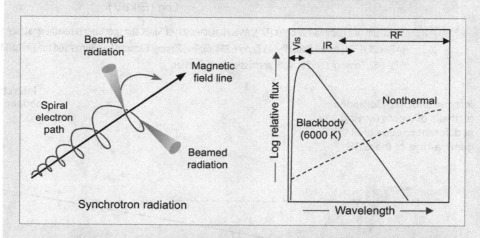

The wavelength of emitted radiation depends on how fast the charged particle spirals in the field. As the particle emits radiation it slows and emits longer wavelength radiation, which explains the broad distribution in wavelength of synchrotron radiation. Often the distribution of nonthermal emission is well described by a power law (a polynomial in powers of the energy).

Implications of Nonthermal Emission

Nonthermal emission is much less common than thermal emission in astronomy, but a nonthermal component in a spectrum typically signals violent processes and large accelerations of charged particles. High-frequency synchrotron radiation also implies the presence of *strong magnetic fields* because the frequency increases with tighter electron spirals, which are characteristic of strong fields.

Fig. 21.7, where deposition of energy by some central engine initiates a fireball expanding at relativistic velocities that is responsible for the afterglows.

Ultrarelativistic jets are a central piece of the GRB mechanism, so we shall now discuss the reason that this is believed to be so.

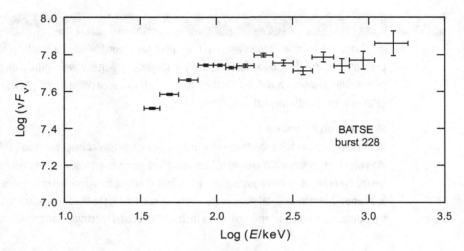

Fig. 21.6 Spectrum of a gamma-ray burst [171]. As is characteristic of GRBs, the spectrum is nonthermal (see the discussion in Box 21.1). Reprinted from *Physics Report*, **314**(6), Tsvi Piram, "Gamma-Ray Bursts and the Fireball Model," 575–667, Copyright (1999), with permission from Elsevier.

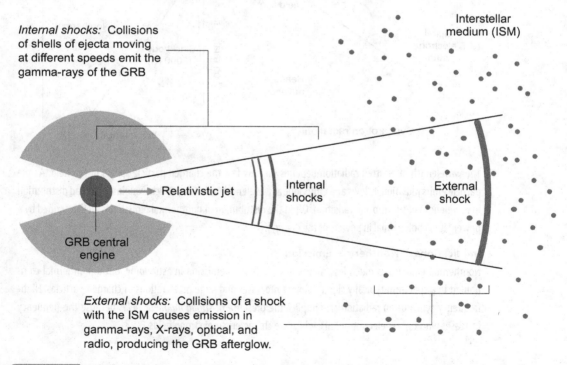

Fig. 21.7 Relativistic fireball model for afterglows following gamma-ray bursts. Internal shocks in the ultrarelativistic jet produce the gamma-rays; the external shocks resulting from the jet impacting the interstellar medium produce the afterglows.

21.4 The Importance of Ultrarelativistic Jets

As was illustrated in Example 17.1, the *Lorentz γ-factor* defined by

$$\gamma \equiv \left(1 - \frac{v^2}{c^2}\right)^{-1/2} \tag{21.1}$$

is a measure of how relativistic the particles in a jet are. Ultrarelativistic particles (those for which the rest mass energy can be neglected relative to the kinetic energy) will have $\gamma \gg 1$. Because the spectrum is nonthermal (see Box 21.1 and Fig. 21.6), we will now argue that gamma-ray bursts must be produced by jets that are ultrarelativistic. This can be understood in terms of the opacity of the medium with respect to formation of electron–positron pairs through $\gamma\gamma \to e^+ e^-$. A more extensive discussion is given in Section 15.7 of Ref. [100] but the essential features of the argument presented there are summarized below.

21.4.1 Optical Depth for a Nonrelativistic Burst

We first assume that the burst involves nonrelativistic velocities. The number of counts $N(E)$ as a function of gamma-ray energy can be approximated for particular ranges of energy as a power law,

$$N(E)\,dE \propto E^{-\alpha}\,dE, \tag{21.2}$$

where the *spectral index α* may be assumed to be ~ 2 for present purposes [171, 172]. The observed nonthermal spectrum requires the medium to be *optically-thin* (small optical depth; see Box 10.1 and Problem 21.4), because an optically-thick medium would thermalize the photons. Energy conservation for $\gamma\gamma \to e^+ e^-$ requires the two photons with energies E_1 and E_2, respectively, to satisfy $(E_i E_2)^{1/2} \gtrsim m_e c^2$, where m_e is the electron mass. If f is the fraction of photon pairs having sufficient energy, the optical depth τ_0 is [171]

$$\tau_0 = \frac{f\sigma_T F D^2}{R^2 m_e c^2} \simeq \frac{f\sigma_T F D^2}{\delta t^2 m_e c^4}, \tag{21.3}$$

where σ_T is the Thomson scattering cross section (7.10), F is the burst fluence, D is the distance to the source and R is its size, with R related to the period δt for time structure in the burst by $R = c\delta t$, using the arguments of Box 17.1. You are asked to show in Problem 21.4 that an optical depth estimated using this formula is *enormous* ($\tau_0 \sim 10^{14}$), and therefore inconsistent with the low optical depth required by the nonthermal GRB spectrum.

21.4.2 Optical Depth for an Ultrarelativistic Burst

If the burst is instead ultrarelativistic (Lorentz factor $\gamma \gg 1$), the relative motion with $v \sim c$ between source and observer will modify Eq. (21.3) in two essential ways:

1. The blueshift of the emitted radiation will change the fraction f of photon pairs that have sufficient energy to make electron–positron pairs. This multiplies the factor f in Eq. (21.3) by a factor of $\sim \gamma^{-2\alpha}$, where α is the spectral index.
2. The effective size R of the emitting region will be altered by relativistic effects, which will multiply R by a factor of γ^2.

Incorporating these corrections, the ultrarelativistic modification of Eq. (21.3) is

$$\tau \simeq \frac{\tau_0}{\gamma^{4+2\alpha}}, \qquad (21.4)$$

where τ_0 is the nonrelativistic limit evaluated from Eq. (21.3). Thus, the medium will be optically thin if γ is sufficiently large. Taking the estimate from Problem 21.4 that $\tau_0 \sim 10^{14}$ and assuming $\alpha = 2$ for the spectral index, τ will be less than one provided that $\gamma > 56$. Hence consistency with the observed nonthermal spectrum requires that the GRB involve an ultrarelativistic jet with a Lorentz γ of order 100 or more.

21.4.3 Confirmation of Large Lorentz Factors

Observational confirmation that gamma-ray bursts are indeed associated with large values of γ is supplied by the location of "breaks" in the lightcurves for afterglows, which indicate the time when the initially relativistic afterglow begins to slow rapidly through interactions with the interstellar medium. This information can in turn be related to the opening angle of the jet that produced the afterglow. Such analyses typically find small jet opening angles $\Delta\theta$, suggesting large Lorentz factors for many GRB because from relativistic kinematics $\Delta\theta \propto \gamma^{-1}$. Because of this beaming, a fixed observer sees only a fraction of all gamma-ray bursts. The beaming also solves a potential energy-conservation problem. If the energy from detected bursts were assumed to be emitted isotropically, total energies exceeding 10^{54} erg would be inferred for some strong gamma-ray bursts, which is comparable to the rest mass energy of the Sun! But if gamma-ray bursts are assumed to be emitted as collimated jets, then the total energy released would be much smaller than that inferred from the measured energy assuming isotropic emission.

21.5 Association of GRBs with Galaxies

The generic features of gamma-ray bursts summarized in Section 21.3 provide substantial clues to their nature, but are not sufficient to allow a specific mechanism to be identified. They describe the phenomenology of depositing locally a large amount of energy produced by ultrarelativistic jets from some central engine, but do not specify the exact nature of the central engine. The first steps in identifying the central engine will be to take seriously the suggestion that there are two classes of gamma-ray bursts (long-period and short-period) that are likely produced in different ways, and to associate gamma-ray bursts observationally with particular kinds of galaxies. An afterglow is connected only indirectly to the gamma-ray burst central engine, but it permits *localization on the celestial sphere* of the gamma-ray burst that produced it. This localization has permitted a number of GRB to be associated with specific distant galaxies, with the following conclusions:

1. Long-period (softer) bursts are strongly correlated with *star-forming regions*.
2. Short-period (harder) bursts typically *do not occur in star-forming regions*.
3. Some evidence indicates that long-period bursts favor star-forming regions having *low metallicity*.

These observations provide further evidence that long-period and short-period bursts are initiated by different mechanisms because they occur preferentially in star populations of very different ages.

21.6 Mechanisms for the Central Engine

Based on the observational evidence, an acceptable model for the central engines that produce gamma-ray bursts and their afterglows must embody at least the following features:

1. Highly relativistic, strongly focused jets.

    (a) Lorentz γ-factors of at least 100, perhaps larger, are required by observations.[2]
    (b) Jets with opening angles of only $1°–10°$, associated with a total energies of order 10^{52} erg.
    (c) As will be described below, long-period bursts may be associated with a supernova explosion, so the central engine also must (at least sometimes) deliver $\sim 10^{52}$ erg to a much larger angular range (~ 1 rad) to produce the accompanying supernova, and it must operate for 10 seconds or longer in these long-period bursts to account for their duration.

2. The large radiated power over an extended period, particularly for long-period bursts, points strongly to a mechanism involving accretion onto a compact object.

Almost the only way known to explain such a rapid release of that amount of energy is from a collapse involving a compact gravitational source. Two general classes of models are now thought to account for GRB.

1. A *hypernova* in which a spinning massive star collapses to a Kerr black hole and jet outflow from this collapsed object produces a burst of gamma-rays (see the collapsar model discussed in Section 21.8). This is the favored mechanism for long-period bursts.

---

[2] These are very large Lorentz factors by astrophysical standards. Most jets observed from quasars and active galactic nuclei do not exceed $\gamma \sim 10$. On the other hand, one of the counter-circulating proton beams of the Large Hadron Collider (LHC) at CERN has $\gamma \sim 7000$. This translates to a velocity only about $3\,\mathrm{m\,s^{-1}}$ less than the speed of light (and a special relativity time dilation factor of γ, so time passes about 7000 times more slowly for an LHC beam particle than for a stationary observer beside the beam line).

2. The merger of two neutron stars, or a neutron star and a black hole, with jet outflow perpendicular to the merger plane producing a burst of gamma-rays as the two objects collapse to a Kerr black hole. This is the favored mechanism for short-period bursts.

The unifying theme for both mechanisms is the collapse of stellar amounts of spinning mass to a Kerr black hole central engine that powers the burst. In the next section we describe observations that suggest a specific mechanism powering the central engine of a long-period gamma-ray burst that is partial to star-forming regions, as required by data.

21.7 Long-Period GRB and Supernovae

As suggested by Fig. 21.8(a), long-period GRB afterglow spectra have been observed to evolve into spectra resembling those of supernovae, hinting strongly that the underlying mechanism for long-period bursts may be a particular type of supernova.

21.7.1 Types Ib and Ic Supernovae

As discussed in Section 20.1, supernovae may be classified observationally by their spectra and lightcurves, or theoretically by their explosion mechanism. The supernovae that are thought to be associated with long-period GRB are classified in Fig. 20.3 and Section 20.1.2 as Types Ib and Ic, for which the explosion mechanism is core collapse in a rapidly rotating, 15–30 $M_\odot$ *Wolf–Rayet star* (see Section 14.3.1). These stars can shed their hydrogen and even helium envelopes before their cores collapse, and are likely to collapse directly to a Kerr black hole instead of to a neutron star because they are so massive. Figure 21.8(b) shows a Wolf–Rayet star shedding its outer layers. It is thought that a Type Ib supernova occurs in a Wolf–Rayet star for which the H shell has been ejected, and that a Type Ic supernova occurs in a Wolf–Rayet star for which the H and part of the He shells have been ejected, before the stellar core collapses and triggers the supernova.

21.7.2 Role of Metallicity

The occurrence of long-period gamma-ray bursts in star-forming regions is not surprising if they are associated with core collapse supernovae, since these occur only for young, massive stars. The possible affinity of long-period GRB with star-forming regions of low metallicity has been interpreted in terms of a model of long-period gamma-ray bursts resulting from the core collapse of a Wolf–Rayet star [230]. Low metallicity makes it harder for radiation pressure to eject matter because it decreases the surface photon opacity. The suppression of mass loss has two favorable effects on the *collapsar model* to be described below: (1) High mass favors a collapse directly to a black hole instead of to a neutron star. (2) Suppression of mass loss disfavors angular momentum loss, leading to higher spin rates at collapse. This in turn aids in the creation of a substantial accretion disk

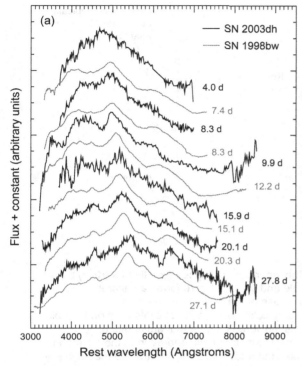

Fig. 21.8 (a) Time evolution in the optical spectrum of SN2003dh (GRB 030329) in black, compared with a reference supernova SN1998bw in gray [121]. The initial rather featureless spectrum of the GRB 030329 afterglow develops bumps similar to supernova SN1998bw over time, suggesting that as the afterglow of the gamma-ray burst energy fades an underlying supernova explosion is revealed. Hence GRB 030329 is also denoted as the supernova SN2003dh. Reprinted by permission from Springer Nature: *Nature*, "A Very Energetic Supernova Associated with the γ-Ray Burst of 29 March 2003," Hjorth, E. et al. COPYRIGHT (2003). (b) Wolf–Rayet star (tip of black arrow) surrounded by gas that it has emitted. These massive, high-mass-loss, rapidly-spinning stars may be progenitors of Type Ib and Type Ic core collapse supernovae, and of long-period GRB.

around the black hole as it forms, which can be tapped as a source of the extended power output needed for a long-period gamma-ray burst.

21.8 Collapsar Model of Long-Period Bursts

An overview of the collapsar model is given in Fig. 21.9. Simulations of ultrarelativistic jets breaking out of a Wolf–Rayet star are shown in Fig. 21.10, and Fig. 21.11(a) displays a simulation of a black hole formed from a Wolf–Rayet star, 20 seconds after core collapse. The GRB is powered by an ultrarelativistic jet driven by rotating magnetic fields or neutrino–antineutrino annihilation, but the supernova is powered by the disk wind of

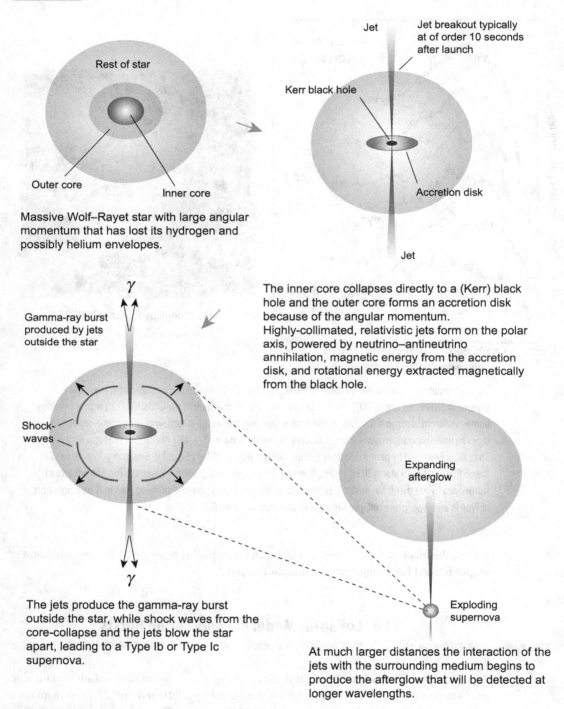

Fig. 21.9 Collapsar model for a long-period GRB and accompanying Type Ib or Ic supernova [230].

21.8 Collapsar Model of Long-Period Bursts

(a)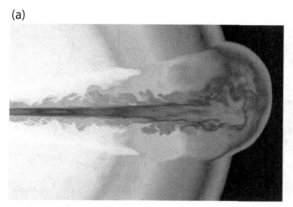

(b)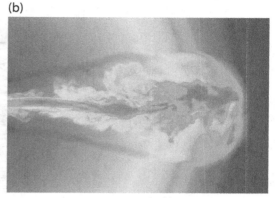

Fig. 21.10 Simulations of ultrarelativistic jets breaking out of Wolf–Rayet stars [144, 230]. Breakout of the jet is eight seconds after launch from the center of a 15 $M_\odot$ Wolf–Rayet star. The Lorentz γ for the jet is about 200. (a) Reprinted from "Long Gamma-Ray Bursts," Andrew Macfadyen, *Science*, **303**(5654), 45–46, 2004. DOI: 10.1126/science.1091764. (b) Reprinted from "The Supernova–Gamma-Ray Burst Connection," S. E. Woosely and J. S. Bloom, *Annual Review of Astronomy and Astrophysics*, **44**(1), 507–556, 2006.

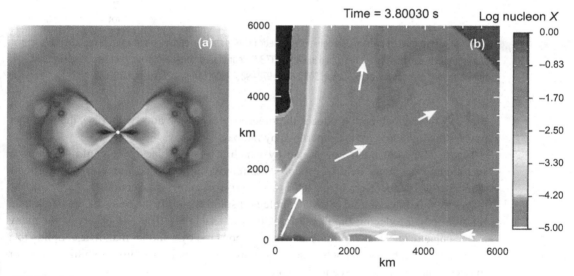

Fig. 21.11 (a) A rapidly spinning Wolf–Rayet star of 14 solar masses, about 20 seconds after core collapse (the polar axis is vertical). The density scale is logarithmic and the 4.4 $M_\odot$ Kerr black hole has been accreting at about 0.1 $M_\odot$ per second for 15 seconds at this point in the calculation [230]. (a) Reproduced from "Long Gamma-Ray Bursts," Andrew Macfadyen, *Science*, **303**(5654), 45–46, 2004. DOI: 10.1126/science.1091764. (b) Simulation of the nucleon wind blowing off the accretion disk in a collapsar model [230]. Gray-scale contours represent the log of the nucleon mass fraction X and arrows indicate the general flow. (b) Reproduced from "The Supernova–Gamma-Ray Burst Connection," S. E. Woosely and J. S. Bloom, *Annual Review of Astronomy and Astrophysics*, **44**(1), 507–556, 2006.

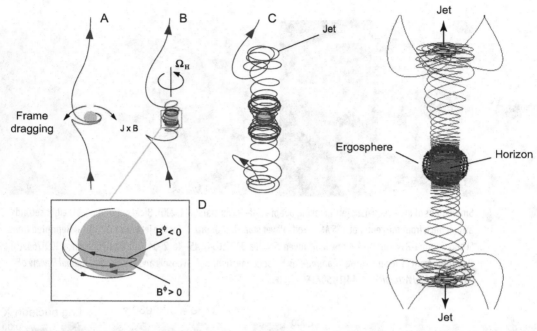

Fig. 21.12 Relativistic jets produced by frame dragging of magnetic fields in the spacetime around a Kerr black hole [199]. The ergosphere and horizon are described in Section 17.3.5. Reproduced from, "Simulations of Jets Driven by Black Hole Rotation," Semenov, V. et al. *Science*, **305**(5686), 978–980, 2004, DOI:10.1126/science.1100638.

Fig. 21.11(b), which both produces the supernova explosion and synthesizes the ^{56}Ni that powers the lightcurve of the supernova by radioactive decay.

Figure 21.12 illustrates one model by which a rotating black hole could couple to a surrounding magnetic field to produce ultrarelativistic jets. The frame-dragging effects associated with the Kerr black hole (see Kerr spacetimes in Section 17.3.5) wind the magnetic flux lines around the black hole and spiral them off the poles of the black hole rotation axis, producing bipolar ultrarelativistic jets [199]. The jets observed for many active galactic nuclei and quasars also may be powered by a similar magnetic coupling to a Kerr black hole, but on a much larger scale with supermassive (millions to billions of solar masses) black holes as the central engines.

21.9 Neutron Star Mergers and Short-Period Bursts

The core collapse of a Wolf–Rayet star represents a plausible mechanism for long-period gamma-ray bursts that associates them naturally with star-forming regions. On the other hand, there is little observational evidence that short-period bursts are associated either with star-forming regions or supernovae, suggesting that the mechanism responsible for

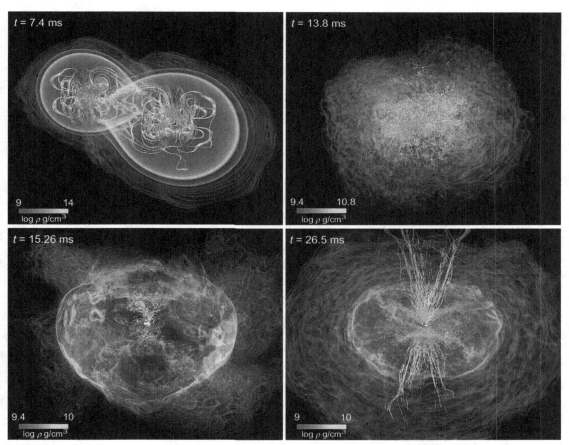

Fig. 21.13   Neutron star merger simulation with strong magnetic fields [83, 184]. Successive panels show the evolution in time of the mass density with magnetic field lines superposed; see text for further description. Reproduced from Rezzolla et al., "The Missing Link: Merging Neutron Stars Naturally Produce Jet-Like Structures and Can Power Short Gamma-Ray Bursts, © 2011. The American Astronomical Society. All rights reserved. *Astrophysical Journal Letters*, **732**(1). Reproduced with permission from the authors.

them must be something other than the core collapse of Wolf–Rayet stars. The favored mechanism for short-period bursts also involves the formation of an accreting Kerr black hole, but one produced by the merger of two neutron stars, or merger of a neutron star and a black hole, rather than by the core collapse of a massive star.

A simulation of a neutron-star merger to form a Kerr black hole with strong magnetic fields is shown in Fig. 21.13 [83, 184], The first panel shows the state shortly after initial contact and the second displays a high-mass neutron star configuration (one too massive to remain a neutron star for long). In the bottom two panels a Kerr black hole has formed in the center with a disk around it, and the magnetic field is wound up by the disk to a strength of order 10^{15} gauss, with an opening angle for the field lines in the polar direction of about $30°$. These simulations indicate the propensity to develop very high magnetic fields in neutron star mergers, enabling a possible magnetically powered gamma-ray burst. It is also expected that emitted neutrinos could provide some of the power for such a gamma-ray burst.

21.10 Multimessenger Astronomy

The large asymmetric mass distortion, high velocities generated by revolution on millisecond timescales, and highly compact mass distribution, imply that neutron-star mergers will be a strong source of gravitational waves. Likewise, core collapse events involve asymmetric motion of dense matter at high velocity and are expected to be strong gravitational wave sources. The frequencies for gravitational waves emitted from neutron star mergers (as well as mergers of neutron stars with black holes) and core collapse supernovae are expected to lie in the range accessible to gravitational-wave interferometers like LIGO (see Fig. 22.8). Thus, in addition to being potential sources of gamma-ray bursts, neutron star mergers and core collapse events could be excellent candidates for producing gravitational waves of sufficient strength to be observable in earth-based gravitational wave detectors. This possibility will be discussed in Chapter 22 and raises the intriguing prospect of *multimessenger astronomy,* where traditional astronomy, neutrino, or gamma-ray burst signals might be detected in coincidence with gravitational waves.

Background and Further Reading

General overviews of gamma-ray bursts have been given by Piran [171, 172]. An overview of gamma-ray bursts with emphasis on collapsar models may be found in Woosley and Bloom [230]. Neutron star merger models for gamma-ray bursts are discussed in Price and Rosswog [177], and in Rosswog [189].

Problems

21.1 What is the velocity of the particles in a gamma-ray burst jet exhibiting a Lorentz γ-factor of 200? In the first upgrade of the Large Hadron Collider each of the colliding proton beams was designed to reach an energy of 7 TeV (7×10^{12} eV), which corresponds to a Lorentz γ-factor of 7460. What is v/c for a proton beam with this energy?

21.2 Verify that γmc^2, where m is the rest mass and γ is the Lorentz γ-factor, behaves like the total energy in the limit of low velocity. *Hint*: The total energy is the potential (rest mass) energy plus the kinetic energy.

21.3 In a particular gamma-ray burst the intensity is observed to vary with a period of approximately 10 ms. Estimate an upper limit for the size of the central engine powering the burst.

21.4 Assume for a typical gamma-ray burst that the distance is ~ 3000 Mpc, the fluence is $\sim 10^{-7}$ erg cm^{-2}, the fraction of photon pairs with sufficient energy for $\gamma\gamma \to e^+e^-$ to occur is of order one, and that the burst intensity exhibits fluctuations on a 10 ms timescale. If the burst is taken (contrary to fact) to not be relativistic, show that the optical depth given by Eq. (21.3) is inconsistent with the observed nonthermal spectrum for gamma-ray bursts. Show that in principle this problem is alleviated if the kinematics are ultrarelativistic and the value of the Lorentz γ-factor is large enough. *Hint*: See Eq. (21.4) and the discussion preceding it.***

22 Gravitational Waves and Stellar Evolution

Detection of the gravitational wave event GW150914 in late 2015 by the LIGO (Laser Interferometer Gravitational-Wave Observatory) collaboration, and its interpretation as resulting from the merger of two $\sim 30\,M_\odot$ black holes, may be of as much importance for stellar physics as for gravitational physics. Certainly the confirmation that gravitational waves exist and can (through monumental technical ingenuity!) be detected, coming some 100 years after Einstein's prediction of such ripples in the fabric of spacetime, was a remarkable achievement for gravitational physics and the theory of general relativity.[1] But it is also arguably the most direct evidence yet for black holes, and begins to place strong new constraints on theories of massive-star evolution. Of even broader significance for stellar evolution was the detection in 2017 of gravitational waves from a neutron star merger in coincidence with a GRB and accompanied by electromagnetic signals observed at multiple wavelengths. This chapter introduces the new field of gravitational wave astronomy and its potentially large implications for understanding the evolution of stars.

22.1 Gravitational Waves

Gravitational waves, the requisite general relativity background, and details of the first gravitational wave events are covered more thoroughly in Ref. [100]. This chapter will draw heavily on the discussion in that book, introducing only the bare minimum of mathematics and instead concentrating on the potential implications of gravitational wave observation for understanding of stellar evolution. It will be useful for later discussion to summarize some basic principles without getting too deeply into the mathematical weeds. The essential idea is that the Einstein equations that were introduced in Eqs. (17.3) and (17.4) admit solutions that are *wavelike and propagate at the speed of light* (which could be termed more precisely in this context, the *speed of gravity*). These gravitational wave solutions have many similarities with the corresponding electromagnetic wave solutions of the Maxwell equations, but there are some essential differences. The most fundamental concerns the question "what is waving"? Electromagnetic waves are propagating ripples in the electric and magnetic fields, which are defined *in spacetime*; gravitational waves are ripples propagating in the metric of spacetime, so it is *spacetime itself*, not some field defined in spacetime, that is "waving."

---

[1] As illustrated in Fig. 16.14, there already was strong indirect evidence for the existence of gravitational waves, but GW150914 was the first *direct* confirmation.

22.1 Gravitational Waves

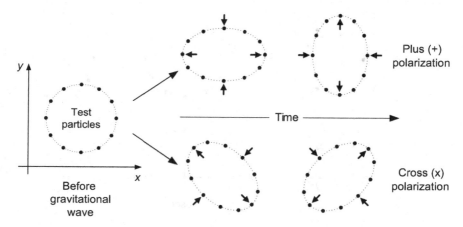

Fig. 22.1 Effect of a gravitational wave incident along the z axis on test masses in the x–y plane. The top pattern is called plus (+) polarization (test masses oscillate in a + pattern) and the bottom pattern is called cross (×) polarization (the test masses oscillate in a × pattern).

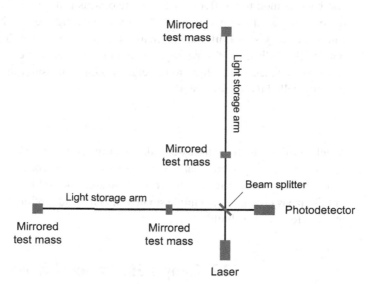

Fig. 22.2 Laser interferometer gravitational wave detector. In the storage arms of actual detectors light typically is multiply reflected, greatly increasing the effective length of the arms.

As for electromagnetic waves, gravitational waves are transverse and have two states of polarization, commonly denoted *plus* (+) and *cross* (×). Gravity acts on mass so gravitational wave polarization may be illustrated by considering the effect of a polarized gravitational wave on a circular array of test masses, as shown in Fig. 22.1. These wave patterns in spacetime may be detected using Michelson laser interferometers with kilometer or longer arms, as illustrated in Fig. 22.2 [185]. Because the gravitational wave causes

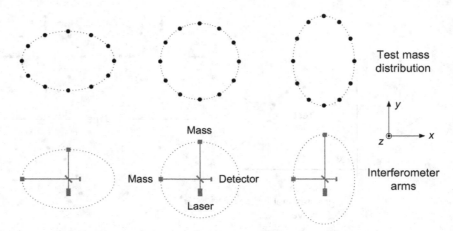

Fig. 22.3 Analogy between interaction of a gravitational wave with a test mass distribution and with an interferometer.

periodic fluctuations in the spacetime metric, the distance that light travels down an arm and back is modified differently for the two arms if a gravitational wave passes through the detector, as illustrated in Fig. 22.3. By comparing the two beams, the interferometer can detect very small differential changes in the light travel distances for the two arms, potentially indicating the passage of a gravitational wave. The fractional change in effective distance for the light to travel $\delta L(t)/L_0$ is measured in terms of a dimensionless quantity called the *strain h*, with

$$\frac{\delta L(t)}{L_0} \simeq \tfrac{1}{2} h(t, 0), \tag{22.1}$$

which oscillates with the time dependence of the gravitational wave. Exquisite precision is required because gravitational waves from expected astronomical sources require strains $\sim 10^{-21}$ to be measured. As you are asked to show in Problem 22.1, $\delta L \sim h L_0$ for a strain of this size is orders of magnitude smaller than the width of nuclei in the atoms from which the interferometer is built!

22.2 Sample Gravitational Waveforms

We begin the discussion with an overview of some computer simulations indicating the varied waveforms and potential astrophysical information that gravitational waves may carry. At least four kinds of events involving objects from late stellar evolution are expected to produce detectable gravitational waves: (1) merger of two black holes, (2) merger of a black hole and neutron star, (3) merger of two neutron stars, and (4) a core collapse supernova explosion. Simulations indicate that the corresponding waveforms will carry signatures of the event that produced the gravitational wave, and that these may encode detailed information about the objects involved. Some computed gravitational waveforms

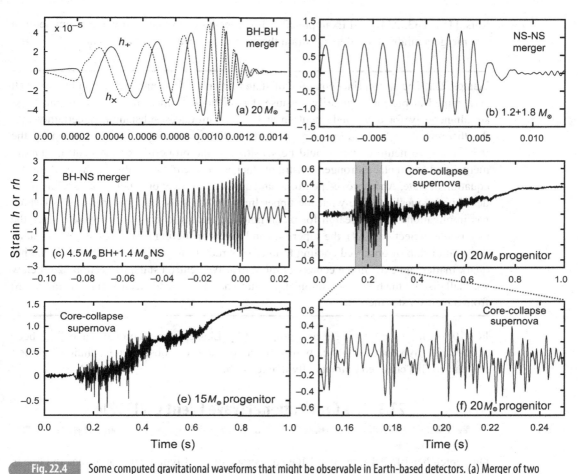

Fig. 22.4 Some computed gravitational waveforms that might be observable in Earth-based detectors. (a) Merger of two 20 $M_\odot$ black holes (BH–BH) [10, 34, 35, 61]. (b) Merger of 1.2 $M_\odot$ + 1.8 $M_\odot$ (all masses are baryonic) neutron stars (NS–NS) at distance of 15 Mpc [191]. (c) 4.5 $M_\odot$ black hole and 1.4 $M_\odot$ neutron star merger (BH–NS) at 15 Mpc [10, 141]. (d)–(f) Supernova at 15 kpc for two progenitor masses; time measured from bounce [234]. Panel (f) displays the initial burst of panel (d) at higher resolution. In panel (a) rh is shown, where r is the distance to the source in cm. In panels (b)–(f) strain is given in dimensionless units of 10^{-21} by assuming a distance to the source. All waves are h_+ polarization except for in (a), where both h_+ and $h_\times$ are shown. Further details may be found in the references. Figure plotted from data available at https://astrogravs.gstc.nasa.gov/docs/catalog.html and from Figure 1, Kanstantin N. Yakunin et al., *Physical Review D*, 084040, published October 19, 2015, Copyright (2015) by the American Physical Society. DOI:https://doi.org/10.1103/PhysRevD.92.084040.

for various events of the type described above are displayed in Figs. 22.4. As we may see by comparing these examples, the gravitational waveform is very dependent on the nature of the objects participating in formation of the wave, and hence should be sensitive to their detailed physics. For example, a discussion of how supernova microphysics influences the form of gravitational waves emitted in a core collapse supernova may be found in

Refs. [157, 233, 234], and Example 22.1 discusses how gravitational waves from neutron star mergers might be used to constrain the equation of state for neutron stars.

Example 22.1 The appropriate equation of state to employ for neutron stars is not very well constrained at present, primarily because it is difficult to measure the radius and mass simultaneously for any single neutron star. This introduces substantial uncertainty into the theoretical understanding of neutron stars. The gravitational waves emitted by the merger of two neutron stars would be sensitive to the properties of the neutron stars at merger and could place stronger constraints than are presently available on the neutron star equation of state. An improved neutron-star equation of state would permit answering more definitively questions like what the upper limit for the mass of a neutron star is (which has implications for the search for black hole candidates in binary star systems; see the discussion in Section 17.4), the superfluid and superconducting properties of neutron stars, the relationship of observed cooling to internal structure for the neutron star, and whether quark matter can exist in the centers of more massive neutron stars. In Section 22.6 below we shall discuss the first observation of gravitational waves (and electromagnetic radiation) from a neutron star merger.

For 100 years after they were first proposed by Einstein, gravitational waves had been a primarily hypothetical issue, with only a few indirect observations indicating their existence. This changed dramatically in late 2015.

22.3 The Gravitational Wave Event GW150914

On September 14, 2015, the two LIGO detectors, one in Livingston, Louisiana, and one in Hanford, Washington, observed simultaneously[2] a transient signal lasting about a quarter of a second that was flagged almost immediately as a strong gravitational wave candidate. Extensive analysis confirmed with significance greater than five standard deviations that the transient labeled GW150914 (with the numbers a reference to the date of discovery) was indeed a gravitational wave that was produced by the merger of two $\sim 30\,M_\odot$ black holes at a distance of more than 400 Mpc [12, 13, 14]. This was a milestone event in general relativity because gravitational waves were the last of Einstein's major predictions not tested by direct observation. But of equal import, in particular for the subject matter of this book, was that GW150914 marked the opening of a new observational window on the Universe for "dark events" that might not be seen easily in traditional astronomy observing modes. As this book goes to press in late 2017, gravitational waves from five confirmed binary black hole mergers and one neutron star binary merger have been reported (see Fig. 17.7 and Section 22.6). Thus prospects appear to be bright for studying stellar evolution through dark events observed by LIGO and future gravitational wave detectors such as Virgo in Italy, which came online in 2017 following an upgrade with capabilities comparable to one of the LIGO detectors.

[2] The detection times were actually separated by about 7 ms because of the finite light (and gravitational wave) travel time between Livingston and Hanford.

22.3.1 Observed Waveforms

The waveforms observed by the LIGO detectors in the GW150914 event are shown in the upper panel of Fig. 22.5. The gravitational wave arrived first at the Livingston detector (L1) and then 6.9 ms later at the Hanford detector (H1). In the top-right image the H1 wave is also shown superposed and shifted by 6.9 ms. The third row displays the result of subtracting a numerical relativity waveform for black hole merger from the observed waveform. The last row shows a time-frequency representation, with the grayscale contours indicating strain. The signal swept upward in frequency from about 35 to 250 Hz ("the chirp," indicative of the final rapid inspiral of a merger event), with a measured peak strain $\sim 1.0 \times 10^{-21}$.

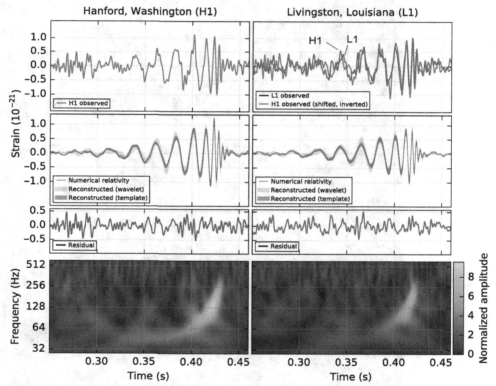

Fig. 22.5 LIGO gravitational wave event GW150914 [12]. Left panels correspond to data from the Hanford detector (H1) and right panels to data from the Livingston detector (L1). Top row is measured strain in units of 10^{-21}. In the top right panel the Hanford signal has been superposed on the Livingston signal. The second row shows numerical relativity simulations [156] of the waveform assuming a binary black hole merger event. The third row shows residuals after subtracting the numerical relativity waveform (second row) from the detector waveform (first row). The fourth row shows frequency versus time for the strain data, with grayscale contours indicating strain amplitude. The rapidly rising pattern (chirp) is indicative of a binary merger. For visualization purposes the data and the simulations have been filtered, as described in more detail in Ref. [12].

22.3.2 The Black Hole Merger

In Fig. 22.6 a computer simulation of what the black holes might have looked like from up close during the merger is shown (at least in a high-speed snapshot; by this point the black holes would have been whirling around each other many times each second). The black, sharply defined objects are shadows of the black holes that block all light

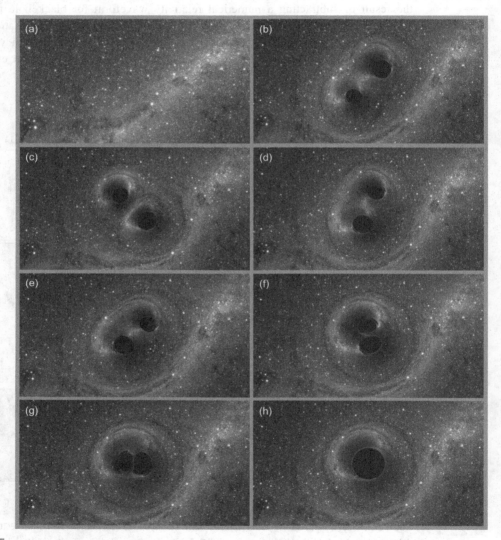

Fig. 22.6 Computer simulation of the GW150914 merger. (a) The undistorted background field of stars in the absence of the black holes. (b)–(g) Successively later times in the merger sequence. (h) The final Kerr black hole. Notice the strong gravitational lensing effects near the black holes. The background stars, of course, are fixed in position as in (a) for each panel, but the gravitational lensing completely distorts their apparent positions. The ring around the black holes is an *Einstein ring*, which results from strong focusing of light from stars behind the black holes by gravitational lensing. Images extracted from video in [73].

Table 22.1 Properties of the black-hole merger event GW150914

Quantity	Value†
Primary black hole mass	$36^{+5}_{-4}\ M_\odot$
Secondary black hole mass	$29^{+4}_{-4}\ M_\odot$
Final black hole mass	$62^{+4}_{-4}\ M_\odot$
Final black hole spin	$0.67^{+0.05}_{-0.07}$
Mass radiated as gravitational waves	$3.0^{+0.5}_{-0.5}\ M_\odot$
Peak gravitational wave luminosity (erg s^{-1})	$3.6^{+0.5}_{-0.4} \times 10^{56}$
Peak gravitational wave luminosity ($M_\odot$ s^{-1})	200^{+30}_{-20}
Source redshift z	$0.09^{+0.03}_{-0.04}$
Source luminosity distance	410^{+160}_{-180} Mpc

†Masses in source frame. Multiply by $(1+z)$, where z is redshift, for mass in detector frame. Spin given in units of spin for an extreme Kerr black hole of that mass; see Eq. (17.14).

from behind. Flattened dark features around them and the marked displacement of the apparent background star images [compare with the undistorted background star field in Fig. 22.6(a)] are caused by pronounced gravitational lensing effects described in the figure caption that arise from the strongly curved space near the event horizons of the black holes.

Extensive analysis comparing simulations of the merger with data measured for the gravitational wave yields quantitative information about the two black holes that merged, and the final Kerr black hole that resulted from the merger. These parameters for GW150914 are displayed in Table 22.1, along with uncertainty estimates that typically are in the 10% to 20% range. For example, the initial masses of the merging black holes were determined to be 36 $M_\odot$ and 29 $M_\odot$, respectively, the mass of the final black hole was 62 $M_\odot$ (implying from the difference of initial and final masses that $\sim$ 3 solar masses were radiated as gravitational waves),[3] and the redshift and corresponding distance to the source were $z = 0.09$ and 410 Mpc, respectively. The spin of the final black hole was determined to be 67% of that for an extremal Kerr black hole, lending support to the conjecture of Section 17.3.5 that near-extremal Kerr black holes may be common. (Spins of the two initial black holes were estimated, but with present limited data the uncertainties were of order 100%.)

The direction to the source was determined also. Since the gravitational wave was observed by only two detectors, tracking the wave back to its source entailed considerable uncertainty. The analysis was able to localize the source to an error box of about 230 square degrees in the Southern Hemisphere near the Large Magellanic Cloud. As more

---

[3] The 3 solar masses were converted to gravitational waves over a period of less than half a second, with a peak gravitational wave luminosity of an astonishing $\sim 200\ M_\odot$ s^{-1}! This translates through $E = mc^2$ to well over 10^{56} erg s^{-1}, which is about 23 orders of magnitude greater than the Sun's photon luminosity and about 5 orders of magnitude brighter than the photon luminosity of a supernova.

gravitational wave observatories come online and a signal can be triangulated from more than two detectors, this uncertainty will be decreased substantially (see an example in Fig. 22.10), but gravitational wave detectors will always have lower intrinsic angular resolution than traditional astronomy instruments. On the other hand, gravitational wave interferometers see essentially the entire sky at all times, not just a narrow field as for traditional telescopes.

22.4 A New Probe of Massive-Star Evolution

Notice from Fig. 17.7 that each of the two initial black holes for GW150914 had at least a factor of two more mass than the most massive black holes that have been inferred from X-ray binary data. Thus GW150914 provided the first conclusive evidence that such massive black holes can exist, that they can occur in binary pairs, and that these binaries can form with sufficiently compact orbits that they can merge within the age of the Universe through gravitational wave emission. Understanding this is likely to have implications for understanding the evolution of massive stars, in particular for those in binary systems.

22.4.1 Formation of Massive Black Hole Binaries

The formation of massive black hole binaries implied by the merger event GW150914 requires a sequence of four events to occur in the course of stellar evolution.

1. Stars must form with very large masses (probably in the vicinity of 100 $M_\odot$).
2. These stars must not lose too much of their mass to stellar winds while evolving to core collapse.
3. These massive stars must collapse to black holes, so they must avoid collapsing to neutron stars and they must avoid being destroyed by the pair instability discussed in Box 20.2.
4. The black holes thus formed must end up as part of a binary star system.

The factors determining whether massive stars collapse to neutron stars or to black holes when they exhaust their core fuel are not completely understood, but most simulations indicate that the more massive a star is the more prone it is to produce a black hole than a neutron star. In addition, the most massive stars are more likely to accrete their envelope rather than eject it when they collapse to black holes. Thus, the formation of 30 or greater solar mass black holes probably involves core collapse of the most massive stars. Two general pictures for the formation of massive binary black holes have been proposed:

1. Evolution of massive galactic binaries in relative isolation from other stars.
2. Capture of single black holes into binary orbits in dense star clusters.

In the first case a binary with two massive stars must form and eventually undergo successive core collapse of each star without unbinding the binary system and without too much mass loss to stellar winds prior to collapse. Specific models in this category often involve one or more accretion episodes and periods of common-envelope evolution. In the second case it is hypothesized that in a dense star cluster black holes formed by core collapse of massive stars quickly will become the most massive objects by merger and accretion, and sink to the dense center of the cluster. There they can form binaries and be ejected from the cluster by dynamical interactions. In either case, the black hole binaries must form with orbits such that inspiral times from gravitational wave emission are less than the age of the Universe. Reviews of these mechanisms may be found in Refs. [40, 174].

22.4.2 Gravitational Waves and Massive Binary Evolution

A discussion of the implications of GW150914 for the binary black hole formation mechanism is given in Ref. [15], where it is concluded that binary black hole masses as large as those found for GW150914 are consistent with a range of models for isolated formation of binary black holes provided that

1. Compositions with metal content much lower than that of the Sun are employed, which allows more massive stars to form.
2. Stellar models with newer prescriptions favoring reduced stellar winds for massive stars are used. The lower winds lead to less mass loss and hence to higher masses at the end of main sequence evolution.

They conclude also that the masses for GW150914 are consistent with models for formation of black hole binaries in dense star clusters if the metal content is less than solar, though further study is needed to determine whether details are consistent in such models.

On the other hand, comprehensive simulations performed after the LIGO discovery of gravitational waves [39, 208] suggest that the binary black holes responsible for the gravitational wave events observed thus far by LIGO could have formed through a single mechanism for formation and merger of isolated binary black holes. There are three crucial assumptions underlying this mechanism:

1. Massive binary stars are formed in low-metallicity environments with initial orbital periods of hundreds of days or less.
2. The subsequent evolution of the binary involves episodes of both Roche-overflow accretion and common-envelope evolution.
3. The cores of the massive stars in the binary collapse directly to black holes, with no ejection of a supernova remnant and with little of the natal kick (see Section 20.3.7) that is common for neutron stars in normal core collapse events. Thus, the black holes form without the parent stars undergoing a supernova explosion in the normal sense, which keeps the binary from unbinding gravitationally.

It is argued that this mechanism gives a much higher merger rate than a mechanism based on capture in dense star clusters, with a predicted rate consistent with the (still highly uncertain) observed black hole merger rate for LIGO events.

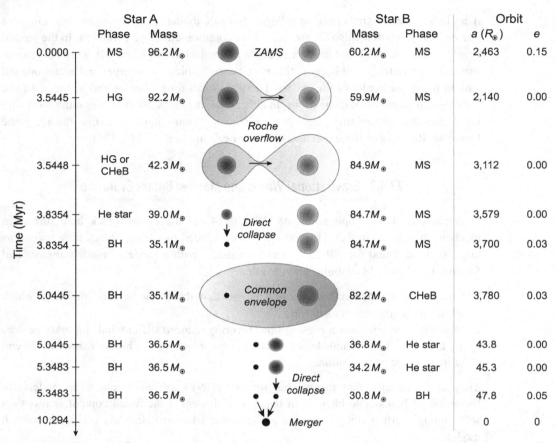

Fig. 22.7 A scenario for evolution of the massive black hole binary leading to GW150914 [39]. ZAMS means zero age main sequence, MS means main sequence, HG means a star evolving through the Hertzsprung gap, CHeB means core helium burning, a He star is a star exhibiting strong He and weak H lines (indicating loss of much of its outer envelope), and BH indicates a black hole. Time is measured from formation of the binary, about 2 billion years after the big bang, and the scale is *highly nonlinear*. The separation of the pair is a and the eccentricity of the orbit is e. Adapted by permission of Springer Nature: *Nature*, **534**(7608), "The First Gravitational-Wave Source from the Isolated Evolution of Two Stars in the 40–100 Solar Mass Range," Bekzynski, K. et al., Copyright (2016).

In Fig. 22.7 a possible specific scenario is sketched for the generation of GW150914 [39]. In this simulation the massive binary formed about 2 billion years after the big bang (redshift $z \sim 3.2$), with initial main sequence masses of $96.2\,M_\odot$ and $60.2\,M_\odot$, respectively, metal fraction of $Z = 0.03\,Z_\odot$, average separation of $a \sim 2500\,R_\odot$, and orbital eccentricity $e = 0.15$. The initially more massive star (A) evolved to fill its Roche lobe and transferred more than half of its mass to the other star (B) by Roche lobe overflow, as the orbit was circularized and star A evolved through the Hertzsprung gap and possibly to core helium burning. The core of star A then collapsed directly to a black hole of mass $35.1\,M_\odot$, with no ejection of a supernova remnant and with minimal natal kick from the collapse, leaving a bound, slightly eccentric orbit. (However, in the formation of black holes by direct collapse in this simulation it was assumed that 10% of the mass is carried off by neutrinos during a collapse.)

Star B had grown by accretion to 84.7 $M_\odot$ by this point and it evolved quickly off the main sequence to core helium burning, initiating a common envelope phase with the black hole formed from star A. During the short common envelope phase the orbit was again circularized and the average separation shrank rapidly from $a \sim 3800\ R_\odot$ to $\sim 45\ R_\odot$. At the end of the common envelope phase the mass of the black hole formed from star A was 36.5 $M_\odot$ and star B was now a helium star of mass 36.8 $M_\odot$, as a consequence of accretion and wind loss. Star B then collapsed directly to a black hole of mass 30.8 $M_\odot$, leaving a binary black hole system with masses of 36.5 $M_\odot$ and 30.8 $M_\odot$, respectively, orbital separation 47.8 $R_\odot$, and eccentricity 0.05. This system then spiraled together through gravitational wave emission over a period of 10.3 billion years, merging about 1.1 billion years ago ($z \sim 0.09$) to produce GW150914.

Note that the timescale on the left vertical axis of Fig. 22.7 is highly nonlinear. It took only about 5 million years from birth of the binary to form two black holes with a separation of about 48 $R_\odot$, but then an additional ~ 10 *billion* years was required for them to merge because of gravitational wave emission. The common envelope phase was instrumental in allowing the merger on a timescale less than the age of the Universe. In this simulation both stars lose their outer envelopes through binary interactions and the final black holes result from direct collapse of massive Wolf–Rayet stars to black holes.

Although the simulations described above are plausible and indicate that massive black hole binaries such as those observed in gravitational wave events to date can form through standard binary evolution channels, they are subject to large uncertainties. In particular, common envelope evolution may be essential to forming massive black hole binaries that merge on observable timescales, and this is the least understood aspect of binary evolution. As more black hole merger events are accumulated by gravitational wave detectors, we may expect increasingly strong constraints on stellar population models for binary star evolution that were not possible before the advent of gravitational wave astronomy.

22.4.3 Formation of Supermassive Black Holes

An important question is whether there is any connection between the formation of stellar-size black holes and the formation of supermassive black holes found often in the centers of galaxies. Two pictures for the formation of supermassive black holes have been proposed.

1. Supermassive black holes may have formed by successive merger of intermediate-mass black holes created by core collapse of massive first-generation stars,[4] or they may have formed directly from the collapse of large clouds.
2. The seeds for the growth of supermassive black holes may instead have been massive (say greater than 25 $M_\odot$) stellar black holes such as those responsible for GW150914.

---

[4] The big bang produced essentially no metals, so the first stars had near-zero metallicity. Simulations indicate that such stars could have had masses as large as hundreds or even thousands of solar masses.

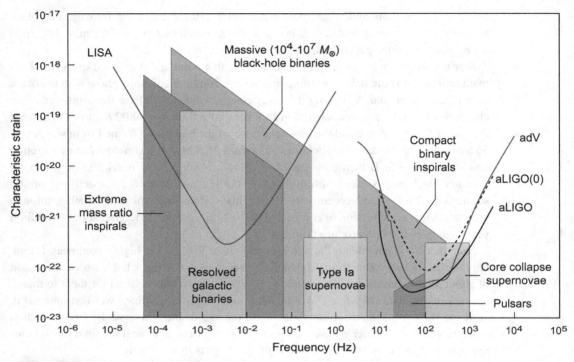

Fig. 22.8 Strain amplitude and frequency ranges expected for gravitational waves from various astronomical sources [153]. Minimum strain detection bounds for advanced LIGO (aLIGO) at full design capacity (∼2020), advanced Virgo (adV) at full design capability (∼2020), advanced LIGO in the first observing run after the upgrade [aLIGO(0), indicated by the dashed curve], during which the gravitational wave GW150914 was observed in 2015, and the proposed space-based array LISA are indicated.

In either case, it is possible that the evolution of massive stars leading to the creation of massive stellar black holes also has implications for the origin of supermassive black holes. The merger of supermassive black holes in galaxy collisions cannot be studied with Earth-based gravitational wave observatories like LIGO because the gravitational wave frequency is too low and the background noise level too high, but they could be studied in large space-based gravitational wave arrays; see Fig. 22.8.

22.5 Listening to Multiple Messengers

The prospects are good for the systematic accumulation of gravitational wave events from binary black hole mergers, binary neutron star mergers, mergers of neutron star–black hole binaries, and core collapse supernovae by present and future gravitational wave observatories. Even more interesting is the possibility of *multimessenger astronomy*, where, for example, a short-period gamma-ray burst might be observed in coincidence with gravitational waves from a neutron star merger, or a neutrino burst observed in coincidence with gravitational waves from the accompanying supernova.[5] With the Advanced Virgo gravitational wave detector now online in Italy, detection of a gravitational wave by three detectors is routine, greatly decreasing the uncertainty in localizing a GW source. However, the situation is likely similar to that of gamma-ray bursts initially, where the error box for the gravitational wave location will be large enough that many galaxies lie within it. Thus, some amount of cleverness may be required to establish a definite correlation between a gravitational wave signal and other signals in the general case.

Implementation of multimessenger astronomy with gravitational waves presents huge technical challenges but could have a large impact on the understanding of neutron-star structure, the core collapse supernova mechanism, and the detailed properties of central engines powering gamma-ray bursts and gravitational wave emission, as well as providing unprecedented tests of general relativity under strong-gravity conditions. Since these events involve various aspects of late stellar evolution, systematic gravitational wave astronomy and multimessenger astronomy have the potential to revolutionize our understanding of how stars evolve. In Section 22.6 we shall describe briefly the first multimessenger event observed: the coincidence of a gravitational wave with a gamma-ray burst and the subsequent electromagnetic transient.[6]

22.6 Gravitational Waves from Neutron Star Mergers

On August 17, 2017 the LIGO–Virgo collaboration detected gravitational wave GW170817, which would be quickly interpreted as originating in the merger of two neutron stars [16]. Approximately 1.7 seconds after the gravitational wave both the Fermi Gamma-ray Space Telescope (Fermi) and the International Gamma-Ray Astrophysics Laboratory (INTEGRAL) observed a gamma-ray burst of two seconds duration in the same direction as the source of the gravitational wave, and that night various observatories alerted to the approximate location of these events discovered a new point source in the galaxy NGC 4993 lying within the position error box for the gravitational wave and gamma-ray burst. In the ensuing weeks a multitude of observatories studied the transient afterglow in NGC 4993 (named officially AT 2017gfo) intensively at various wavelengths. Thus was the discipline of multimessenger gravitational-wave astronomy born (see Section 22.5).

The coincidence of the gravitational wave and the gamma-ray burst is illustrated in Fig. 22.9 and the sky localization of the event is illustrated in Fig. 22.10. The final combined LIGO–Virgo sky position localization corresponded to an uncertainty area of 28 deg^2, illustrating the dramatically improved angular resolution with a third detector

---

[5] SN 1987A may be viewed as a multimessenger astronomy event, since it was observed in neutrinos and at various electromagnetic wavelengths. However, in current usage "multimessenger" typically has come to mean that one of the messengers is a gravitational wave. We will use it in that restricted sense here.

[6] The discussion in Section 22.6 is a synopsis of a more extensive discussion in Ref. [101] and Section 24.7 in Ref. [100].

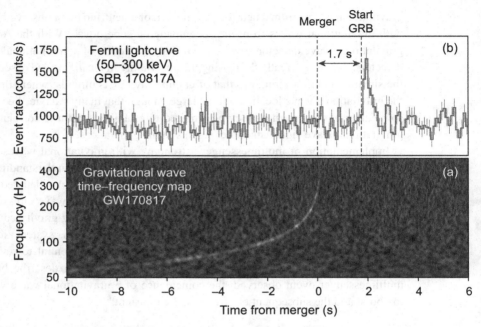

Fig. 22.9 (a) Gravitational wave GW170817 (LIGO) and (b) gamma-ray burst GRB 170817A (Fermi satellite) [17]. The source was at a luminosity distance of 40 Mpc (130 Mly).

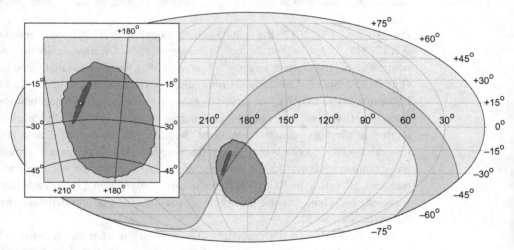

Fig. 22.10 Localization of gravitational wave GW170817 and gamma-ray burst GRB 170817A [17]. The 90% contour for LIGO–Virgo localization is shown in the darkest gray. The 90% localization for the gamma-ray burst is shown in intermediate gray. The 90% annulus from triangulation using the difference in GRB arrival time for Fermi and INTEGRAL is the lighter gray band. The zoomed inset shows the location of the transient AT 2017gfo (small white star). Axes correspond to right ascension and declination in the equatorial coordinate system.

augmenting the LIGO detectors. Rapid localization of the gravitational wave source and the coincidence with the gamma-ray burst allowed electromagnetic observers to locate quickly the afterglow about 2 kpc from the center of the lenticular/elliptical galaxy NGC 4993, in the southern constellation Hydra. The location of the afterglow is indicated by the small white star in the error box of Fig. 22.10. The luminosity distance was 40^{+8}_{-14} Mpc, which is consistent with the known distance to the host galaxy NGC 4993.

22.6.1 New Insights Associated with GW170817

The multimessenger nature of GW170817 led to a number of discoveries having fundamental importance in astrophysics, the physics of dense matter, gravitation, and cosmology:

Viability of multimessenger gravitational-wave astronomy: The event confirmed that gravitational wave detectors could see and distinguish events that did not correspond to merger of two black holes, which had been the interpretation of all previous gravitational waves, and demonstrated that electromagnetic signals could be detected in coincidence with a confirmed gravitational wave event.

Mechanism for short-period GRBs: The interpretation of the gravitational wave as originating in the merger of binary neutron stars and the coincident (short-period) gamma-ray burst provided the first conclusive evidence that short-period gamma-ray bursts are produced in the merger of neutron stars (see Section 21.9). The gamma-ray burst was relatively weak, which was interpreted provisionally as evidence that the gamma-ray burst was not aimed directly at Earth. (See the discussion of ultrarelativistic beaming for gamma-ray bursts in Section 21.4.2.) Confirmation came two weeks later when radio waves and X-rays characteristic of a gamma-ray burst were detected.

Site of the r-process: The signature of heavy-element production in the event demonstrated that neutron star mergers are one (or perhaps the dominant) source of the rapid neutron capture or r-process thought to make many of the heavy elements (see Fig. 20.21 and the discussion in Section 20.5). Now we have a quantitative way to investigate the relative importance of the two primary candidate sites for the r-process: core collapse supernovae, and neutron star mergers, but already it is clear that the once-common view that the r-process occurs primarily in core-collapse supernovae is probably not correct.

Observation of a kilonova: The expanding radioactive debris was observed at UV, optical, and IR wavelengths, giving the first direct evidence for the *kilonova* (also termed a *macronova*) predicted to occur as a result of radioactive heating by newly synthesized r-process nuclei. That the gamma-ray burst was emitted off-axis may have been essential in allowing the kilonova associated with the radioactivity of heavy elements produced in the merger to be observed, as will be described in Fig. 22.11.

Nuclei far from stability: The r-process runs far to the neutron-rich side of the β-stability valley in the chart of the isotopes shown in Fig. 20.21. The lightcurves for the kilonova are a statistical blend of contributions from many neutron-rich nuclei with no sharp lines because of the high velocities (as large as $v \sim 0.3c$) for the ejecta. However, they carry

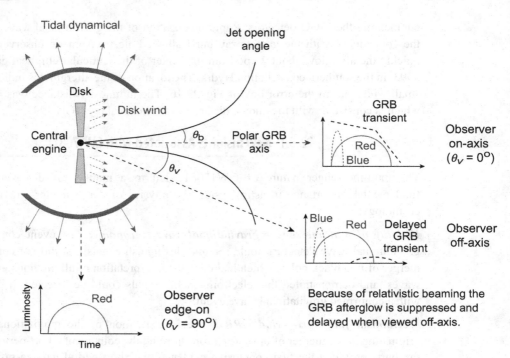

Fig. 22.11 Geometry of GW170817 afterglows [131, 170, 216]. The neutron-rich ejected matter labeled "Tidal dynamical" emits a kilonova peaking in the IR (solid arrows and solid curves labeled "Red" in the time–luminosity diagrams) associated with production of heavy r-process nuclei and high opacity (the *red kilonova*). Additional mass is emitted by winds along the polar axis (dotted arrows and dotted curves labeled "Blue") that is processed by neutrinos emitted from the hot central engine, giving matter less rich in neutrons and a kilonova peaking in the optical that is associated with production of light r-process nuclides and lower opacity (the *blue kilonova*). The usual GRB afterglow is indicated by dashed curves in the plots. It dominates all other emission when viewed on-axis but when viewed off-axis it appears as a low-luminosity component delayed by days or weeks (until $\theta_v < \theta_b$), which permits the kilonova to be seen.

information about the average decay rates and other general properties of these largely unknown r-process nuclei that could provide future constraints on theories of nuclear structure far from β stability.

The speed of gravity: General relativity predicts that gravity propagates at the speed of light. Arrival of the GRB within 1.7 seconds of the gravitational wave from a distance of 40 Mpc established conclusively that the difference of the speed of gravity and the speed of light is no larger than 3 parts in 10^{15} [17]. Thus, alternatives to general relativity for which gravity does not propagate at c are now excluded.

Neutron-star equation of state: The multimessenger nature of the event indicates that neutron star mergers will provide an opportunity to make much more precise statements about the neutron-star equation of state because – for example – the merger wave signature is sensitive to the tidal deformability of the neutron star matter.

Demographics of neutron-star binaries: The observation provides quantitative information about the probability that neutron star binaries form in orbits that can lead to merger in a time less than the age of the Universe. The rate currently inferred corresponds to 0.8×10^{-5} mergers per year in a galaxy the size of the Milky Way, but this number should become more precise with future neutron star merger observations. This has large implications for our understanding of stellar evolution, and also for the expected rate of gravitational wave detection from such events.

Determination of the Hubble constant: Multimessenger gravitational-wave astronomy provides an independent way to determine the Hubble constant H_0 governing the rate of expansion of the Universe, by comparing the distance inferred from the gravitational wave signal with the redshift of the electromagnetic signal [198]. Analysis of the GW170817 multimessenger event suggests a value $H_0 \sim 70^{+12}_{-8}$ km s^{-1}Mpc^{-1} [18], which is consistent with the value from other methods but the error bars are large. It has been estimated that 100 independent gravitational wave detections with host galaxy identified as in GW170817 would reduce the error in H_0 to about 5% [18].

Off-axis gamma-ray bursts: The initial observation of the kilonova followed two weeks later by observation of X-ray and radio emission provides corroborating evidence for the highly beamed nature of gamma-ray bursts [216] and represents the first clear detection of a weak, off-axis GRB and its slowing in the interstellar medium. Systematic studies of such events should greatly enrich our understanding of gamma-ray bursts.

We will not discuss all of these in any detail but let us elaborate further on the kilonova powered by the production of radioactive r-process nuclei.

22.6.2 The Kilonova Associated with GW170817

Simulations of neutron star mergers identify two mechanisms for mass ejection [131]. (1) On millisecond timescales, matter may be expelled dynamically by tidal forces during the merger itself and as surfaces come into contact shock heating may squeeze matter into the polar regions. (2) On a longer ($\sim$ 1 second) timescale, matter in an accretion disk around the merged objects can be blown away by winds. Heavy elements may be synthesized by the r-process in the ejected material. If the matter is highly neutron-rich, repeated neutron captures form the *heavy r-process nuclei* ($58 \leq Z \leq 90$), while if the ejecta is less neutron-rich, *light r-process nuclei* ($28 \leq Z \leq 58$) are formed. The matter in the tidal tails is neutron-rich and tends to form heavy r-process nuclei. The disk winds and ejecta squeezed dynamically into the polar regions may be subject to neutrino irradiation from the central engine, which converts some neutrons to protons, making the winds less neutron-rich and favoring the light r-process.

The photon opacity of the r-process ejecta is generated largely by transitions between bound atomic states (*bound–bound transitions*). For light r-process nuclei the valence electrons typically fill atomic d shells but a substantial fraction of heavy r-process species (often 1–10% by mass) are *lanthanides* ($58 \leq Z \leq 71$), for which valence electrons fill the f shells. These have densely spaced energy levels and an order of magnitude more line transitions than for the d shells in light r-process species. As a consequence, the opacity of heavy r-process nuclei is roughly a factor of 10 larger than the corresponding opacity for light r-process species, and they have correspondingly long photon diffusion times [131].

Hence the cloud of light r-process species is considerably less opaque with shorter diffusion times, and tends to radiate in the optical and fade over a matter of days. In contrast, the cloud of heavy r-process species radiates in the IR with lightcurves that may last for weeks because of the high opacity and long diffusion times. This accounts for the observed characteristics of the transient AT 2017gfo, which differed essentially from all other astrophysical transients that have been observed: it brightened quickly in the optical and then faded but a rapidly-growing IR emission remained strong for weeks, and only after a period of weeks did X-ray and RF signals begin to emerge.

The above considerations suggest a general picture of the geometry of GW170817 that is sketched in Fig. 22.11. The kilonova transient AT 2017gfo that followed the gravitational wave GW170817 and associated gamma-ray burst had two distinct components. The tidal dynamical ejection flung out on millisecond timescales very neutron-rich matter at high velocities $v \sim 0.3c$ that underwent extensive neutron capture to produce heavy r-process species and extremely high opacity because of the lanthanide content. In addition, winds ejected matter from the disk region on a timescale of seconds. This matter was subject to irradiation by neutrinos from the hot center, which increased the proton to neutron ratio.

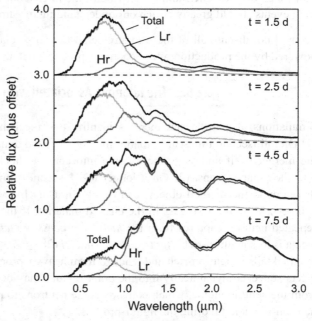

Fig. 22.12 Evolution of different components of the GW170817 kilonova [131]. The total flux is a sum of two spatially separated components: the dominantly optical emission from light r-process isotopes (the "blue kilonova," labeled Lr) and the dominantly infrared emission from heavy r-process isotopes (the "red kilonova," labeled Hr). Adapted by permission of Springer Nature: *Nature*, **551**(7678), "Origin of the Heavy Elements in Binary Neutron-Star Mergers from a Gravitational-Wave Event, Kasen, D. et al. Copyright (2017).

Nucleosynthesis in this less neutron-rich matter was likely to produce light r-process matter of lower opacity, since there weren't enough neutrons to produce lanthanides and other heavy r-process nuclei.

This picture is supported by model calculations that are displayed in Fig. 22.12. These simulations exhibit clearly the early emergence and rapid decay of the optical component associated with the light r-process (the *blue kilonova*), followed by the longer-lived IR component associated with the heavy-r process (the *red kilonova*). The color evolution, spectral continuum shape, and IR spectral peaks of this composite model resemble the observed time evolution of AT 2017gfo. In the model suggested by Fig. 22.11, the kilonova components are visible only because the gamma-ray burst was seen off-axis, which suppressed the GRB afterglow because of relativistic beaming so that the underlying kilonova could be seen.

22.7 Gravitational Wave Sources and Detectors

Let us conclude with a brief overview of the prospects for detecting gravitational waves in different frequency ranges from astrophysical sources. Amplitude and frequency ranges for operating and proposed gravitational wave observatories, along with corresponding ranges expected for some important astrophysical sources of gravitational waves, were illustrated earlier in Fig. 22.8. Earth-based detectors like LIGO are prime instruments for elucidating the physics of neutron stars, black holes, and core collapse supernovae. Space-based arrays could probe the gravitational waves emitted from merger of supermassive black holes in the collisions of galaxies, but also those emitted from ordinary binary stars within the galaxy. The proposed space-based LISA array of Fig. 22.8 fell victim to budgetary constraints but variations are being resurrected under the leadership of the European Space Agency.

Background and Further Reading

See Ref. [100] for a more extensive discussion of gravitational waves and GW150914. See also the LIGO discovery paper [12], the comments on the discovery paper in [42], and the paper on astrophysical implications of the discovery [15] for an overview of GW150914. Numerical relativity references include Refs. [23, 34, 35, 37, 156], the supernova mechanism and emission of gravitational waves are discussed in Refs. [129, 136, 137, 150, 157, 162, 234], and neutron star mergers are reviewed in [83]. Short-period gamma-ray bursts are reviewed in Berger [41] and an overview of the r-process in neutron star mergers may be found in Thielemann, Eichler, Panov, and Wehmeyer [213].

Problems

22.1 Assume interference arms of length $L = 2$ km for a gravitational-wave detector. At the peak strain of the gravitational wave displayed in Fig. 22.5, how large would be the change in path length for the laser light for one pass through the arm? Compare that with the size of the protons in the atoms making up the detection system (a rough estimate for the radius of a proton is about 10^{-13} cm).***

22.2 In the binary black hole merger leading to the gravitational wave event GW150914, the total mass of the black holes was $\sim 70\,M_\odot$ and the observed frequency at peak amplitude was 150 Hz for the gravitational wave. Assuming for a rough estimate that the merger orbits were still described approximately by Keplerian trajectories at peak frequency, what is the lower limit on the sum of the Schwarzschild radii for the two colliding black holes and what was the separation between centers for the two black holes when emitting gravitational waves at peak frequency? *Hint*: Because of symmetry, for gravitational wave emission from a binary the gravitational wave frequency is *twice* the orbital frequency for the binary pair.

Appendix A Constants

Fundamental constants

Gravitational constant: $G = 6.67408 \times 10^{-8}$ dyn cm^2 g^{-2}
$= 6.67408 \times 10^{-8}$ g^{-1} cm^3 s^{-2}
$= 6.67408 \times 10^{-8}$ erg cm g^{-2}
$= 2.960 \times 10^{-4}\, M_\odot^{-1}$ AU^3 days^{-2}
$= 1.327 \times 10^{11}\, M_\odot^{-1}$ km^3 s^{-2}

Speed of light: $c = 2.99792458 \times 10^{10}$ cm s^{-1}

Planck's constant: $h = 2\pi\hbar = 6.6261 \times 10^{-27}$ erg s
$= 4.136 \times 10^{-21}$ MeV s
$\hbar = 1.0546 \times 10^{-27}$ erg s $= 6.5827 \times 10^{-22}$ MeV s
$\hbar c = 197.3$ MeV fm $= 197.3 \times 10^{-13}$ MeV cm

Electrical charge unit: $e = 4.8032068 \times 10^{-10}$ esu
$= 4.8032068$ erg$^{1/2}$cm$^{1/2}$
$= 4.8032068$ g$^{1/2}$ cm$^{3/2}$s^{-1}

Fine structure constant: $\alpha = (137.036)^{-1} = 0.0073$

Weak (Fermi) constant: $G_\text{F} = 8.958 \times 10^{-44}$ MeV cm^3
$= 1.16637 \times 10^{-5}$ GeV^{-2} $[G_\text{F}/(\hbar c)^3;\, \hbar = c = 1]$

Mass of electron: $m_\text{e} = 9.1093898 \times 10^{-28}$ g
$= 5.4858 \times 10^{-4}$ amu
$= 0.5109991$ MeV/c^2

Mass of proton: $m_\text{p} = 1.6726231 \times 10^{-24}$ g
$= 1.00727647$ amu
$= 938.27231$ MeV/c^2

Mass of neutron: $m_\text{n} = 1.6749286 \times 10^{-24}$ g
$= 1.0086649$ amu
$= 939.56563$ MeV/c^2

Atomic mass unit (amu): $M_u = 1.6605390 \times 10^{-24}$ g $= 931.49411$ MeV/c^2

Avogadro's constant: $N_\text{A} = 6.0221409 \times 10^{23}$ mol^{-1}

Boltzmann's constant: $k = 1.38065 \times 10^{-16}$ erg K^{-1}
$= 8.617389 \times 10^{-5}$ eV K^{-1}

Ideal gas constant: $R_\text{gas} \equiv N_\text{A} k = 8.314511 \times 10^7$ erg K^{-1} mole^{-1}

Stefan–Boltzmann constant: $\sigma = 5.67051 \times 10^{-5}$ erg cm^{-2} K^{-4} s^{-1}

Radiation density constant: $a \equiv 4\sigma/c = 7.56591 \times 10^{-15}$ erg cm^{-3} K^{-4}
$= 4.7222 \times 10^{-9}$ MeV cm^{-3}K^{-4}

Planck mass: $M_P = 1.2 \times 10^{19}$ GeV/c^2

Planck length: $\ell_P = 1.6 \times 10^{-33}$ cm

Planck time: $t_P = 5.4 \times 10^{-44}$ s

Planck temperature: $T_P = 1.4 \times 10^{32}$ K

Solar quantities

Solar (photon) luminosity: $L_\odot = 3.828 \times 10^{33}$ erg/s

Solar absolute magnitude $M_v = 4.83$

Solar bolometric magnitude $M_{\text{bol}}^\odot = 4.74$

Solar mass: $M_\odot = 1.989 \times 10^{33}$ g

Effective surface temperature: $T_\odot^{\text{eff}} = 5780$ K

Solar radius: $R_\odot = 6.96 \times 10^{10}$ cm

Central density: $\rho_\odot^{\text{core}} \simeq 160$ g/cm^3

Central pressure: $P_\odot^{\text{core}} \simeq 2.7 \times 10^{17}$ dyn cm^{-2}

Central temperature: $T_\odot^{\text{core}} \simeq 1.6 \times 10^7$ K

Color indices: $B - V = 0.63 \qquad U - B = 0.13$

Solar constant: 1.36×10^6 erg cm^{-2} s^{-1}

General quantities

1 tropical year (yr) = 3.1556925×10^7 s = 365.24219 d

1 parsec (pc) = 3.0857×10^{18} cm = $206,265$ AU = 3.2616 ly

1 lightyear (ly) = 9.4605×10^{17} cm

1 astronomical unit (AU) = 1.49598×10^{13} cm

Energy per gram from H $\rightarrow$ He fusion = 6.3×10^{18} erg/g

Thomson scattering cross section: $\sigma_T = 6.652 \times 10^{-25}$ cm^2

Mass of Earth $M_\oplus = 5.98 \times 10^{27}$ g

Radius of Earth $R_\oplus = 6.371 \times 10^8$ cm

Useful conversion factors

1 eV = $1.60217733 \times 10^{-12}$ ergs = $1.60217733 \times 10^{-19}$ J

1 J = 10^7 ergs = 6.242×10^{18} eV

1 amu = $1.6605390 \times 10^{-24}$ g

1 fm = 10^{-13} cm

0 K = -273.16 Celsius

1 atomic unit (a_0) = 0.52918×10^{-8} cm

1 atmosphere (atm) = 1.013250×10^6 dyn cm^{-2}
1 pascal (Pa) = 1 N m^{-2} = 10 dyn cm^{-2}
1 arcsec = $1''$ = 4.848×10^{-6} rad = 1/3600 deg
1 Å = 10^{-8} cm
1 barn (b) = 10^{-24} cm^2
1 newton (N) = 10^5 dyn
1 watt (W) = 1 J s^{-1} = 10^7 erg s^{-1}
1 gauss (G) = 10^{-4} tesla (T)
1 g cm^{-3} = 1000 kg m^{-3}
Opacity units: 1 m^2 kg^{-1} = 10 cm^2 g^{-1}

Conversion between normal and geometrized units: see Appendix B.

Appendix B Natural Units

In astrophysics it is common to use the CGS (centimeter–gram–second) system of units. However, in many applications it is more convenient to define new sets of units where fundamental constants such as the speed of light or the gravitational constant may be given unit value. Such units are sometimes termed *natural units* because they are suggested by the physics of the phenomena being investigated. For example, the velocity of light c is clearly of fundamental importance in problems where special relativity is applicable. In that context, it is far more natural to use c to set the scale for velocities than to use an arbitrary standard (such as the length of some king's foot divided by a time unit that derives from the apparent revolution of the heavens!) that has arisen historically in nonrelativistic science and engineering. Defining a set of units where c takes unit value is equivalent to making velocity a dimensionless quantity that is measured in units of c, as illustrated below, thus setting a "natural" scale for velocity.

The introduction of a natural set of units has the advantage of more compact notation, since the constants rescaled to unit value need not be included explicitly in the equations, and the standard "engineering" units like CGS may be restored easily by dimensional analysis if they are required to obtain numerical results. This appendix outlines the use of such natural units for problems encountered in astrophysics.

B.1 Geometrized Units

In gravitational physics it is useful to employ a natural set of units called *geometrized units* or $c = G = 1$ units that give both the speed of light and the gravitational constant unit value. Setting

$$1 = c = 2.9979 \times 10^{10} \text{ cm s}^{-1} \qquad 1 = G = 6.6741 \times 10^{-8} \text{ cm}^3 \text{ g}^{-1} \text{ s}^{-2}, \tag{B.1}$$

one may solve for standard units like seconds in terms of these new units. For example, from the first equation

$$1 \text{ s} = 2.9979 \times 10^{10} \text{ cm}, \tag{B.2}$$

and from the second

$$1 \text{ g} = 6.6741 \times 10^{-8} \text{ cm}^3 \text{ s}^{-2}$$
$$= 6.6741 \times 10^{-8} \text{ cm}^3 \left(\frac{1}{2.9979 \times 10^{10} \text{ cm}}\right)^2$$
$$= 7.4261 \times 10^{-29} \text{ cm}. \tag{B.3}$$

So both time and mass have the dimension of length in geometrized units. Likewise, one may derive from the above relations

$$1 \text{ erg} = 1 \text{ g cm}^2 \text{ s}^{-2} = 8.2627 \times 10^{-50} \text{ cm} \tag{B.4}$$

$$1 \text{ g cm}^{-3} = 7.4261 \times 10^{-29} \text{ cm}^{-2} \tag{B.5}$$

$$1 M_\odot = 1.477 \text{ km}, \tag{B.6}$$

and so on. Velocity is dimensionless in these units since cm s^{-1} is dimensionless (that is, v is measured in units of v/c).

In geometrized units, all explicit instances of G and c are dropped in the equations. When quantities need to be calculated in standard units, appropriate combinations of c and G must be reinserted to give the right standard units for each term. For example, in geometrized units the Schwarzschild radius for a spherical black hole is

$$r_S = 2M,$$

so both sides of this equation have dimensions of length in geometrized units. What is the Schwarzschild radius of the Sun in normal units? The result is obtained by inspection since from above the mass of the Sun is 1.477 km in geometrized units. Thus,

$$r_S^\odot = 2M_\odot = 2 \times 1.477 \text{ km} = 2.95 \text{ km}.$$

More formally, to convert this equation to CGS units note that $r_S = 2M$ implies that the right side must be multiplied by a combination of G and c having the units of cm g^{-1} to make it dimensionally correct in the CGS system. This clearly requires the combination G/c^2, so in CGS units the Schwarzschild radius is

$$r_S = \frac{2GM}{c^2}.$$

Inserting the mass of the Sun in grams and the CGS values for G and c then gives the same answer as found above:

$$r_S^\odot = \frac{2(6.6741 \times 10^{-8} \text{ cm}^3 \text{ g}^{-1} \text{ s}^{-2})(1.989 \times 10^{33} \text{ g})}{(2.9979 \times 10^{10} \text{ cm s}^{-1})^2} = 2.95 \times 10^5 \text{ cm} = 2.95 \text{ km}.$$

As a second example, the escape velocity from the radial coordinate R outside a spherical black hole is

$$v_{esc} = \sqrt{\frac{2M}{R}}$$

in geometrized units. Thus, the escape velocity at the event horizon ($R = r_S = 2M$) is

$$v_{esc}(R = r_S) = \sqrt{\frac{2M}{2M}} = 1$$

(that is, $v = c$). In the CGS system velocity has the units of cm s^{-1}, so to convert to the CGS system the right side of the above equation must be multiplied by a combination of G and c to give this dimensionality. The required factor is clearly $\sqrt{G}$ and

$$v_{esc} = \sqrt{\frac{2GM}{R}}$$

in CGS units. Working the preceding problem in these units,

$$v_{\text{esc}}(R = r_{\text{S}}) = \sqrt{\frac{2GM}{R}} = \sqrt{\frac{2GM}{2GM/c^2}} = c,$$

which is the same result as before.

B.2 Natural Units in Particle Physics

In relativistic quantum field theory the explicit role of gravity in the interactions can be ignored (except on the Planck scale), but the equations expressed in standard units are populated by a multitude of the fundamental constants c (expressing the importance of special relativity) and $\hbar$ (expressing the importance of quantum mechanics). It is convenient in this context to define natural units where $\hbar = c = 1$. Using the notation $[a]$ to denote the dimension of a and using $[L]$, $[T]$, and $[M]$ to denote the dimensions of length, time, and mass, respectively, for the speed of light c,

$$[c] = [L][T]^{-1}. \tag{B.7}$$

Setting $c = 1$ then implies that $[L] = [T]$, and since $E^2 = p^2 c^2 + M^2 c^4$,

$$[E] = [M] = [p] = [k] \tag{B.8}$$

where $p = \hbar k$. Furthermore, because

$$[\hbar] = [M][L]^2[T]^{-1} \tag{B.9}$$

one has

$$[M] = [L]^{-1} = [T]^{-1} \tag{B.10}$$

if $\hbar = c = 1$. These results then imply that $[M]$ may be chosen as the single independent dimension of our set of $\hbar = c = 1$ natural units. This dimension is commonly measured in MeV (10^6 eV) or GeV (10^9 eV). Useful conversions are

$$\hbar c = 197.3 \text{ MeV fm} \quad 1 \text{ fm} = \frac{1}{197.3} \text{ MeV}^{-1} = 5.068 \text{ GeV}^{-1}$$
$$1 \text{ fm}^{-1} = 197.3 \text{ MeV} \quad 1 \text{ GeV} = 5.068 \text{ fm}^{-1}. \tag{B.11}$$

where 1 fm = 10^{-13} cm (one fermi or one femtometer). For example, the Compton wavelength of the pion is

$$\lambda_\pi = \frac{1}{M_\pi} \simeq (140 \text{ MeV})^{-1}$$

in $\hbar = c = 1$ units. This may be converted to standard units through

$$\lambda_\pi = \left(\frac{1}{140} \text{ MeV}^{-1}\right) \times 197.3 \text{ MeV fm} = 1.41 \text{ fm},$$

where Eq. (B.11) was used.

Appendix C Mean Molecular Weights

In this appendix the formula for the mean molecular weight quoted in Section 3.4.2 is derived. The fraction by number f_i of each ionic species in the gas is

$$f_i = \frac{n_i^I}{\sum_j n_j} = \frac{\rho N_A Y_i}{\rho N_A \sum_j Y_j} = \frac{Y_i}{\tilde{Y}}, \tag{C.1}$$

where n_i^I is the ionic number density in Eq. (3.18) and $\tilde{Y} \equiv \sum_j Y_j$. The average mass μ_I of an ion in the gas in atomic units is then

$$\mu_I = \sum_i f_i A_i = \sum_i \frac{Y_i A_i}{\tilde{Y}} = \frac{\sum_i X_i}{\tilde{Y}} = \tilde{Y}^{-1}, \tag{C.2}$$

where Eq. (3.18) and $\sum_i X_i = 1$ have been used. Upon ionization, the average number of electrons produced per ion is $\bar{z} = \sum_i f_i y_i Z_i$, where y_i is the *fractional ionization* of the species i ($y_i = 0$ for no ionization and $y_i = 1$ if the species i is completely ionized). Thus the electron number density n_i^e associated with ionization of the species i is given by

$$n_i^e = y_i Z_i n_i^I = \rho N_A y_i Z_i Y_i. \tag{C.3}$$

The *total number of particles* produced by the ionization of an average atom i is then one for the residual ion plus $\bar{z}$ for the ionized electrons and

$$\begin{aligned}
1 + \bar{z} &= 1 + \sum_i f_i y_i Z_i = 1 + \tilde{Y}^{-1} \sum_i y_i Z_i Y_i \\
&= \tilde{Y}^{-1} \sum_i Y_i + \tilde{Y}^{-1} \sum_i y_i Z_i Y_i \\
&= \tilde{Y}^{-1} \sum_i (1 + y_i Z_i) Y_i,
\end{aligned} \tag{C.4}$$

where the identity $1 = \tilde{Y}/\tilde{Y} = \sum_i Y_i/\tilde{Y}$ has been employed in the second line. Then the average mass of a particle (atoms, ions, and electrons) in the gas is given by

$$\mu = \frac{\mu_I}{1 + \bar{z}} = \frac{\tilde{Y}^{-1}}{\tilde{Y}^{-1} \sum_i (1 + y_i Z_i) Y_i} = \left(\sum_i (1 + y_i Z_i) Y_i \right)^{-1}, \tag{C.5}$$

where the contribution of electrons to the mass has been ignored and in the final expression the first term is contributed by the ions and the second by electrons.

With this formalism the actual gas (a mixture of electrons and different atomic, possibly molecular, and ionic species) has been replaced with a gas containing a single kind of fictitious particle having an effective mass μ (often termed the *mean molecular weight*) that is given by Eq. (3.19). As discussed in Section 3.11.1, in very hot stars the momentum and energy density carried by photons is non-trivial and this will further modify the effective mean molecular weight of the gas.

Appendix D Reaction Libraries

This appendix outlines two parameterizations and associated libraries for the nuclear reaction rates required in practical stellar physics calculations (see Section 5.8). A more comprehensive discussion of these and other rate compilations, and tools for dealing with reaction rates, may be found in Ref. [11].

D.1 Caughlan–Fowler Rates

The Gamow peak is not quite gaussian (the peak is asymmetric about the maximum with a high energy tail) so the result (5.31) is only approximately correct. By using expansions to characterize the deviation from gaussian behavior of the realistic curve and allow a slow variation of $S(E)$ around $S(E_0)$, correction terms may be derived for the simple result (5.31) that give a more accurate representation of the thermally averaged cross section (see Problem 5.6 and the discussion in chapter 4 of Ref. [188], for example). One parameterization that incorporates such correction terms and is used often in reaction rate compilations is

$$\langle \sigma v \rangle = a \left(f_0 + f_1 T^{1/3} + f_2 T^{2/3} + f_3 T + f_4 T^{4/3} + f_5 T^{5/3} \right) \frac{e^{-bT^{-1/3}}}{T^{2/3}}, \qquad (D.1)$$

where a, b, and f_n are parameters. The nonresonant rates for the Caughlan and Fowler compilation [88] used in some problems have been parameterized in this manner.

D.2 The ReacLib Library

Another commonly used parameterization of the nuclear reaction rates important in astrophysics is the ReacLib Library [182], which is a systematic compilation of measured rates and theoretical rates where data are not available. At the time of this writing the ReacLib library contains the parameters necessary to compute rates for more than 60,000 reactions, with the rates expected to be reliable in most cases over a range of temperatures from 10^7 to 10^{10} K. Many rates used in this book have been derived from the ReacLib library.

Class	Reaction	Description and examples
1	a → b	β-decay or e^- capture: $^{13}N \to {}^{13}C + e^+ + \nu$
2	a → b + c	Photodisintegration: $(\gamma +)\ ^{28}Si \to {}^{24}Mg + \alpha$
3	a → b + c + d	Example: $^{12}C \to 3\alpha$ (Inverse triple-α)
4	a + b → c	Capture: $p + {}^{12}C \to {}^{13}N\ (+\gamma)$
5	a + b → c + d	Exchange reactions: $^1H + {}^{13}C \to {}^4He + {}^{10}Be$
6	a + b → c + d + e	Example: $^2H + {}^7Be \to {}^1H + {}^4He + {}^4He$
7	a + b → c + d + e + f	Example: $^3He + {}^7Be \to {}^1H + {}^1H + {}^4He + {}^4He$
8	a + b + c → d (+ e)	Effective 3-body reactions: $3\alpha \to {}^{12}C$

Table D.1 Reaction classes in the ReacLib library

D.2.1 Reaction Classes

In ReacLib reactions are sorted into eight reaction classes according to the number of *nuclear species* before and after the reaction, as illustrated in Table D.1. Only nuclear species are displayed explicitly in the labeling. For example, the photodisintegration $\gamma + {}^{28}Si \to {}^{24}Mg + \alpha$ is classified as a → b + c (reaction class 2) because the photon is ignored in the labeling. Likewise neutrinos, antineutrinos, electrons, and positrons are suppressed in the reaction labeling. In Table D.1 classes 1–3 are termed 1-body reactions, classes 4–7 are termed 2-body reactions, and class 8 consists of 3-body reactions, according to the number of nuclear species on the left side of the reaction equation.

Expressed in nuclear physics notation, the various reactions correspond to the vectors in the neutron–proton (N–P) plane illustrated in Fig. D.1. For example, a (p, α) reaction generally decreases the proton number by one and decreases the neutron number by two for the "target" nucleus: in $p + {}^{15}N \to \alpha + {}^{12}C$, which can be expressed compactly as $^{15}N(p,\alpha)^{12}C$, the proton number of ^{15}N (with seven protons and eight neutrons) is decreased by one and the neutron number decreased by two, converting it into ^{12}C with six protons and six neutrons, and an α-particle is emitted.

D.2.2 Parameterization of Rates

The rates in ReacLib are then expressed in terms of a sum over components R_k

$$R = \sum_k R_k, \tag{D.2}$$

where each component R_k has a temperature dependence that is parameterized in the form

$$R_k = \exp\left(p_1 + \frac{p_2}{T_9} + \frac{p_3}{T_9^{1/3}} + p_4 T_9^{1/3} + p_5 T_9 + p_6 T_9^{5/3} + p_7 \ln T_9\right), \tag{D.3}$$

where T_9 is the temperature in units of 10^9 K and the parameters p_n are tabulated constants. In some cases a single component R_k is sufficient to parameterize a rate over the full

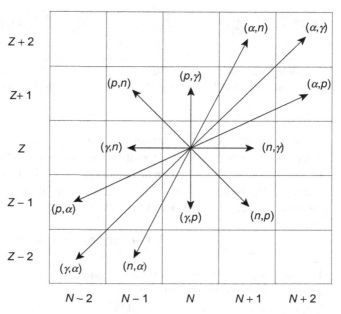

Fig. D.1 Vectors describing some nuclear reactions on an isotope with proton number Z and neutron number N that may play a role in stellar structure and evolution studies.

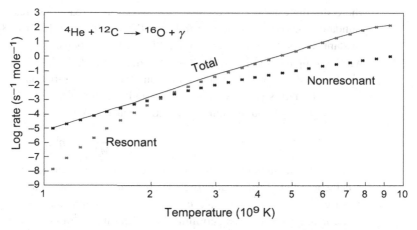

Fig. D.2 A reaction rate expressed in terms of two components in the ReacLib parameterization of Eqs. (D.2)–(D.3). Parameters for the two components of this reaction are given in Table D.2.

temperature range of interest, but in many cases several terms are required in the sum over R_k to adequately parameterize the temperature dependence. If there are two terms, typically one is labeled as resonant and the other as nonresonant, but this is sometimes only a formal distinction in the compilation and not necessarily a statement about reaction mechanism.

An example of a reaction rate parameterized by two components is shown in Fig. D.2.

Table D.2 ReacLib parameters for the reaction $^{12}C(\alpha, \gamma)^{16}O$

	p_1	p_2	p_3	p_4	p_5	p_6	p_7
Nonresonant:	18.4977	0.0048	−33.2522	3.3352	−0.7017	0.0782	−2.8075
Resonant:	142.191	−89.1608	2204.35	−2380.31	108.931	−5.3147	1361.18

The corresponding ReacLib parameters for the two components are shown in Table D.2. Notice the extremely large temperature dependence exhibited for the rate plotted in Fig. D.2: a change of one order of magnitude in T yields a change in the reaction rate of seven orders of magnitude. This illustrates the necessity of a parameterization that can accommodate very large changes of rates with temperature.

D.2.3 Rates in Reaction Networks

The coupled set of ordinary differential equations describing the evolution of nuclear species in stellar burning environments that was described schematically in Box 6.1 is often expressed in the specific form [119]

$$\dot{Y}_i = \sum_j \eta^i_j \lambda_j Y_j + \sum_{jk} \eta^i_{jk} \rho N_A \langle jk \rangle Y_j Y_k + \sum_{jk\ell} \eta^i_{jk\ell} \rho^2 N_A^2 \langle jk\ell \rangle Y_j Y_k Y_\ell, \quad (D.4)$$

where the three terms represent 1-, 2-, and 3-body reactions, respectively, that alter the abundance of species i, the abundance Y_i is defined in Eq. (3.18), N_A is Avogadro's constant, λ_j is a 1-body decay rate, $\langle jk \rangle$ is the velocity-averaged 2-body reaction rate [with a compact notation $\langle ij \rangle \equiv \langle \sigma v \rangle_{ij}$] defined in Eq. (5.18), $\langle jk\ell \rangle$ is a corresponding velocity-averaged 3-body reaction rate, and the η factors are constants that account for the signs of the terms (positive for sources and negative for sinks for Y_i) and for counting particles (without double counting identical particles). Specifically,

$$\eta^i_j = \text{sgn} \times N_i \qquad \eta^i_{jk} = \text{sgn} \times \frac{N_i}{M!} \qquad \eta^i_{jk\ell} = \text{sgn} \times \frac{N_i}{M!}, \quad (D.5)$$

where N_i is the number of particles of species i created or destroyed in a single reaction, M is the number of identical particles in the entrance channel, and sgn is $+1$ if the reaction increases Y_i and -1 if it decreases Y_i. The denominator factors $M!$ prevent overcounting of interacting particles in determining the reaction rate,[1] and the numerator factors N_i keep track of how many particles of species i are created or destroyed in each reaction.

The sums in Eq. (D.4) represent contributions from all reactions that can increase or decrease Y_i, with the reactions distinguished by the indices. The parameterized rates R defined in Eq. (D.2) generally correspond to

---

[1] The product $n_\alpha n_X v \sigma_{\alpha\beta}(v)$ is the rate per unit volume for the 2-body reaction (5.12) and $n_\alpha n_X$ is the number of unique particle pairs (α, X) contained in the unit volume. But for the collision of identical particles (so that $\alpha = X$), the number of independent particle pairs (α, α) is not $N_\alpha(N_\alpha - 1) \simeq N_\alpha^2$ but $N_\alpha^2/2$. Therefore, for two identical particles the rate expression must be multiplied by a factor of $1/(1 + \delta_{\alpha X}) = 1/2$ to avoid double counting. More generally, one finds that for M identical particles a factor $1/M!$ is required to prevent overcounting.

1. $\lambda_j \equiv 1/\tau_j$ (where τ_j is the decay or photodisintegration mean life) in units of s^{-1} for 1-body reactions,
2. $N_A \langle jk \rangle$ in units of cm^{3}mol^{-1}s^{-1} for 2-body reactions, and
3. $N_A^2 \langle jk\ell \rangle$ in units of cm^{6}mol^{-2}s^{-1} for 3-body reactions

in the terms of Eq. (D.4). The following example illustrates explicitly equations (D.4) for a small thermonuclear network.

Example D.1 Consider a simple network including only the three species $\{\alpha, {}^{12}\text{C}, {}^{16}\text{O}\}$ connected by the four reactions $3\alpha \rightleftarrows {}^{12}\text{C}$ and $\alpha + {}^{12}\text{C} \rightleftarrows {}^{16}\text{O} + \gamma$. (There are other reactions that these species can undergo, but for simplicity let us assume only these four to be operative.) There are three equations (D.4) corresponding to the index $i = \{\alpha, {}^{12}\text{C}, {}^{16}\text{O}\}$,

$$\dot{Y}_\alpha = -\tfrac{1}{2}\rho^2 3R_{3\alpha} Y_\alpha^3 + 3R_{12} Y_{12} - \rho 3R_{\alpha 12} Y_\alpha Y_{12} + 3R_{16} Y_{16}$$

$$\dot{Y}_{12} = \tfrac{1}{2}\rho^2 3R_{3\alpha} Y_\alpha^3 - 3R_{12} Y_{12} - \rho 3R_{\alpha 12} Y_\alpha Y_{12} + 3R_{16} Y_{16}$$

$$\dot{Y}_{16} = \rho 3R_{\alpha 12} Y_\alpha Y_{12} - 3R_{16} Y_{16}$$

where a compact notation $Y_\alpha \equiv Y_{{}^4\text{He}}, Y_{12} \equiv Y_{{}^{12}\text{C}}, \ldots$ has been employed, the index $\alpha 12$ is shorthand for the reaction $\alpha + {}^{12}\text{C} \to {}^{16}\text{O} + \gamma$, the index 3α is shorthand for the reaction $3\alpha \to {}^{12}\text{C}$, and the quantities

$$3R_{3\alpha} \equiv N_A^2 \langle \sigma v \rangle_{3\alpha \to {}^{12}\text{C}} \qquad 3R_{\alpha 12} \equiv N_A \langle \sigma v \rangle_{\alpha + {}^{12}\text{C} \to {}^{16}\text{O}}$$

$$3R_{12} = \lambda_{12} \equiv 3R_{{}^{12}\text{C} \to 3\alpha} \qquad 3R_{16} = \lambda_{16} \equiv 3R_{{}^{16}\text{O} \to \alpha + {}^{12}\text{C}}$$

are the reaction rates for the corresponding reactions evaluated from Eqs. (D.2) and (D.3). These rates, and a numerical solution of the preceding equations for constant temperature $T_9 = 5$ and density $\rho = 10^8$ g cm^{-3}, are shown in the following figure.

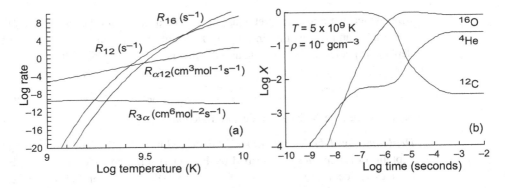

This example exhibits many features found in more realistic thermonuclear networks (which can contain as many as thousands of species with tens of thousands of reactions): it contains 1-body, 2-body, and 3-body reactions, and the equations are coupled and nonlinear. For example, from the three differential equations the change in abundance of any isotope depends non-trivially on the abundance of other isotopes in the network.

In very simple cases the coupled set of nonlinear ordinary differential equations defined in Eq. (D.4) may be solved analytically but generally they must be solved numerically. Their solution presents serious computational issues associated with the rate parameters in the equations often differing by many orders of magnitude. Numerically, equations with this property are said to be *stiff* and their efficient solution requires special numerical techniques that are beyond the scope of our discussion. An overview of the methods used to solve these equations numerically may be found in Refs. [26, 99, 119, 215].

D.2.4 Example: Mean Life for a Species

It is sometimes useful to consider the mean life for destruction of a species through a reaction or set of reactions by generalizing the law of radioactive decay.

Radioactive decay: The change in abundance Y with time caused by decay of some radioactive species is governed by the differential equation

$$\frac{dY}{dt} = -\lambda Y = -\frac{1}{\tau} Y = -\frac{\ln 2}{t_{1/2}} Y, \tag{D.6}$$

where λ is the *decay constant*, the *mean life* denoted by τ and *half-life* denoted by $t_{1/2}$ are related to λ by

$$\tau \equiv \frac{1}{\lambda} \qquad t_{1/2} \equiv \frac{\ln 2}{\lambda} = (\ln 2)\,\tau, \tag{D.7}$$

and $\ln 2 \sim 0.693$ denotes the natural (base-e) logarithm of 2. For radioactive decay these are all independent of time so the solution of Eq. (D.6) is a decaying exponential,

$$\frac{Y(t)}{Y_0} = e^{-\lambda t} = e^{-t/\tau} \qquad Y_0 \equiv Y(t=0), \tag{D.8}$$

from which we may interpret λ as the rate of decay, τ as the time for Y/Y_0 to fall by a factor e^{-1}, and $t_{1/2}$ as the time for Y/Y_0 to fall by a factor $\frac{1}{2}$. If N different decay paths deplete a species, the decay rate and the inverse mean life are additive

$$\lambda_{\text{total}} = \lambda_1 + \lambda_2 + \cdots + \lambda_N \qquad \frac{1}{\tau_{\text{total}}} = \frac{1}{\tau_1} + \frac{1}{\tau_2} + \cdots + \frac{1}{\tau_N}, \tag{D.9}$$

provided that the N decay modes are independent.

Depletion by 1-body reactions: The preceding considerations can be generalized to depletion of a species i described by Eq. (D.4). First consider a single 1-body term that reduces Y_i according to $\dot{Y}_i = -\lambda_i Y_i = -(1/\tau_i) Y_i$. Two general classes of reactions would give an equation of this sort:

1. Radioactive decay of species i; for example, the β-decay $^{17}\text{F} \to {}^{17}\text{O} + \beta^+ + \nu_e$, which corresponds to reaction class 1 in Table D.1.

2. Photodisintegration of species i; for example $\gamma + {}^{28}\text{Si} \to {}^{4}\text{He} + {}^{24}\text{Mg}$ which corresponds to reaction class 2 in Table D.1. (This is considered 1-body because there is only one nuclear species on the left side and the photon number isn't tracked since it is not conserved.)

The first case has already been discussed. The second case looks formally like radioactive decay except that τ_i is no longer necessarily constant because the rate of photodisintegration depends on the temperature, which may change with time. Nevertheless, if the photodisintegration rate does not change too rapidly with time it is useful to define an approximate mean life for decay by photodisintegration $\tau_i = \lambda_i^{-1}$ set by Eq. (D.4).

Example D.2 From Fig. 6.9 the rate-determining step $\gamma + {}^{28}\text{Si} \to {}^{4}\text{He} + {}^{24}\text{Mg}$ in silicon burning has a rate $\lambda \simeq 10^{-4}\,\text{s}^{-1}$ at $T_9 = 3$, implying a mean life for silicon photodisintegration of $\tau = \lambda^{-1} \sim 10^4$ s at that temperature. In the simulation of silicon burning at constant $T_9 = 3$ in Fig. 6.10 the ${}^{28}\text{Si}$ begins to be noticeably depleted at $t \sim 10^4$ s (about 10% of the original silicon has been consumed by that time). In this simulation reactions such as ${}^{4}\text{He} + {}^{24}\text{Mg} \to {}^{28}\text{Si} + \gamma$ are included that replenish ${}^{28}\text{Si}$, so its mass fraction does not fall to e^{-1} of the initial value until $t \sim 4 \times 10^6$ s. Nevertheless, this example illustrates that $\tau \sim 10^4$ s is a useful rough estimate of the timescale for silicon burning under these conditions.

Depletion by 2-body reactions: Next, consider depletion of species i by a 2-body reaction $i + k \to X$, where X denotes some number of isotopic products. From Eq. (D.4) the depletion of i by this single reaction will be governed by the differential equation

$$\dot{Y}_i = \eta_{ik}^i \rho N_\text{A} \langle ik \rangle Y_i Y_k = -\left(\frac{N_i}{M!} \rho R_{ik} Y_k\right) Y_i \equiv -\frac{1}{\tau_{ik}} Y_i, \quad (\text{D.10})$$

where N_i is the number of times species i appears on the left side of the reaction equation, M is the number of identical particles on the left side of the reaction equation, and R_{ik} is the reaction rate in units of cm^3 mol^{-1} s^{-1} for $i + k \to X$ determined from Eqs. (D.2) and (D.3). The quantity

$$\tau_{ik} = \frac{1}{(N_i/M!)\rho R_{ik} Y_k} \quad (\text{D.11})$$

is not constant because ρ, R_{ik}, and Y_k will evolve with time. But if τ_{ik} varies slowly it may be interpreted as an approximate mean life for depletion of i by the reaction $i + k \to X$.

Example D.3 Let $Y_i = Y_\text{p}$ denote the proton abundance and consider how it changes because of the single reaction $p + p \to \beta^+ + \nu_e + d$. In Eq. (D.10) one has $i = k = \text{p}$ and $N/M! = 2/2! = 1$, and sgn $= -1$ since the reaction reduces Y_p, so

$$\dot{Y}_\text{p} = -(\rho R_{\text{pp}} Y_\text{p}) Y_\text{p} \equiv -\frac{1}{\tau_{\text{pp}}} Y_\text{p} \quad (\text{D.12})$$

where R_{pp} is Eq. (D.2) evaluated for the reaction $p + p \to \beta^+ + \nu_e + d$. Thus τ_{pp} may be interpreted as the approximate mean life for destruction of a proton in the reaction $p + p \to \beta^+ + \nu_e + d$. This relation is used in Section 6.1.1 to estimate the timescale for hydrogen burning in the PP chains.

D.3 Representative Applications

Rates computed from reaction libraries are useful for a quantitative understanding of various stellar processes that are discussed in the text. For example, the rates compared in Example 5.7 to examine the competition between β-decay in the CNO cycle and the breakout reaction ^{13}N$(p,\gamma)^{14}$O leading to the hot-CNO cycle, the CNO-cycle network integrations shown in Box 6.1, Fig. D.3 illustrating some rates important in the advanced burning stages discussed in Section 6.5, and Fig. D.4 displaying some competing α-capture and photodisintegration rates important in silicon burning, were all computed using rates from the ReacLib library.

The α-capture rates and photodisintegration rates shown in Fig. D.4 have different units, so to compare them (for example, to examine their competition in silicon burning) they must be converted to the same units. In general, for the reactions $a+b \to c$ and $c \to a+b$, assuming only these reactions to contribute, Eq. (D.4) may be used to write

$$\dot{Y}_a = \dot{Y}_b = -\dot{Y}_c = \eta_a^c \lambda_c Y_c + \eta_{ab}^a \rho N_A \langle \sigma v \rangle_{ab} Y_a Y_b$$
$$= \lambda_c Y_c - \rho N_A \langle \sigma v \rangle_{ab} Y_a Y_b$$
$$= R_c Y_c + R_{ab} \rho Y_a Y_b,$$

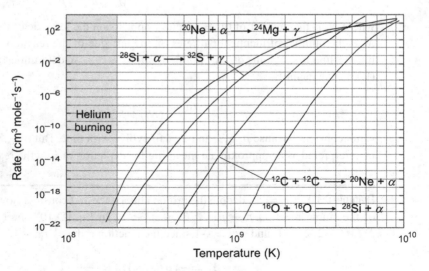

Fig. D.3 Some rates important in advanced burning stages for stellar evolution.

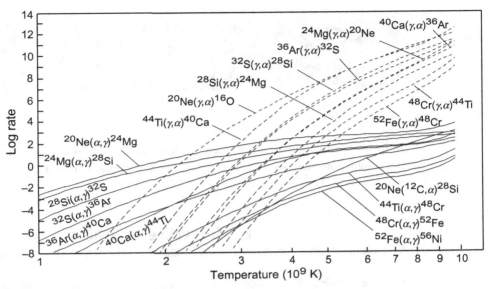

Fig. D.4 Some rates for competing capture reactions $A(\alpha, \gamma)B$ and photodisintegration reactions $A(\gamma, \alpha)B$ that are important for silicon burning. Photodisintegration rates are in units of s^{-1} and α-capture rates are in units of $cm^3\, mol^{-1}\, s^{-1}$.

where R_c is computed from Eq. (D.3) for $c \rightarrow b + c$ and is equal to λ_c, and R_{ab} is computed from Eq. (D.3) for $a + b \rightarrow c$ and is equal to $N_A \langle \sigma v \rangle_{ab}$. Thus to compare the two contributions to $\dot{Y}_i$ on the same footing, one should compare $R_c Y_c$ with $R_{ab} \rho Y_a Y_b$, both of which will have units of s^{-1}. For example, in comparing the relative rates for

$$\alpha + {}^{20}\text{Ne} \rightarrow {}^{24}\text{Mg} + \gamma \qquad {}^{24}\text{Mg} + \gamma \rightarrow \alpha + {}^{20}\text{Ne}$$

in Fig. D.4, the plotted rate for ${}^{24}\text{Mg} + \gamma \rightarrow \alpha + {}^{20}\text{Ne}$ should be multiplied by Y_{24} (where a shorthand notation subscript 24 is used to stand for ${}^{24}\text{Mg}$) and the plotted rate for $\alpha + {}^{20}\text{Ne} \rightarrow {}^{24}\text{Mg} + \gamma$ should be multiplied by $\rho Y_\alpha Y_{20}$ to compare in common units of s^{-1}.

Appendix E A Mixing-Length Model

In this appendix a simple mixing-length model of convection is constructed according to the assumptions described in Section 7.9. The model is based primarily on the discussion in Carrol and Ostlie [68].

E.1 Convective Velocities

Consider a rising blob in a convective region described by an ideal gas equation of state that is assumed to be in pressure equilibrium. By differentiating both sides of the ideal gas law $P = \rho kT/\mu$,

$$\frac{\delta P}{P} = \frac{\delta \rho}{\rho} + \frac{\delta T}{T}, \tag{E.1}$$

where δx indicates the difference in the variable x evaluated inside and outside the blob. But pressure equilibrium between the bubble and surroundings implies that the left side vanishes and

$$\delta \rho \simeq -\rho \frac{\delta T}{T}. \tag{E.2}$$

The buoyancy force per unit volume f acting on the blob is

$$f = -g\delta\rho = g\rho \frac{\delta T}{T}. \tag{E.3}$$

Initially the temperature difference δT between the blob and surroundings is zero, since the blob begins with the same temperature as its environment. Thus, the force averaged over the motion of the blob may be approximated by

$$\bar{f} \simeq \tfrac{1}{2} g\rho \frac{\delta T_{\rm f}}{T}, \tag{E.4}$$

where $\delta T_{\rm f}$ is the final temperature difference between the blob and surroundings. The work W done by the buoyancy force goes into kinetic energy $E_{\rm kin}$ of the blob (any viscous forces are neglected in this discussion). These quantities are given by

$$E_{\rm kin} = \tfrac{1}{2}\rho \bar{v}^2 \qquad W = \bar{f}\ell, \tag{E.5}$$

where ℓ is the mixing length and $\bar{v}$ is the average velocity of the blob. Therefore, equating the kinetic energy and the work gives $\tfrac{1}{2}\rho\bar{v}^2 = \bar{f}\ell$, which may be solved for the average velocity, giving

$$\bar{v} = \sqrt{\frac{2\bar{f}\ell}{\rho}} = \sqrt{g\ell\frac{\delta T_{\mathrm{f}}}{T}}. \tag{E.6}$$

The mixing length ℓ is critical to the entire formulation but has not been specified yet. We may expect that the pressure scale height H_{p} is the most relevant length scale for the overall problem: it determines the distance over which there is a substantial change in gas pressure and the assumption of minimal difference in properties between convective blobs and the surrounding medium would likely not be justified if the mixing length were large measured on this scale. Therefore, we parameterize the mixing length in units of the scale height,

$$\ell = \alpha H_{\mathrm{p}}, \tag{E.7}$$

where the constant of proportionality α is termed the *mixing-length parameter*. It will be taken as adjustable and we expect that it should be of order unity to be consistent with our assumptions. It will be necessary to justify, after the fact in specific cases, whether this choice implies violation of the other assumptions such as convective velocities being small compared with the local speed of sound. From Eqs. (E.5)–(E.7), (7.35), and (7.46), and using $\delta T \sim \delta(dT/dr)\,dr \sim \delta(dT/dr)\,\ell$,

$$\bar{v} = \frac{\alpha k}{\mu M_u}\sqrt{\frac{T}{g}\,\delta\left(\frac{dT}{dr}\right)} \tag{E.8}$$

for the average convective velocity [68].

E.2 Convective Energy Fluxes

Our primary concern is energy transport, so let us now ask what the convective energy flux associated with the velocity (E.8) would be. The convective flux F_{c} is given by a product of the temperature difference δT between the blob and its surrounding environment, the specific heat at constant pressure c_{p} (because of our pressure equilibrium assumptions), the density ρ, and the average convective velocity $\bar{v}$:

$$F_{\mathrm{c}} = \tfrac{1}{2}\rho\bar{v}c_{\mathrm{p}}\delta T. \tag{E.9}$$

The factor of $\tfrac{1}{2}$ is included to account for the assumption of balanced upward and downward motion of blobs, but such numerical factors will not be very important because of the phenomenological nature of the mixing-length model. Substituting the expression (E.8) for the average convective velocity in Eq. (E.9) gives

$$F_{\mathrm{c}} = \frac{\rho c_{\mathrm{p}} k\alpha}{2\mu M_u}\sqrt{\frac{T}{g}\,\delta\left(\frac{dT}{dr}\right)}\,\delta T. \tag{E.10}$$

But from Equations (E.7) and (7.46),

$$\delta T = \delta\left(\frac{dT}{dr}\right)\ell = \delta\left(\frac{dT}{dr}\right)\alpha H_{\mathrm{p}} = \frac{\alpha k T}{\mu g M_u}\,\delta\left(\frac{dT}{dr}\right), \tag{E.11}$$

and the convective flux is

$$F_c = \tfrac{1}{2}\rho\alpha^2 c_p \left(\frac{k}{\mu M_u}\right)^2 \left[\frac{T}{g}\delta\left(\frac{dT}{dr}\right)\right]^{3/2}. \tag{E.12}$$

Equation (E.12) gives us an approximate expression for the convective flux, but to use it a value of the phenomenological parameter α must be chosen and the difference between the temperature gradient of the blob and its surroundings $\delta(dT/dr)$ must be determined. Solving Eq. (E.12) for this difference gives

$$\delta\left(\frac{dT}{dr}\right) = \frac{g}{T}\left[2\left(\frac{\mu M_u}{k}\right)^2 \frac{F_c}{\rho c_p \alpha^2}\right]^{2/3}. \tag{E.13}$$

If the critical temperature gradient is exceeded the energy transport through a region could involve a combination of radiative and convective transport. Therefore, it is necessary to determine the relative contribution of radiative and convective energy fluxes in convective regions. Assume for the sake of argument that *all* flux is being carried by convection. Then

$$F_c = \frac{L(r)}{4\pi r^2}, \tag{E.14}$$

where $L(r)$ is the luminosity evaluated at the radius r. For this special case of pure convection, substituting (E.14) into (E.13) gives

$$\delta\left(\frac{dT}{dr}\right) = \frac{g}{T}\left[\left(\frac{\mu M_u}{k}\right)^2 \frac{L(r)}{2\pi r^2 \rho c_p \alpha^2}\right]^{2/3}. \tag{E.15}$$

How superadiabatic does the temperature gradient have to be in order for the preceding equation to be correct (that is, for all flux to be carried by convection)? The ratio of the superadiabatic gradient to adiabatic gradient is obtained by dividing Eq. (E.15) by the expression (7.34) for the adiabatic gradient. These equations are used to analyze subsurface convection for the Sun in Example 7.4, where it is found that the temperature gradient in the convective zone is only slightly steeper than the adiabatic gradient. This then justifies the prescription that we adopted in Chapter 7: assume purely radiative transport unless the actual temperature gradient becomes steeper than the adiabatic temperature gradient, in which case the transport is assumed to be purely convective with a temperature gradient equal to the adiabatic gradient.

Appendix F Quantum Mechanics

In modern astrophysics we often have to deal with concepts from quantum mechanics and relativistic quantum field theories. For example, equations of state for dense matter, neutrino oscillations, and the resolution of the solar neutrino problem through the MSW effect are all essentially quantum phenomena, and lie at the heart of many modern topics for which students display keen interest. However, many students at the advanced undergraduate level, particularly those who are not physics majors, have had a basic introduction to quantum concepts in elementary courses but may not yet have had a rigorous course in these matters at the time that they encounter the material in this book. A conceptual summary of quantum mechanics in dense-matter applications has been given in Box 3.6, but this appendix gives a brief overview in more formal terms. It is no substitute for a more substantial course in quantum mechanics and quantum field theory, but may provide some orientation to concepts and terminology for students without a solid quantum background who wish to understand at least the basic ideas for topics like dense-matter equations of state and solar neutrino oscillations. The author has found that even students who have not yet had a full course in quantum mechanics are able to grasp the essence of say neutrino oscillations and the MSW resonance if they understand the basic concepts contained in this appendix.

F.1 Wavefunctions and Operators

Quantum mechanics describes the world using *wavefunctions* ψ and *operators* $\hat{O}$. A useful realization is a matrix representation for the $\hat{O}$ operating on column vectors for ψ and row vectors for the complex conjugate ψ^*. In the general case operators don't commute ($\hat{P}\hat{Q} \neq \hat{Q}\hat{P}$) and those corresponding to observables are *hermitian*: $\hat{O} = \hat{O}^\dagger$, where $\dagger$ denotes Hermitian conjugation: complex-conjugation of elements and transposition of rows and columns in a matrix representation.

F.2 Wave Equations

If there is no time dependence, the wavefunction ψ and associated energy E result from solving a partial differential equation,

$$\hat{H}\psi = E\psi \tag{F.1}$$

(*Schrödinger equation* or *wave equation*), where $\hat{H}$ is the *Hamiltonian operator*. Solutions ψ are termed the *eigenstates* or *eigenvectors* and energies E the *eigenvalues* of the problem. The time-dependent Schrödinger equation is

$$i\hbar \frac{\partial \psi}{\partial t} = \hat{H}\psi \tag{F.2}$$

and its solution defines a *time-evolution operator* $\hat{T}$ that propagates a wavefunction at time $t = 0$ into the corresponding one at time t according to

$$\psi(t) = \hat{T}\psi(0) = \exp(-iEt/\hbar)\psi(t_0), \tag{F.3}$$

where E is the energy. The Schrödinger equation is nonrelativistic but a propagator of the form (F.3) applies also to ultrarelativistic particles like neutrinos if their spin structure is neglected.

F.3 Calculation of Observables

The state is specified by ψ, while an operator $\hat{O}$ applied to ψ extracts information or causes transitions, but ψ and $\hat{O}$ are *not observable*. Observables are related to *matrix elements*

$$M_{ij} \equiv \langle \psi_i | \hat{O} | \psi_j \rangle \equiv \int \psi_i^* \hat{O} \psi_j \, d\tau, \tag{F.4}$$

where τ denotes parameters, and in *Dirac notation* a "ket" $|\psi\rangle$ is a wavefunction, a "bra" $\langle \psi |$ is its complex conjugate, and a "bra–ket" $\langle \psi_i | \cdots | \psi_j \rangle$ implies integration over all variables. For example, the probability that $\hat{O}$ operating on $|\psi_j\rangle$ causes a transition to $|\psi_i\rangle$ is

$$|M_{ij}|^2 = |\langle \psi_i | \hat{O} | \psi_j \rangle|^2. \tag{F.5}$$

A special case is the *overlap*, where $\hat{O}$ is set to the unit operator $\hat{1}$ and

$$M_{ij} = \langle \psi_i | \hat{1} | \psi_j \rangle = \langle \psi_i | \psi_j \rangle. \tag{F.6}$$

Then $\langle \psi_i | \psi_j \rangle$ is the *probability amplitude* that ψ_j is a component of the state ψ_i and $|\langle \psi_i | \psi_j \rangle|^2$ is the probability that it is.

F.4 Unitary Transformations

Unitary transformations are of central importance in quantum mechanics because they preserve all quantum-mechanical observables. Thus everything of consequence that is true about a quantum system remains true after a unitary transformation. In a matrix representation unitary transformations are implemented by *unitary matrices* U, which have the properties that

$$U = U^\dagger \qquad UU^\dagger = U^\dagger U = I, \tag{F.7}$$

where I denotes the unit matrix and the dagger denotes the hermitian conjugate. A unitary transformation for a quantum system then consists of the following simultaneous operations on the operators $\hat{O}$ and wavefunction $|\psi\rangle$

$$\hat{O} \to U\hat{O}U^{\dagger} \qquad |\psi\rangle \to U|\psi\rangle \qquad \langle\psi| \to \langle\psi|U^{\dagger}. \tag{F.8}$$

In the special case that the transformation matrices contain $\hat{O} \equiv U$ only real entries the unitarity condition reduces to the *orthogonality condition*, $O^{\mathsf{T}}O = OO^{\mathsf{T}} = I$, where O^{T} denotes the transpose of O. Hence, unitary transformations are a complex generalization of orthogonal transformations. For example, the transformation matrix in Eq. (11.2) is both unitary and orthogonal (Problem 11.6).

F.5 Quantum Field Theory

If $|\psi\rangle$ is a many-body state it is useful to generalize to a *quantum field theory*. Similar concepts apply but now wavefunctions are *fields* defined over all space and particles are *quanta of the field*. This is an instance of quantum particle–wave duality: the system may be described in terms of fields, or in terms of localized quanta of those fields. If $v \sim c$, Lorentz invariance is required, giving a *relativistic quantum field theory*. The machinery is more complex than for basic quantum mechanics but the principles are the same. In particular, *observables are specified in terms of matrix elements*. Construction of a quantum field theory typically starts with a *Lagrangian density* $\mathscr{L}$, which generalizes the classical Lagrangian to a continuous field. For quantum field theory a graphical representation of matrix elements called *Feynman diagrams* becomes extremely useful (see Box 11.3).

Appendix G Using arXiv and ADS

Some journal articles referenced in this book are published in journals with limited public availability. They will likely be available from university libraries but readers without immediate access to a university library may still be able to access many of these papers free of charge by using the *arXiv preprint server* or the *ADS Astronomy Abstract Service*. Where possible, references to journal articles in the bibliography include sufficient information to allow arXiv and/or ADS to be used to retrieve copies of the articles according to the following instructions.

arXiv access: An arXiv reference will be of the general form *arXiv: xxxx*. Typing the string *xxxx* into the Search or Article-id field at http://arxiv.org and clicking the search icon will return an abstract with links to the article in PDF form, and a more general search on arXiv can be implemented by clicking the Form Interface button.

ADS access: The ADS interface may be found at http://adsabs.harvard.edu/bib_abs.html.

  (i) If a DOI number of the form *DOI: yyyy* is given for a reference, the article often can be accessed through the ADS interface by putting the string *yyyy* into the Bibliographic Code Query box and clicking Send Query.
 (ii) If a BibCode reference *BibCode: zzzz* is given, the article can be accessed through the ADS interface by typing the BibCode string *zzzz* into the Bibliographic Code Query box and clicking Send Query. Alternatively, the BibCode string can be used directly in a Web browser. For example, *BibCode: 1971Natur.232..246B* can be accessed as http://adsabs.harvard.edu/abs/1971Natur.232..246B.
(iii) A search for a general article may be implemented with the ADS interface by giving the Journal Name/Code (there is a link on the page to the list of codes for standard journals; for example *The Astrophysical Journal* is ApJ), Year, Volume, and beginning Page of the article, and clicking Send Query.

Articles in ADS are scanned so the quality is not high, but they are generally quite readable.

Example G.1 Reference [219] corresponds to an article on a Wolf–Rayet star published by Tuthill et al. in the *Astrophysical Journal*. If you do not have access to that journal,

1. Reference [219] also gives an arXiv reference *0712.2111*. Putting 0712.2111 into the Search or Article-id field at http://arxiv.org and clicking the search icon should return an abstract with links to a PDF version of the preprint for the journal article.

2. Ref. [219] also lists the DOI number *DOI: 10.1086/527286*; putting 10.1086/527286 into the Bibliographic Code Query box at http://adsabs.harvard.edu/bib_abs.html and clicking Send Query should return links to a scanned PDF version of the journal article.

Other references that list a DOI, BibCode, or arXiv number may be retrieved in a similar way as for this example.

References

[1] **Journal Access** Some journal articles referenced in this bibliography are published in journals with limited public access. Readers with university affiliations often will have access through the university library but to ensure broad availability for the general reader, references to the *arXiv preprint server* or to the *ADS Astronomy Abstract Service* are included where available for journal articles. Instructions for using these arXiv and ADS references to retrieve journal articles are given in Appendix G.

[2] The SIMBAD Astronomical Database (http://simbad.u-strasbg.fr/simbad/).

[3] http://zebu.uoregon.edu/spectra.html.

[4] http://sci.esa.int/education/35774-stellar-radiation-stellar-types/?fbodylongid=1703.

[5] Eta Carinae General Information (http://etacar.umn.edu/etainfo/basic/).

[6] Particle Data Group at http://pdg.lbl.gov/.

[7] www.ligo.caltech.edu/image/ligo20160615e.

[8] www.aavso.org/v1974-cyg-nova-cygni-1992 (American Association of Variable Star Observers).

[9] "The Physics and Astrophysics of Type Ia Supernova Explosions"; special section of *Frontiers of Physics,* **8**, April, 2013, 111–216, Mike Guidry and Bronson Messer, editors.

[10] http://astrogravs.gsfc.nasa.gov/docs/catalog.html.

[11] A comprehensive compilation of tools and reaction rates relevant for astrophysics may be found at www.nucastrodata.org/datasets.html.

[12] Abbott, B. P., et al. 2016a. *Phys. Rev. Lett.*, **116**, 161102 (DOI: 10.1103/PhysRevLett.116.061102).

[13] Abbott, B. P., et al. 2016b. *Phys. Rev.*, **D93**, 122004 (arXiv: 1602.03843).

[14] Abbott, B. P., et al. 2016c. *Phys. Rev. Lett.*, **116**, 221101 (arXiv: 1602.03841).

[15] Abbott, B. P., et al. 2016d. *Astrophys. J. Lett.*, **818**, L22 (DOI: 10.3847/2041-8205/818/2/L22).

[16] Abbott, B. P., et al. 2017a. *Phys. Rev. Lett.*, **119**, 161101 (arXiv: 1710.05832).

[17] Abbott, B. P., et al. 2017b. *Astrophys. J. Lett.*, **848**, L13 (DOI: 10.3847/2041-8213/aa920c).

[18] Abbott, B. P., et al. 2017c. *Nature*, **551**, 85 (DOI: 10.1038/nature24471).

[19] Abe, S., et al. 2008. *Phys. Rev. Lett.*, **100**, 221803 (DOI: 10.1103/PhysRevLett.100.221803).

[20] Adams, S. M., Kochanek, C. S., Gerke, J. R., Stanek, K. Z., and Dai, X. 2017. *Mon. Not. R. Astr. Soc.*, **468**, 4968 (DOI: 10.1093/mnras/stx816).

[21] Ahmad, Q. R., et al. 2002a. *Phys. Rev. Lett.*, **89**, 011301 (DOI: 10.1103/PhysRevLett.89.011301).

[22] Ahmad, Q. R., et al. 2002b. *Phys. Rev. Lett.*, **89**, 011302 (DOI: 10.1103/PhysRevLett.89.011302).

[23] Alcubierre, M. 2008. *Introduction to 3 + 1 Numerical Relativity*. Oxford University Press.

[24] Annett, J. F. 2004. *Superconductivity, Superfluids and Condensates*. Oxford University Press.

[25] Aprahamian, A., Langanke, K., and Wiescher, M. 2005. *Prog. Part. and Nuc. Phys.*, **54**, 535 (DOI: 10.1016/j.ppnp.2004.09.002).

[26] Arnett, D. 1996. *Supernovae and Nucleosynthesis*. Princeton University Press.

[27] Arnett, W. D., Bahcall, J. N., Kirshner, R. P., and Woosley, S. E. 1989. *Annu. Rev. Astron. Astrophys.*, **27**, 629 (BibCode: 1989ARA&A..27..629A).

[28] Asplund, M., Grevesse, N., Sauval, A. J., and Scott, P. 2009. *Annu. Rev. Astron. Astrophys.*, **47**, 481 (DOI: 10.1146/annurev.astro.46.060407.145222).

[29] Audi, G., Wapstra, A. H., and Thibault, C. 2003. *Nuc. Phys.*, **A729**, 337 (DOI: 10.1016/j.nuclphysa.2003.11.003).

[30] Bahcall, J. N. 1989. *Neutrino Astrophysics*. Cambridge University Press.

[31] Bahcall, J. N. and Pinsonneault, M. 1995. *Rev. Mod. Phys.*, **67**, 781 (DOI: 10.1103/RevModPhys.67.781).

[32] Bahcall, J. N., Pinsonneault, M. H., and Basu, S. 2001. *Astrophys. J.*, **555**, 990 (arXiv: astro–ph/0010346).

[33] Bahcall, J. N. 1969. *Scientific American*, **221**, 28 (DOI: 10.1038/scientificamerican0769-28).

[34] Baker, J. G., Centrella, J., Choi, D., Koppitz, M., and van Meter, J. 2006. *Phys. Rev.*, **D73**, 104002 (arXiv: gr–qc/0602026; DOI: 10.1103/PhysRevD.73.104002).

[35] Baker, J. G., Campanelli, M., Pretorius, F., and Zlochower, Y. 2007. *Class. Quant. Grav.*, **24**, S25 (arXiv: gr–qc/0701016).

[36] Barrat, J. L., Hansen, J. P., and Mochkovitch, R. 1988. *Astron. Astrophys.*, **199**, L15 (BibCode: 1988A&A...199L..15B).

[37] Baumgarte, T. W. and Shapiro, S. L. 2010. *Numerical Relativity: Solving Einstein's Equations on the Computer*. Cambridge University Press.

[38] Beaudet, G., Petrosian, Vahé, and Saltpeter, E. E. 1967. *Astrophys. J.*, **150**, 979 (DOI: 10.1086/149398).

[39] Belczynski, K., Holz, D. E., Bulik, T., and O'Shaughnessy, R. 2016. *Nature*, **534**, 512 (DOI: 10.1038/nature18322; arXiv: 1602.04531).

[40] Benacquista, M. J. and Downing, J. M. B. 2013. *Living Rev. Rel.*, **16**, 4 (DOI: 10.12942/lrr–2013–4).

[41] Berger, E. 2014. *Annu. Rev. Astron. Astrophys.*, **52**, 43 (DOI: 10.1146/annurev–astro–081913–035926).

[42] Berti, E. 2016. *Physics*, **9**, 17.

[43] Bethe, H. 1986. *Phys. Rev. Lett.*, **56**, 1305 (DOI: 10.1103/PhysRevLett.56.1305).
[44] Bethe, H. 1990. *Rev. Mod. Phys.*, **62**, 801 (DOI: 10.1103/RevModPhys.62.801).
[45] Bethe, H. and Wilson, J. R. 1985. *Astrophys. J.*, **295**, 14 (DOI: 10.1086/163343).
[46] Binney, J. and Tremaine, S. 2008. *Galactic Dynamics*. Princeton University Press.
[47] Bionta, R. M., et al. 1987. *Phys. Rev. Lett.*, **58**, 1494 (DOI: 10.1103/PhysRevLett.58.1494).
[48] Blandford, R., and Gehrels, N. 1999. *Physics Today*, **52**, 40 (DOI: 10.1063/1.882697).
[49] Blennow, M. and Smirnov, A. Yu. 2013. *Adv. High Energy Phys.*, **2013**, www.hindawi.com/journals/ahep/2013/972485/.
[50] Blundell, S. J. and Blundell, K. M. 2010. *Concepts in Thermal Physics, 2nd edn.* Oxford University Press.
[51] Bodenheimer, P., Laughlin, G. P., Różyczka, M., and Yorke, H. W. 2007. *Numerical Methods in Astrophysics*. Boca Raton: Taylor and Francis.
[52] Böhm-Vitense, E. 1992. *Introduction to Stellar Astrophysics, Vols. 1–3*. Cambridge University Press.
[53] Bolton, C. T. 1972. *Nature*, **235**, 271 (DOI: 10.1038/235271b0).
[54] Bonolis, L. 2005. *Am. J. Phys.*, **73**, 487 (DOI: 10.1119/1.1852540).
[55] Bowers, R. L. and Deeming, T. 1984. *Astrophysics, Vols. I and II*. Boston: Jones and Bartlett.
[56] Bowyer, S., Byram, E. T., Chubb, T. A., and Friedman, H. 1965. *Science*, **147**, 394 (DOI: 10.1126/science.147.3656.394).
[57] Braes, L. L. E, and Miley, G. K. 1971. *Nature*, **232**, 246 (BibCode: 1971Natur.232..246B).
[58] Brassard, P. and Fontaine, G. 2005. *Astrophys. J.*, **622**, 572 (DOI: 10.1086/428116).
[59] Brown, G. E. 1988. *Phys. Rep.*, **163**.
[60] Bruenn, S. 1993. In Guidry, M. W. and Strayer, M. R. (eds), *Proceedings of the First Symposium on Nuclear Physics in the Universe*. Elsevier.
[61] Buonanno, A., Cook, G. B., and Pretorius, F. 2007. *Phys. Rev.*, **D75**, 124018 (arXiv: gr–qc/0610122).
[62] Burbidge, E. M., Burbidge, G. R., Fowler, W. A., and Hoyle, F. 1957. *Rev. Mod. Phys.*, **29**, 547 (DOI: 10.1103/RevModPhys.29.547).
[63] Burgay, M., et al. 2003. *Nature*, **426**, 531 (DOI: 10.1038/nature02124).
[64] Busso, M., Gallino, R., and Wasserburg, G. J. 1999. *Annu. Rev. Astron. Astrophys.*, **37**, 239 (DOI: 10.1146/annurev.astro.37.1.239).
[65] Cameron, A. G. W. 1957. *Stellar Evolution, Nuclear Astrophysics, and Nucleogenesis*. Atomic Energy of Canada Report: https://fas.org/sgp/eprint/CRL-41.pdf.
[66] Cardall, C. Y. 2007. Supernova Neutrinos, from Back of the Envelope to Supercomputer. arXiv: astro-ph/0701831.
[67] Carpenter, J. R. and Timmermans, M.-L. 2012. *Physics Today*, **65**, 66 (DOI: 10.1063/PT.3.1485).
[68] Carrol, B. W. and Ostlie, D. A. 2007. *An Introduction to Modern Astrophysics*. San Francisco: Pearson/Addison–Wesley.

[69] Caughlan, G. R. and Fowler, W. A. 1988. *Atomic Data Nuc. Data Tables*, **40**, 283 (DOI: 10.1016/0092-640X(88)90009-5).

[70] Champagne, A. and Wiescher, M. 1992. *Annu. Rev. Nucl. Part. Sci.*, **42**, 39 (DOI: 10.1146/annurev.ns.42.120192.000351).

[71] Clayton, D. D. 1983. *Principles of Stellar Evolution and Nucleosynthesis*. University of Chicago Press.

[72] Colgate, S. A. and White, R. A. 1966. *Astrophys. J.*, **143**, 626 (DOI: 10.1086/148549).

[73] Collaboration, SXS. 2016. What the First LIGO Detection Would Look Like Up Close. www.ligo.caltech.edu/video/ligo20160211v3.

[74] Collins, P. D. B., Martin, A. D., and Squires, E. J. 1989. *Particle Physics and Cosmology*. New York: Wiley Interscience.

[75] Cox, A. N. 1999. *Allen's Astrophysical Quantities*. New York: AIP (Springer-Verlag).

[76] Cox, J. P. 1974. *Rep. Prog. Phys.*, **37**, 563 (DOI: 10.1088/0034-4885/37/5/001).

[77] Cox, J. P. 1980. *The Theory of Stellar Pulsations*. Princeton University Press.

[78] de Gouvea, A., Kayser, B., and Mohapatra, R. 2003. *Phys. Rev. D*, **67**, 053004 (DOI: 10.1103/PhysRevD.67.053004).

[79] Dexheimer, V. and Schramm, S. 2008. *Astrophys. J.*, **683**, 943 (DOI: 10.1086/589735).

[80] Eguchi, K., et al. 2003. *Phys. Rev. Lett.*, **90**, 021802 (DOI: 10.1103/PhysRevLett.90.021802).

[81] Einstein, A. 1916. *Sitzungsber. K. Preuss. Akad. Wiss.*, **1**, 688.

[82] Einstein, A. 1918. *Sitzungsber. K. Preuss. Akad. Wiss.*, **1**, 154.

[83] Faber, J. A. and Rasio, F. A. 2012. *Living Rev. Rel.*, **15**, 8 (arXiv: 1204.3858).

[84] Federrath, C. 2018. *Physics Today*, **71 no. 6**, 38 (arXiv: 1806.05132).

[85] Figer, D. F. 2005. *Nature*, **434**, 192 (DOI: 10.1038/nature03293).

[86] Filippenko, A. V. 1997. *Annu. Rev. Astron. Astrophys.*, **35**, 309 (DOI: 10.1146/annurev.astro.35.1.309).

[87] Foster, J. and Nightingale, J. D. 2006. *A Short Course in General Relativity*. New York: Springer.

[88] Fowler, W. A., Caughlan, G. R., and Zimmerman, B. A. 1967. *Annu. Rev. Astron. Astrophys.*, **5**, 525 (DOI: 10.1146/annurev.aa.05.090167.002521).

[89] Frank, J., King, A., and Raine, D. 2002. *Accretion Power in Astrophysics*. Cambridge University Press.

[90] Frebl, A., and Beers, T. C. 2018. *Physics Today*, **71 no. 1**, 30 (arXiv: 1801.01190).

[91] Frieman, J. A., Turner, M. S., and Huterer, D. 2008. *Annu. Rev. Astron. Astrophys.*, **46**, 385 (DOI: 10.1146/annurev.astro.46.060407.145243).

[92] Fukuda, Y., et al. 1998. *Phys. Rev. Lett.*, **81**, 1562 (DOI: 10.1103/PhysRevLett.81.1562).

[93] Gehrels, N., et al. 2005. *Nature*, **437**, 851 (DOI: 10.1038/nature04142).

[94] Gillessen, S., Eisenhauer, F., Fritz, T. K., Bartko, H., Dodds-Eden, K., Pfuhl, O., Ott, T., and Genzel, R. 2009. *Astrophys. J. Lett.*, **707**, L114 (DOI: 10.1088/0004–637X/707/2/L114).

[95] Glendenning, N. K. 1997. *Compact Stars*. Springer.

[96] Gou, L., et al. 2011. *Astrophys. J.*, **742**, 85 (arXiv: 1106.3690; DOI: 10.1088/0004–637X/742/2/85).

[97] Guidry, M. W. 1991. *Gauge Field Theories: An Introduction with Applications*. New York: Wiley Interscience.

[98] Guidry, M. W. 1998. Neutrino Transport and Large-Scale Convection in Core-Collapse Supernovae. In Hirsch, J. G. and Page, D. (eds), *Nuclear and Particle Astrophysics*. Cambridge University Press, p. 115.

[99] Guidry, M. W. 2012. *J. Comp. Phys.*, **231**, 5206 (arXiv: 1112.4778; DOI: 10.1016/j.jcp.2012.04.026).

[100] Guidry, M. W. 2019. *Modern General Relativity: Black Holes, Gravitational Waves, and Cosmology*. Cambridge University Press.

[101] Guidry, M. W. 2018 (DOI: 10.1016/j.scib.2017.11.021). *Sci. Bull.*, **63**

[102] Guidry, M. W. and Billings, J. J. 2018 arXiv: 1812.00035

[103] Guidry, M. W., and Messer, B. 2013. *Front. Phys.*, **8**, 111 (DOI: 10.1007/s11467–013–0317–9).

[104] Hajduk, M., et al. 2005. *Science*, **308**, 231 (DOI: 10.1126/science.1108953).

[105] Hamuy, M., et al. 1996. *Astronomical J.*, **112**, 2408 (DOI: 10.1086/118192).

[106] Handler, G. 2012. *Asteroseismology*. arXiv: 1205.6407.

[107] Hansen, C. J., Kawaler, S. D., and Trimble, V. 2004. *Stellar Interiors: Physical Principles, Structure, and Evolution, 2nd edn*. New York: Astronomy and Astrophysics Library, Springer–Verlag.

[108] Harpaz, A. 1994. *Stellar Evolution*. Wellesley: A. K. Peters.

[109] Hartle, J. B. 2003. *Gravity, An Introduction to Einstein's General Relativity*. Addison-Wesley.

[110] Hartmann, L. 1998. *Accretion Processes in Star Formation*. Cambridge University Press.

[111] Haswell, C. A., Robinson, E. L., Horne, K., Steinig, R. F., and Abbott, T. M. C. 1993. *Astrophys. J.*, **411**, 802 (DOI: 10.1086/172884).

[112] Haxton, W. C. 1986. *Phys. Rev. Lett.*, **57**, 1271 (DOI: 10.1103/PhysRevLett.57.1271).

[113] Haxton, W. C. and Holstein, B. R. 2000. *Am. J. Phys.*, **68**, 15 (DOI: 10.1119/1.19368).

[114] Haxton, W. C. and Holstein, B. R. 2004. *Am. J. Phys.*, **72**, 18 (DOI: 10.1119/1.1619142).

[115] Haxton, W. C., Robertson, R. G. H., and Serenelli, A. M. 2013. *Annu. Rev. Astron. Astrophys.*, **51**, 21 (DOI: 10.1146/annurev–astro–081811–125539).

[116] Hessels, J. W. T., et al. 2006. *Science*, **311**, 1901 (DOI: 10.1126/science.1123430).

[117] Hilditch, R. W. 2001. *An Introduction to Close Binary Stars*. Cambridge University Press.

[118] Hirata, H., et al. 1987. *Phys. Rev. Lett.*, **58**, 1490 (DOI: 10.1103/PhysRevLett.58.1490).
[119] Hix, W. R. and Meyer, B. S. 2006. *Nuc. Phys.*, **A777**, 188 (DOI: 10.1016/j.nuclphysa.2004.10.009).
[120] Hjelling, R. M. and Wade, C. M. 1971. *Astrophys. J. Lett.*, **168**, L21 (DOI: 10.1086/180777).
[121] Hjorth, J., et al. 2003. *Nature*, **423**, 847 (DOI: 10.1038/nature01750).
[122] Ho, W. C. G., Andersson, N., Espinoza, C. M., Glampedakis, K., Haskell, B., and Heinke, C. 2013. *P.S (confinement X)*, **260**. arXiv: 1303.3282.
[123] Hobson, M. P., Efstathiou, G., and Lasenby, A. N. 2006. *General Relativity: An Introduction for Physicists*. Cambridge University Press.
[124] Holberg, J. B., et al. 1998. *Ap. J.*, **497**, 935 (DOI: 10.1086/305489).
[125] Iben, I. 1967. *Ann. Rev. Astron. Ap.*, **5**, 571 (DOI: 10.1146/annurev.aa.05.090167.003035).
[126] Iben, I. 1985. *Quart. J. Roy. Astron. Soc.*, **26**, 1 (BibCode: 1985QJRAS..26....1I).
[127] Iglesias, C. A., Rogers, R. J., and Wilson, B. G. 1992. *Ap. J.*, **397**, 717 (DOI: 10.1086/171827).
[128] Iliadis, C. 2012. *Nuclear Physics of Stars*. Weinheim: Wiley-VCH.
[129] Janka, H.-T. 2012. *Annu. Rev. Nucl. Part. Sci.*, **62**, 407 (DOI: 10.1146/annurev-nucl-102711-094901).
[130] Jr., Icko Iben. 2013. *Stellar Evolution Physics: (I) Physical Processes in Stellar Interiors; (II) Advanced Evolution of Single Stars*. Cambridge University Press.
[131] Kasen, D., Metzger, B., Barnes, J., Quataert, E., and Ramirez-Ruiz, E. 2017. *Nature*, **551**, 80 (DOI: 10.1038/nature24453).
[132] Kayser, B. 1981. *Phys. Rev.*, **D24**, 110 (DOI: 10.1103/PhysRevD.24.110).
[133] King, S. F. and Luhn, C. 2013. *Rep. Prog. Phys.*, **76**, 056201 (DOI: 10.1088/0034-4885/76/5/056201).
[134] Kippenhahn, R., Weigert, A., and Weiss, A. 2012. *Stellar Structure and Evolution, 2nd edn*. Berlin: Springer-Verlag.
[135] Kiziltan, B., Baumgardt, H., and Loeb, A. 2017. *Nature*, **542**, 203 (DOI: doi:10.1038/nature21361; arXiv:1702.02149).
[136] Kotake, K., Ohnishi, N., and Yamada, S. 2007. *Astrophys. J.*, **655**, 406 (DOI: 10.1086/509320).
[137] Kotake, K., Ohnishi, N., and Yamada, S. 2013. *Comptes Rendus Physique*, **14**, 318 (arXiv: 1110.5107).
[138] Kuo, T. K. and Pantaleone, J. 1989. *Rev. Mod. Phys.*, **61**, 937 (DOI: 10.1103/RevModPhys.61.937).
[139] Langanke, K. and Martínez-Pinedo, G. 2003. *Rev. Mod. Phys.*, **75**, 819 (DOI: 10.1103/RevModPhys.75.819).
[140] Lattimer, J. and Prakash, M. 2004. *Science*, **304**, 536 (DOI: 10.1126/science.1090720).
[141] Lee, W. H. 2001. *Mon. Not. R. Astr. Soc.*, **328**, 583 (DOI: 10.1046/j.1365-8711.2001.04898.x).

[142] Leff, H. S. 2002. *Am. J. Phys.*, **70**, 792 (DOI: 10.1119/1.1479743).
[143] Leighton, R. B., Noyes, R. W., and Simon, G. W. 1962. *Astrophys. J.*, **135**, 474 (DOI: 10.1086/147285).
[144] MacFadyen, A. 2004. *Science*, **303**, 45 (DOI: 10.1126/science.1091764).
[145] Marcy, G. W. and Butler, R. P. 1998. *Annu. Rev. Astron. Astrophys.*, **36**, 57 (DOI: 10.1146/annurev.astro.36.1.57).
[146] Marsh, T. R., Robinson, E. L., and Ward, J. H. 1994. *Mon. Not. R. Astr. Soc.*, **266**, 137 (DOI: 10.1093/mnras/266.1.137).
[147] McClintock, J. and Remillard, R. 1986. *Astrophys. J.*, **308**, 110 (DOI: 10.1086/164482).
[148] McClintock, J. E., Petro, L. D., Remillard, R. A., and Ricker, G. R. 1983. *Astrophys. J.*, **266**, L27 (DOI: 10.1086/183972).
[149] Metcalfe, T. S., Mongomery, M. H., and Kanaan, A. 2004. *Astrophys. J.*, **605**, L133 (DOI: 10.1086/420884).
[150] Mezzacappa, A. 2005. *Annu. Rev. Nucl. Part. Sci.*, **55**, 467 (DOI: 10.1146/annurev.nucl.55.090704.151608).
[151] Mikheyev, S. P. and Smirnov, A. Yu. 1986. *Nuovo Cimento*, **9C**, 17 (DOI: 10.1007/BF02508049).
[152] Montgomery, M. H. and Winget, D. E. 1999. *Astrophys. J.*, **526**, 976 (DOI: 10.1086/308044).
[153] Moore, C. J, Cole, R. H., and Berry, C. P. L. 2015. *Class. Quantum Grav.*, **32**, 015014 (arXiv: 1408.0740; DOI: 10.1088/0264–9381/32/1/015014).
[154] Morii, T., Lim, C. S., and Mukherjee, S. N. 2004. *The Physics of the Standard Model*. Singapore: World Scientific.
[155] Moro-Martín, A., Garnavich, P. M., and Noreiga-Crespo, A. 2001. *Astronomical J.*, **121**, 1636 (DOI: 10.1086/319387).
[156] Mroué, A., et al. 2013. *Phys. Rev. Lett.*, **111**, 241104 (DOI: 10.1103/PhysRevLett.111.241104).
[157] Müller, B., Janka, H.-T., and Marek, A. 2013. *Astrophys. J.*, **766**, 43 (DOI: 10.1088/0004–637X/766/1/43).
[158] Nugis, T. and Lamers, H. J. G. L. M. 2000. *Astron. and Astrophys.*, **360**, 227–244 (BibCode: 2000A&A...360..227N).
[159] Oda, M., Gorenstein, P., Gursky, H., Kellogg, E., Schreier, E., Tananbaum, H., and Giaconni, R. 1971. *Astrophys. J.*, **166**, L1 (DOI: 10.1086/180726).
[160] Oke, J. B. 1977. *Astrophys. J.*, **217**, 181 (DOI: 10.1086/155568).
[161] Orosz, J. A., J. E, McClintock, Aufdenberg, J. P., Remillard, R. A., Reid, M. J., Narayan, R., and Gou, L. 2011. *Astrophys. J.*, **742**, 84 (DOI: 10.1088/0004–637X/742/2/84).
[162] Ott, C. D. 2009. *Class. Quantum Grav.*, **26**, 063001 (arXiv: 0809.0695).
[163] Padmanabhan, T. 2001. *Theoretical Astrophysics, Vol. II: Stars and Stellar Systems*. Cambridge University Press.
[164] Page, D., Prakash, M., Lattimer, J. M., and Steiner, A. W. 2011. Superfluid Neutrons in the Core of the Neutron Star in Cassiopeia A. arXiv: 1110:5116.

[165] Page, D., Lattimer, J. M., Prakash, M., and Steiner, A. W. 2013. *Stellar Superfluids*. arXiv: 1302.6626.
[166] Pagel, B. E. J. 1997. *Nucleosynthesis and Chemical Evolution of Galaxies*. Cambridge University Press.
[167] Percy, J. R. 2007. *Understanding Variable Stars*. Cambridge University Press.
[168] Peterson, B. 1997. *An Introduction to Active Galactic Nuclei*. Cambridge University Press.
[169] Phillips, A. C. 1994. *The Physics of Stars*. Chichester: John Wiley and Sons, Ltd.
[170] Pian, E., et al. 2017. *Nature*, **551**, 67 (DOI: 10.1038/nature24298).
[171] Piran, T. 1999. *Phys. Rep.*, **314**, 575 (DOI: 10.1016/S0370-1573(98)00127-6).
[172] Piran, T. 2004. *Rev. Mod. Phys.*, **76**, 1143 (DOI: 10.1103/RevModPhys.76.1143).
[173] Popper, D. M. 1980. *Annu. Rev. Astron. Astrophys.*, **18**, 115 (DOI: 10.1146/annurev.aa.18.090180.000555).
[174] Postnov, K. A. and Yungelson, L. R. 2014. *Living Rev. Rel.*, **17**, 3 (DOI: 10.12942/lrr-2014-3).
[175] Press, W. H., Teukolsky, S. A., Vetterling, W. T., and Flannery, B. P. 2006. *Numerical Recipes, 3rd edn.* Cambridge University Press.
[176] Prialnik, D. 2010. *An Introduction to the Theory of Stellar Evolution and Stellar Structure, 2nd edn.* Cambridge University Press.
[177] Price, D. J. and Rosswog, S. 2006. *Science*, **312**, 719 (DOI: 10.1126/science.1125201).
[178] Qin, Y., et al. 2000. *Publ. Astron. Soc. Jpn.*, **52**, 759 (DOI: 10.1093/pasj/52.5.759).
[179] Raine, D. and Thomas, E. 2005. *Black Holes: An Introduction*. London: Imperial College Press.
[180] Rana, N. C. 1983. *Astron. and Astrophys.*, **184**, 104 (BibCode: 1987A&A...184..104R).
[181] Ransom, S. M., et al. 2014. *Nature*, **505**, 520 (DOI: 10.1038/nature12917).
[182] Rauscher, T. and Thielemann, F. K. 2000. *At. Data Nuclear Data Tables*, **75**, 1 (DOI: 10.1006/adnd.2000.0834).
[183] Reifarth, R, Lederer, C., and Käppeler, F. 2014. *J. Phys. G*, **41**, 053101 DOI: 10.1088/0954-3899/41/5/053101.
[184] Rezzolla, L, et al. 2012. *Astrophys. J. Lett.*, **732**, L6 (arXiv: 1101.4298; DOI: 10.1088/2041-8205/732/1/L6).
[185] Riles, K. 2013. *Prog. Part. Nuc. Phys.*, **68**, 1 (arXiv: 1209.0667).
[186] Robson, I. 1996. *Active Galactic Nuclei*. Chichester: Wiley.
[187] Rogers, F. J., and Iglesias, C. A. 1992. *Ap. J. Suppl.*, **79**, 507 (DOI: 10.1086/191659).
[188] Rolfs, C. E. and Rodney, W. S. 1988. *Cauldrons in the Cosmos*. University of Chicago Press.
[189] Rosswog, S. 2004. *Science*, **303**, 46 (DOI: 10.1126/science.1091767).
[190] Rozsnyai, B.F. 2000. *Solar Opacities*. Preprint UCRL-JC-140161; https://e-reports-ext.llnl.gov/pdf/238718.pdf.
[191] Ruffert, M., Ruffert, H. Th., and Janka, H. Th. 2006. *Astron. Astrophys*, **380**, 544 (arXiv: astro-ph/0106229).

[192] Ryan, S., and Norton, A. 2010. *Stellar Evolution and Nucleosynthesis*. Cambridge University Press.
[193] Ryden, B., and Peterson, B. 2010. *Foundations of Astronomy*. San Francisco: Addison-Wesley.
[194] Salaris, M. and Cassisi, S. 2005. *Evolution of Stars and Stellar Populations*. Chichester: Wiley.
[195] Salpeter, E. E. 1955. *Astrophys. J.*, **121**, 161 (DOI: 10.1086/145971).
[196] Schmidt-Kaler, T. 1982. In Schaifers, K. and Voigt, H. H. (eds), *Landolt-Börnstein: Numerical Data and Functional Relationsips in Science and Technology, New Series, Vol. 2, Astronomy and Astrophysics*, p. 1.
[197] Schödel, R., et al. 2002. *Nature*, **419**, 694 (DOI: 10.1038/nature01121).
[198] Schutz, B. F. 1986. *Nature*, **323**, 310 (DOI: 10.1038/323310a0).
[199] Semenov, V., Dyadechkin, S., and Punsly, B. 2004. *Science*, **305**, 978 (DOI: 10.1126/science.1100638).
[200] Shapiro, S. L. and Teukolsky, S. A. 1983. *Black Holes, White Dwarfs, and Neutron Stars: The Physics of Compact Objects*. New York: Wiley Interscience.
[201] Silbar, R. R. and Reddy, S. 2004. *Am. J. Phys.*, **72**, 892 (DOI: 10.1119/1.1703544).
[202] Silva, D. R. and Cornell, M. E. 1992. *Ap. J. Suppl.*, **81**, 865 (DOI: 10.1086/191706).
[203] Smirnov, A. Yu. 2003. *The MSW Effect and Solar Neutrinos*. arXiv: hep-ph/0305106.
[204] Smith, N. 2014. *Annu. Rev. Astron. Astrophys.*, **52**, 487 (DOI: 10.1146/annurev-astro-081913-040025).
[205] Sobral, D., et al. 2015. *Astrophys. J.*, **808**, 139 (DOI: 10.1088/0004-637X/808/2/139).
[206] Stahler, S. W. and Palla, F. 2004. *The Formation of Stars*. Weinheim: Wiley-VCH.
[207] Stehle, M., Mazzali, P. A., Benetti, S., and Hillebrandt, W. 2005. *Mon. Not. R. Astr. Soc.*, **360**, 1231 (DOI: 10.1111/j.1365-2966.2005.09116.x).
[208] Stevenson, S., et al. 2017. *Nature Comm.*, **8**, 14906 (DOI: 10.1038/ncomms14906; arXiv:1704.01352).
[209] Suntzeff, N. B., Phillips, M. M., Elias, J. H., DePoy, D. L., and Walker, A. R. 1992. *Astrophys. J. Lett.*, **384**, L33 (DOI: 10.1086/186256).
[210] Tauris, T. M. and van den Heuvel, E. P. J. 2014. *Astrophys. J. Lett.*, **781**, L13 (DOI: 10.1088/2041-8205/781/1/L13).
[211] Tayler, R. J. 1994. *The Stars: their Structure and Evolution*. Cambridge University Press.
[212] Taylor, J. H., and Weisberg, J. M. 1982. *Astrophys. J.*, **253**, 908 (DOI: 10.1086/159690).
[213] Thielemann, F.-K., Eichler, M., Panov, I.V., and Wehmeyer, B. 2017. *Annu. Rev. Nucl. Part. Sci.*, **67** (arXiv: 1710.02142).
[214] Tielens, A. G. G. M. 2008. *Annu. Rev. Astron. Astrophys.*, **46**, 289 (DOI: 10.1146/annurev.astro.46.060407.145211).
[215] Timmes, F. X. 1999. *Astrophys. J. Supp.*, **124**, 241 (DOI: 10.1086/313257).
[216] Troja, E., et al. 2017. *Nature*, **551**, 71 (DOI: 10.1038/nature24290).

[217] Troy, T. P. and Ahmed, M. 2015. *Physics Today*, **68**, 62 (DOI: 10.1063/PT.3.2729).

[218] Tuthill, P. J., Monnier, J. D., and Danchi, W. C. 1999. *Nature*, **398**, 487 (DOI: 10.1038/19033; arXiv: astro–ph/9904092).

[219] Tuthill, P. J., Monnier, J. D., Lawrance, N., Danchi, W. C., Owocki, S. P., and Gayley, K. G. 2008. *Astrophys. J.*, **675**, 698 (DOI: 10.1086/527286; arXiv: 0712.2111).

[220] Usov, V. V. 1991. *Mon. Not. R. Astr. Soc.*, **252**, 49 (DOI: 10.1093/mnras/252.1.49).

[221] Wallace, D. J., Moffat, A. F. J., and Shara, M. M. 2002. Hubble Space Telescope Detection of Binary Companions Around Three WC9 Stars: WR 98a, WR 104, and WR 112. Page 407 (BibCode: 2002ASPC..260..407W) of: Moffat, A. F. J., and St-Louis, N. (eds), *Interacting Winds from Massive Stars. ASP Conference Proceedings, Vol. 260*. San Francisco: Astronomical Society of the Pacific.

[222] Waltham, C. 2004. *Am. J. Phys.*, **72**, 742 (DOI: 10.1119/1.1646132).

[223] Webster, B. L. and Murdin, P. 1972. *Nature*, **235**, 37 (DOI: 10.1038/235037a0).

[224] Weiss, A., Hillebrandt, W., Thomas, H.-C., and Ritter, H. 2004. *Cox & Giuli's Principles of Stellar Structure: Extended Second Ed*. Cambridge University Press.

[225] Williams, P. M., van der Hucht, K. A., and Thé, P. S. 1987. *Astron. and Astrophys.*, **182**, 91 (BibCode: 1987A&A...182...91W).

[226] Wilson, J. R. 1985. In Centrella, J., LeBlanc, J., and Bowers, R. (eds), *Numerical Astrophysics*. Boston: Jones and Bartlett, p. 422.

[227] Wilson, J. R. and Mayle, R. W. 1988. *Phys. Rep.*, **163**, 63 (DOI: 10.1016/0370-1573(88)90036–1).

[228] Winget, D. E., Sullivan, D. J., Metcalfe, T. S., Kawaler, S. D., and Montgomery, M. H. 2004. *Astrophys. J. Lett.*, **602**, L109 (DOI: 10.1086/382591).

[229] Wolfenstein, L. 1978. *Phys. Rev.*, **D17**, 2369 (DOI: 10.1103/PhysRevD.17.2369).

[230] Woosley, S. E. and Bloom, J. S. 2006. *Annu. Rev. Astro. Astrophysics*, **44**, 507 (DOI: 10.1146/annurev.astro.43.072103.150558).

[231] Woosley, S. E., Heger, A., and Weaver, T. A. 2002. *Rev. Mod. Phys.*, **74**, 1015 (DOI: 10.1103/RevModPhys.74.1015).

[232] Yakovlev, D. G. and Pethick, C. J. 2004. *Annu. Rev. Astron. Astrophys.*, **42**, 169 (DOI: 10.1146/annurev.astro.42.053102.134013).

[233] Yakunin, K. N., Marronetti, P., Mezzacappa, A., Bruenn, S. W., Lee, C.-T., Chertkow, M. A., Hix, W. R., Blondin, J. M., Lentz, E. J., Messer, O. E. B., and Yoshida, S. 2010. *Class. Quant. Grav.*, **27**, 194005 (arXiv: 1005.0779).

[234] Yakunin, K. N., Mezzacappa, A., Marronetti, P., Yoshida, S., Bruenn, S. W., Hix, W. R., Lentz, E. J., Messer, O. E. B., Harris, J. A., Endeve, E., Blondin, J. M., and Lingerfelt, E. J. 2015. *Phys. Rev.*, **D92**, 084040 (arXiv: 1505.05824).

Index

β-decay, 78, 178
β-stability valley, see valley of stability
η Carinae, 218, 327, 435
30 Doradus, see Tarantula Nebula

absorption
    continuum, 40
    line, 40
abundance, 59
accretion
    accretion-induced phenomena, 415
    disks, 406, 408
    Eddington luminosity, 218
    efficiency, 414
    limiting rate, 413
    nova, 422
    power, 412
    Roche lobe overflow, 372, 373, 401, 406, 408, 421, 487, 488
    streams, 406
    temperature, 413
    wind-driven, 401, 410
    X-ray burst, 425
accretion disks
    as elevators, 409
    as energy-storage reservoirs, 415
    formation, 222, 408
    heating, 408
    internal angular momentum transfer, 222, 409
    luminosities, 409
active galactic nuclei, 416, 469
adiabatic index, 57
ADS astronomy abstract service, 522
advection, 90
AGB stars, 308
    and planetary nebulae, 316
    carbon stars, 318
    deep convection, 308, 314
    dredge-up episodes, 317
    ejection of envelope, 315
    mass loss, 308, 314
    sources of PAHs, 318
    superwind, 317
    thermal pulses, 308
AGN, see active galactic nuclei
Algol paradox, 416

Algol variables (algols), 418
angular size of stars, 13
arXiv preprint server, 522
asteroseismology, 233, 338, 358
AT 2017gfo, 491
atomic mass unit, 55
Avogadro constant, 55

Balmer series, 37, 40
BAT, 464
BATSE, 462
BeppoSAX, 463
binary pulsars, 372
    formation, 372
    Hulse–Taylor binary (Binary Pulsar), 373
    tests of general relativity, 373
binary star system, 14
    accretion, 387, 401
    Algol, 20, 416
    and novae, 421
    binary neutron stars, 372
    Castor, 14
    circularization of orbits, 387, 406
    classification, 405
    common-envelope evolution, 373, 486
    contact binary, 405
    detached binary, 405
    eclipsing, 14, 19, 25
    evolution, 405
    formation of massive, 486
    geometry, 15
    mass function, 387
    primary star, 406
    secondary star, 406
    semidetached binary, 405
    Sirius, 15
    spectroscopic, 17, 387
    tidal interactions, 387, 406
    tilt angle, 18, 387
    true orbit, 18
    visual, 18
    X-ray binaries, 387, 411
binding energy
    Q-values, 108
    curve of, 105, 106
    nuclear, 105, 106, 108
bipolar outflows, 201, 221

Index

black hole central engines
  causality and source size, 390
  for gamma-ray bursts, 465, 469
black holes, 378
  and frame dragging, 386, 474
  and gravitational waves, 392, 396, 478, 482, 487, 488
  and magnetic fields, 386, 474
  as central engines, 470
  as endpoint of stellar evolution, 297, 330, 378, 487, 488
  at center of Milky way (Sgr A*), 392
  candidates in X-ray binaries, 391, 392
  Cygnus X-1, 389, 411
  event horizons, 378
  evidence for stellar-mass black holes, 386, 387
  extremal Kerr, 386, 485
  formation of supermassive, 489
  formed by direct stellar collapse, 331
  Hawking, 393
  imaging event horizons, 394
  in X-ray binaries, 387, 396
  intermediate-mass, 393
  jets, 416
  Kerr, 470
  known masses of, 396
  mini, 393
  natal kick, 450
  negative heat capacity of, 98
  supermassive rotating, 416, 474
blackbodies, 9–11, 13, 78
blue stragglers, 418
Boltzmann equation, 32, 36
Bose–Einstein distribution, 69
BPM 37093, 358
bremsstrahlung radiation, 158, 179
broken symmetries, 255
brown dwarfs, 215

carbon–nitrogen–oxygen cycle, see CNO cycle
cataclysmic variables, 415
causality and source size, 390, 468
Chandrasekhar limit, 303, 350, 352, 353
Cherenkov radiation, 240
chirality, 181, 274
chromosphere, 32
clusters
  age of, 44
  globular, 3, 43
  open or galactic, 3, 43
  turnoff point, 43, 44, 418
CNO cycle, 110–113, 135, 136, 451
collapsar model of gamma-ray bursts, 471
  and long-period bursts, 471
  and Wolf–Rayet stars, 471
  jets, 471
color index, 12
  and surface temperature, 12, 13

B, 12
$B-V$, 12
  of Antares, 13
  of Spica, 13
$U-B$, 12
common envelope evolution, 372, 373, 405, 487–489
Compton Gamma-Ray Observatory (CGRO), 462
Compton scattering, 157, 179
concentration variables
  abundance, 59
  mass density, 59
  mass fraction, 59
  number density, 59
convection
  actual temperature gradients, 170, 189
  adiabatic temperature gradient, 167, 189
  and helium core burning, 306
  convective fluxes, 517
  convective instability, 163
  convective velocity, 516
  critical temperature gradient, 167
  fully convective stars, 212
  in core collapse supernova, 447, 448
  in red giant stars, 303, 308, 314
  in stellar cores, 175, 306
  in the Sun, 174
  Ledoux, 165
  macroconvection, 162
  microconvection, 162
  mixing-length theory, 171, 173, 516
  pressure scale height, 172
  radiative temperature gradients, 171
  role of adiabatic index, 168
  role of pressure gradient, 169
  salt-finger, 166, 167
  Schwarzschild, 164
  standing accretion shock instability (SASI), 449
  surface ionization zones, 177
coordinates
  Eulerian, 88
  Lagrangian, 88, 94
Coulomb barrier, 117
Coulomb forces, 107
Crab Pulsar, 369
Cygnus X-1, 389, 411

Dalton's law of partial pressures, 58
dark ages, 330
de Broglie wavelength, 67, 71, 78
deflagration, 437
deflection of light by gravity
  gravitational curvature radius, 381
  gravitational lensing, 395, 484
degeneracy
  lifting of, 305, 423
  of white dwarfs, 348, 349
degenerate matter, 67
  electron, 66, 74, 423

degenerate matter (cont.)
  helium flash, 305
  neutron, 66
  thermonuclear runaway in, 305, 423, 426
  white dwarfs, 349
density exponents, *see* nuclear reactions, density exponents
detonation, 437
deuterium burning in protostars, 211, 247
differentials, exact and inexact, 57
Dirac neutrinos, 257
distance
  Cepheid distance scale, 26
  parallax, 6
  parsec, 6
  spectroscopic parallax, 48
  to α Centauri by spectroscopic parallax, 48
  to Arcturus by spectroscopic parallax, 48
  to stars in a cluster, 41
distance modulus, 8
DOI number, 522
Doppler spectroscopy method, 223
double pulsars, 372
  formation, 372
  tests of general relativity, 373
doubly-diffusive instability, *see* salt-finger convection
dredge-up episodes, 303

Eddington luminosity, 218, 314
Eddington model (of star), 81
effective temperature
  and blackbody radiators, 10
  and stellar radii, 11
  of Sun, 11
Einstein equation, 382
electron capture, 178, 180, 182, 360
energy conservation for stars, 95
energy production rate, 123, 137
  CNO cycle, 137
  triple-α reaction, 142
energy transport, 153
  bound–bound absorption, 158
  bound–free absorption, 158
  conduction, 153, 156
  convection, 153, 162–171, 173
  diffusion, 154
  free–free absorption, 158
  in degenerate matter, 158
  Kramer's law, 160
  mean free path, 155
  neutrino emission, 153, 178, 327, 329
  photon absorption, 158
  photon opacities, 159
  radiation, 153, 157
  Thomson scattering, 157
energy widths, 116

equation of state, 53
  nonrelativistic gas, 70
  adiabatic, 65
  and de Broglie wavelength, 67, 71
  and quantum mechanics, 67, 353
  and quantum statistics, 67
  and special relativity, 353
  Bose–Einstein distribution, 69
  classical gas, 69, 70
  closing stellar equations, 190
  degenerate, 66, 72, 74, 78
  density of quantum states, 71
  Fermi–Dirac distribution, 69
  ideal gas, 54
  isothermal helium cores, 304
  matter and radiation, 79
  Maxwell distribution, 70
  neutron star, 482
  nonrelativistic degenerate gases, 75
  perfect gas, *see* ideal gas
  photons, 78
  polytropic, 63, 64, 76, 192
  pressure integral, 54
  quantum critical density, 70
  quantum gas, 69, 70
  transition from classical to quantum gas, 72, 78
  ultrarelativistic degenerate gases, 76, 353
  ultrarelativistic gases, 72
  white dwarf, 347, 353
equivalence principle, 379
Euler equation, 406
Eulerian coordinates, 88
Eulerian derivative, 90
event horizons, 378
exascale computers, 449
exoplanets, 222
  Doppler spectroscopy method, 223
  transits, 224
extensive properties, 92
extrasolar planets, *see* exoplanets

Fermi energy, 74
Fermi momentum, 74, 78
Fermi–Dirac distribution, 69
Feynman diagrams, 258
first law of thermodynamics, 56, 94
fluence, 461, 462
function of state, *see* state functions

Gaia mission, 7, 48
gamma-ray bursts, 460
  afterglows, 464, 491, 495
  and energy conservation, 468
  and gravitational waves, 491
  and multimessenger astronomy, 491
  and star-forming regions, 468

association with core collapse supernovae,
  469, 470
association with galaxies, 468
association with neutron star mergers, 469,
  474, 491
beamed emission, 464, 468, 491, 495
causality and source size, 468
central engine, 469
characteristics, 464
collapsar model, 471
cosmological origin, 464
discovery, 460
duration, 466
fireball model, 464
fluence, 462
hardness of spectrum, 462
isotropic distribution on sky, 462, 464
localization, 463, 491
long-period, 416, 461, 464, 469
long-period bursts and Wolf–Rayet stars, 470
Lorentz γ-factor, 469
nomenclature, 464
nonthermal emission, 464
off-axis emission, 495
optical depth, 467
power-law spectrum, 467
short-period, 461, 464, 469, 491
the gamma-ray sky, 460
Gamow window, 120
general covariance, 379
general relativity, see gravity
geometrized units, 502
giant stars, 43
GPS and relativistic corrections, 379
gravitational curvature radius
  and deflection of light, 381
  and strength of gravity, 381
gravitational lensing, see deflection of
  light by gravity
gravitational stability, 78, 80, 96
  and adiabatic index, 354
  collapse of protostar, 207, 208
  in adiabatic approximation, 206, 207
  Jeans density, 204
  Jeans length, 203
  Jeans mass, 203
gravitational waves, 478
  and equations of state, 482
  and late stellar evolution, 482, 487, 489
  and multimessenger astronomy, 476, 490
  as standard sirens, 495
  determining the Hubble parameter, 495
  frequency, 490
  from binary pulsars, 373
  from black hole formation, 331
  from merger of black hole binaries, 482, 487, 488

from neutron star mergers, 476, 491
from supernovae, 476
GW150914, 482, 483, 487, 488
luminosity, 485
multimessenger astronomy, 491
strain, 481, 483, 490
template waveforms, 480
test of general relativity in strong gravity, 476
gravity
  and curved spacetime, 380, 381
  general covariance, 379
  general relativity, 78
  Newtonian, 78, 86
  Newtonian versus general relativity, 378
  Poisson equation, 86
  principle of equivalence, 379
  strength of (table), 381
GRB, see gamma-ray bursts
GW150914
  detection, 482
  luminosity, 485
  properties of (table), 485
  waveform, 483
GW170817, 491

Hayashi tracks,
  dependence on composition and mass, 214
  forbidden zone, 212
heat
  and inexact differentials, 57
  and work in non-adiabatic pulsations, 340
  change is not a state function, 57
  flow though mass shell, 95
Heisenberg uncertainty principle
  and crystallization of white dwarfs, 358
  and delocalized electrons, 318
  and momentum, 78
  and neutrino oscillations, 259
  and quantum pressure, 77
  energy widths, 116
  statement of, 67
  virtual particles, 258
helicity, 180, 181
helioseismology, 233
  and asteroseismology, 358
  and speed of sound in Sun, 236
  g-modes, 233, 358
  p-modes, 233
helium burning
  and anthropic principle, 146
  energy production rate, 142, 143
  equilibrium population of ^{8}Be, 139
  ground state of ^{12}C, 141
helium flash, 140
helium ignition on the RGB branch, 304
helium main sequence, 306

helium burning (cont.)
  Hoyle resonance, 140
  NSE, 140
  outcome, 146
  production of neon, 143
  production of oxygen, 143
  simulated, 145
  temperature sensitivity, 142, 306
  triple-α, 138, 139
helium flash, 140, 305
helium main sequence, 306
helium shell flashes, *see* thermal pulses
Herbig–Haro objects, 201
Hertzsprung gap, 303, 488
Hertzsprung–Russell (HR) diagram, 32
  evolution of, 49
  excitation and ionization, 32
  for globular clusters, 43
  for open clusters, 43, 44
  for stars near the Sun, 41, 42
  from Hipparcos data, 42
  giant stars, 43, 46
  instability strip, 337
  ionization and Saha equations, 33
  luminosity classes, 45
  main sequence, 42, 46
  Morgan–Keenan (MK) system, 45
  spectral classes, 32, 38
  supergiant stars, 43, 47
  temperature boundaries of instability strip, 343
  white dwarfs, 43
hierarchical triple-star system, 371
Hipparcos satellite, 7, 42, 48
homology, 208
hot CNO cycle, 423
Hoyle resonance in ^{12}C, 140
HR diagram, *see* Hertzsprung–Russell (HR) diagram
Hubble parameter, 495
hydrogen burning
  CNO catalysis, 112
  CNO cycle, 110–113
  competition of PP chains and CNO cycle, 113
  efficiency of, 109
  in earliest stars, 113
  PP chains, 110, 113
hydrostatic equilibrium, 86
  and radiation pressure, 435
  conditions for, 86
  contrasting Eulerian and Lagrangian descriptions, 89
  Eulerian description, 88, 188
  for polytropes, 192
  for Sun, 229
  in general relativity, 365
  Lagrangian description, 88, 188

onset in protostar collapse, 209
Oppenheimer–Volkov equations, 366
Poisson equation, 86
virial theorem, 92

IMF, *see* initial mass function (IMF)
initial mass function (IMF), 201, 219, 335
instability
  convective, 163–166, 168–170, 230, 448
  doubly diffusive, 167
  gravitational, 80, 203, 207, 354
  of iron core, 324
  pair-instability supernova, 434, 435
  pulsational, 27, 337, 342, 343
  Schönberg–Chandrasekhar, 303, 304
  thermal pulses, 308
  thermonuclear runaway in degenerate matter, 305, 415, 416, 421–423, 426, 430, 434, 435
  thin-shell, 308, 310
instability strip,
  and κ-mechanism, 342
  temperature boundaries, 343
intensive properties, 92
interstellar reddening, 41
ionization
  and mean molecular weight, 59
  and Saha equations, 33, 35, 68
  by pressure, 66, 68, 77
  by temperature, 33, 38, 68
  helium ionization zones, 343
  hydrogen ionization zones, 343
  in solar surface, 36
  of helium, 35, 343
  of hydrogen, 35, 343
  of metals, 62
iron peak, 107

jets
  collapsar model, 471
  from gamma-ray bursts, 464
  Lorentz γ-factor, 469

Kelvin–Helmholtz timescale, *see* timescale
Kepler's laws, 16, 18, 387
Kerr spacetime
  metric, 385
  vacuum solution, 385
kilonova, 493, 495, 496
kinetic energy for stars, 95
Kramer's law, 160

Lagrange points, 403, 404
Lagrangian coordinates, 88, 94
Lagrangian derivative, 90
Lane–Emden equation, 193, 195, 349
lanthanides and opacity, 495

Large Magellanic Cloud (LMC), 440, 451
LBV, see luminous blue variables
lifecycles, see stellar lifecycles
LIGO, 373, 392, 476, 478, 482
limiting lower mass for stars, 214
limiting upper mass for stars, 217, 218
line element, see metric
Lorentz factor
   definition, 384, 467
   for gamma-ray bursts, 468
   for LHC, 469
   for quasars, 469
luminosity
   and magnitudes, 3
   expression for, 171, 189
   for mass shell, 95
   of stars, 3
   of Sun, 11
   photon, 4
   relationship to density, 28
   relationship to mass, 22
luminosity classes, 45
   pressure broadening of spectral lines, 46
   spectroscopic parallax, 48
   surface densities, 47
luminous blue variables, 327

macronova, see kilonova
magnetars, 374
magnitude, 3, 12
   absolute, 8, 28
   apparent, 5
   blue sensitive (B), 5
   bolometric, 8
   normalizations, 6
   of the Sun, 8
   photovisual (V), 5
   ultraviolet (UV), 5
   visual, 5
main sequence stars, 42
Majorana neutrinos, 257
mass density, 59
mass fraction, 59
mass loss
   and black hole formation, 486
   and massive black hole binaries, 487
   evidence from white dwarf masses, 4
   from AGB stars, 314
   from RGB stars, 304
   from Wolf–Rayet stars, 326
   from young stars, 326, 487
   soot and interstellar dust, 318
   source of luminosity, 4
mass–luminosity relationship, 20, 22
masses
   atomic, 106

   binding energy, 108
   effective neutrino mass in medium, 273
   effective plasmon mass in medium, 179
   for binary stars, 387
   gravitational, 379
   inertial, 379
   mass excess, 107
   mass matrix, 274
   nuclear, 106
Maxwell distribution, 70, 115
mean free path, 155
mean molecular weight, 55, 58–60
metal-poor stars, 24, 330, 487
metallicity, 62
   and binary black holes, 487
   and gamma-ray bursts, 469
   and opacity, 470
   and the r-process, 457
   enrichment in galaxies, 334
   massive stars as primary source, 335
   of Large Magellanic Cloud, 451
   of Pop III stars, 330
metals, 61, 330
metric
   and line element, 382
   Kerr, 385
   Minkowski, 383
   Schwarzschild, 384
metric tensor
   and geometry of spacetime, 381
   Kerr, 385
   Minkowski, 383
   Schwarzschild, 384
   source of gravitational field, 380
millisecond pulsars, 370, 371
Minkowski spacetime
   metric, 383
   vacuum solution, 383
mirror principle, 299, 307
mixing-length theory, see convection
moles, 55
Morgan–Keenan (MK) system, 45
MSP recycling, see millisecond pulsars
MSW effect, see neutrino oscillations
multimessenger astronomy, 476, 490, 491

natal kick, 450, 487
natural units, 502, 504
Navier–Stokes equations, 406
negative heat capacity
   black holes, 96, 98
   stars, 96, 98
neutrino cooling, 178, 183, 327, 329, 356, 357
neutrino oscillations
   adiabatic condition, 286
   adiabatic resonance flavor conversion, 283

neutrino oscillations (cont.)
    adiabaticity parameter ξ, 286
    antineutrino matter resonance, 281
    CP violation, 266
    effective neutrino mass in medium, 273
    energy dependence of flavor conversion, 290
    interaction of neutrinos with matter, 271
    mass matrix, 274
    matter mixing angle, 277
    matter oscillation length, 278
    MSW effect in matter, 271
    MSW flavor conversion, 279, 285, 287
    MSW resonance condition, 280
    MSW solutions in matter, 276
    neutrino mass hierarchy, 267
    oscillations with three flavors, 265
    propagation of left-handed neutrinos, 274
    Super-K evidence for oscillations, 291
    time-average or classical probabilities, 263
    vacuum oscillation length, 262
    vacuum oscillations, 260
neutrinoless double β-decay, 257
neutrinos
    and cooling of massive stars, 183, 329
    and parity, 181
    bremsstrahlung, 179
    chirality, 181, 274
    coherent scattering, 184
    cooling by neutrino emission, 153, 178, 183, 304, 327, 329, 356, 357
    cooling rates, 182
    cross section, 178
    Dirac, 257
    emission, 4, 178, 327, 329
    emission during silicon burning, 183
    finite mass, 260
    flavor mixing, 242, 259
    flavor production, 182
    helicity, 180, 181
    Majorana, 257
    mass hierarchy, 267
    masses, 256
    masses in Standard Model, 256
    pair annihilation, 179
    photoneutrinos, 179
    plasma neutrinos, 179, 180, 356, 357
    production mechanisms, 178
    recombination, 180
    reheating in core collapse supernova, 446, 447
    solar, 236, 357
    solar neutrino spectrum, 236
    trapping in core collapse supernova, 444
    Urca process, 180, 182
neutron capture reactions
    no Coulomb barrier, 125
    rapid neutron capture (r-process), 125, 456, 457
    slow neutron capture (s-process), 125, 310
neutron stars
    as dense quantum liquids, 365
    as endpoint of stellar evolution, 297, 330
    basic properties, 359
    binary neutron stars, 372
    Cas A, 364
    cooling, 362
    equation of state, 482
    internal structure, 361
    luminosity, 360
    magnetars, 374
    magnetic field, 374
    mass, 360
    natal kick, 450
    neutronization, 360
    Oppenheimer–Volkov equations, 366
    prediction of, 359
    protoneutron star, 362
    pulsars, 367
    size, 360
    superfluidity in, 364
    Urca process, 362
    X-ray emission, 362
neutronization, 360
nonthermal emission, 465
    from gamma-ray bursts, 464, 467
    implications of, 465
    polarization, 465
    requires optically thin medium, 467
    synchrotron radiation, 465
nova, 421
    breakout from CNO to hot CNO, 126
    hot CNO cycle, 126, 423
    lightcurve, 422
    Nova Cygni 1992, 421
    nucleosynthesis, 425
    recurrence, 421, 425
    RS Ophiuchi, 425
NSE, 140
nuclear forces
    Coulomb repulsion, 107
    pairing interactions, 107
    quadrupole interactions, 107
    saturation, 106
    shell effects, 107
    surface effects, 106
    symmetry energy, 107
    volume effects, 106
nuclear reactions
    advanced burning, 147
    angular momentum, 127
    barrier penetration factor, 118
    carbon burning, 147
    Caughlan–Fowler rates, 507

CNO cycle, 135, 136
competition of capture and β-decay, 126
Coulomb barrier, 117
cross section parameterization, 116, 123
cross sections, 114, 115, 121
density exponents, 123
deuterium burning in protostars, 247
deuterium burning in stars, 211
energy widths, 116
Gamow window, 120, 121
Hoyle resonance, 140
isotopic abundance changes, 189
isotopic spin, 127
neon burning, 147
neutron capture, 125
neutron reactions, 124
nonresonant, 116, 117
nuclear statistical equilibrium (NSE), 140, 141
oxygen burning, 147
parity, 127
PP chains, 131
rate libraries, 123, 507
rate of energy production, 123, 137
rates, 78, 114, 115
ReacLib library, 507
reaction networks, 510, 511
resonant, 116, 121
rp-process, 425
S-factor, 119
selection rules, 127
silicon burning, 148, 149, 514
Sommerfeld parameter, 118
temperature exponents, 123, 133, 142
thermal averaging, 115
timescales for advanced burning, 151
triple-α, 138, 140, 141, 304
weak interactions, 124
nuclear statistical equilibrium, *see* NSE
nucleosynthesis
big bang, 139, 211
CNO cycle, 135, 136
hot CNO cycle, 425
isotopic abundance changes, 189
nova, 425
PP chains, 131
r-process, 310, 314, 456
rp-process, 425, 426
s-process, 308, 310, 314, 315
triple-α, 138
X-ray burst, 426
number density, 59

OB associations, 201
opacity, 159
and metallicity, 469, 470, 487
and optical depth, 230, 231, 236

contributions to, 160
dominant contributions, 161
Kramer's law, 160
Rosseland mean, 159, 161, 230, 344
solar opacity, 160
temperature and density dependence, 78, 161
Thomson opacity, 160
Oppenheimer–Volkov equations, 366
comparison with Newtonian gravity, 366
hydrostatic equilibrium in general relativity, 366
optical depth
altered by relativity, 468
and solar surface, 230, 231, 236
definition, 230, 231, 236
in gamma-ray burst, 467

P Cygni, 327
P Cygni profiles, 201
PAH (polycyclic aromatic hydrocarbon), 318
pair-instability supernova, *see* supernova
parallax
α Centauri, 7
measured by Gaia, 7
measured by ground-based telescopes, 7
measured by Hipparcos, 7
parallax angle, 6
range of, 26
spectroscopic, 48
parity
and chirality of neutrinos, 181
and nuclear reactions, 127
nonconservation in weak interactions, 181
parsec, 6, 7
partition function, 33
phase transition
quantum, 77
thermal, 77
photoneutrinos, 179
Planck law, 9, 78
planetary nebulae, 297, 330
plasmons, 179
as heavy photons, 180
decay of, 180
in condensed matter, 179
source of neutrino–antineutrino pairs, 180
Poisson equation, 86
Pop I, Pop II, Pop III, *see* stellar populations
PP chains
competition with CNO cycle, 113
dominant energy source for Sun, 110
minimum temperature for, 214
reactions of, 110, 131
pressure
broadening, 46
for ideal gases, 54, 60
from electrons, 60

pressure (cont.)
  from ions, 60
  integral, 54
  ionization, 66, 68
  quantum, 77
  radiation, 435
  scale height, 172
proper motion, 22
proton–proton chains, *see* PP chains
protoplanetary disks, 221
protostar
  contraction to main sequence, 207, 247
  deuterium burning, 247
  initial free-fall collapse, 208
  onset of hydrostatic equilibrium, 209
PSR J0337+1715, 371
pulsars, 367
  and tests of general relativity, 373
  as spinning neutron stars, 367
  basic properties, 367
  binary pulsars, 372
  binary spinup, 371
  Crab Pulsar, 369
  discovery, 369
  double pulsars, 372
  gamma-ray emission, 461
  glitches, 369
  lighthouse mechanism, 367
  magnetars, 374
  magnetic fields, 368, 374
  millisecond pulsars, 370, 371
  nomenclature, 370
  pulsar–WD–WD triplet PSR J0337+1715, 371
  spindown, 369
pulsation, 337
  ε-mechanism and stability of massive stars, 344
  adiabatic radial, 337
  and heat engines, 340
  fundamental mode, 339
  nearly adiabatic, 340
  non-adiabatic radial, 340
  non-radial, 344
  opacity and κ-mechanism, 342
  overtones, 339
  partial ionization zones and instability strip, 342
  radial acoustic modes, 338
  role of radiative opacity, 342
  thermodynamics of sustained, 340
  work done, 340

quantum mechanics, 519
  de Broglie wavelength, 67
  exclusion principle and degeneracy, 67
  matrix elements, 520
  observables, 520
  operators, 519
  quantum field theory, 521
  quantum statistics, 67

  relativistic quantum field theory, 521
  uncertainty principle, 67
  unitary transformations, 520
  wave equations, 519
  wavefunctions, 519
quantum statistics, 67
quasars, 330, 416, 469
Q-values, 108, 134

r-process
  and Solar System abundances, 310
  and the origin of heavy elements, 456
  competition with s-process, 314
  heavy r-process nuclei, 495–497
  in neutron star mergers, 495–497
  kilonova, 495–497
  light r-process nuclei, 495–497
  time delay, 457
  timescales, 457
radial velocities (stars), 22
radiation laws, 9
  Planck law, 9
  Stefan–Boltzmann law, 9
  Wien law, 9
random walk, 100
recombination transition, 330
recycling of stellar material, *see* stellar lifecycles
red giant evolution, 297
  asymptotic giant branch (AGB), 300, 307
  crossing the Hertzsprung gap, 303
  deep convection, 303, 308, 314
  development of isothermal core, 303
  evolution away from main sequence, 246
  helium flash, 305
  helium ignition, 304
  horizontal branch (HB), 300, 306
  mass loss, 308
  mirror response of mass shells, 299, 307
  red giant branch (RGB), 300, 302
  Schönberg–Chandrasekhar limit, 303, 304
  shell burning, 298
  thermal pulses, 308
reionization transition, 330
restricted 3-body problem, 402
Roche lobes, 372, 373, 404
Roche potential, 402
Rosseland mean, *see* opacity
rp-process, 426

s-process, 310, 314
S0-2, 392
Saha equations, 33, 36, 140
Sakurai's Object (V4334 Sgr), 311
Sanduleak −69 202, *see* Sk −69 202
Schönberg–Chandrasekhar limit, 303, 304
Schwarzschild spacetime
  metric, 384
  vacuum solution, 384

selection rules, 127
  conservation of angular momentum, 127
  conservation of isospin, 127
  conservation of parity, 127
Sgr A*
  and the star S0-2, 392
  evidence for a supermassive black hole, 393
  mass, 393
shell burning, 298
  helium shell source, 307
  hydrogen shell source, 138, 301
  mirror principle, 299, 307
shell sources
  helium shell source, 307, 308
  hydrogen shell source, 307, 308
  thermal pulses, 308
  thin-shell instability, 308, 310
shockwave
  excluded in mixing-length models, 172
  in core collapse supernova, 444
silicon burning, 148, 149, 336
Sk −69 202, 440, 451
SN 1987A, see Supernova 1987A
soft gamma-ray repeater (SGR), see magnetars
solar neutrino problem
  astrophysics versus particle physics, 241
  Davis chlorine experiment, 238
  gallium experiments, 239
  KamLAND constraints on mixing angles, 292
  large mixing angle and MSW, 294
  neutrino deficit, 238
  neutrino flavor mixing, 259
  resolution of, 290
  SNO mixing solution, 292
  SNO observation of flavor mixing, 291
  Super Kamiokande results, 239
  total SNO solar neutrino flux, 292
sound speed
  and expansion timescale, 91
  in Sun, 236
space velocities (stars), 22
special relativity
  Lorentz γ-factor, 469
  relationship with general relativity, 379
spectral classes, 32
  (table), 38
  and absorption spectra, 38
  as color index sequence, 39
  as temperature sequence, 39
  carbon stars (C), 39
  hot blue stars with strong emission, 39
  IR objects (L, T, Y), 39
  old R, N, S classes, 39
  special classes, 39
  white dwarfs (D), 39
spectral index, 467

spectroscopic binaries, 17
spectroscopic parallax, 48
spectrum–luminosity diagram, see Hertzsprung–Russell (HR) diagram
speed of sound, see sound speed
stability valley, see valley of stability
Standard Model of elementary particle physics, 254
  beyond the Standard Model, 268
  CP symmetry, 266
  CPT symmetry, 266
  electroweak interactions, 182
  Feynman diagrams, 258
  finite neutrino mass, 260
  leptonic flavor mixing, 259
  neutrino masses, 256
  quark flavor mixing, 259
standard sirens, 495
Standard Solar Model, 228
  assumptions of, 228
  constraints and solution, 230
  testing with helioseismology, 233
  testing with neutrinos, 236, 237
standardizable candle, 431, 436
star formation, 201
  brown dwarfs, 215
  collapse of protostar, 207
  dependence on composition, 205
  development of radiative core, 213
  exoplanets, 222
  fragmentation, 204
  fully convective stars, 212
  gravitational stability in adiabatic approximation, 206, 207
  Hayashi forbidden zone, 212
  Hayashi tracks, 212
  Herbig–Haro objects, 201
  homology, 208
  in the early Universe, 464
  initial free-fall collapse, 208
  initial mass function, 201, 219
  Jeans criteria for gravitational collapse, 203
  limiting lower mass, 201, 214
  limiting upper mass, 201, 217, 218
  OB associations, 201
  onset of hydrostatic equilibrium, 209
  protoplanetary disks, 201, 221
  T Tauri stars, 201
  termination of fragmentation, 212
starbursts, 334
state functions, 57
Stefan–Boltzmann law, 9, 10
stellar lifecycles, 333
stellar luminosities, 22
stellar masses, 13, 17, 21, 22
stellar populations, 23
  features of, 24

stellar populations (cont.)
  Pop I, 23, 24
  Pop II, 23, 24
  Pop III, 24, 113, 330
stellar radial velocities, 17
stellar radii, 11, 13, 21, 22
stellar structure equations
  boundary conditions, 196
  general solutions, 190, 196
  Lane–Emden solution, 192, 195
  stiffness, 196
  summary, 188
stress–energy tensor, 382
structure formation, 464
subgiant stars, 303
Sun, 228
  5-minute oscillations, 233
  bolometric magnitude, 8
  CNO reactions, 135
  composition, 61, 229
  degeneracy of core, 73
  dynamical timescale, 91
  effective surface temperature, 11
  energy generation, 229
  energy transport, 230
  engulfing Earth, 319, 320
  evolution on the main sequence, 242
  helioseismology, 233
  Kelvin–Helmholtz timescale, 99
  luminosity, 11
  neutrino production, 236
  optical depth, 230, 231, 236
  PP chains, 131
  pressure of atmosphere, 34
  red giant phase, 319, 320
  rotation rate, 236
  solar constant, 11
  solar neutrino spectrum, 236
  speed of sound, 236
  Standard Solar Model, 228
superfluid
  in helium-3, 365
  in neutron stars, 364
  phase transition, 365
supergiant stars, 43
supernova, 429
  1987A, 357
  and formation of black holes, 486
  association with gamma-ray bursts, 470
  convection in, 447
  core collapse, 297, 330, 360, 431, 440, 442
  double-degenerate mechanism, 435
  entropy of core, 442
  failed, 331
  lightcurves, 429, 474
  neutrino emission, 357
  nucleosynthesis, 438, 456
  observational classification, 429

pair-instability, 330, 434, 435
remnants, 439
remnants of core collapse, 449
single degenerate mechanism, 435
standing accretion shock instability (SASI), 449
Supernova 1987A, 440, 450
thermonuclear, 434
Type Ia, 416, 430, 434
Type Ib, 431, 470
Type Ic, 431, 470
Type II, 433
Supernova 1987A, 440, 450, 451
SWIFT, 464
symmetric matter, 61, 348
synchrotron radiation, 465

T Tauri stars, 201
tangential velocities (stars), 22
Tarantula Nebula, 43, 440, 451
temperature
  effective, 10
  effective of Sun, 11
temperature exponents, see nuclear reactions
tensors, 380
thermal adjustment timescale, see Kelvin–Helmholtz timescale
thermal equilibrium, 94
thermal pressure, 77
thermal pulses, 308
  and dredge-up episodes, 318
  duration, 308
  for Sakurai's Object, 311
  number of, 308
  thin-shell instability, 308, 310
thermonuclear reactions, see nuclear reactions
thermonuclear runaway, 305, 423, 426, 439
thin-shell instability, 308, 310
Thomson scattering
  cross section, 157, 467
  opacity, 160
  relation to Compton scattering, 157
timescale
  contraction, 100
  dynamical, 91, 92, 101, 191
  evolutionary timescales, 245
  expansion, 91
  for advanced burning, 151, 192
  free-fall, 28, 91, 101, 208
  hydrodynamical, 92
  Kelvin–Helmholtz, 97, 99–101
  main sequence lifetime, 243
  nuclear burning, 133, 191, 192
  nuclear burning of helium, 306
  random walk, 100
  table of timescales, 191
  thermal adjustment, 97, 191
Tolman–Oppenheimer–Volkov (TOV) equations, see Oppenheimer–Volkov equations

total energy of a star, 95
triple-α reaction, *see* nuclear reactions and helium burning
triple-star systems, 372, 373

uncertainty principle, *see* Heisenberg uncertainty principle
unitary transformations, 520
Urca process, 180, 182

valley of stability, 125, 313
variable stars, 337
   δ Sct, 338
   ε-mechanism, 344
   κ-mechanism, 342
   as heat engines, 340
   Cepheid, 25, 28, 338, 343
   eclipsing, 25, 337
   eruptive, 25, 337
   instability strip, 337
   long-period red, 28, 338
   Mira, 28, 338
   nomenclature, 26
   non-adiabatic radial pulsations, 340
   non-radial pulsating, 344
   partial ionization zones in pulsating, 342
   period–luminosity relations, 25, 26, 28
   pulsating, 25, 27, 28, 337, 338
   role of radiative opacity in pulsating, 342
   RR Lyra, 26, 28, 338
   W Vir, 338
   white dwarf, 39, 357
   ZZ Ceti, 39, 338, 358
virial theorem
   and negative heat capacities, 96
   for ideal gas, 92
VLBA, 394

weak interactions, 106, 124
   and neutrino physics, 254
   charged currents, 257
   electron capture, 134
   neutral currents, 257
white dwarf, 346
   and degeneracy, 348
   and novae, 421
   and polytropic equation of state, 76
   and special relativity, 347
   as AGB core, 316
   as endpoint of stellar evolution, 297, 330
   as Sun's destiny, 319
   asteroseismology, 233, 358
   asteroseismology of BPM 37093, 358
   Chandrasekhar limiting mass, 350, 352, 353
   cooling, 356
   cooling curves as age indicators, 356, 358
   crystallization, 358
   density and gravity, 347
   equation of state, 347, 349
   high-mass, 76
   in HR diagram, 43
   insulating blanket model of cooling, 356
   internal structure, 355
   ions in, 348
   Lane–Emden equation, 349
   low-mass, 76, 349
   masses as evidence for mass loss, 4
   photons in, 348
   properties, 346
   Sirius B, 15, 346
   spectral classes, 39
   temperature profile, 355
Wien law, 9
Wolf–Rayet star
   and gamma-ray bursts, 470, 471, 474
   and Type Ib and Ic supernovae, 433
   characteristics, 326
   collapsar model, 473
   direct collapse to black holes, 489
   mass emission, 327, 328, 335, 470, 471
   spectral classification, 39
work
   and inexact differentials, 57
   done by mass shell, 95
   done by pulsation, 340
   is not state function, 57

X-ray binaries
   and black holes, 389, 396
   classification, 411
   Cygnus X-1, 389
   high-mass X-ray binaries, 387, 411, 415
   low-mass X-ray binaries, 372, 373, 411, 415
X-ray burst, 415, 425
X-rays, 426

ZAMS, *see* zero age main sequence (ZAMS)
zero age main sequence (ZAMS), 61, 219, 246, 297, 301, 488

Printed in the United States
by Baker & Taylor Publisher Services